LEARNING MASTERCAM X2 MILL 2D Step by Step

James Valentino
Joseph Goldenberg

Valentino, James.
  Learning Mastercam X2 Mill 2D step by step / James Valentino, Joseph Goldenberg.
      p. cm.
  Includes index.
  ISBN 978-0-8311-3353-5
  1. CAD/CAM systems. I. Goldenberg, Joseph. II. Title.
  TS155.6.V3463 2008
  670'.285--dc22

                                    2008055865

Industrial Press Inc.
989 Avenue of the Americas
New York, NY 10018

First Edition

Learning Mastercam X2 Mill 2D
Step by Step

1 2   3   4   5   6 7 8 9 10

DEDICATION

To my wife Barbara and to my children, Sarah and Andrew.

-James Valentino

To my students past, present and future.

-Joseph Goldenberg

.

ACKNOWLEDGEMENTS

The authors would like to express their thanks to CNC Software, Inc, especially Mr Ben Mund for his continued support.

We would also like to express our appreciation to Mr Bernard Hunter, the laboratory technician in the manufacturing processes laboratory at Queesnborough Community College for proofing many portions of the manuscript.

CONTENTS

CHAPTER-3   EDITING 2D GEOMETRY   3-1

CHAPTER-7 EDITING MACHINING OPERATIONS VIA THE OPERATIONS MANAGER   7-1

CHAPTER-8 USING TRANSFORM TO TRANSLATE, ROTATE OR MIRROR EXISTING TOOLPATHS   8-1

CHAPTER-9   USING A LIBRARY TO SAVE OR IMPORT MACHINING OPERATIONS   9-1

CHAPTER-10   USING TABS AND WORK OFFSETS   10-1

CHAPTER-11   CREATING BASIC SOLID MODELS   11-1

CHAPTER-12 EXECUTING 2D MILLING OPERATIONS ON SOLID MODELS   12-1

PREFACE

The CNC programmer now has a powerful tool to assist in the job of creating and verifying part programs. *MasterCam X2* CNC software provides the programmer with a full array of easy to use features. The benefits of using *MasterCAM X2* include: automatic calculation of toolpath coordinates, determination of speeds and feeds, animation of the machining process, off-line, without tying up the CNC machine, and postprocessing the part program.

MasterCAM X2 is a robust PC based package. Its many cababilities must be presented in a clear and logical sequence. This text was written to provide a thorough introduction to *MasterCAM X2*'s MILL package for students with little or no prior experience. Past users of *MasterCAM* will find release *X2* contains several enhancements and new features. LEARNING *MasterCAM X2* MILL-2D Step by Step has been expanded to include chapters on creating basic solid models, executing milling operations on solid models and file tracking and change recognition. Examples of creating user defined tool planes and machining on those planes have also been included. The Appendix contains step by step instructions users can follow to customize *MasterCAM X2*'s toolbars, drop down menus and key strokes.

Several learning aids have been designed throughout.

- Good graphical displays rather than long text and definitions are emphasised.

- An overview of the process of generating a word address program is presented.

- Key definitions are boxed in.

- Examples provide step-by-step instructions with excellent graphical displays.

- Needless cross-referencing has been eliminated. Each example is presented with all explainations appearing on the same page.

- Exercises are presented at the ends of chapters.

- A process plan is provided for many machining exercises to indicate the machining operations to be performed and the tools to be used.

- A CD provided with the text contains:

    - ► *MasterCAM X2*, DEMO version. Students can use the DEMO to practice interactively on their own PC's.

    - ► Files now keyed in sequence to the selected examples. Students can follow interactively when learning the procedure with the concepts presented in the text.

    - ► Files containing CAD parts for machining exercises.

LEARNING *MasterCAM X2* MILL-2D can be used for many different types of training applications; these include:

- Undergraduate one-semester or two semester CNC programming courses.
- Computer assisted component of a CNC programming course.
- Indistrial training courses.
- Trade school courses on computer assisted CNC programming.
- Seminar on computer assisted CNC programming.
- Adult education courses.
- Reference text for self-study.

This text is designed to be used in many types of educational institutions such as:

- Four-year engineering schools.
- Four-year technology schools.
- Community colleges.
- Trade schools.
- Industrial training centers.

CHAPTER - 1

INTRODUCTION TO *Mastercam X2*

1-1 Chapter Objectives

After completing this chapter you will be able to:

1. State the system requirements for installing *Mastercam X2*
2. Describe the general process of generating a word address program via *Mastercam X2*
3. Know the types of files created by *Mastercam X2*
4. Understand how to start *Mastercam X2*
5. Describe the nine general elements of *Mastercam X2*'s interface window
6. State how to set the system's working parameters
7. Understand the basic concepts of Gview, View, WCS, Cplane and Tplane
8. Know how to use the **Help**, **Save**, and **Exit** commands

1-2 *Mastercam X2* CNC Software

One of the most popular CNC software packages available today is *Mastercam X2* from CNC Software, Inc located at 671 Old Post Road, Tolland, CT 06084. CNC Software can also be reached at 860-875-5006. For information on the latest product developments and downloads of enhancements and patches, visit their web site at *www.mastercam.com*.

This software has a short learning curve. It presents the user with an easy to follow menu system that works fully with the Windows 00, NT4.0 or higher, ME or XP operating systems. Part geometry can be easily created with *Mastercam*'s CAD package. The CAM package enables the operator to quickly select the part material, machining operations and cutting tools. The software allows the operator to identify the CAD geometry to be selected for a machining operation then quickly generates the required tool path. The operator uses the machine and control definitions managers to specify the features of the CNC machine to be used. The appropriate postprocessor is also selected from the system's library. *Mastercam* is then directed to generate the corresponding word address part program. A very powerful feature of the software is its ability to verify the part program by animating the entire machining process. *Mastercam* automatically checks for any tool collisions when verification is running.

1-3 System Requirements for Version X2

Mastercam X2 is a 32-bit CNC software package.
The following minimum system hardware and software must be installed.

◆ Windows® XP or Windows 2000 with the latest service packs and updates; .NET 2.0 framework and DirectX ® version 9. Oc.

◆ A 1.5 GHz Intel processor

◆ 1024 x 768 resolution (minimum)

◆ 64 Mbytes of OpenGL-compatible(minimum), 128Mbytes(recommended)

◆ 512 Mbytes RAM, 1GB available hard disk space

1-4 Conventions Used Throughout the Text

The following conventions are used throughout this text.

DISPLAY	MEANING	PICTORIAL
Enter ⏎	Directs the operator to *press* the ⌸Enter⌸ key on the *keyboard*.	
Space Bar	Directs the operator to *press* the ⌸Space Bar⌸ key on the *keyboard*.	
Click Ⓝ	Means to move the mouse cursor to position Ⓝ and press the *LEFT* mouse button	position Ⓝ 〔Mouse cursor〕 **Depress left mouse button after moving the cursor to position Ⓝ** Monitor Mouse
Bold	Commands to be *typed at the keyboard* appear in **bold**	

Ⓞ Ⓝ Ⓒ Ⓓ	When placed next to examples and exercises, this icon indicates a *file by the same name is on the enclosed CD*. The student can *get the file and follow the work interactively*.
	
	◇ Place the DEMO+EXERCISES CD in the drive
	◇ When the *Mastercam X2* graphic auto-displays:
	
	    Click the (Exit) button.
	
	◇ *Double* Click on the Mastercam X2 Demo icon 〔X2 Mastercam X2 Demo〕
	
	◇ Click ⌸File⌸ ; Click 🗁 Open
	
	◇ Click the drive containing the CD Look in: 📁 [E:] ▽
	
	◇ *Double* Click on the chapter folder to open it and see all the chapter's files.
	        🗀 CHAPTER2
	        🗀 CHAPTER3
	        • •	

1-5 Installation of *Mastercam* X2 Demo CD Software for Student Use

➤ Place the DEMO X2 CD in the CD drive

The *Mastercam X2* graphic shown below will auto-display

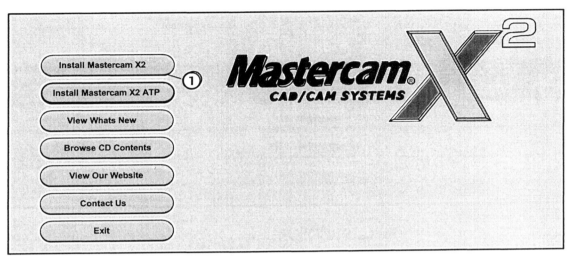

➤ Click ① the (Install Mastercam X2) button

➤ Follow the prompts *automatically triggered* by the software on the CD to complete the installation.

1-6 An Overview of Generating a Word Address Program Via *Mastercam X2*

Any part to be machined using *Mastercam X2* software must first be drawn using either the *Mastercam X2* computer aided drafting (CAD) package or imported from another CAD package such as AutoCAD or Solidworks. The part geometry created by the CAD package is used directly by the computer aided machining (CAM) package in specifying the location of machining cycles and in the determination of the corresponding tool paths. Ultimately, the CAM package produces a complete word address program for machining a specified part on a particular CNC machine tool (to learn more about word address programming refer to Introduction to Computer Numerical Control 4th Ed by J.Valentino and J.Goldenberg, published by Prentice Hall

   The sequence of steps to be followed to direct *Mastercam X2* to generate a word address program for a milling CNC machine tool are shown in Figure 1-1.

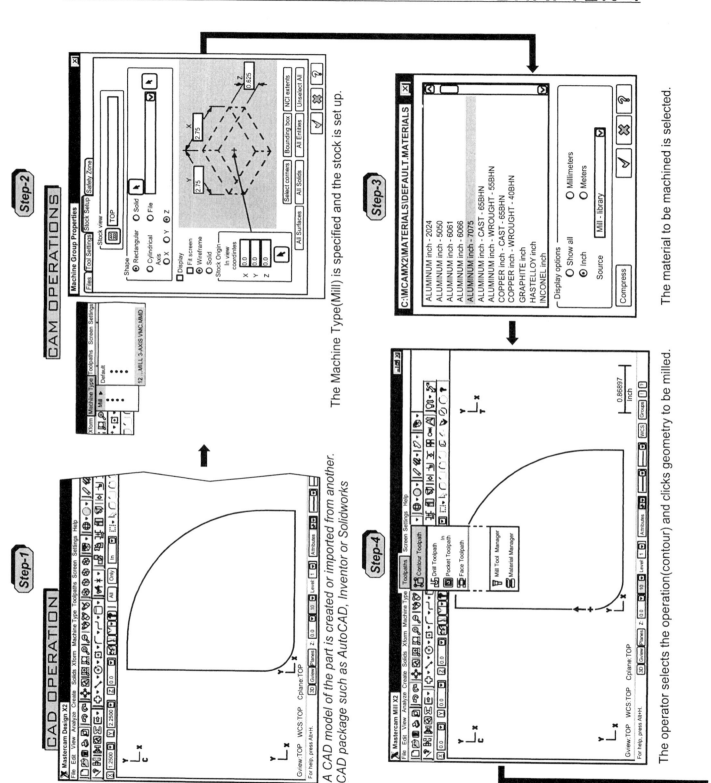

Step-1

CAD OPERATION

A CAD model of the part is created or imported from another. CAD package such as AutoCAD, Inventor or Solidworks

Step-2

CAM OPERATIONS

The Machine Type(Mill) is specified and the stock is set up.

Step-3

The material to be machined is selected.

Step-4

The operator selects the operation(contour) and clicks geometry to be milled.

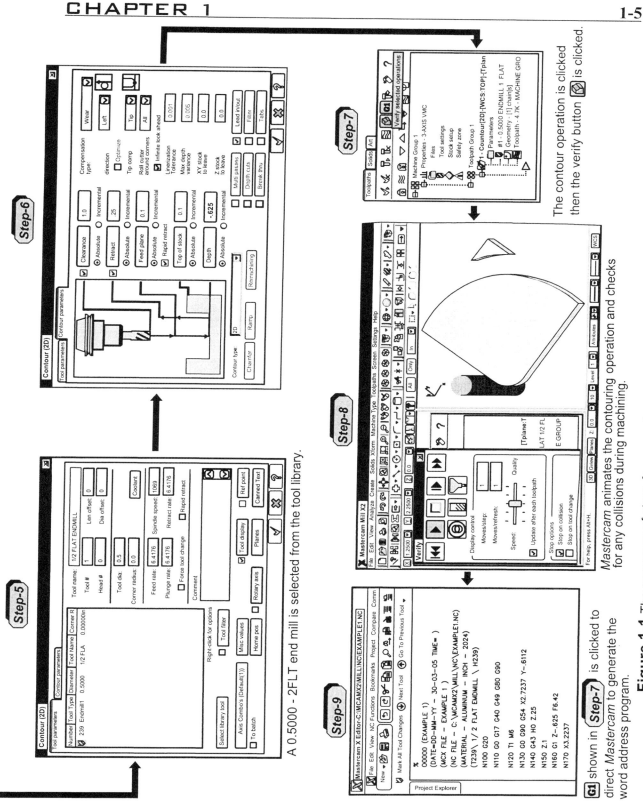

Figure 1-1 The sequence of steps for generating a part program with *Mastercam's* mill package.

1-7 Types of Files Created by *Mastercam X2*

The operator will encounter various file extension names in the course of working with *Mastercam X2* software. The extension names and their meanings are listed below in one place for quick reference.

a) CONFIG File

The system default values such as *units,allocations, tolerances, NC settings, screen and CAD settings* are stored in the configuration file that has the *extension* **.CONFIG**

b) MCX Files

All the *part geometry* generated in DESIGN mode and *tool path information* created in MILL *mode* is stored in files that have the *extension* **.MCX**

c) MATERIALS Files

Mastercam X2 has an extensive library of various materials. The information on a particular *material* is stored in the materials file with extension **.MATERIALS** *Mastercam X2* uses this information to *automatically set the recommended speeds and feeds* for a particular cutting tool used in a machining operation. The operator can also *manually* enter desired speeds and feeds.

d) TOOLS Files

The *tool library* files with the extension **.TOOLS** contain a comprehensive set of the most common *tools* to be selected in order to execute the machining of a part.

e) PST Files

The *post processors* for various CNC control units are stored in files that have the *extension* **.PST**

f) NCI Files

NCI(Numerical Control Intermediate) files contain the tool path coordinate values as well as speeds, feeds and other important machining information for a job. These files have the extension **. NCI**

f) NC Files

These files contain the *word address part programs*. *Mastercam X2* uses a particular **.PST** file and a selected **.NCI** file to generate the corresponding **.NC** file. The **.NC** file is sent to the CNC machine tool for producing the part.

1-8 Starting *Mastercam X*2

*Mastercam X*2 is started by double clicking ① on the *MastercamX2* icon appearing on the Windows desktop. See Figure 1-2.

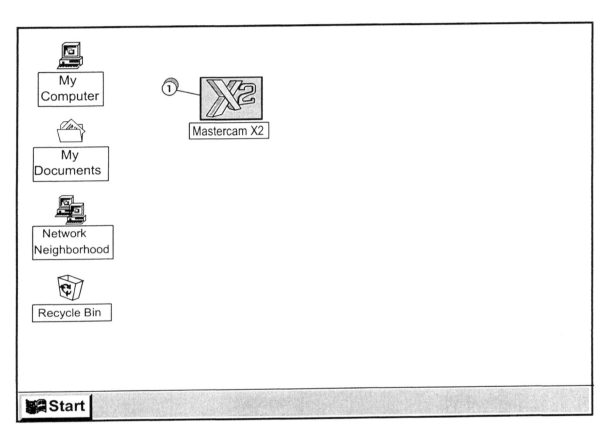

Figure 1-2 Starting *MastercamX2* from the Windows Desktop

1-9 Entering the *MastercamX2* Mill Package

To enter the Mill package the operator must select a milling machine from the machine type drop down menu.

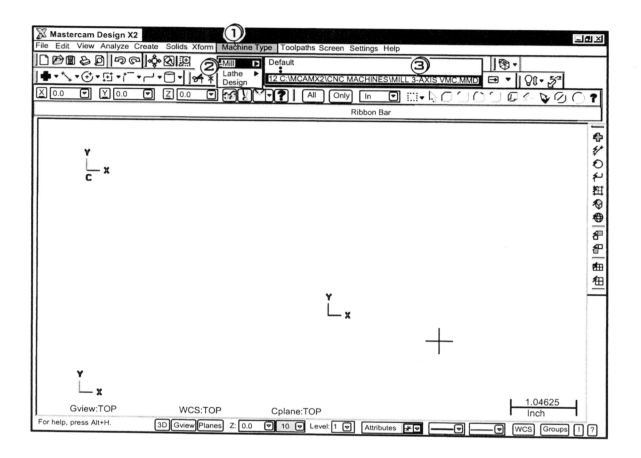

➤ Click ① the   Machine Type   drop down menu

➤ Click ② the   Mill   package

➤ Click ③ the   MILL 3-AXIS VMC.MMD   general vertical milling machine with a

Fanuc type controller.

Mastercam will then *change* the main interface window listing from **Design** to **Mill**

1-10 A Description of the *Mastercam X2* Mill Main Interface Window

The Mastercam X2 Mill main interface window is similar to any other window that is used in the Windows operating system. The interface window consists of *nine* general elements:

- menu bar
- tool bars
- autocursor ribbon bar
- general selection ribbon bar
- current function ribbon bar
- operations manager
- graphics window
- status bar
- MRU(most recently used) function bar

These elements are shown in Figure 1-3

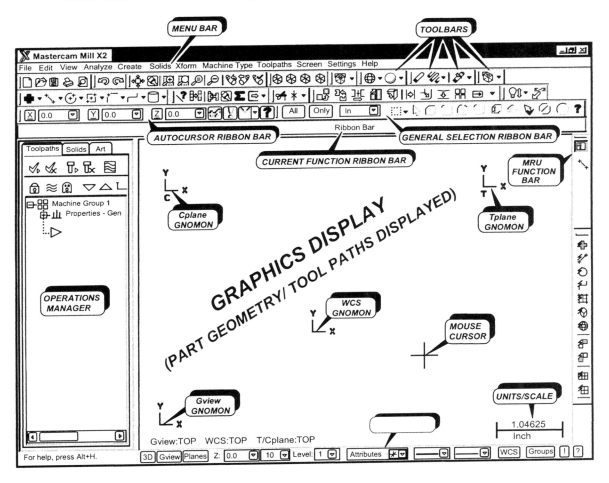

Figure 1-3 *Mastercam* X2 design/mill main interface window.

MENU BAR

This element contains *twelve* drop down menus. Several of the functions have icons next to them so the operator gets a graphic picture of what the command accomplishes. A function is *executed* by moving the mouse cursor *over* it and pressing the *left* mouse button. The operator also has the ability to create customized drop down menus. Refer to Chapter 11.

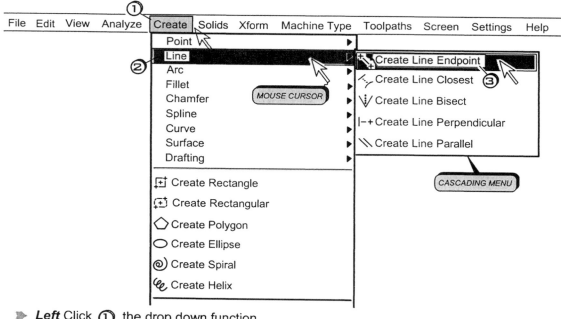

➤ *Left* Click ① the drop down function

➤ move the cursor down to the desired command ②

➤ The symbol ▶ indicates a cascading sub-menu relating to the command exists

➤ move the cursor across to the desired command in the sub-menu and *left* click ③
                                                                    to *execute* it

TOOLBARS

Toolbars are sets of buttons that execute Mastercam functions. Toolbars are normally docked but they can be undocked and relocated by clicking on their "grab" handle. Toolabrs can also be reshaped and customized. Refer to Chapter 11 for a presentation on toolbar customization.

EXECUTING A FUNCTION IN A TOOLBAR

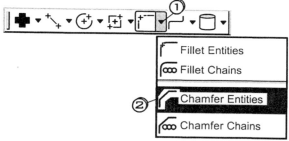

➤ *Left* Click ① the down arrow

➤ move the cursor down to the desired command; *Left* click ②

RELOCATING A TOOLBAR

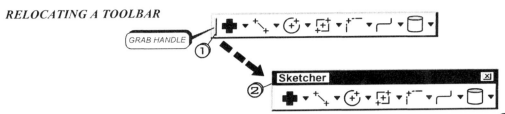

>> move the mouse cursor on the grab bar and press the left mouse button ①

>> *keeping the left button depressed* move the toolbar to its new location ②; release.

RSHAPING AN UNDOCKED TOOLBAR

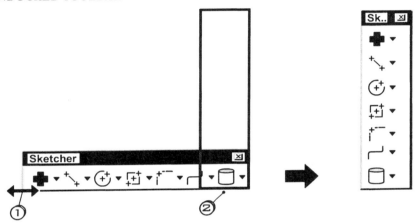

>> move the cursor *near* the edge of the toolbar until the double headed arrow appears

>> *Depress the left* mouse button and *keeping it depressed* drag it to position ② ; release.

REMOVING or ADDING TOOLBARS TO THE INTERFACE DISPLAY

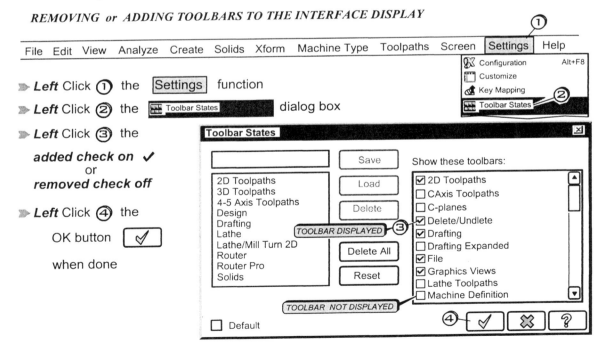

>> *Left* Click ① the [Settings] function

>> *Left* Click ② the [Toolbar States] dialog box

>> *Left* Click ③ the

added check on ✓
or
removed check off

>> *Left* Click ④ the

OK button [✓]

when done

AUTOCURSOR RIBBON BAR

Autocursor is used to **automatically detect and snap to** certian point locations on graphic entities displayed on the screen. This is done as the operator moves the mouse cursor over the entities. Snap locations include endpoints, centers of arcs, intersections, midpoints, etc. Autocursor is activated every time Mastercam prompts the operator to select a location on the screen. Upon detecting a snapped point location autocursor displays a "visual cue" to the right of the cursor. Cues include:

                   endpoint snap

                ✕   intersection point snap

                ⊕   center of arc snap

If Autocursor does not detect any points to snap to it automatically defaults to the **Sketcher** function. The operator then enters the point location using coordinates.

SETTING AND UNSETTING AUTOCURSOR SNAP LOCATIONS

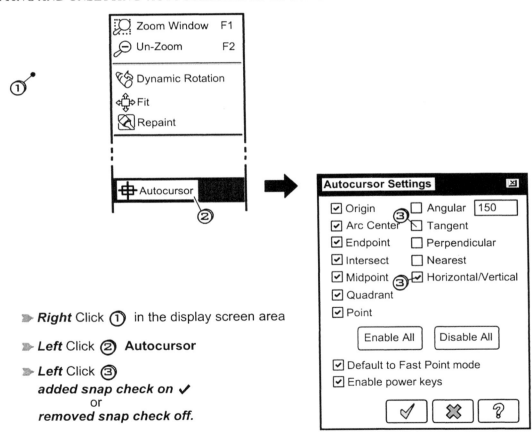

➤ **Right** Click ① in the display screen area

➤ **Left** Click ② **Autocursor**

➤ **Left** Click ③
     added snap check on ✔
           or
     removed snap check off.

Autocursor has been enhanced in *X2* to include two *new* default settings:

☑ Default to Fast Point mode

Sets fastpoint mode as the default method of inputting XYZ corrdinates .

☑ Enable power keys

Enables the operator to specify point locations to snap to *by using single keystrokes* .

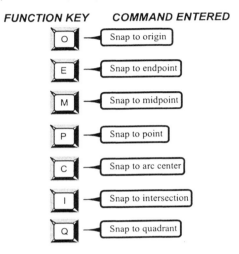

FUNCTION KEY   COMMAND ENTERED

O	Snap to origin
E	Snap to endpoint
M	Snap to midpoint
P	Snap to point
C	Snap to arc center
I	Snap to intersection
Q	Snap to quadrant

GENERAL SELECTION RIBBON BAR

The General Selection ribbon bar is used to select entities displayed in the graphics window. General Selection *is active any time the operator is not in an active function* such as Sketcher, Analyze or View Manipulation. General Selection is activated by *Mastercam X2* functons that prompt the operator to select entities. For graphics, these include Xform and for milling: toolpath creation.

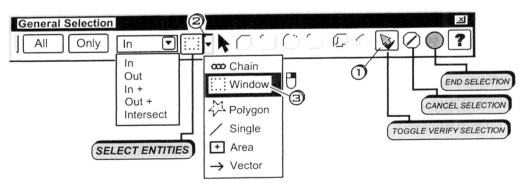

ACTIVATING THE SELECT ENTITIES MENU

➤ **Left** Click ① Toggle Verify Selection button

➤ **Left** Click ② the down arrow

➤ **Left** Click ③ the entity selection method

LOCKING A SELECTION BY RIGHT CLICKING ON IT

Mastercam X2 has a new left/right click selection feature. If the operator *left* clicks on the selection it is *in effect for a single event*. If the operator *right* clicks on the selection it *remains locked in that mode until the function is cancelled* .

CURRENT FUNCTION RIBBON BAR

The current function ribbon bar is used to create and edit CAD geometry. Ribbon bars have a close appearance to toolbars and work very much like dialog boxes. They can be docked or undocked and relocated in a convenient area of the graphics screen. If a ribbon bar for a function exists it is automatically activated and displayed when the function is selected. A blank ribbon bar is displayed by *Mastercam X2* above the graphics window when no ribbon bar is activated. When a function with a ribbon bar is selected *Mastercam X2* replaces the blank with the ribbon bar for the function.

ACTIVATING THE CURRENT FUNCTION RIBBON BAR

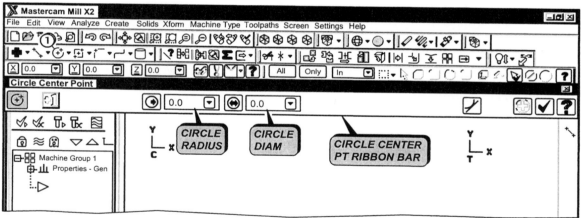

⏩ *Left* Click ① The Circle Center pt Button.

RELOCATING THE CURRENT FUNCTION RIBBON BAR

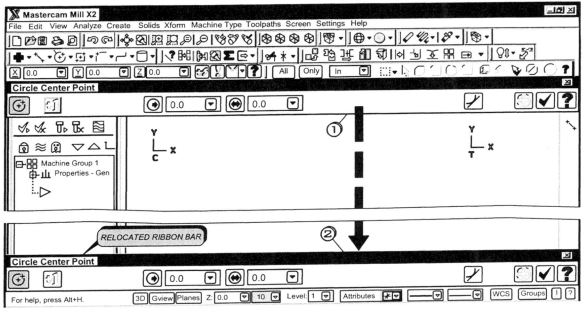

➤ move the mouse cursor on the boundary of the ribbon bar and press the left mouse button ①

➤ *keeping the left button depressed* move the ribbon bar to its new location ② and release.

THE OPERATIONS MANAGER

The Operations Manager is displayed to the left of the graphics window. It contains both the Toolpath Manager , the Solids Manager and Art. The Operations Manager is *Mastercam X2*'s control center for a job. Some of its functions include:

- listing all machining operations in the order which they will occur
- backplotting and verifying toolpaths for listed machining operations
- editing any of the machining parameters of listed operations
- re-sequencing the order of listed machining operations
- deleting listed machining operations
- creating new machining operations by copying existing ones and editing
- postprocessing a word address part program

RESIZING THE PANE WIDTH OF THE OPERATIONS MANAGER

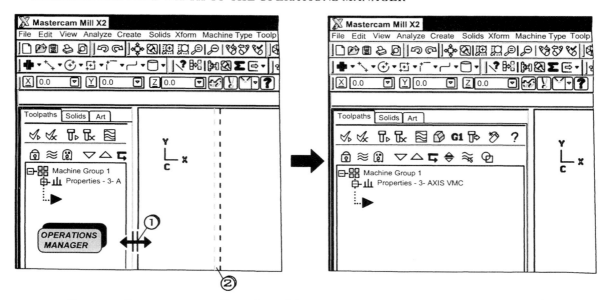

➤ move the cursor *near* the edge of the toolbar until the double headed arrow appears

➤ *Depress the left* mouse button and *keeping it depressed* drag it to position ② ; release

TURNING THE OPERATIONS MANAGER DISPLAY ON/OFF

The Operations Manager is display is turned on, by default. The operator can toggle the display *off* or *on* by using the short-cut keys on the keyboard:

`Alt` **+** `O` ◀— toggle Operations Manager display On/Off

THE GRAPHICS WINDOW

Part geometry is created, edited and viewed in the working area called the Graphics window. Backplotted toolpaths and verification of machining operation is also displayed in this area. The Graphics window can be resized. Doing so also resizes the width of the Operations Manager window. When the Graphics Window is resized Mastercam X2 remembers the setting and will maintain it when the software is opened again.

STATUS INDICATORS DISPLAYED IN THE GRAPHICS WINDOW

Mastercam X2 displays the following indicators in the graphics window to keep the operator updated with important parameters relating to the job at hand. These include :

> *THE CURRENT WORKING UNITS AND SCALE*
> *inch or metric* and the number of units
> displayed per inch or millimeter. This
> type of scaling provides a feel for the
> *actual* size of the part.

|———— 1.04625 ————|
Inch

THE COORDINATE AXIS ICON("GNOMON")

Gview Gnomon

Mastercam X2 displays the orientation of the current Gview in the lower left corner of the Graphics Window. The X axis is shown red, Y axis green and Z axis blue. Below the gnomon Mastercam X indicates that the Gview is aligned with TOP and views WCS, Cplane and Tplane are also aligned TOP or with the current Gview.

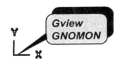

Gview:TOP   WCS:TOP   T/Cplane:TOP

Cplane Gnomon

Mastercam X2 displays the Cplane gnomon in the *upper left* corner of the Graphics Window. This indicates the orientation of the Cplane.

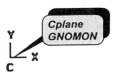

Tplane Gnomon

Mastercam X2 displays the Tplane gnomon in the *upper right corner* of the Graphics Window to show the Tplane's orientation.

WCS Gnomon

The Tplane gnomon is displayed at the *origin of the current WCS* and indicates its current orientation. The color of the WCS gnomon is set in the View Manager dialog box.

Refer to Section 1-12 for the steps to be taken in to direct Mastercam X to display these gnomons at all times.

When the operator works in *Mastercam X2* Design the Tplane gnomon *is not displayed* in the Graphics Window.

A brief discussion of Gview, View, WCS, Cplane and Tplane is given in Section 1-10.

MOUSE CURSOR

The Graphics cursor is used as a *pointing/identifying* device. The *location* of a point can be inputted by moving the cursor to a location in the Graphics Window and pressing the mouse button. The cursor is also used to *identify* which objects in the Graphics Window are to be operated on by the current command.

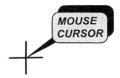

STATUS BAR

Mastercam X2's status bar is positioned at the bottom of the main interface window. The *left* side of the status bar displays *status messages* and the *right* side enables the operator to specify *important parameter settings*. These include entity creation types(2D,3D) , orientations (Gview and Planes), depth (Z), entity color, levels on which information is placed, line types/widths and point types(attributes), view settings(WCS) and Groups.

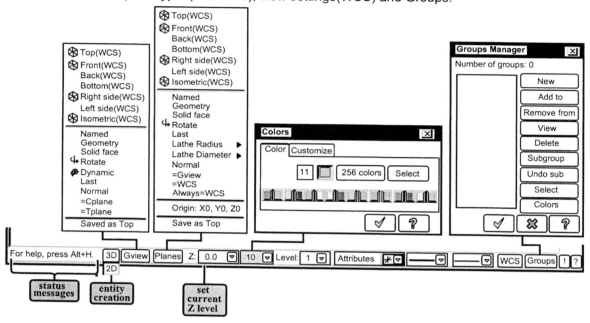

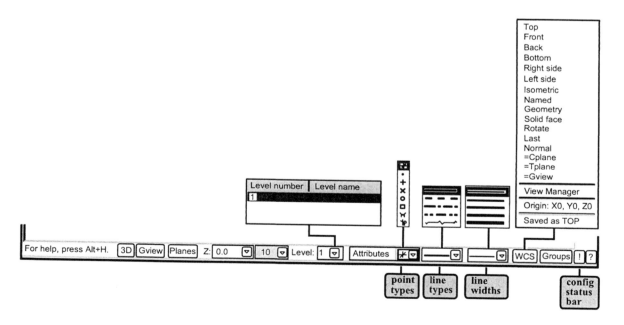

MRU FUNCTION BAR

The MRU(Most Recently Used) function bar is located to the right of the main interface window. *Mastercam X2* keeps track of the *last function the operator selected and creates a button for it in the MRU toolbar*. The MRU makes the most recently used functions easily accessible for use again in one location.

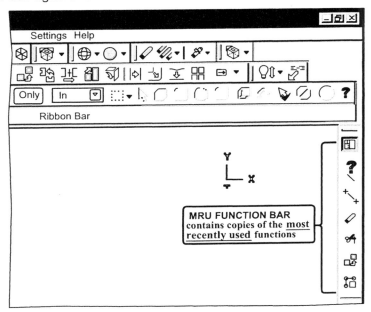

1-11 A Brief Explanation of the Terms Gview, VIEW, WCS, Cplane and Tplane

Proper use of important tools such as Gview, VIEW, WCS, Cplane and Tplane is key to insuring *Mastercam X2* produces correct CAD models, toolpaths, and word address programs.

Gview

A Gview is the *display* that *Mastercam X2* produces in the graphics window when it *points its camera along a line of sight* on the CAD model or part. The operator can instruct the system to present displays by looking along any of the **seven** standard default lines of sight: TOP, FRONT, BACK, BOTTOM, RIGHT SIDE, LEFT SIDE, ISO, or create a named line of sight.

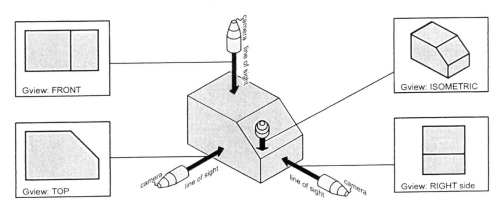

VIEW

A view is the orientation of a plane. A view has a zero point. The **six** standard default views, TOP, FRONT, BACK, BOTTOM, RIGHT SIDE and LEFT SIDE are aligned with the faces of a cube. An additional *ISO* view is also provided. If necessary, the operator can create *other* views with *different orientations or origins*. The operator uses views as **devices to set the orientation and origins** of WCS's, Cplane and Tplanes.

WORK COORDINATE SYSTEM - WCS

The WCS is the *active working XYZ coodrinate system* currently in use by *Mastercam X2*.

In *Mastercam X2* Design: the WCS origin is the **origin(0,0)** from which *absolute X,Y coordinates* are measured.

In *Mastercam X2* Mill: the WCS origin is the **part or program zero (0,0,0)**

Mastercam X2's default WCS specifies the orientation and origin of the TOP view, TOP Cplane and TOP Tplane as well as the orientation and origin's of the set of **six** default views Cplanes and Tplanes that key off TOP. An additional *ISO* view is also provided. The WCS is key to presenting the **top of the part to the CNC machine tool.**

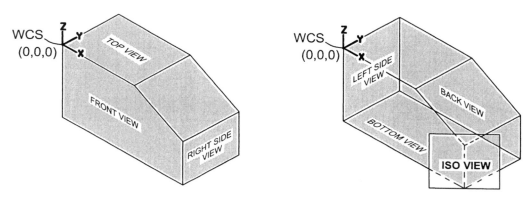

CONSTRUCTION PLANE- CPLANE

All **CAD geometry** is created on a Cplane. Cplanes are used to **orient geometric entities** such as 2D lines, arcs and circles. When created these entities **lie in the XY plane of the selected Cplane.** For 2D applications a single Cplane is usually adequate. Normally it is the default Cplane **TOP**(XY). The Z depth is always taken relative to the current Cplane.

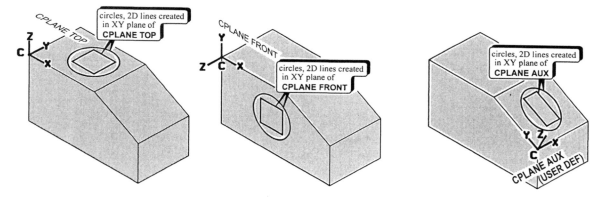

TOOL PLANE- TPLANE

The tool plane *sets the orientation of the cutting tool.* The tool will *align itself along the -Z axis of the current tool plane* and will approach and retract from the part along the Tplane's Z axis. The (-Z)depth of cut in *absolute coordinates* is taken from the *origin of the tool plane.* Rotary motion (A,B,C) word address code in the part program is triggered by changing the Tplane. An example would be rotating the tool plane 45° about the Y axis could rotate the tool axis on a 5-axis machining center or cause rotary index table rotation on another type of machine.

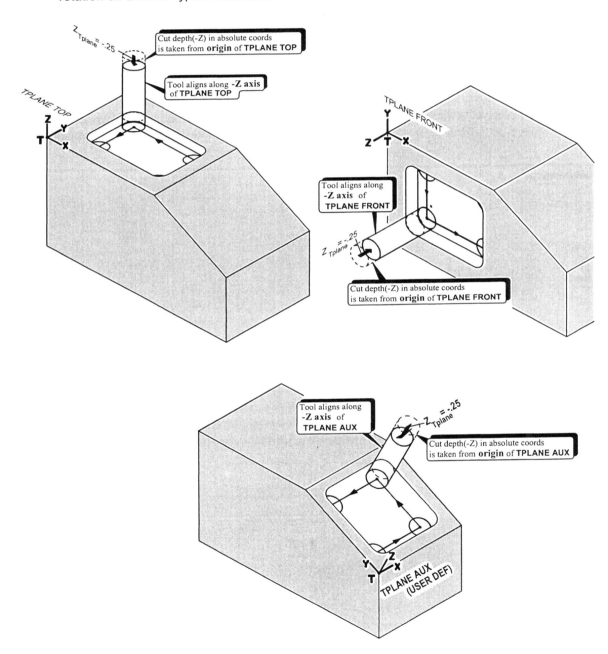

1-12 *Mastercam X2*'s Short-Cut Keys for Entering Functions

In addition to using toolbar and drop down menus, the operator can also use short-cut keys for quickly entering functions. Key use reduces the need for clicking into sub-menus thereby dramatically lowering the time it takes to complete operations in *Mastercam X2*. Keys can provide additional flexibility. For example, the keys that cause pan and zoom can be used *while* a line function is being executed.

FUNCTION KEYS FUNCTION ENTERED   *FUNCTION KEYS   FUNCTION ENTERED*

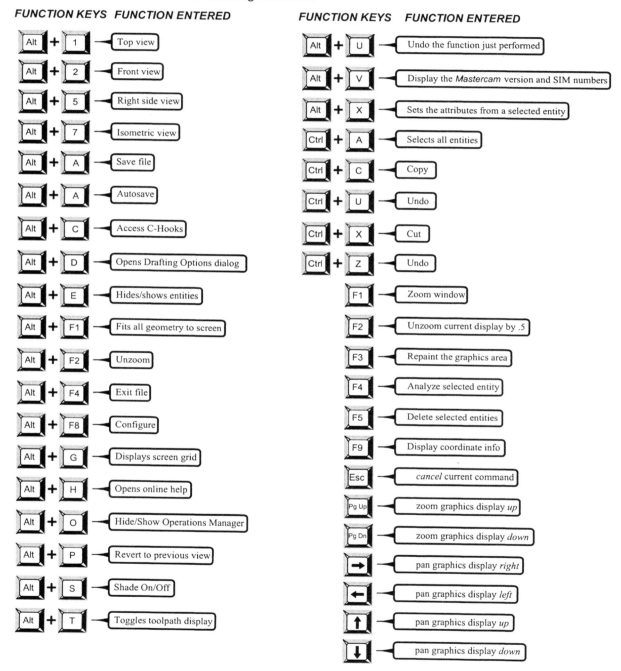

Alt + 1	Top view
Alt + 2	Front view
Alt + 5	Right side view
Alt + 7	Isometric view
Alt + A	Save file
Alt + A	Autosave
Alt + C	Access C-Hooks
Alt + D	Opens Drafting Options dialog
Alt + E	Hides/shows entities
Alt + F1	Fits all geometry to screen
Alt + F2	Unzoom
Alt + F4	Exit file
Alt + F8	Configure
Alt + G	Displays screen grid
Alt + H	Opens online help
Alt + O	Hide/Show Operations Manager
Alt + P	Revert to previous view
Alt + S	Shade On/Off
Alt + T	Toggles toolpath display

Alt + U	Undo the function just performed
Alt + V	Display the *Mastercam* version and SIM numbers
Alt + X	Sets the attributes from a selected entity
Ctrl + A	Selects all entities
Ctrl + C	Copy
Ctrl + U	Undo
Ctrl + X	Cut
Ctrl + Z	Undo
F1	Zoom window
F2	Unzoom current display by .5
F3	Repaint the graphics area
F4	Analyze selected entity
F5	Delete selected entities
F9	Display coordinate info
Esc	*cancel* current command
Pg Up	zoom graphics display *up*
Pg Dn	zoom graphics display *down*
→	pan graphics display *right*
←	pan graphics display *left*
↑	pan graphics display *up*
↓	pan graphics display *down*

1-13 Setting Working Parameters Via the System Configuration Dialog Box

Important working parameters in *Mastercam X2* are specified in the System Configuration dialog box. These include tolerance setting, file subdirectory creation, screen settings, NC settings, CAD settings, etc. The configuration file is saved as **MCAMX.CONFIG** for English units or **MCAMXM.CONFIG** for metric units.

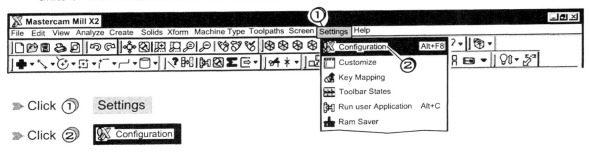

➤ Click ① Settings

➤ Click ② Configuration

SPECIFYING WORKING UNITS(ENGLISH OR METRIC)

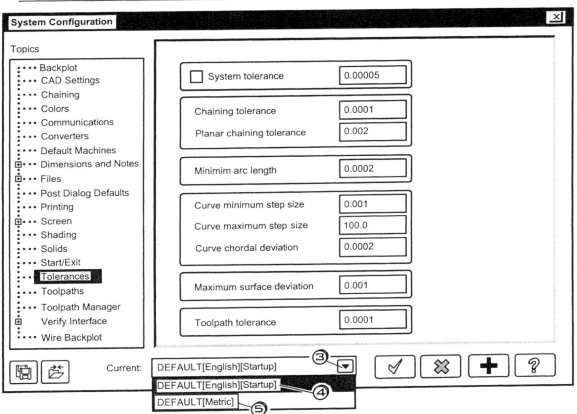

➤ Click ③ the Current configuration file down button ▼

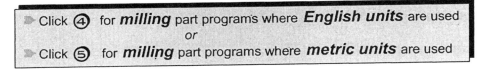

➤ Click ④ for *milling* part programs where *English units* are used

or

➤ Click ⑤ for *milling* part programs where *metric units* are used

CREATING A DIRECTORY FOR STORING AND GETTING .MCX STUDENT EXERCISE FILES

➤ Click ⑥ **Files**

➤ Click ⑦ Mastercam Parts[MCX] in the Data paths area

➤ Click ⑧ in the Selected item's Data path name

➤ Keep the direction *right* key ➡ depressed until the cursor | is positioned to the right of the text

➤ Type (YOUR INITIALS)-**MILL**

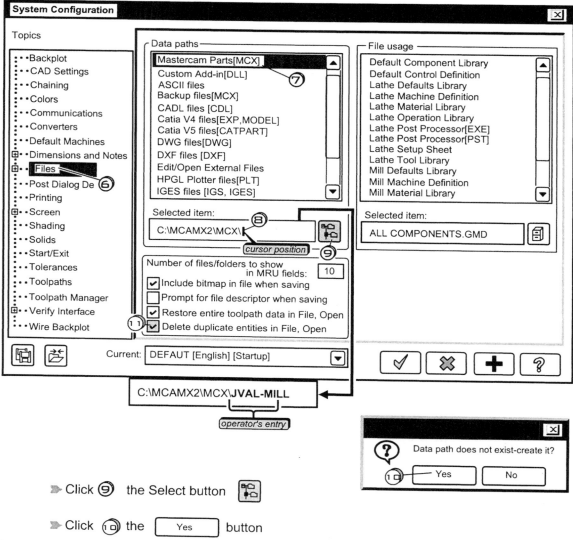

➤ Click ⑨ the Select button

➤ Click ⑩ the [Yes] button

Your mill jobs will now be saved and retrieved from the directory named
MCAMX\MCX2\JVAL-MILL when the edited configuration file is saved.

Configure *Mastercam* to *automatically find and delete duplicate entities* every time a file is opened.

➤ Click ⑪ check on for ☑ Delete duplicate entities in File, Open

DISPLAYING THE WCS GNOMON AND LEARNING MODE PROMPTS

By default *Mastercam X2* has the *continuous display* for the WCS gnomon set to *off*. The operator can change this setting to *on* so the location of the **working origin** is *displayed at all times*. Beginners also find it helpful to have the Learning Mode prompts turned **on** when executing functions.

⟫ Click ⑫ [Screen]

⟫ Click ⑬ the check **on** ☑ for Display WCS XYZ axes

⟫ Click ⑭ the check **on** ☑ for Use Learning Mode prompts

⟫ Click ⑮ the check **off** ☑ for Enable Ribbon modality

> Note: If Ribbon Bar modality is *on* ☑ , all settings previously entered *will remain* when it is opened again. This feature is intended to save time for entering the same function in production work but should be **turned off** ☐ when learning different features of the same function.

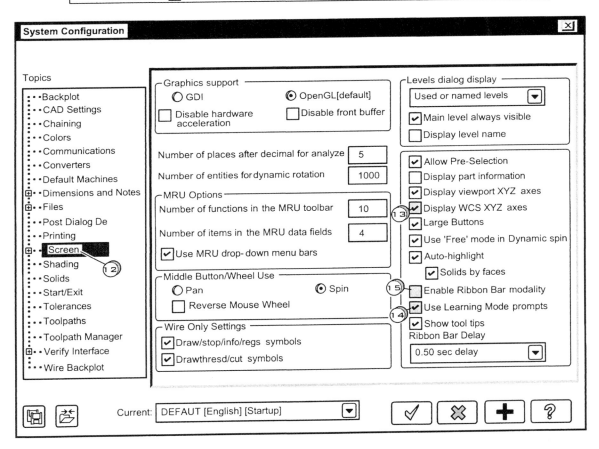

SETTING THE SYSTEM DEFAULT COLORS

➤ Click ⑯   `Colors`

➤ Click ⑰ the *Mastercam X2* interface element   `Graphics backround color`

➤ Click ⑱   the color setting for the element

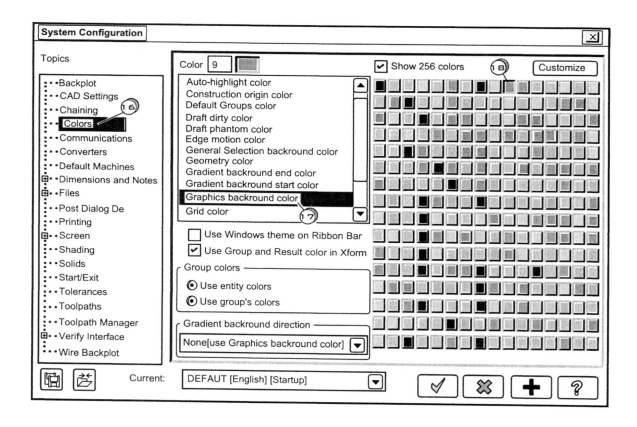

SPECIFYING DEFAULT SETTINGS FOR POINT AND LINE STYLES AND LINE WIDTHS

➤ Click ⑲ **Cad Settings**

➤ Click ⑳ the Point Style down arrow ▾

➤ Click ㉑ the desired point style

➤ Click ㉒ the Line Width down arrow ▾

➤ Click ㉓ the desired line width

➤ Click ㉔ the Line Style down arrow ▾

➤ Click ㉕ the desired line style

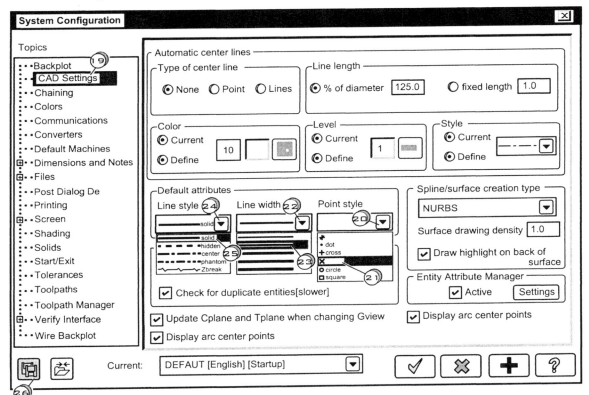

SAVING THE CONFIGURATION SETTINGS TO YOUR OWN CONFIGURATION FILE   NAME

➤ Click ㉖ the Save As button

➤ Click ㉗ in the name box; enter the name of the new configuration file **MYCONFIG1**

➤ Click ㉘ the OK button ✓

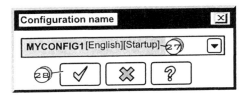

Note: it is advisable to save your *customized* configuration file(s) with extension (.**CONFIG**) to a floppy,FLASH or ZIP diskette so that they are readily avalable if restoration is needed.

1-14 Using On Line Help

Mastercam provides quick access to information regarding basic concepts, commands, tools,and information about the latest release of the software.

General Use Of Help

➽ Press the [Alt] + [H] keys

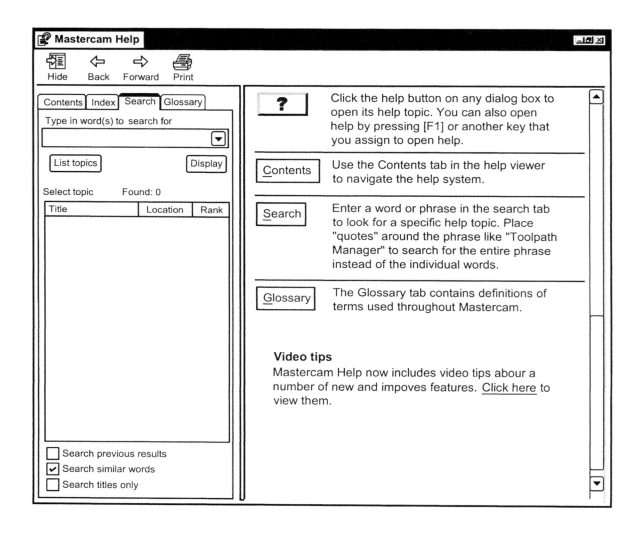

USING THE CONTENTS TAB TO NAVIGATE HELP

➤ Click ① the [Contents] tab

➤ Click ② open the contents of Mastercam Help

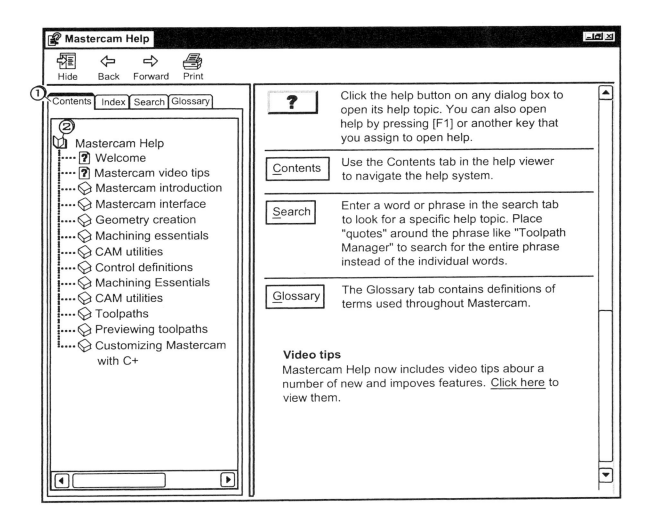

➤ **_double_** Click ③ open the contents of Mastercam Interface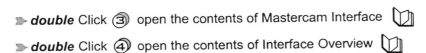

➤ **_double_** Click ④ open the contents of Interface Overview

➤ Click ⑤ Common Mastercam buttons

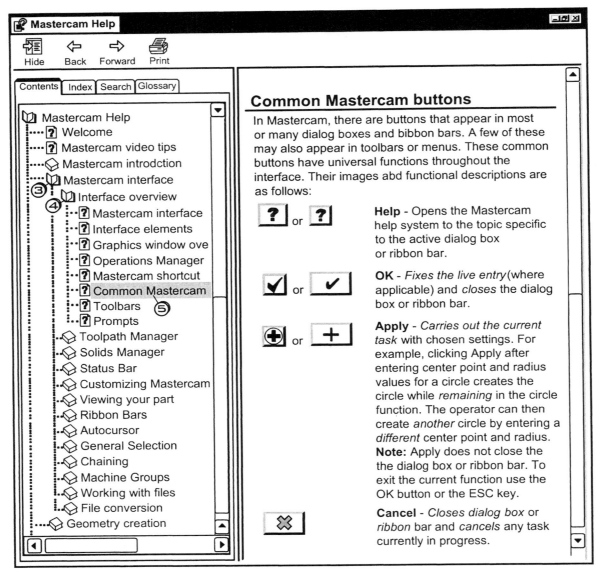

Common Mastercam buttons

In Mastercam, there are buttons that appear in most or many dialog boxes and bibbon bars. A few of these may also appear in toolbars or menus. These common buttons have universal functions throughout the interface. Their images abd functional descriptions are as follows:

Help - Opens the Mastercam help system to the topic specific to the active dialog box or ribbon bar.

OK - _Fixes the live entry_ (where applicable) and _closes_ the dialog box or ribbon bar.

Apply - _Carries out the current task_ with chosen settings. For example, clicking Apply after entering center point and radius values for a circle creates the circle while _remaining_ in the circle function. The operator can then create _another_ circle by entering a _different_ center point and radius. **Note:** Apply does not close the the dialog box or ribbon bar. To exit the current function use the OK button or the ESC key.

Cancel - _Closes dialog box_ or _ribbon_ bar and _cancels_ any task currently in progress.

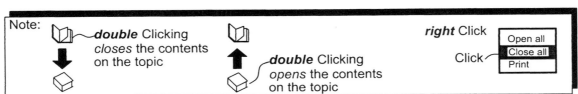

Note: **_double_** Clicking _closes_ the contents on the topic

double Clicking _opens_ the contents on the topic

right Click

Click — Open all / Close all / Print

➤ Click ⑥ the [Search] tab

➤ Click ⑦ in the key word box and enter the search word: **SLOT MILL**

➤ Click ⑧ the [List topics] button

➤ Click ⑨ on [Slot mill toolpaths]

➤ Click ⑩ the [Display] button

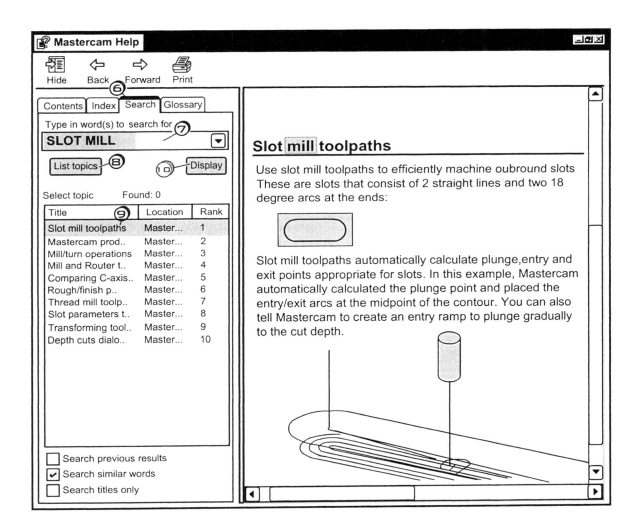

1-15 Saving a File

TO SAVE A FILE

➤ Click ① File pull down menu

➤ Click ② [▣ Save As]

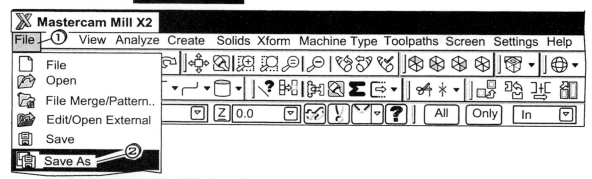

➤ Click ③ in the file name box and enter the name of the **.MCX** file to be saved,
for example file, **EX1-1JV**

➤ Click ④ the [Options] button

Note:
The descriptor 11= *Mastercam X2, 10 = Mastercam X*

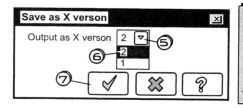

Note:
Choosing Version 1 saves the file in the format used by the first release of *Mastercam X,* as well as releases *MR1* and *MR2.*
Only geometry will be saved when choosing Version 1 Toolpath operations *will not be saved.*

Choosing Version 2 saves the file in the format used by *Mastercam X2.*

➤ Click ⑤ the down button

➤ Click ⑥ the version [2]

➤ Click ⑦ ⑧ the OK buttons [✓]

1-16 Opening a File

TO OPEN A FILE

➤ Click ① the file open icon 📂 in the **File** toolbar

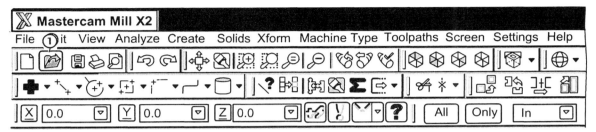

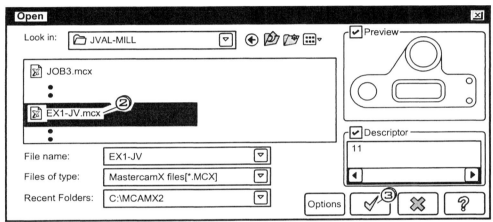

➤ Click ② on the desired file to open **EX1-JV**

➤ Click ③ the OK ☑ button

TO OPEN A PREVIOUSLY STORED FILE ON THE ENCLOSED CD

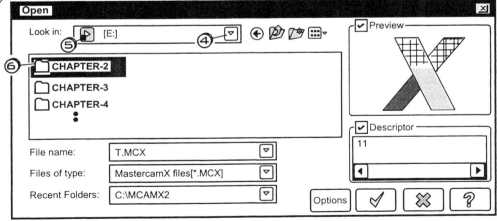

➤ Click ④ on the down button ▽ and ⑤ on the CD drive icon 💽

➤ ***Double*** Click ⑥ on the file folder of the desired chapter 📁

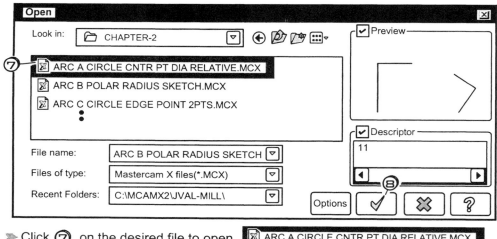

➤ Click ⑦ on the desired file to open

[ARC A CIRCLE CNTR PT DIA RELATIVE.MCX]

➤ Click ⑧ the OK [✓] button

1-16 Using the Zip2Go Utility

The Zip2Go utility is used to gather and compress the current Mastercam part data and other files into a .Z2G file. This is especially useful for sending files over the internet .Other utlities such as WinZip can then be used to unzip all the files.

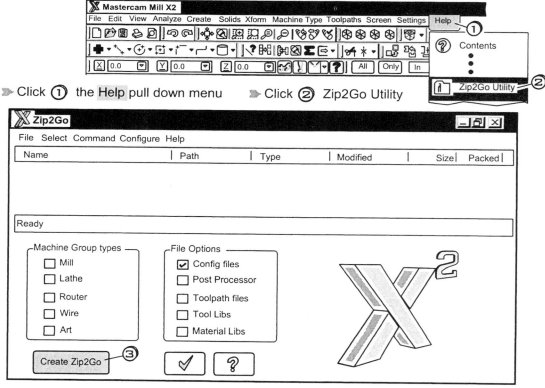

➤ Click ① the Help pull down menu   ➤ Click ② Zip2Go Utility

➤ Click ③ the [Create Zip2Go] button

1-17 Creating a New File

TO CREATE A NEW FILE WHILE IN **Mastercam X2**

➣ Click ① the file new icon [ ] in the **File** toolbar

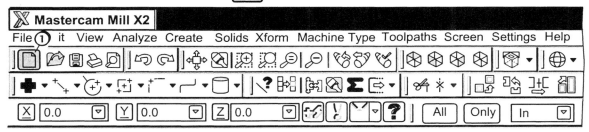

1-17 Converting Files from Previous Releases of *Mastercam X* to *X2*

Users of *Mastercam X* , *MR1* and *MR2* must convert the following files before working
with them in *Mastercam X2*.

- Toolpath and operation defaults ***.DEFAULTS**
- Tool libraries ***.TOOLS**
- Material libraries ***.MATERIALS**
- Operation libraries ***.OPERATIONS**
- Machine definitions ***.MMD**
- Component libraries ***.GMD**
- Part files ***.MCX**

To convert files using *Mastercam X2* select **File** pull down menu then click **Open**
Open the individual file and save it using the proper extension such as ***.TOOLS**
or ***.PST** etc.

The reader is encouraged to consult the *MastercamX2* Transition Guide
that comes with the installation software for further information.

1-18 Exiting the *Mastercam X2* Design/Mill package

TO EXIT Mastercam X2

➡ Click ① the File drop down menu

➡ Click ② Exit

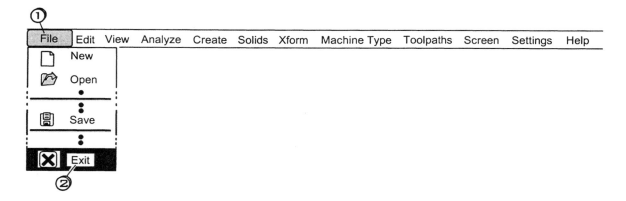

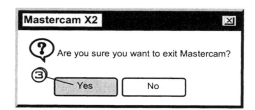

➡ Click ③ [Yes]

If any *changes* to the current **.MCX** file were *not saved* the system
will display the dialog box below

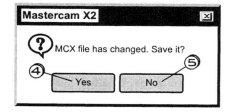

➡ Click ④ [Yes] to *save* the changes and exit *Mastercam X2*

➡ Click ⑤ [No] to *discard* the changes and exit *Mastercam X2*

EXERCISES

1-1) Execute the steps necessary to turn on the computer and start the *Mastercam X2* Mill package

1-2) What information is contained in the following files: a) **CONFIG** b) **MCX** c) **NC**

1-3) Identify and describe the use of each of the areas *A, B, C, D, E, F, G.... O* of *Mastercam X2's* main interface window shown in Figure 1p-1.

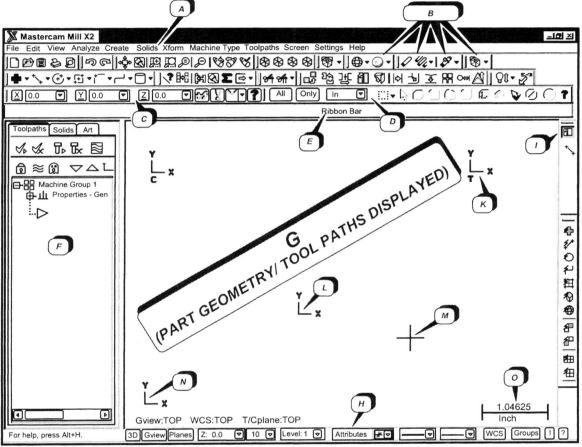

Figure 1-p1

1-4) What is the significance of using the [Esc] key

1-5) What are some advantages of using the short-cut keys

1-6) What functions are executed by the following short-cut keys

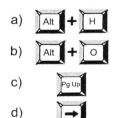

a) [Alt] + [H]

b) [Alt] + [O]

c) [Pg Up]

d) [→]

1-7) Describe the steps required to set the system to *metric* part programming.

1-8) Describe the steps required to create the subdirectory **SPEEDY-MILL** for saving and getting *MastercamX2* **.MCX** files.

1-9) A Gview _____

    A. defines the active working coordinate system   C. is a line of sight on the part

    B. is a orientation plus an origin   D. controls 2D arc/line orientations

1-10) A Cplane _____

    A. defines the active working coordinate system   C. is a line of sight on the part

    B. is a orientation plus an origin   D. controls 2D arc/line orientations

1-11) A View _____

    A. defines the active working coordinate system   C. is a line of sight on the part

    B. is a orientation plus an origin   D. controls 2D arc/line orientations

1-11) A Tplane _____

    A. defines the active working coordinate system   C. is a line of sight on the part

    B. sets the orientation of the cutting tool   D. controls 2D arc/line orientations

1-12) Use the Contents tab in on line help to find and print information about the topic *ribbon bars*

1-13) Use the Search tab in on line help to find and print information about the topic *Configuration files.*

1-14) Explain the steps for saving the file **EX1-2JV** in the sub-directory **JVAL-MILL** and exiting *Mastercam X2.*

CHAPTER - 2

CREATING 2D WIREFRAME MODELS

2-1 Chapter Objectives

After completing this chapter you will be able to:

1. Know how to construct basic 2D wireframe geometric entities such as points, lines, arcs, fillets and splines.
2. Execute rectangle, chamfer and letter constructions in 2D space.
3. Know how to construct 2D ellipse and polygon geometric entities.
4. Understand how to use *Mastercam*'s point positioning commands to quickly and accurately locate points on existing geometric entities.
5. Explain how to pan and zoom screen displays.
6. State the methods used for repainting and regenerating the screen displays.

2-2 Generating a Wireframe CAD Model of a Part

This chapter as well as Chapter 3 presents the commands and techniques for generating 2D wireframe CAD models using *Mastercam X2* Design. A Wireframe model is the most basic method of representing a real 2D part as a collection of lines, arcs, points and splines. A wireframe defines the *boundaries of the part only*. The space enclosed by the boundaries is void of points and undefined. Only parts containing *no surface intersections or flat surface planes between the boundaries* can be modelled by wireframe.

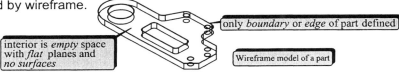

only *boundary* or *edge* of part defined

interior is *empty* space with *flat* planes and *no surfaces*

Wireframe model of a part

2-3 Point Constructions in 2D Space

For rectangular coordinates, *Mastercam X2* employs the *Cartesian coordinate system* for specifying the location of a point in space. This system consists of *three directional numbered* lines, called *axes*, that mutually intersect at an angle of 90°. The point of intersection is called the *origin*. The XY coordinate plane is broken up into *four quadrants*. Direction is determined as follows:

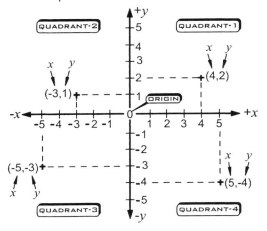

<u>along the *x*- axis</u> - numbers to the *right* of the origin are given a *positive (+) sign*

numbers to the *left* of the origin a *negative(-)sign*.

<u>along the *y*- axis</u> - numbers *above* the origin are given a *positive(+) sign*

numbers *below* the origin a *negative(-) sign*

The location of a point is determined by *first* entering its *signed x-distance*, *then signed y-distance* and *finally signed z-distance* from the *origin*.

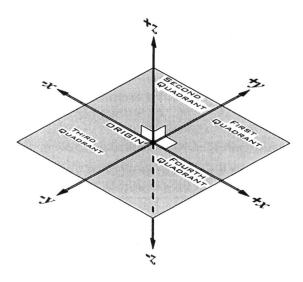

A point is displayed on the screen as the symbol (✘) . Points can be used as references to specify locations for subsequent geometric constructions involving lines, circles, arcs, splines, etc

Create points   **Locations/Snaps**

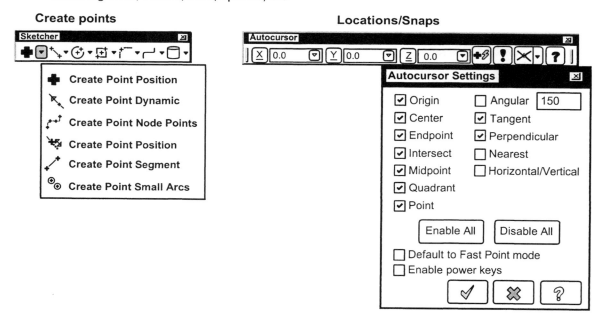

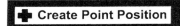 **Create Point Position** ▸ *MANUAL ENTRY* ◂

A *point* is created at any Cartesian (X,Y,Z) coordinates entered.

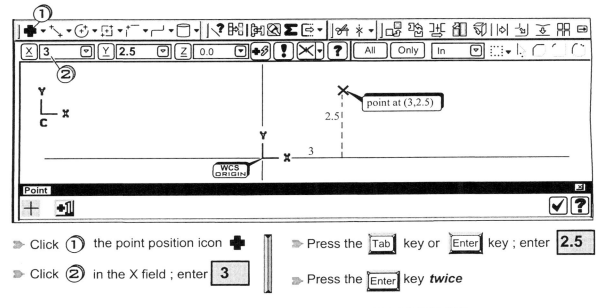

➤ Click ① the point position icon ✚

➤ Click ② in the X field ; enter **3**

➤ Press the [Tab] key or [Enter] key ; enter **2.5**

➤ Press the [Enter] key *twice*

MANUAL ENTRY WITH AUTOCURSOR IN FAST POINT MODE

To default Autocursor to Fastpoint mode

➤ Click ③ the check on for

☑ Default to Fast Point mode

➤ Click ④ the OK [☑] button

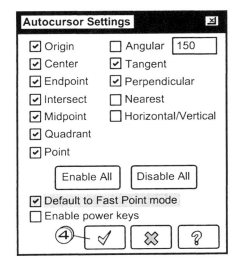

Autocursor-Fast Point active

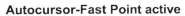

➤ Enter **3,2.5**

Note:
◆ Other permissible entries include: **X3,2.5** or **3,Y2.5**

◆ Fast Point is *modal*. The operator must remain *inside* the value entry *while entering coordinates* and press [Enter] *before* moving outside to select any other part of a current function.

◆ Pressing [Esc] before [Enter] cancels any values entered

◆ *Mastercam X2 stores the values* of the XYZ coordinates last entered. The operator can make a new entry having a single coordinate value change simply by typing the *letter of the coordinate and its new value. Mastercam* will *automatically* use the *last values entered* for the other coordinates.

BUILT IN CALCULATOR FEATURE

The operator can use any of the calculator functions: + add
- subtract
* multiply
/ divide

in fields that take normal input . The following are acceptable input values:

X.75   or   **X3/4**

Y1.25   or   **Y1 + .25**   or   **Y1 + 1/4**

X2.125   or   **X2.25 - .125**   or   **X2+1/4 - 1/8**

X1.875∗.5, Y.875 + 1   or   **X.9375,Y1.875**

X((2-.625)/2 + 1/4),Y((4∗.375 + .5)∗.25 + 1)

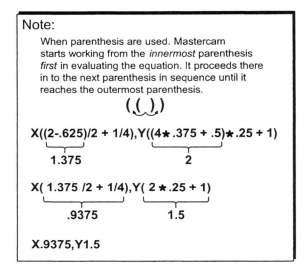

LOCKING VALUE FIELDS WITH AUTOCURSOR IN DEFAULT MODE

The X,Y,Z value fields can be in any one of **three** states when Autocursor is in default mode with Fast Point *inactive.*: **Unlocked**, **Soft-locked** or **Hard-locked.**

Unlocked

This is the default or normal field state. The values in the XYZ fields will be *dynamically* updated as the mouse cursor moves across the graphics screen. The operator can sketch the location of a point by pointing to a location on the screen and clicking.

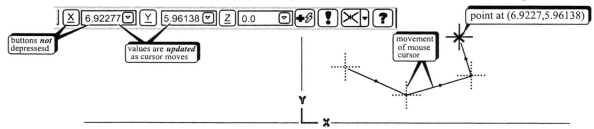

Soft-locked

This is the state that exists when values are being entered in X,Y,Z fields. Autocursor. *returns* to its default unlocked state *after* a point has been created or the operator cancels an entries by hitting the [Esc] key .

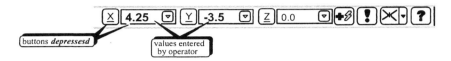

Hard-locked

When the field is hard-locked the values in it *cannot be changed* until the operator *manually unlocks it*.

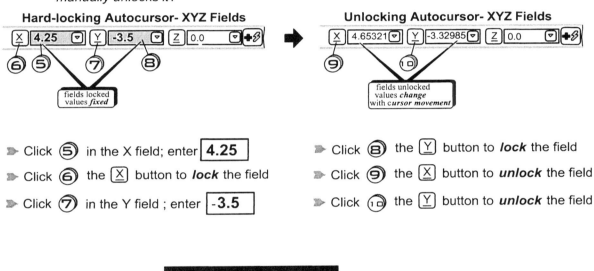

> Click ⑤ in the X field; enter **4.25**

> Click ⑥ the [X] button to *lock* the field

> Click ⑦ in the Y field ; enter **-3.5**

> Click ⑧ the [Y] button to *lock* the field

> Click ⑨ the [X] button to *unlock* the field

> Click ⑩ the [Y] button to *unlock* the field

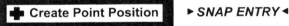

➤ *SNAP ENTRY* ◄

A *point* is created at a specific location on an existing geometric entity. Autocursor **automatically snaps to the location** when the operator moves the cursor near the area on the entity. The setting of snap locations was discussed in Chapter 1, Section 1-9. Autocursor uses the following order to detect and snap to points:

> **Order of Snap** :
> ◆ Point entities - *searched first and snapped to*
> ◆ Endpoints of curves or lines - *searched second and snapped to*
> ◆ Midpoints of curves or lines - *searched third and snapped to*
> ◆ Quadrant points of arcs - *searched fourth and snapped to*
> ◆ Centers of arcs - *searched fifth and snapped to*
> ◆ Real curve or line intersections - *searched sixth and snapped to*
> ◆ Points on the active selection grid - *searched seventh and snapped to*

☑ Center

A *point* is created at the *center* of an existing *circle* or *arc* entity

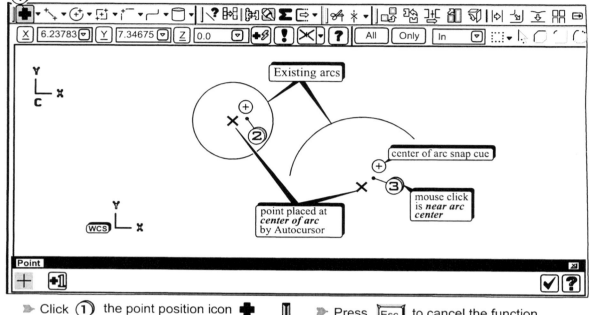

➢ Click ① the point position icon ✚

➢ Click ② ③ *near* the arc *centers*

➢ Press [Esc] to cancel the function

☑ Endpoint

A *point* is created at the *ends* of an existing *line, arc or spline* entity.

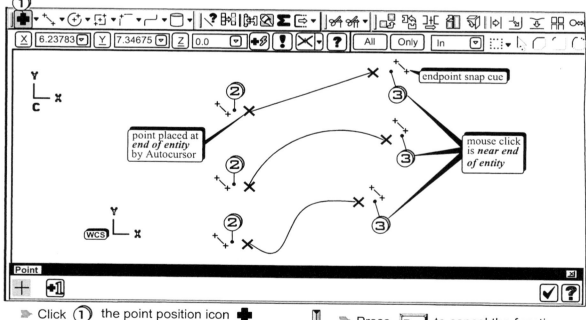

➢ Click ① the point position icon ✚

➢ Click ② ③ *near* the *ends* of the entities

➢ Press [Esc] to cancel the function

☑ Midpoint

A *point* is created at the *midpoint* of an existing *line, arc or spline* entity

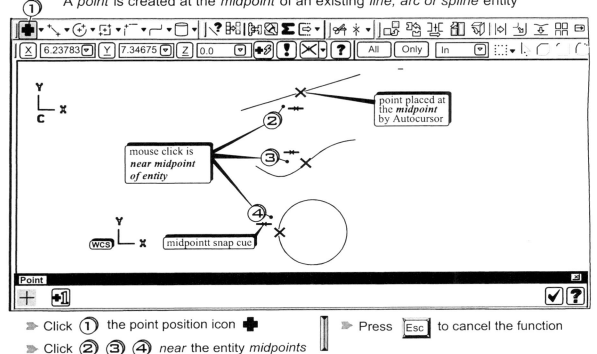

» Click ① the point position icon ✚

» Click ② ③ ④ *near* the entity *midpoints*

» Press Esc to cancel the function

☑ Intersect

A *point* is created at the *intersection* of an existing *line, arc and spline* entities

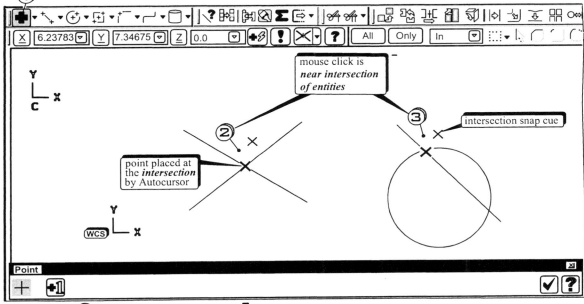

» Click ① the point position icon ✚

» Click ② ③ *near* the *intersection* of the entities

» Press Esc to cancel the function

☑ Quadrant

① A *point* is created at the *0°, 90°, 180° or 270°* angle on an existing *arc*

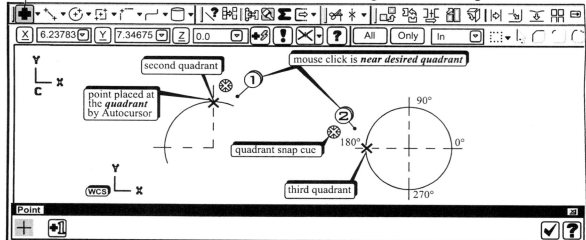

➤ Click ① the point position icon ✚

➤ Click ① ② *near* the *quadrants*

➤ Press [Esc] to cancel the function

⊥ Relative

① A *point* is created at a *relative* distance(*X,Y* or *Rad/Ang*) from an existing point

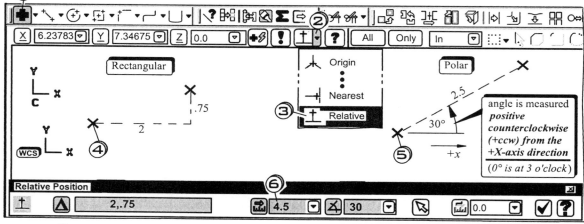

➤ Click ① the point position icon ✚

➤ Click ② the down arrow ▾

➤ Click ③ ⊥ Relative

> Enter a known point or change to Along mode

➤ Click ④ the existing point

➤ Enter *relative* rectang coords **2,.75** [Enter ⏎]

➤ Click ③ ⊥ Relative

➤ Click ⑤ the existing point

➤ Click ⑥ in the *radius* box 🔲 ;

Enter the relative radius [2.5 ▾] [Tab]

➤ In the *angle* box ∡ ;

Enter the relative angle [30 ▾] [Enter ⏎]

Press [Esc] to cancel the function

Create Point Dynamic

Points are created *at any mouse click along* a line, arc or spline entity.

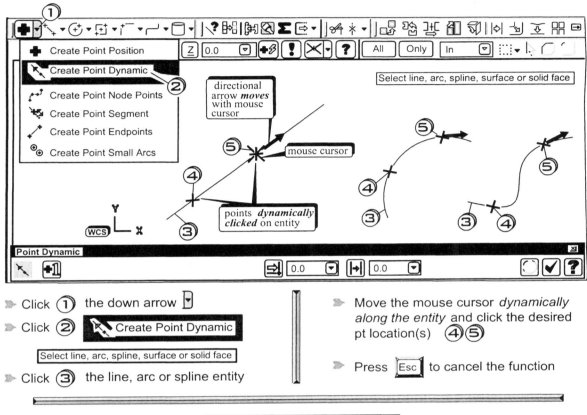

>> Click ① the down arrow ▾

>> Click ② **Create Point Dynamic**

Select line, arc, spline, surface or solid face

>> Click ③ the line, arc or spline entity

>> Move the mouse cursor *dynamically along the entity* and click the desired pt location(s) ④⑤

>> Press Esc to cancel the function

Create Point Node Points   **P-SPLINE**

Points are created *at the node locations of a parametric(P) spline*

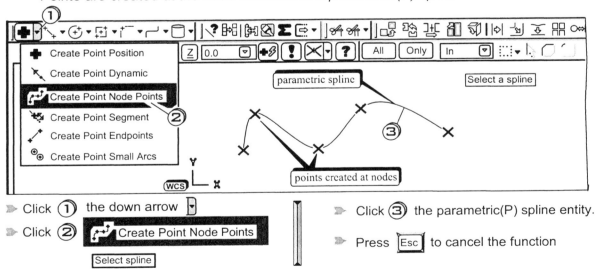

>> Click ① the down arrow ▾

>> Click ② **Create Point Node Points**

Select spline

>> Click ③ the parametric(P) spline entity.

>> Press Esc to cancel the function

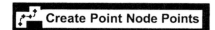 **Create Point Node Points**   **NURBS-SPLINE**

Points are created *at the control locations* of an existing NURBS spline

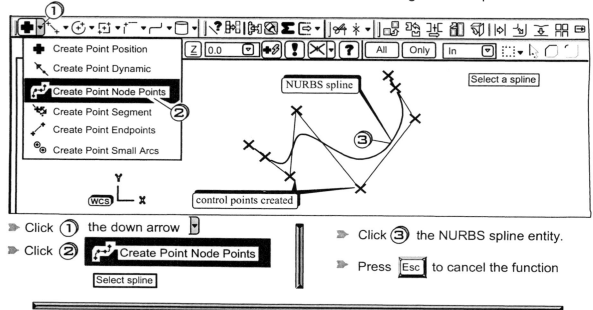

➤ Click ① the down arrow ▾

➤ Click ② 〰 Create Point Node Points

Select spline

➤ Click ③ the NURBS spline entity.

➤ Press Esc to cancel the function

Create Point Segment

A set of *equally* spaced points is created *along* an existing line, arc or
spline entity. The operator can create the point set by specifying either the total
number of points to create ⊞ or the equal spacing between each point ⇥

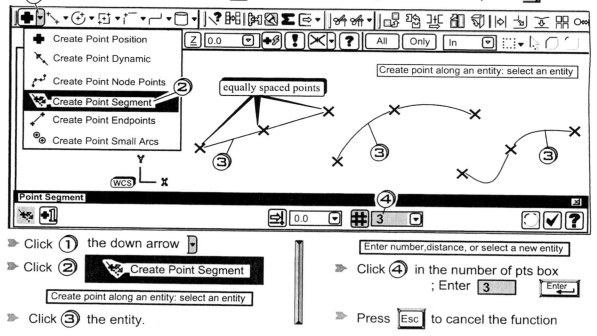

➤ Click ① the down arrow ▾

➤ Click ② 🔻 Create Point Segment

Create point along an entity: select an entity

➤ Click ③ the entity.

➤ Click ④ in the number of pts box
; Enter 3   Enter

➤ Press Esc to cancel the function

Create Point Endpoints

Points are created at the *ends* of existing line, arc or spline entities

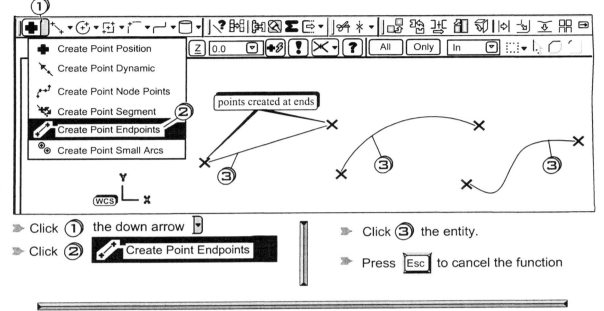

points created at ends

▷ Click ① the down arrow

▷ Click ② **Create Point Endpoints**

▷ Click ③ the entity.

▷ Press Esc to cancel the function

Create Point Small Arcs

Points are created at the *centers* of arcs selected. The operator *specifies* the *maximum radius* of circles to be considered for point creation.

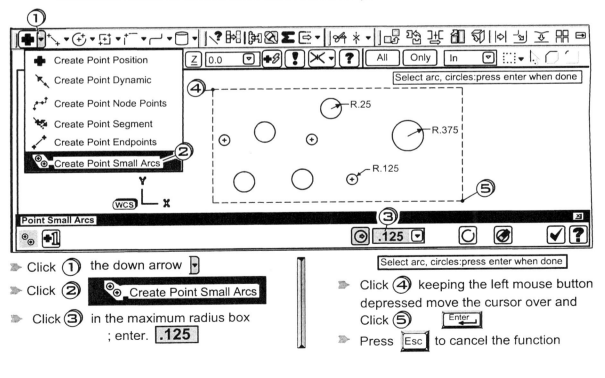

Select arc, circles: press enter when done

▷ Click ① the down arrow

▷ Click ② **Create Point Small Arcs**

▷ Click ③ in the maximum radius box ; enter. **.125**

Select arc, circles: press enter when done

▷ Click ④ keeping the left mouse button depressed move the cursor over and Click ⑤ Enter

▷ Press Esc to cancel the function

2-4 Line Constructions in 2D Space

The various functions for constructing lines via *Mastercam X2* are presented in this section.

Creating Lines

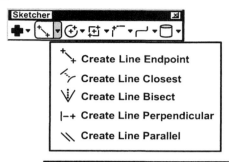

A *horizontal or vertical line* is created between inputted *start* and *end points*. The *horizontal* line can also be *offset* a specified *distance* from the *x*-axis Similarily the *vertical* line can be offset a specified *distance* from the *y-axis*.

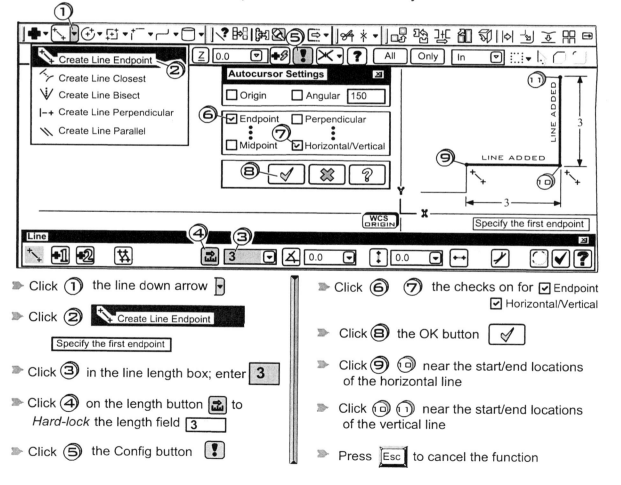

➤ Click ① the line down arrow ▾

➤ Click ② [Create Line Endpoint]
  [Specify the first endpoint]

➤ Click ③ in the line length box; enter [3]

➤ Click ④ on the length button 📊 to *Hard-lock* the length field [3]

➤ Click ⑤ the Config button [!]

➤ Click ⑥ ⑦ the checks on for ☑ Endpoint
                                  ☑ Horizontal/Vertical

➤ Click ⑧ the OK button [✓]

➤ Click ⑨ ⑩ near the start/end locations of the horizontal line

➤ Click ⑩ ⑪ near the start/end locations of the vertical line

➤ Press [Esc] to cancel the function

Create Line Endpoint

A *line* is created *between* inputted *start* and *end points*

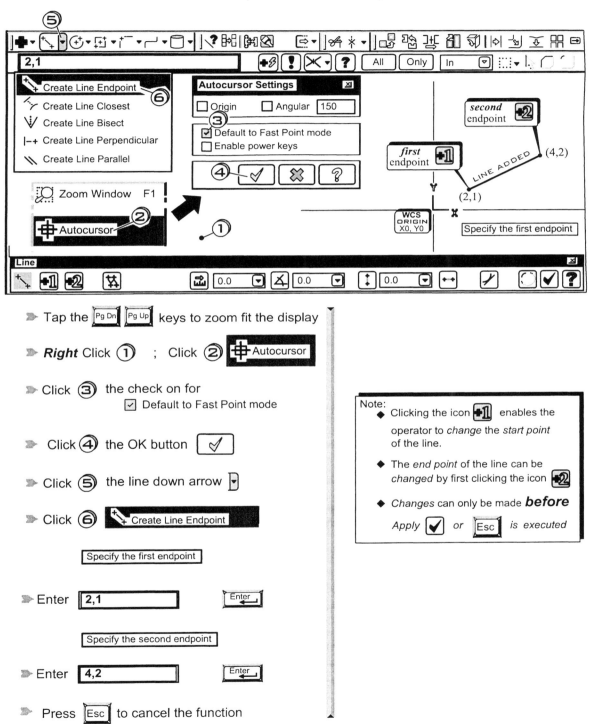

≫ Tap the [Pg Dn] [Pg Up] keys to zoom fit the display

≫ **Right** Click ① ; Click ② ⊞Autocursor

≫ Click ③ the check on for
   ☑ Default to Fast Point mode

≫ Click ④ the OK button ✓

≫ Click ⑤ the line down arrow ⌄

≫ Click ⑥ ⤢ Create Line Endpoint

   Specify the first endpoint

≫ Enter 2,1   [Enter ↵]

   Specify the second endpoint

≫ Enter 4,2   [Enter ↵]

≫ Press [Esc] to cancel the function

Note:
 ◆ Clicking the icon ⊞➊ enables the operator to *change* the *start point* of the line.

 ◆ The *end point* of the line can be *changed* by first clicking the icon ⊞➋

 ◆ *Changes* can only be made **before**

   Apply ☑ or [Esc] *is executed*

Create Line Endpoint ▸ *MULTIPLE* or ◂   M   ☑ Default to Fast Point mode

Multiple lines are created between inputted *start* and *end points*.
*Absolute coordinates are used: each new X,Y is measured from the WCS **origin***

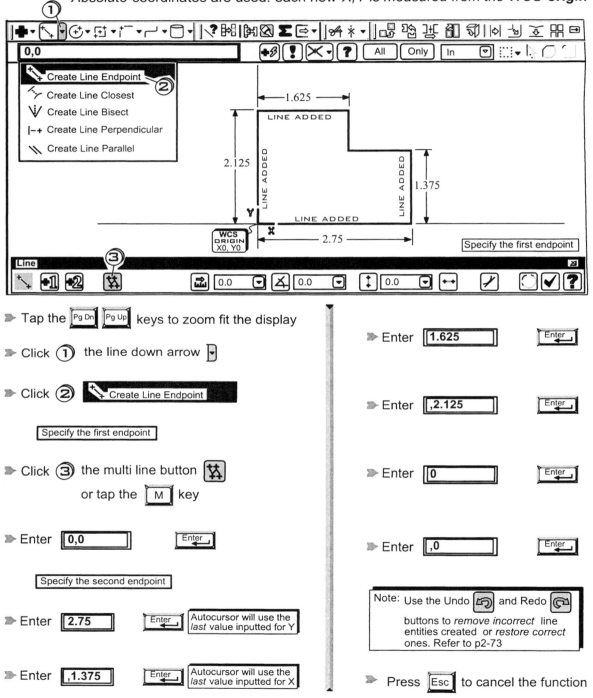

⮞ Tap the Pg Dn Pg Up keys to zoom fit the display

⮞ Click ① the line down arrow ▼

⮞ Click ② ⬛ Create Line Endpoint

 Specify the first endpoint

⮞ Click ③ the multi line button 🗘
       or tap the M key

⮞ Enter 0,0   Enter

 Specify the second endpoint

⮞ Enter 2.75   Enter Autocursor will use the *last* value inputted for Y

⮞ Enter ,1.375   Enter Autocursor will use the *last* value inputted for X

⮞ Enter 1.625   Enter

⮞ Enter ,2.125   Enter

⮞ Enter 0   Enter

⮞ Enter ,0   Enter

Note: Use the Undo 🔄 and Redo 🔁
   buttons to *remove incorrect* line
   entities created or *restore correct*
   ones. Refer to p2-73

⮞ Press Esc to cancel the function

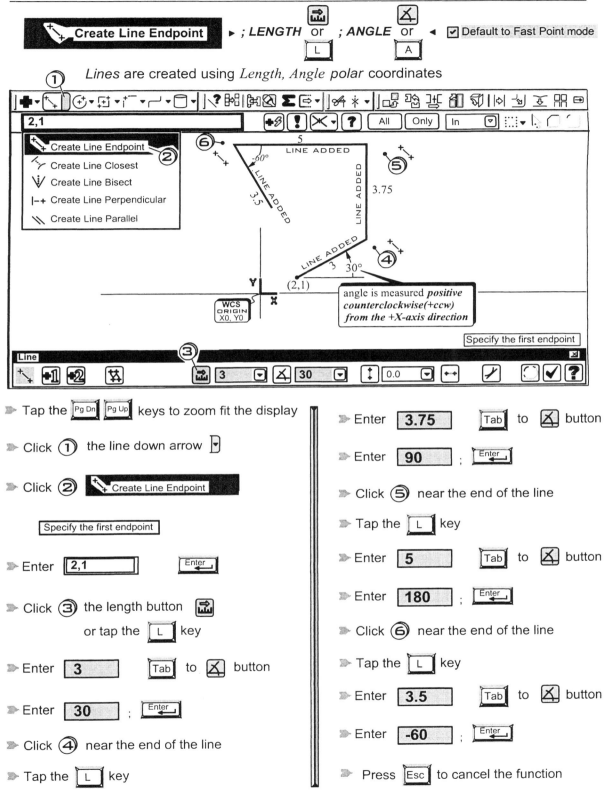

Create Line Endpoint ▸ ; *LENGTH* or ; *ANGLE* or ◂ ☑ Default to Fast Point mode

Lines are created using *Length, Angle polar* coordinates

Create Line Endpoint
Create Line Closest
Create Line Bisect
Create Line Perpendicular
Create Line Parallel

LINE ADDED

-60°
3.5
3.75
LINE ADDED
LINE ADDED
3   30°
(2,1)

angle is measured *positive counterclockwise(+ccw) from the +X-axis direction*

Specify the first endpoint

WCS ORIGIN X0, Y0

Line   3   30   0.0

➤ Tap the `Pg Dn` `Pg Up` keys to zoom fit the display

➤ Click ① the line down arrow ⊡

➤ Click ② **Create Line Endpoint**

Specify the first endpoint

➤ Enter `2,1`   `Enter`

➤ Click ③ the length button 🔲
  or tap the `L` key

➤ Enter `3`   `Tab` to ⊠ button

➤ Enter `30` ; `Enter`

➤ Click ④ near the end of the line

➤ Tap the `L` key

➤ Enter `3.75`   `Tab` to ⊠ button

➤ Enter `90` ; `Enter`

➤ Click ⑤ near the end of the line

➤ Tap the `L` key

➤ Enter `5`   `Tab` to ⊠ button

➤ Enter `180` ; `Enter`

➤ Click ⑥ near the end of the line

➤ Tap the `L` key

➤ Enter `3.5`   `Tab` to ⊠ button

➤ Enter `-60` ; `Enter`

➤ Press `Esc` to cancel the function

Create Line Endpoint ▶ *TANGENT* or ; *LENGTH* or ; *ANGLE* or ◀

T   L   A

A *line* is created starting at a *tangency* to an *arc or spline* and having an inputted *Length* and *Angle*.

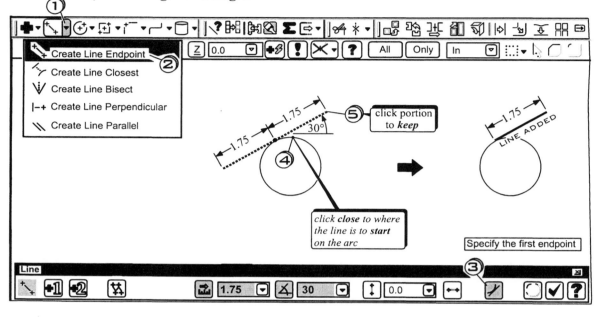

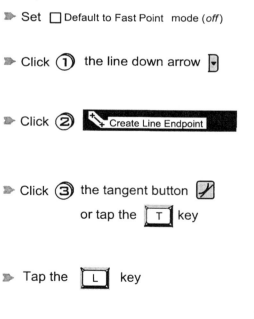

➤ Set ☐ Default to Fast Point mode (*off*)

➤ Click ① the line down arrow ▾

➤ Click ② [Create Line Endpoint]

➤ Click ③ the tangent button 🖉

      or tap the T key

➤ Tap the L key

➤ Enter **1.75** [Tab]

➤ Enter **30**

      Specify the first endpoint

➤ Click ④ the arc or spline entity

      Select which line to keep

➤ Click ⑤ the *portion* of the tangent line to *keep*

➤ Press Esc to cancel the function

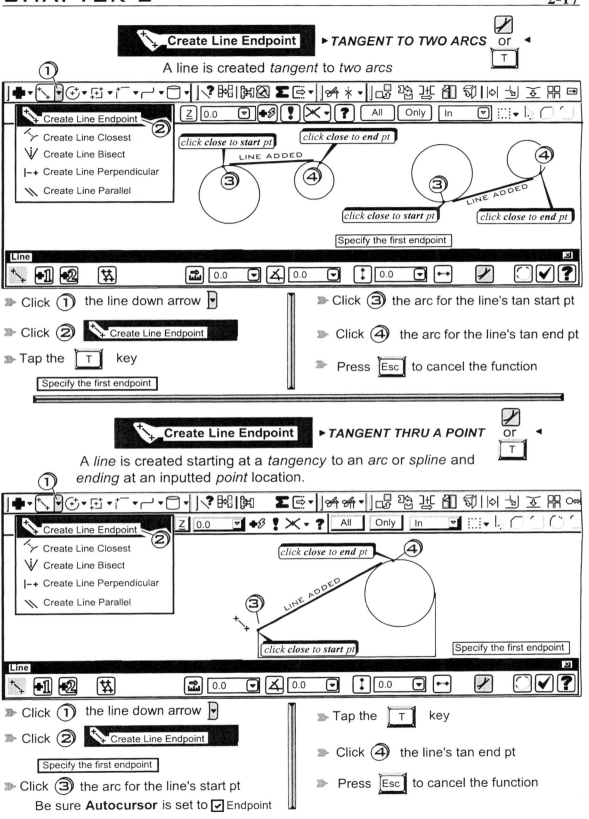

Create Line Endpoint ▸ *TANGENT TO TWO ARCS* or ◂

A line is created *tangent* to *two arcs*

➤ Click ① the line down arrow

➤ Click ② **Create Line Endpoint**

➤ Tap the T key

Specify the first endpoint

➤ Click ③ the arc for the line's tan start pt

➤ Click ④ the arc for the line's tan end pt

➤ Press Esc to cancel the function

Create Line Endpoint ▸ *TANGENT THRU A POINT* or ◂

A *line* is created starting at a *tangency* to an *arc* or *spline* and *ending* at an inputted *point* location.

➤ Click ① the line down arrow

➤ Click ② **Create Line Endpoint**

Specify the first endpoint

➤ Click ③ the arc for the line's start pt
  Be sure **Autocursor** is set to ☑ Endpoint

➤ Tap the T key

➤ Click ④ the line's tan end pt

➤ Press Esc to cancel the function

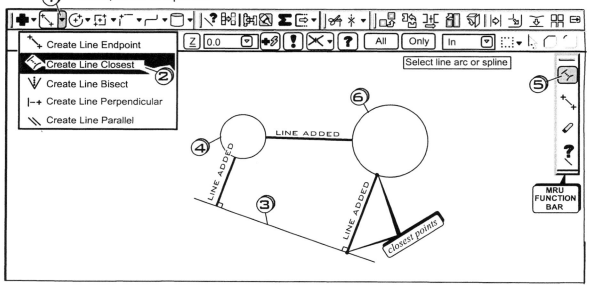

Create Line Closest

A *line* of specified *length* is created *between* the *closest points* of *adjacent* lines, arcs or splines

➤ Click ① the line down arrow ▼

➤ Click ② [Create Line Closest]

Select line arc or spline

➤ Click ③ ④ the line and circle

➤ Click ⑤ the line nearest icon

➤ Click ④ ⑥ the circles

➤ Click ⑤ the line nearest icon

➤ Click ③ ⑥ the line and circle

➤ Press Esc to cancel the function

Create Line Bisect

A *line* of specific *length* is created that *bisects* the *angle* between two *intersecting lines*

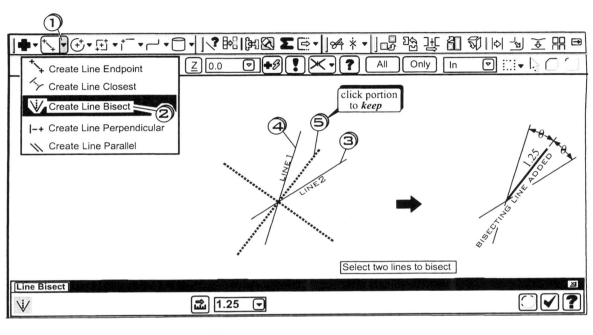

➤ Click ① the line down arrow

➤ Click ② Create Line Bisect

➤ Tap the L key

➤ Enter 1.25

Select two lines to bisect

➤ Click ③ ④ the intersecting line entities

Select which line to keep

➤ Click ⑤ the *portion* of the bisect line to *keep*

➤ Press Esc to cancel the function

|-+ Create Line Perpendicular

A *line* is created *starting* at a *perpendicular* to an existing *line arc, spline* or NURBS curve and *ending* at an inputted *point* location.

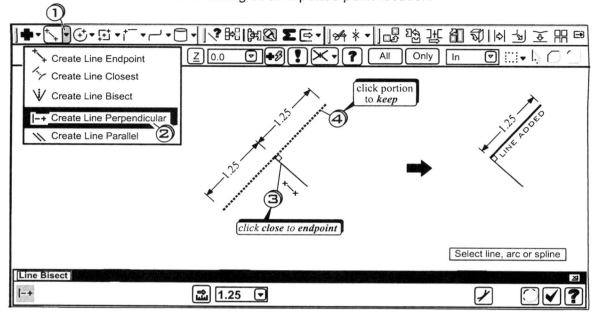

➤ Click ① the line down arrow ▾

➤ Click ②   **|-+ Create Line Perpendicular**

➤ Tap the [L] key

➤ Enter   **1.25**

Select line, arc or spline

➤ Click ③ near the end of the line entity

Select which line to keep

➤ Click ④ the *portion* of the perpendicular line to *keep*

➤ Press Esc to cancel the function

A *line* is created *starting* at a *perpendicular* to an existing *line* and *ending tangent* to an existing *arc*.

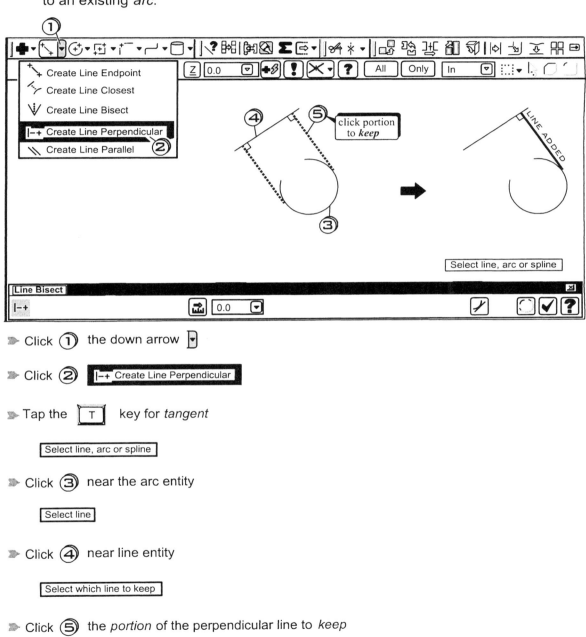

➤ Click ① the down arrow ⊡

➤ Click ② |-+ Create Line Perpendicular

➤ Tap the ⌷ T ⌷ key for *tangent*

   Select line, arc or spline

➤ Click ③ near the arc entity

   Select line

➤ Click ④ near line entity

   Select which line to keep

➤ Click ⑤ the *portion* of the perpendicular line to *keep*

➤ Press ⌷Esc⌷ to cancel the function

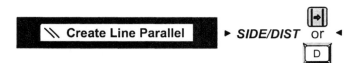

A *line* is created *parallel* to an existing *line* at an *offset distance* inputted by the operator.

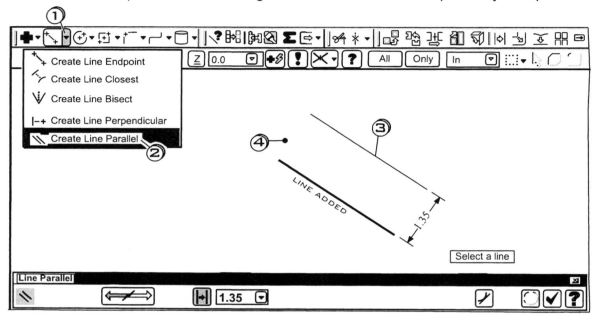

➤ Click ① the down arrow ▾

➤ Click ② \\ Create Line Parallel

➤ Tap the D key for *offset distance*

➤ Enter **1.35**

Select a line

➤ Click ③ near the line entity

Indicate iffset direction

➤ Click ④   the side on which the parallel line is to be created

➤ Press Esc to cancel the function

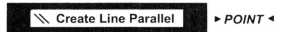

POINT ◄

A *line* is created *parallel* to an existing *line* and *passing through* a specified *point* location

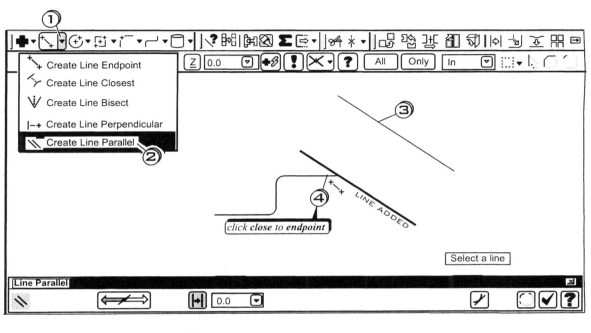

➤ Click ① the down arrow ▾

➤ Click ②

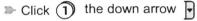

Select a line

➤ Click ③ near the line entity

Select the point to place a parallel line through

Be sure **Autocursor** is set to ☑ Endpoint

➤ Click ④ near the end point of the existing line

➤ Press Esc to cancel the function

A *line* is created *parallel* to an existing *line* and *tangent* to an existing *arc*

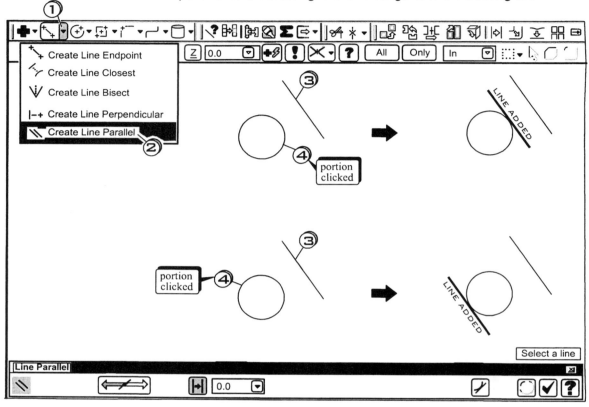

➤ Click ① the down arrow ▾

➤ Click ② ╲╲ Create Line Parallel

➤ Tap the T key for *tangent*

   Select a line

➤ Click ③ near the line entity

   Select an arc to place a parallel line tangent to

➤ Click ④ on the side of the arc *where the tangent line is to be created*

➤ Press Esc to cancel the function

2-5 Arc Constructions in 2D Space

The functions for constructing arcs(including circles) via *Mastercam X2* are considered in this section

Creating Arcs

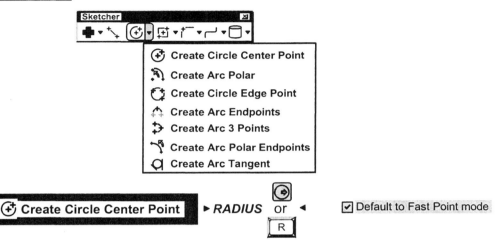

A *circle* is created by entering its *radius* and *center point* values

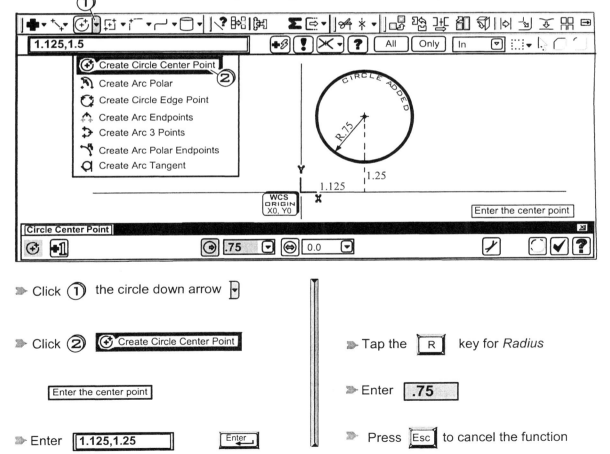

➤ Click ① the circle down arrow

➤ Click ② Create Circle Center Point

Enter the center point

➤ Enter 1.125,1.25   Enter

➤ Tap the R key for *Radius*

➤ Enter .75

➤ Press Esc to cancel the function

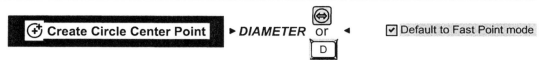

A *circle* is created by entering its *diameter* and *center point* values

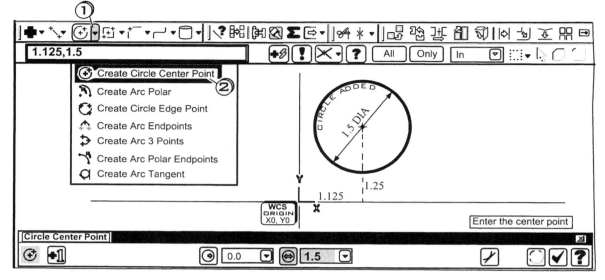

➤ Click ① the circle down arrow ▾

➤ Click ② [⊕ Create Circle Center Point]

[Enter the center point]

➤ Enter [1.125,1.25]   [Enter]

➤ Tap the [D] key for *Diameter*

➤ Enter [1.5]

➤ Press [Esc] to cancel the function

A *circle* is created by entering its *diameter*. Its *center point* value is entered *relative* to an *existing* entity

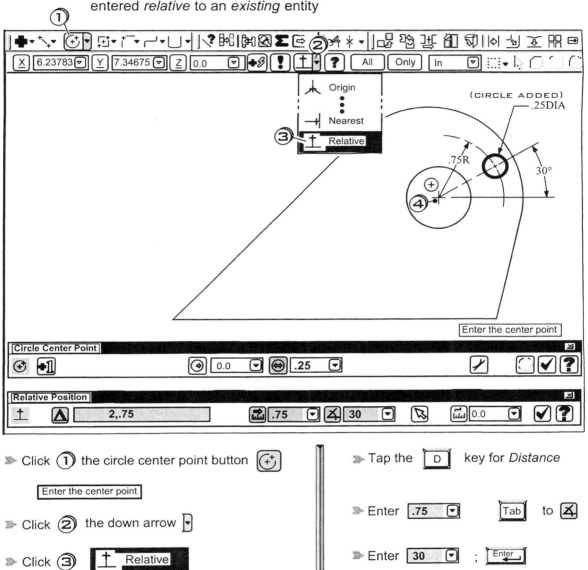

➤ Click ① the circle center point button ⊕

Enter the center point

➤ Click ② the down arrow ▾

➤ Click ③ | ⊥ Relative |

   Be sure **Autocursor** is set to ☑ Center

➤ Click ④ near the center of the circle

   . Move the cursor *into the ribbon bar on the distance icon* and *hold it there*..

➤ Tap the D key for *Distance*

➤ Enter .75 ▾   Tab to X

➤ Enter 30 ▾ ; Enter ↵

➤ Tap the D key for *Diameter*

➤ Enter .25 ▾ ; Enter ↵

➤ Press Esc to cancel the function

An *arc* is created at inputted *center point, radius, starting angle* and *ending angle* values.

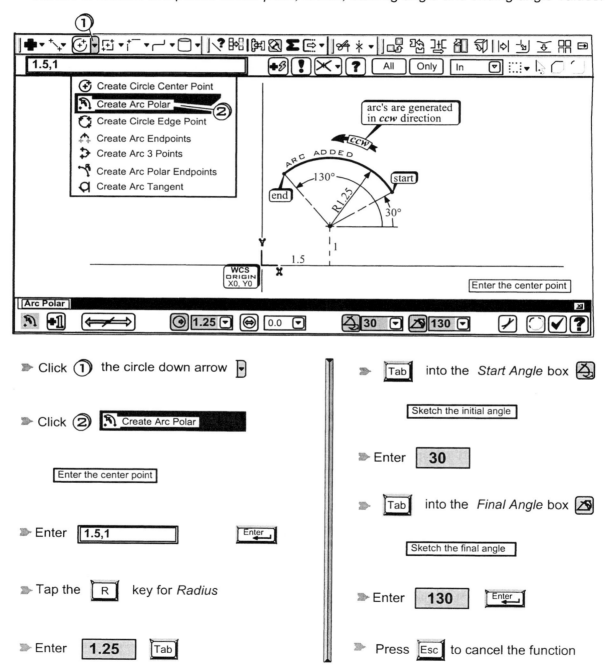

➤ Click ① the circle down arrow ▾

➤ Click ② 🔃 Create Arc Polar

Enter the center point

➤ Enter 1.5,1 Enter↵

➤ Tap the R key for *Radius*

➤ Enter 1.25 Tab

➤ Tab into the *Start Angle* box 🔺

Sketch the initial angle

➤ Enter 30

➤ Tab into the *Final Angle* box 🔻

Sketch the final angle

➤ Enter 130 Enter↵

➤ Press Esc to cancel the function

☑ Default to Fast Point mode

An *arc* is created at inputted *center point and radius values. Mouse clicks* specify the *start* and *end angles*.

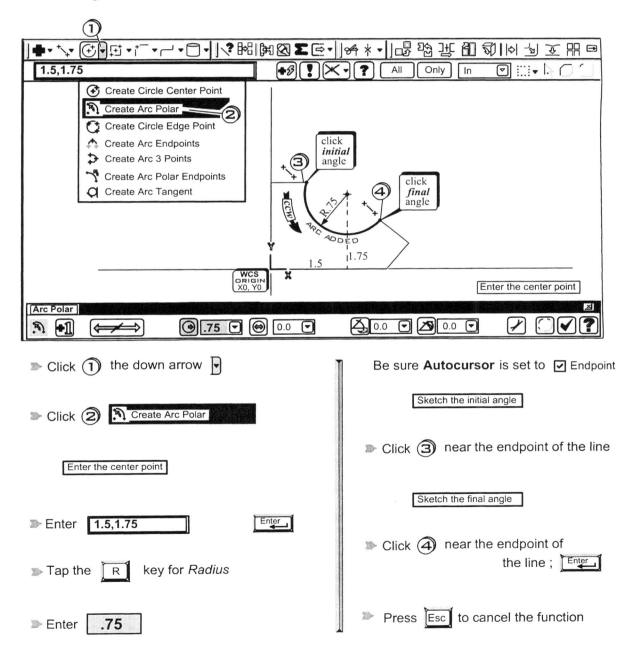

➤ Click ① the down arrow ▾

➤ Click ② 🖋 Create Arc Polar

Enter the center point

➤ Enter **1.5,1.75**   Enter↵

➤ Tap the R key for *Radius*

➤ Enter **.75**

Be sure **Autocursor** is set to ☑ Endpoint

Sketch the initial angle

➤ Click ③ near the endpoint of the line

Sketch the final angle

➤ Click ④ near the endpoint of the line ; Enter↵

➤ Press Esc to cancel the function

A *circle* is created whose *diameter* is defined by two inputted *end point* locations

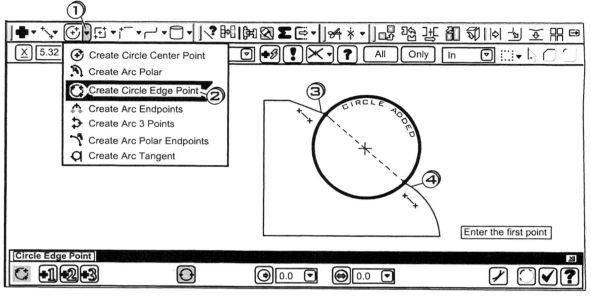

➤ Click ① the circle down arrow 🔽

➤ Click ② ⟳ Create Circle Edge Point

➤ Tap the [W] key for *2-pt circle*

    Be sure **Autocursor** is set to ☑ Endpoint

        Enter the first point

➤ Click ③ near the endpoint of the line

        Enter the second point

➤ Click ④ near the endpoint of the line

➤ Press [Esc] to cancel the function

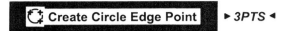

Create Circle Edge Point ► *3PTS* ◄

A *circle* is created whose *circumference* is defined by *three* inputted *end point* locations

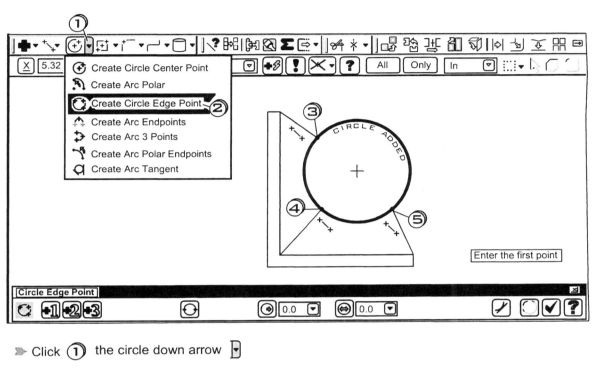

➤ Click ① the circle down arrow ▾

➤ Click ② **Create Circle Edge Point**

　　　Be sure **Autocursor** is set to ☑ Endpoint

　　　Enter the first point

➤ Click ③ near the endpoint of the line

　　　Enter the second point

➤ Click ④ near the endpoint of the line

　　　Enter the third point

➤ Click ⑤ near the endpoint of the line

➤ Press Esc to cancel the function

A *circle* is created whose *circumference* is defined by its *tangency* to *three* clicked entities

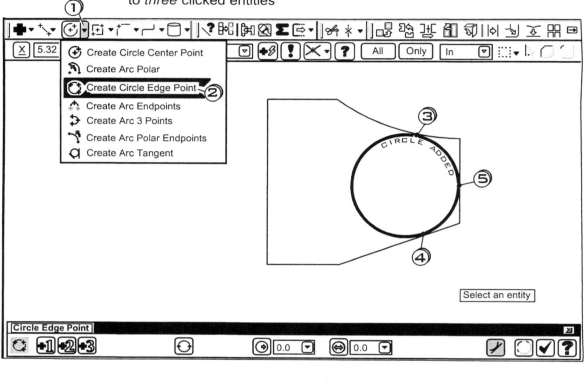

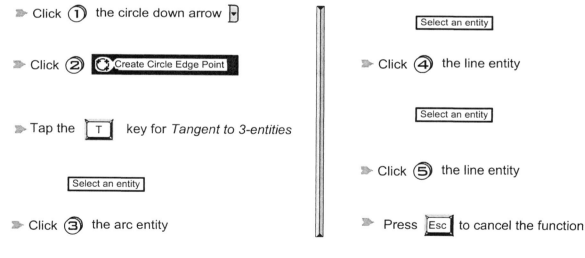

➤ Click ① the circle down arrow

➤ Click ② Create Circle Edge Point

➤ Tap the T key for *Tangent to 3-entities*

Select an entity

➤ Click ③ the arc entity

Select an entity

➤ Click ④ the line entity

Select an entity

➤ Click ⑤ the line entity

➤ Press Esc to cancel the function

A *circle* is created of specified *radius* or *diameter tangent* to *two* other existing *line* or *arc* entities.

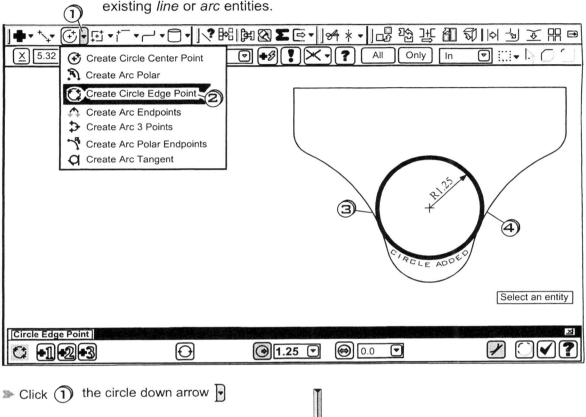

➤ Click ① the circle down arrow ▾

➤ Click ② 🔾 Create Circle Edge Point

➤ Tap the [T] key for *Tangent to 3-entities*

> Select an entity

➤ Tap the [R] key for *Radius*

➤ Enter **1.25**

> Select an entity

➤ Click ③ the arc entity

> Select an entity

➤ Click ④ the line entity

➤ Press [Esc] to cancel the function

Create Arc Endpoints

Arcs are created *starting/ending* at inputted *start/end* points and having
a specified *radius value.* The operator *clicks* the *portion* of the arc to **keep.**

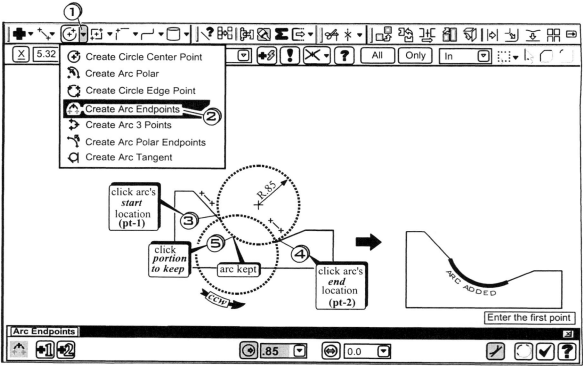

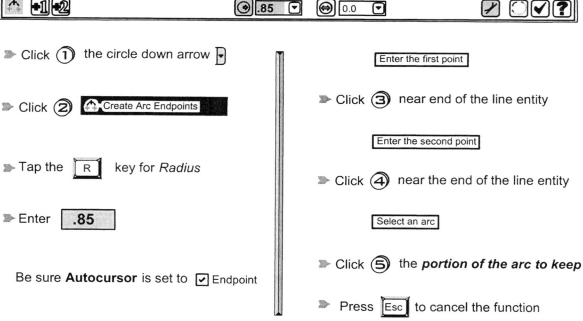

Click ① the circle down arrow

Click ② Create Arc Endpoints

Tap the R key for *Radius*

Enter .85

Be sure **Autocursor** is set to ☑ Endpoint

Enter the first point

Click ③ near end of the line entity

Enter the second point

Click ④ near the end of the line entity

Select an arc

Click ⑤ the *portion of the arc to keep*

Press Esc to cancel the function

Create Arc 3 Points

An *arc* is created passing through *three* inputted points. The *first* point specifies the arc's *start* location the *second* point the arc's *end* location The *third* point clicked is an *intermediate* location *between* the *first* and *second* points through which the arc must also *pass*.

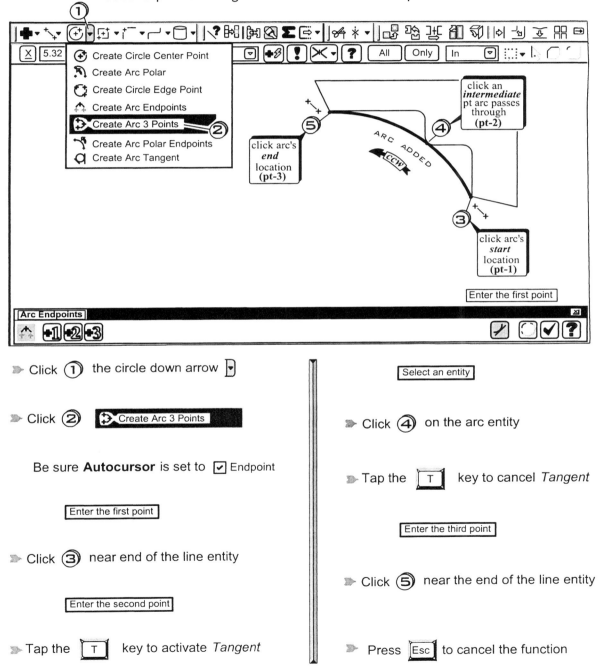

≫ Click ① the circle down arrow 🔽

≫ Click ② **Create Arc 3 Points**

Be sure **Autocursor** is set to ☑ Endpoint

Enter the first point

≫ Click ③ near end of the line entity

Enter the second point

≫ Tap the ⊤ key to activate *Tangent*

Select an entity

≫ Click ④ on the arc entity

≫ Tap the ⊤ key to cancel *Tangent*

Enter the third point

≫ Click ⑤ near the end of the line entity

≫ Press Esc to cancel the function

An *arc* is created *starting* at a inputted *start* point and having specified *values* for its *radius* and *start/end angles*.

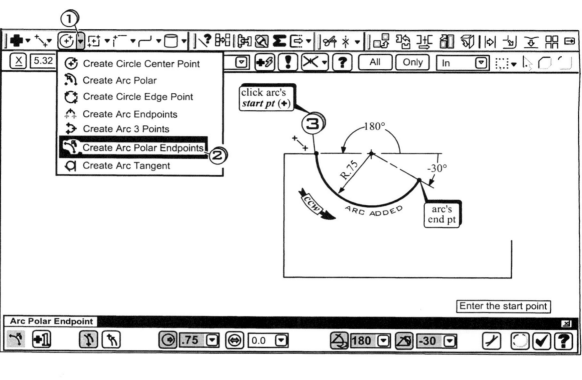

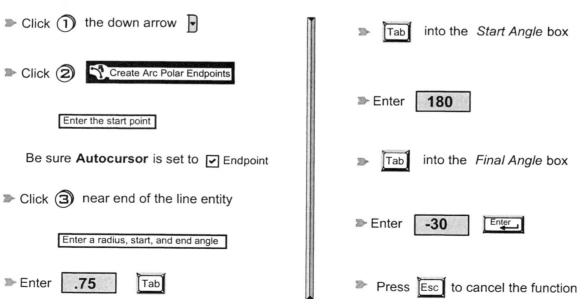

➤ Click ① the down arrow

➤ Click ② Create Arc Polar Endpoints

Enter the start point

Be sure **Autocursor** is set to ☑ Endpoint

➤ Click ③ near end of the line entity

Enter a radius, start, and end angle

➤ Enter **.75** Tab

➤ Tab into the *Start Angle* box

➤ Enter **180**

➤ Tab into the *Final Angle* box

➤ Enter **-30** Enter

➤ Press Esc to cancel the function

An *arc* is created *ending* at a inputted *end* point and having specified *values* for its *radius* and *start/end angles*.

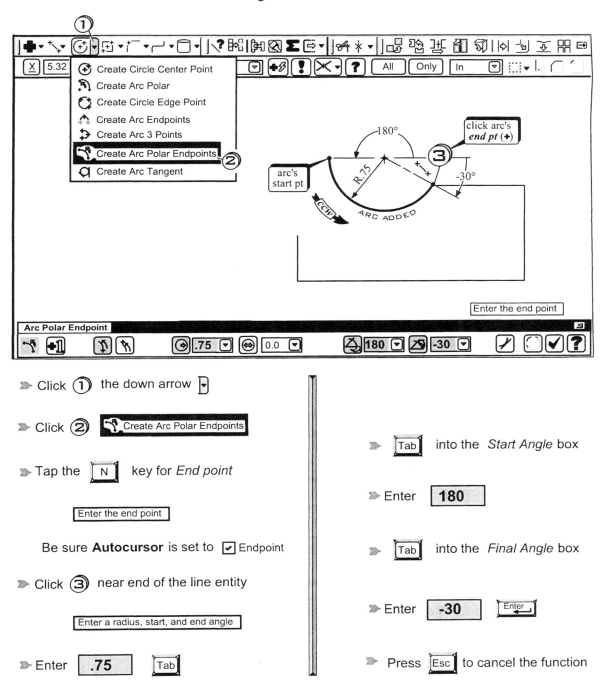

➤ Click ① the down arrow ⊡

➤ Click ② Create Arc Polar Endpoints

➤ Tap the N key for *End point*

Enter the end point

Be sure **Autocursor** is set to ☑ Endpoint

➤ Click ③ near end of the line entity

Enter a radius, start, and end angle

➤ Enter .75 Tab

➤ Tab into the *Start Angle* box

➤ Enter 180

➤ Tab into the *Final Angle* box

➤ Enter -30 Enter

➤ Press Esc to cancel the function

180-degree arcs are created *tangent* to an existing line or arc entity at a specified *point along* the entity. The operator inputs the *radius* and *clicks* the arc *portion to keep*.

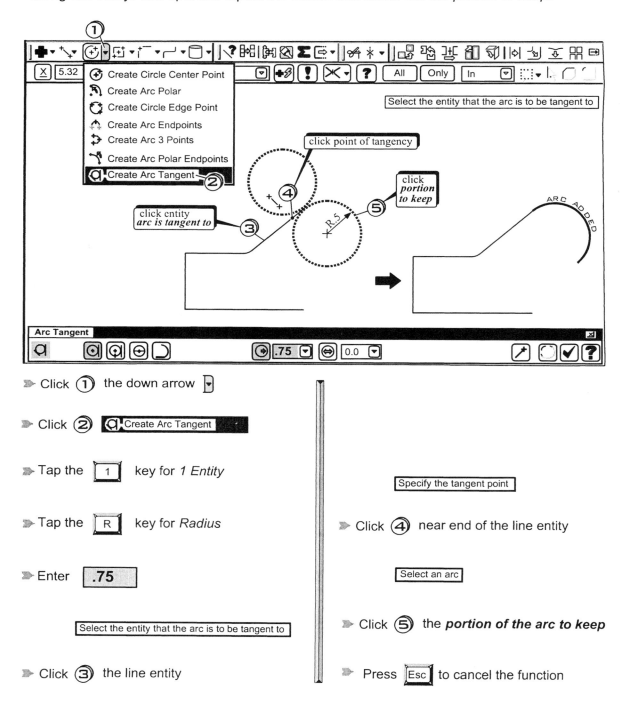

➤ Click ① the down arrow ▾

➤ Click ② 🔲 Create Arc Tangent ▾

➤ Tap the ⊡1 key for *1 Entity*

➤ Tap the ⊡R key for *Radius*

➤ Enter **.75**

Select the entity that the arc is to be tangent to

➤ Click ③ the line entity

Specify the tangent point

➤ Click ④ near end of the line entity

Select an arc

➤ Click ⑤ the *portion of the arc to keep*

➤ Press ⊡Esc to cancel the function

360-degree arcs are created *tangent* to an existing *line* clicked and positioned such that they *pass through a point* inputted. The *radius* is inputted and the operator clicks the *portion* of the *arc to* **keep**.

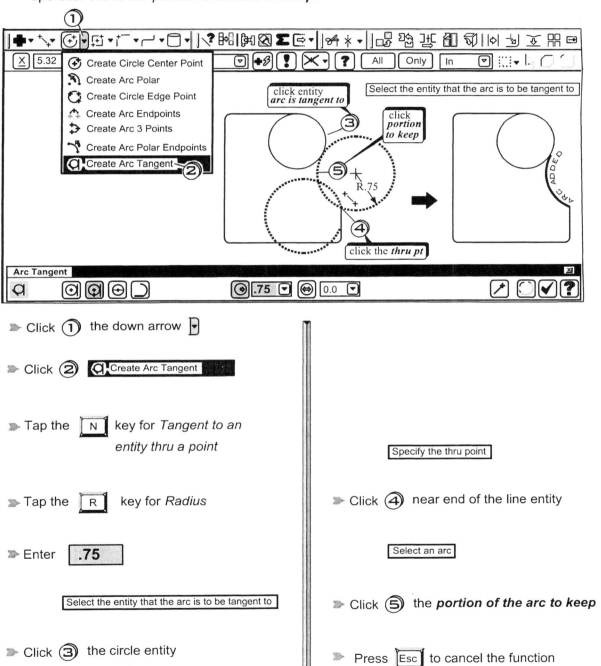

➤ Click ① the down arrow

➤ Click ② Create Arc Tangent

➤ Tap the N key for *Tangent to an entity thru a point*

➤ Tap the R key for *Radius*

➤ Enter **.75**

> Select the entity that the arc is to be tangent to

➤ Click ③ the circle entity

> Specify the thru point

➤ Click ④ near end of the line entity

> Select an arc

➤ Click ⑤ the **portion of the arc to keep**

➤ Press Esc to cancel the function

360-degree arcs are created *tangent* to an existing *line* clicked and positioned such that a *second line* clicked passes through their *centers*. The radius is inputted and the operator clicks the *arc to* **keep**.

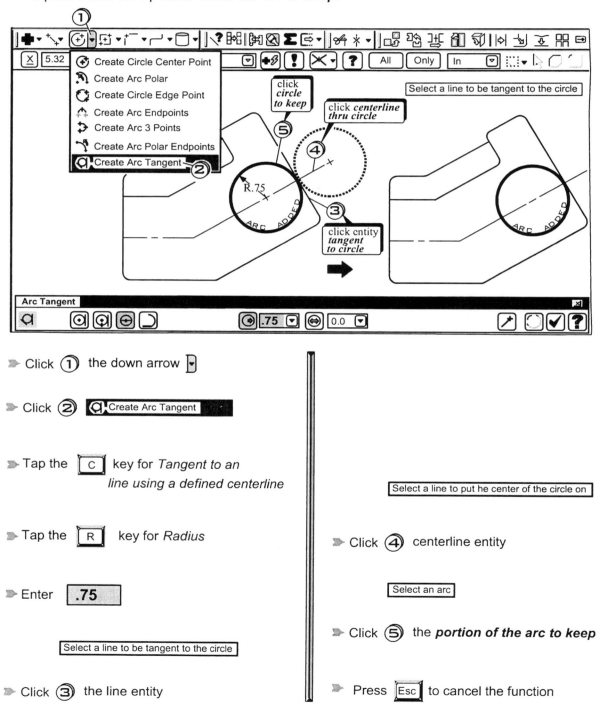

➤ Click ① the down arrow ▾

➤ Click ② [Create Arc Tangent]

➤ Tap the [C] key for *Tangent to an line using a defined centerline*

➤ Tap the [R] key for *Radius*

➤ Enter **.75**

> Select a line to be tangent to the circle

➤ Click ③ the line entity

> Select a line to put he center of the circle on

➤ Click ④ centerline entity

> Select an arc

➤ Click ⑤ the *portion of the arc to keep*

➤ Press [Esc] to cancel the function

An *arc* is created starting at a point *dynamically* clicked on an existing *tangent line* or *arc* and *ending at a point* inputted.

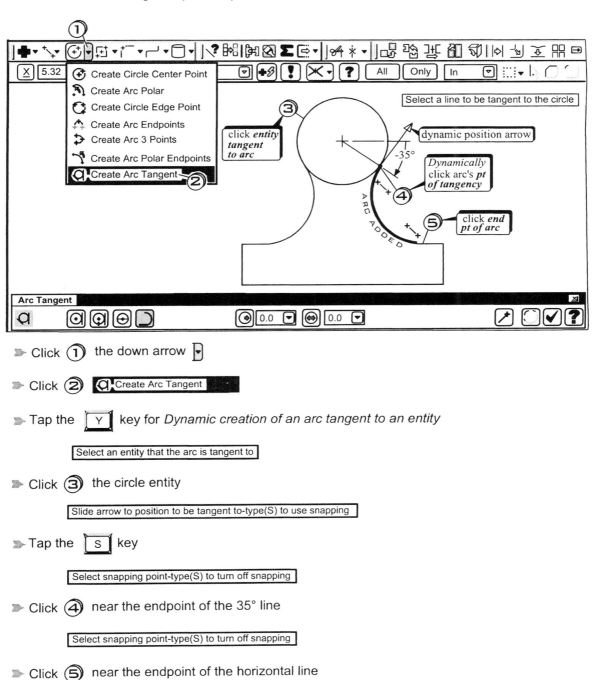

➤ Click ① the down arrow ▾

➤ Click ② ⟨a⟩ Create Arc Tangent ▾

➤ Tap the ⟨Y⟩ key for *Dynamic creation of an arc tangent to an entity*

Select an entity that the arc is tangent to

➤ Click ③ the circle entity

Slide arrow to position to be tangent to-type(S) to use snapping

➤ Tap the ⟨S⟩ key

Select snapping point-type(S) to turn off snapping

➤ Click ④ near the endpoint of the 35° line

Select snapping point-type(S) to turn off snapping

➤ Click ⑤ near the endpoint of the horizontal line

➤ Press ⟨Esc⟩ to cancel the function

2-6 Rectangle Constructions in 2D Space

The create rectangular shapes function enables the operator to create rectangular shapes at any angle having 90° sharp corners or rounded corners. Other shapes that can be created with this function include *Obround*, *Single D*, *Double D* and *Ellipse*.

Creating Rectangular Shapes

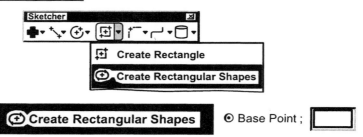

A *rectangle* is created. The operator specifies the *Width*, *Height* and *Anchor position*.

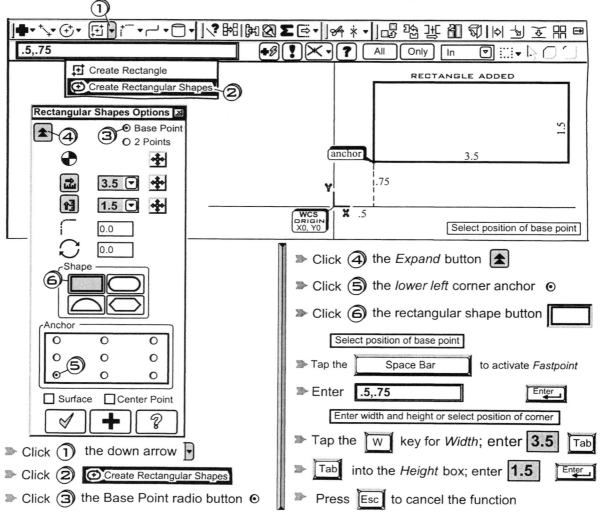

➤ Click ① the down arrow ▾

➤ Click ② ⊕ Create Rectangular Shapes

➤ Click ③ the Base Point radio button ⊙

➤ Click ④ the *Expand* button ⬆

➤ Click ⑤ the *lower left* corner anchor ⊙

➤ Click ⑥ the rectangular shape button ▭

Select position of base point

➤ Tap the Space Bar to activate *Fastpoint*

➤ Enter .5,.75 Enter

Enter width and height or select position of corner

➤ Tap the W key for *Width*; enter 3.5 Tab

➤ Tab into the *Height* box; enter 1.5 Enter

➤ Press Esc to cancel the function

Create Rectangular Shapes   ⊙ 2 Points ;

A *rectangle* is created whose *size* and *location* is determined by inputting its *diagonal endpoints*.

Rectangular Shapes Options

③ ○ Base Point
   ⊙ 2 Points

④ Expand

0.0
0.0

⑤ Shape

☐ Surface   ☐ Center Point

RECTANGLE ADDED

⑥

diagonal

⑦

Select position of first corner

➤ Click ① the down arrow

➤ Click ②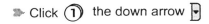

➤ Click ③ the 2 Points radio button ⊙

➤ Click ④ the *Expand* button ⬆

Select position of first corner

➤ Click ⑤ the rectangular shape button

Be sure **Autocursor** is set to

☑ Endpoint

➤ Click ⑥ near the end of the arc

Select position of second corner

➤ Click ⑦ near the end of the line

➤ Press Esc to cancel the function

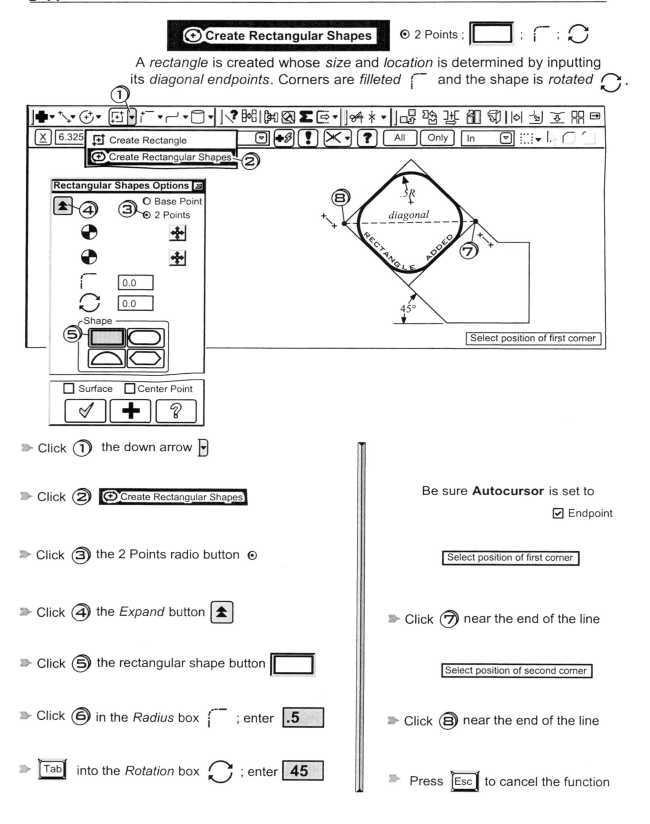

Create Rectangular Shapes   ⊙ 2 Points ; ⬜ ; ⌐ ; ↻

A *rectangle* is created whose *size* and *location* is determined by inputting its *diagonal endpoints*. Corners are *filleted* ⌐ and the shape is *rotated* ↻.

Select position of first corner

➤ Click ① the down arrow ▾

➤ Click ② ⊕Create Rectangular Shapes

➤ Click ③ the 2 Points radio button ⊙

➤ Click ④ the *Expand* button ⬆

➤ Click ⑤ the rectangular shape button ⬜

➤ Click ⑥ in the *Radius* box ⌐ ; enter .5

➤ Tab into the *Rotation* box ↻ ; enter 45

Be sure **Autocursor** is set to

☑ Endpoint

Select position of first corner

➤ Click ⑦ near the end of the line

Select position of second corner

➤ Click ⑧ near the end of the line

➤ Press Esc to cancel the function

2-7 Polygon Constructions in 2D Space
Creating Polygons

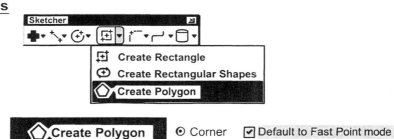

A *polygon* is created. The operator enters the *Number of sides*, *Radius*, *Start angle*(measured *+ccw* from the *+x-axis*). The operator also indicates the *radius* is measured from the *center to the Corner* of the polygon and inputs the polygon's *center* location.

.75,1.125

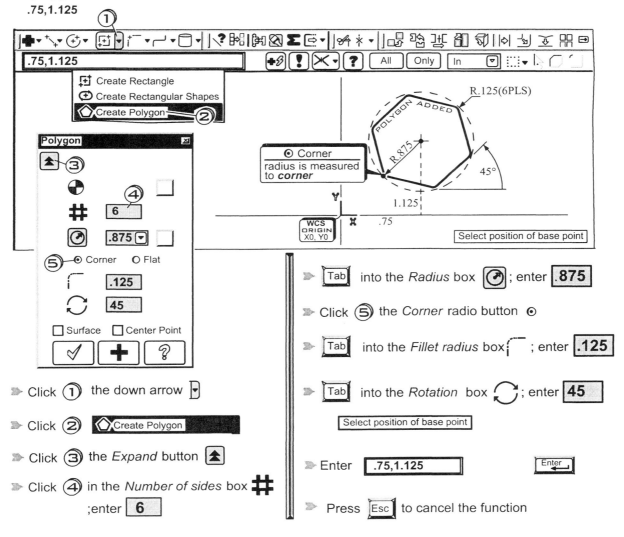

➤ Click ① the down arrow ▾

➤ Click ② Create Polygon

➤ Click ③ the *Expand* button ▲

➤ Click ④ in the *Number of sides* box #
     ;enter **6**

➤ Tab into the *Radius* box ⊘ ; enter **.875**

➤ Click ⑤ the *Corner* radio button ⊙

➤ Tab into the *Fillet radius* box ; enter **.125**

➤ Tab into the *Rotation* box ↻ ; enter **45**

Select position of base point

➤ Enter **.75,1.125**   Enter ↵

➤ Press Esc to cancel the function

⬠ **Create Polygon**   ⊙ Flat   ☑ Default to Fast Point mode

A *polygon* is created. The operator enters the *Number of sides*, *Radius*, *Start angle*(measured *+ccw* from the *+x-axis*). The operator also indicates the *radius* is measured from the *center to the midpoint of the flat side* of the polygon and clicks the polygon's *center* location.

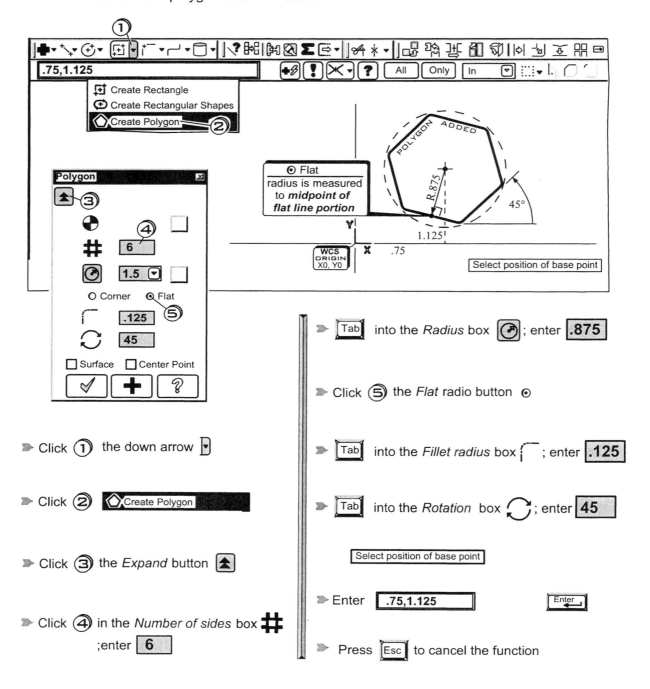

➤ Click ① the down arrow ▾

➤ Click ② ⬠ Create Polygon

➤ Click ③ the *Expand* button ⬆

➤ Click ④ in the *Number of sides* box ⌗
  ;enter 6

➤ Tab into the *Radius* box ⊘; enter .875

➤ Click ⑤ the *Flat* radio button ⊙

➤ Tab into the *Fillet radius* box ⌐; enter .125

➤ Tab into the *Rotation* box ↻; enter 45

Select position of base point

➤ Enter .75,1.125   Enter ↵

➤ Press Esc to cancel the function

2-8 Ellipse Constructions in 2D Space
Creating Ellipses

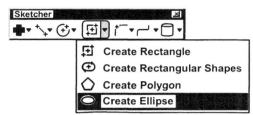

A *full ellipse* is created. The operator specifies the *X Axis Radius, Y Axis Radius* and the *center point* location. The *Start angle* is taken as *0°*, the *End angle* as *360°*.

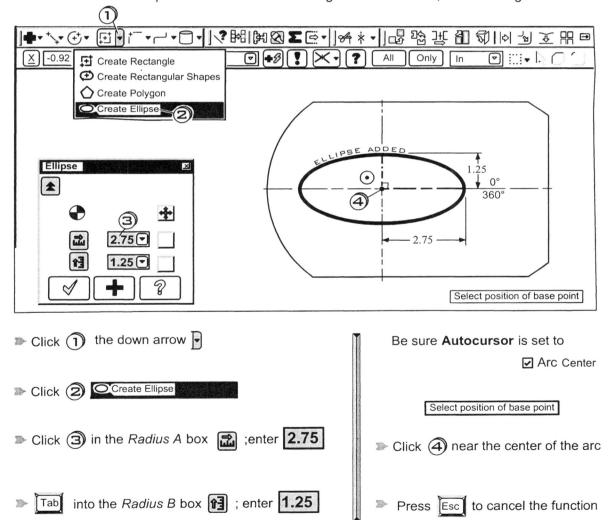

➤ Click ① the down arrow ▾

➤ Click ② ⬭ Create Ellipse

➤ Click ③ in the *Radius A* box 📷 ; enter **2.75**

➤ Tab into the *Radius B* box 🔧 ; enter **1.25**

Be sure **Autocursor** is set to

☑ Arc Center

Select position of base point

➤ Click ④ near the center of the arc

➤ Press Esc to cancel the function

A *portion of an ellipse* is created. The operator specifies the *X Axis Radius, Y Axis Radius* and the *center point* location.The *Start and End angles* for the elliptical portion of the ellipse are inputted. Finally, the operator keys in the desired *Rotation Angle* of the element.

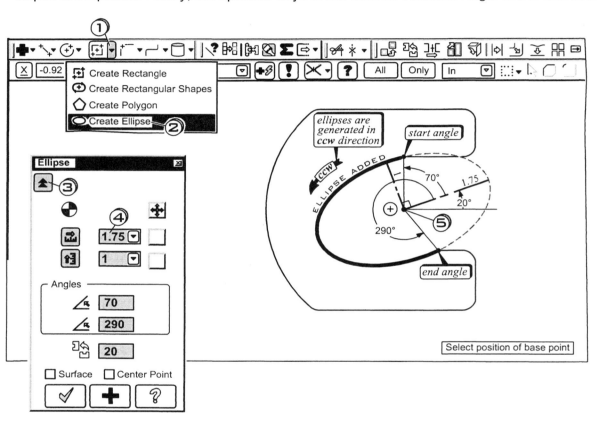

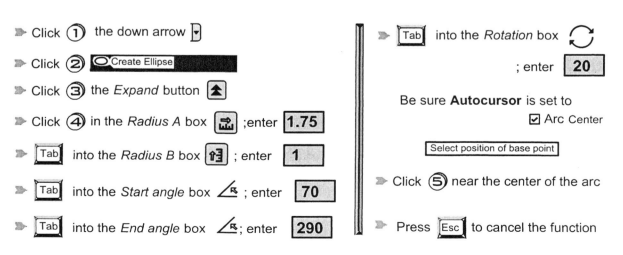

➤ Click ① the down arrow ▾

➤ Click ② ⬭ Create Ellipse

➤ Click ③ the *Expand* button ▲

➤ Click ④ in the *Radius A* box 🔛 ;enter **1.75**

➤ Tab into the *Radius B* box 🔧 ; enter **1**

➤ Tab into the *Start angle* box ∠ ; enter **70**

➤ Tab into the *End angle* box ∠; enter **290**

➤ Tab into the *Rotation* box ↻

             ; enter **20**

Be sure **Autocursor** is set to

            ☑ Arc Center

Select position of base point

➤ Click ⑤ near the center of the arc

➤ Press Esc to cancel the function

2-9 Geometric Letter Constructions in 2D Space

Geometric letters are used for creating text to be engraved(machined) into stock material. Geometric letters are composed of separate line,arc, and NURBS spline entities that are connected end to end. This is different from regular or non-geometric text which is taken as a single entity and cannot be machined.

Creating Geometric Letters

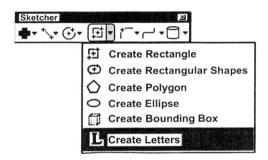

The Create Letters dialog box enables the operator to convert typed alphanumeric text into corresponding geometric alphanumeric characters. *Mastercam X* has pre-defined letter files for four fonts: Block, Box, Roman, and Slant. Mastercam X can create geometric letters from any True Type font listed in the Create Letters dialog box.

Geometric letters are created from *True Type* fonts. The operator clicks the desired *font, font style, size, and alignment*(Horizontal). Then enters the text *height and spacing* parameters. Finally, the operator clicks the starting *location* of the text.

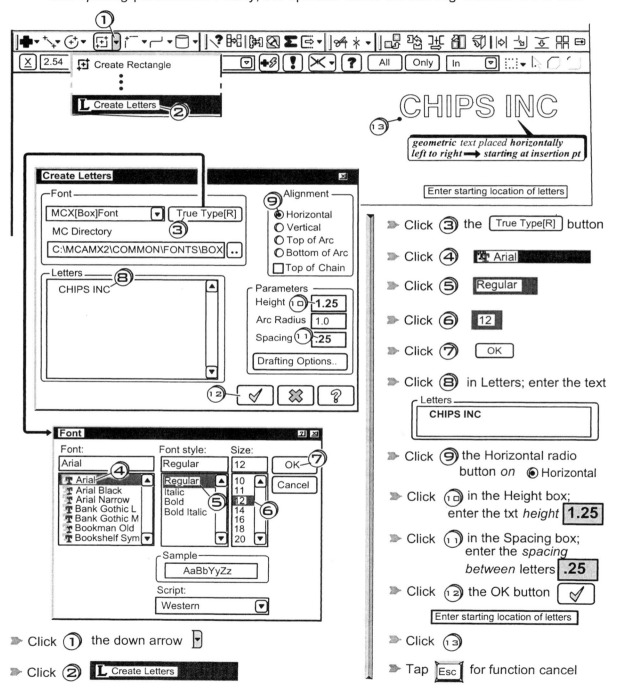

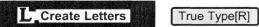

Geometric letters are created from *True Type* fonts. The operator clicks the desired *font, font style, size, and alignment* (Top of Arc) .Then enters the text *height, arc radius and spacing* parameters. Finally, the operator clicks the starting *location* of the text.

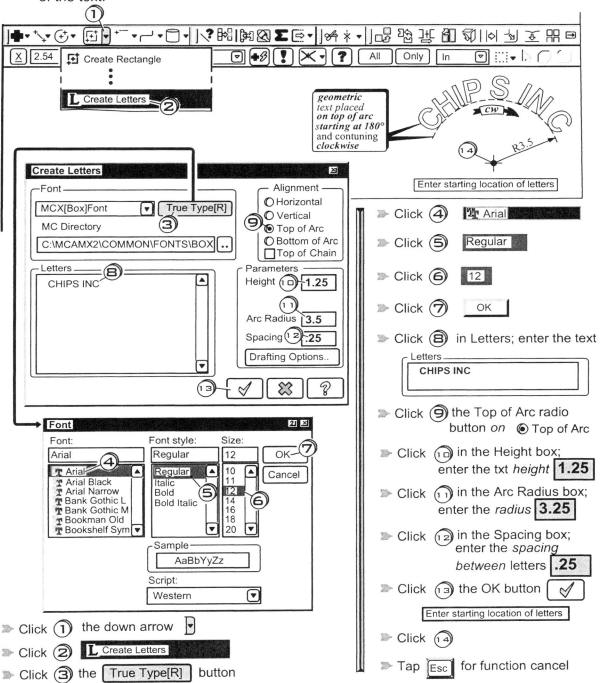

➤ Click ① the down arrow

➤ Click ② **L Create Letters**

➤ Click ③ the **True Type[R]** button

➤ Click ④ **T Arial**

➤ Click ⑤ **Regular**

➤ Click ⑥ **12**

➤ Click ⑦ **OK**

➤ Click ⑧ in Letters; enter the text

```
┌─ Letters ──────────────────────┐
│  CHIPS INC                      │
└────────────────────────────────┘
```

➤ Click ⑨ the Top of Arc radio
button *on* ● Top of Arc

➤ Click ⑩ in the Height box;
enter the txt *height* **1.25**

➤ Click ⑪ in the Arc Radius box;
enter the *radius* **3.25**

➤ Click ⑫ in the Spacing box;
enter the *spacing*
between letters **.25**

➤ Click ⑬ the OK button ✓

Enter starting location of letters

➤ Click ⑭

➤ Tap **Esc** for function cancel

Note Text

The Note Text dialog box enables the operator to set additional text properties. These include: text height, width, spacing, lines and borders, text path or direction, font, horizontal and vertical alignment, mirroring, angle, slant and rotation. Any Font and height settings in the Note Text override those entered in the Create Letters dialog box.

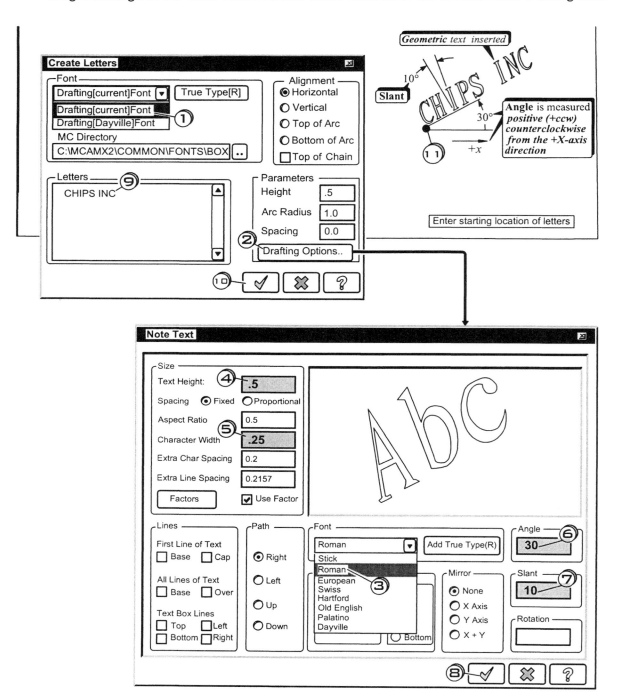

➤ Click ① the font style Drawing[Current]Font

➤ Click ② the Drafting Globals.. button

➤ Click ③ select the font style Roman

➤ Click ④ in the Height box; enter the text height .5

➤ Click ⑤ in the Character Width box; enter the text width .25

➤ Click ⑥ in the Angle box; enter the text angle 30

➤ Click ⑦ in the Slant box; enter the text slant angle 10

➤ Click ⑧ the OK button ✓

➤ Click ⑨ inside Letters and enter the text

Letters
CHIPS INC

➤ Click ⑩ the OK button ✓

Enter starting location of letters

➤ Click ⑪ *Mastercam X* will insert the text as a set of geometric lines,arcs and NURBS splines that can be used for machining

➤ Tap Esc for function cancel

2-10 Fillet Constructions in 2D Space

A fillet is an arc that is fitted *tangent* to two potentially intersecting line, arc or spline entities. The *Mastercam X* fillet ribbon bar provides the operator with several control settings. The operator selects the fillet style(normal, inverse, circle, or clearance) and inputs the fillet radius. The operator also specifies if *Mastercam* is to trim the two entities to their point of tangency with the fillet. The default setting has trimming turned *on* .

Another feature of the fillet ribbon bar is "auto-preview". This allows the operator to click the first entity that meets the fillet and then move the mouse cursor over any other possible intersection of the fillet with another entity. *Mastercam* will display a preview of the fillet as "phantom dotted " . If the preview image is accepticable, the operator can then simply click the second entity to create the fillet.

Creating Fillets

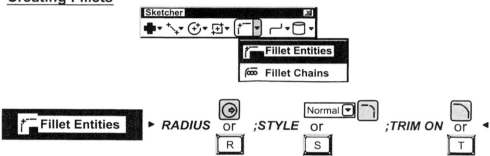

The operator *sets* the fillet *Radius* for *all subsequent* fillets created. The fillet arc *Style* is set to *Normal(*an arc *less than 180°). Trim* is set to *On.*The two entities filleted *will be automatically trimmed* to their *tangency* point with the fillet.

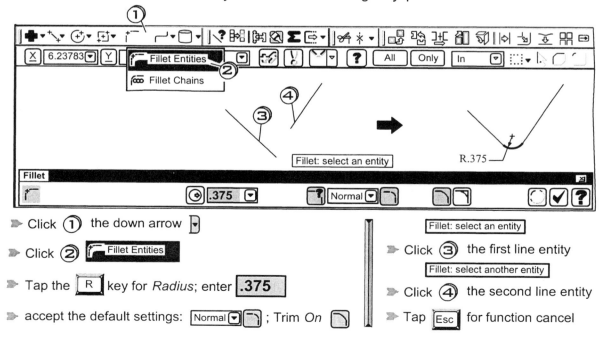

➥ Click ① the down arrow ▾

➥ Click ② **Fillet Entities**

➥ Tap the R key for *Radius*; enter .375

➥ accept the default settings: Normal▾ ⌐ ; Trim *On* ⌐

| Fillet: select an entity |

➥ Click ③ the first line entity

| Fillet: select another entity |

➥ Click ④ the second line entity

➥ Tap Esc for function cancel

The operator *sets* the fillet *Radius* for *all subsequent* fillets created. The fillet *arc Style* is set to *Inverse(*an arc *greater than 180°). Trim* is set to *On.*The two entities filleted *will be automatically trimmed* to their *tangency* point with the fillet.

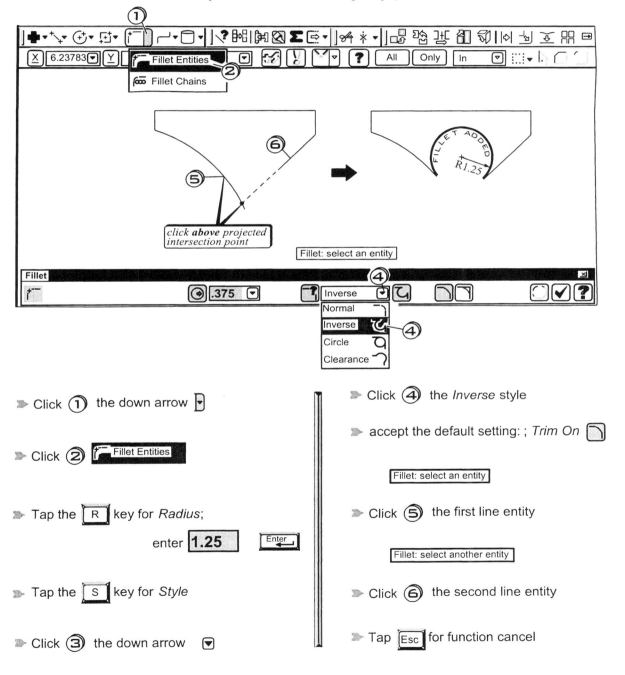

➤ Click ① the down arrow ▾

➤ Click ② [Fillet Entities]

➤ Tap the [R] key for *Radius*; enter **1.25** [Enter]

➤ Tap the [S] key for *Style*

➤ Click ③ the down arrow ▾

➤ Click ④ the *Inverse* style

➤ accept the default setting: ; *Trim On*

Fillet: select an entity

➤ Click ⑤ the first line entity

Fillet: select another entity

➤ Click ⑥ the second line entity

➤ Tap [Esc] for function cancel

▶ **RADIUS** or ⊕ / R   ;**STYLE** or `Circle ▾` / S   ;**TRIM ON** or / T ◀

The operator *sets* the fillet *Radius* for *all subsequent* fillets created. The fillet *arc Style* is set to *Circle(an* arc *equal to 360°). Trim* is set to *On.*The two entities filleted *will be automatically trimmed* to their *tangency* point with the fillet.

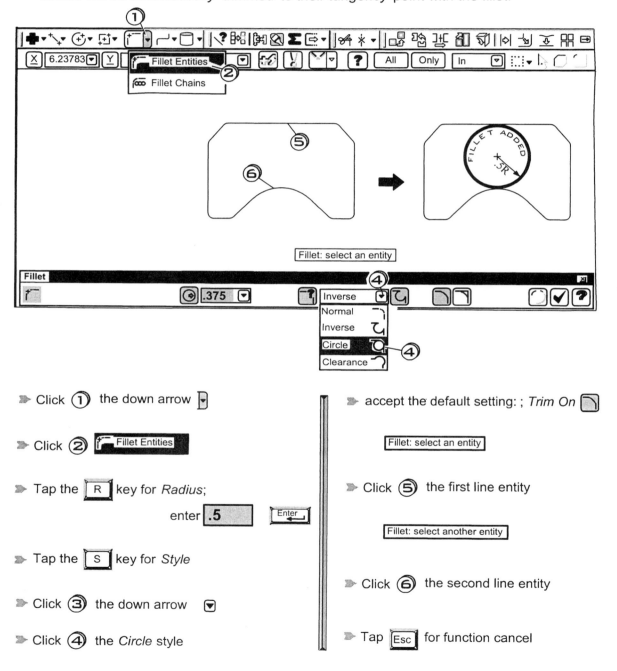

▶ Click ① the down arrow ▾

▶ Click ② `Fillet Entities`

▶ Tap the R key for *Radius*;
　　　　　　　enter `.5` `Enter`

▶ Tap the S key for *Style*

▶ Click ③ the down arrow ▾

▶ Click ④ the *Circle* style

▶ accept the default setting: ; *Trim On*

　　　　　Fillet: select an entity

▶ Click ⑤ the first line entity

　　　　　Fillet: select another entity

▶ Click ⑥ the second line entity

▶ Tap `Esc` for function cancel

The operator *sets* the fillet *Radius* for *all subsequent* fillets created. The fillet *arc Style* is set to *Clearance*. *Mastercam* creates fillets on the inside corners of a contour. This is useful for cases where an inside "clearance cut" is called for. That is the tool must reach completely into the corner and remove material. *Trim* is set to *On*. The two entities fillited *will be automatically trimmed* to their *tangency* point with the fillet.

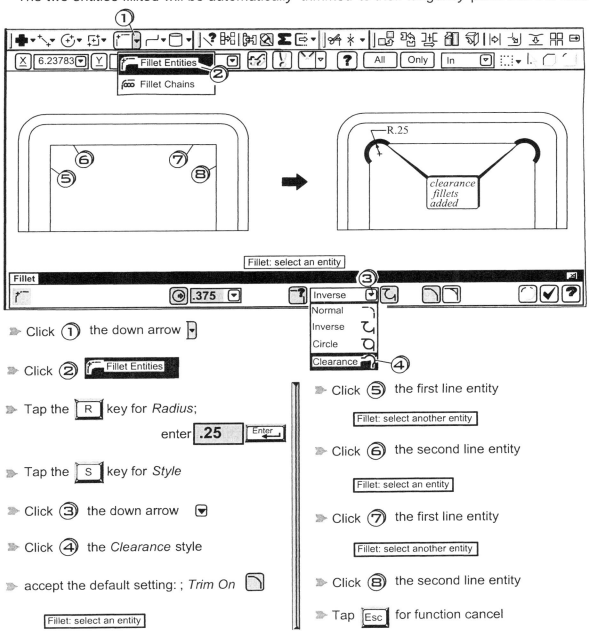

➤ Click ① the down arrow ▾

➤ Click ② [Fillet Entities]

➤ Tap the [R] key for *Radius*;

   enter **.25** [Enter]

➤ Tap the [S] key for *Style*

➤ Click ③ the down arrow ▾

➤ Click ④ the *Clearance* style

➤ accept the default setting: ; *Trim On*

   [Fillet: select an entity]

➤ Click ⑤ the first line entity

   [Fillet: select another entity]

➤ Click ⑥ the second line entity

   [Fillet: select an entity]

➤ Click ⑦ the first line entity

   [Fillet: select another entity]

➤ Click ⑧ the second line entity

➤ Tap [Esc] for function cancel

Fillets are created at the corners of *several* entities that are *connected* end to end in a *chain*

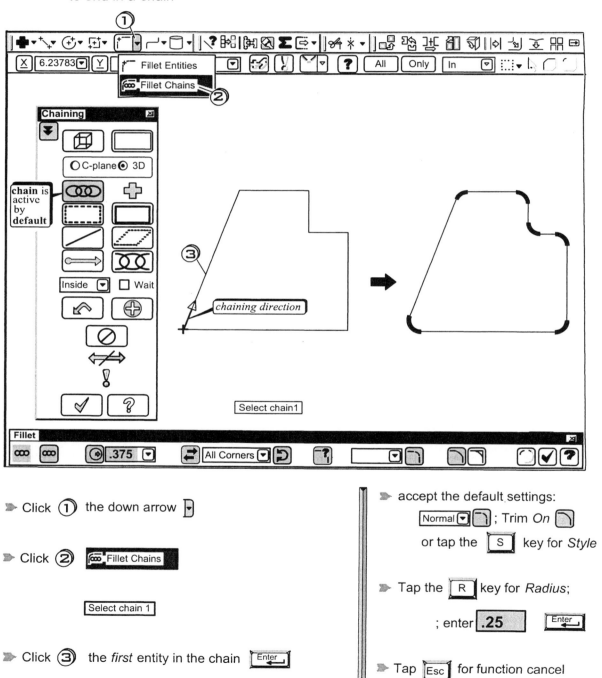

⯈ Click ① the down arrow 🔽

⯈ Click ② 🔲 Fillet Chains

    [Select chain 1]

⯈ Click ③ the *first* entity in the chain [Enter]

⯈ accept the default settings:
   [Normal ▼][🔲] ; Trim *On* [🔲]
   or tap the [S] key for *Style*

⯈ Tap the [R] key for *Radius*;
   ; enter [.25]   [Enter]

⯈ Tap [Esc] for function cancel

2-11 Chamfer Constructions in 2D Space

A chamfer is a *beveled edge* machined to break a sharp corner between two intersecting line or arc entities. The *Mastercam X2* fillet ribbon bar provides the operator with several control settings. The operator selects the chamfer style(1 Distance, 2 Distances, Distance/Angle or Width) and inputs the values appropriate to the style selected. The operator specifies whether *Mastercam* is to trim the two entities to their point of intersection with the chamfer. The default setting has trimming turned *on* . The "auto-preview" feature also works the same way as described for previewing fillets.

Creating Chamfers

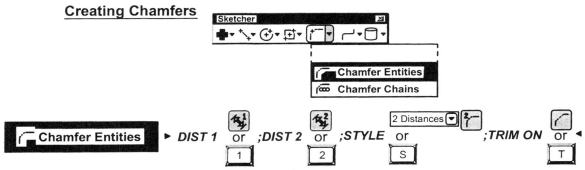

A *chamfer* is created. The operator specifies the *first* and *second* chamfer *distances* then clicks the *first* and *second lines or arcs* to be chamfered.

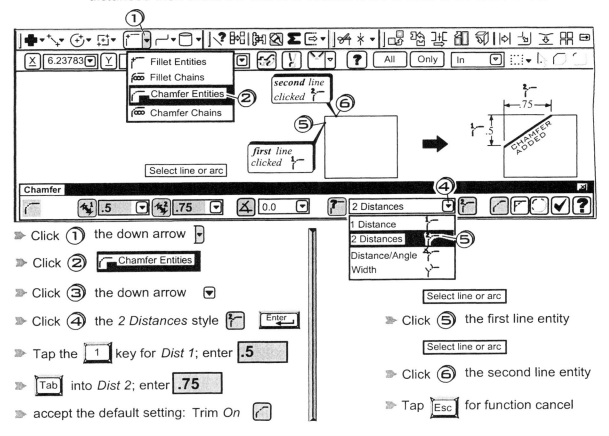

➤ Click ① the down arrow

➤ Click ② Chamfer Entities

➤ Click ③ the down arrow

➤ Click ④ the *2 Distances* style

➤ Tap the [1] key for *Dist 1*; enter **.5**

➤ [Tab] into *Dist 2*; enter **.75**

➤ accept the default setting: Trim *On*

➤ Click ⑤ the first line entity

➤ Click ⑥ the second line entity

➤ Tap [Esc] for function cancel

2-12 Spline Constructions in 2D Space

A spline is a smooth free form curve that passes through a set of operator generated node points

Mastercam can create *two* types of spline curves as follows:

Parametric spline(**P**) - is a curve that is made to pass through a desired set of points. If any of the control points are changed the operator must recreate the spline such that it passes through the new node point locations.

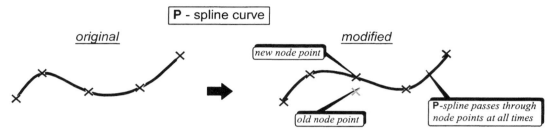

Non-Uniform Rational B-Spline/NURBS(**N**) - is a curve that is made to pass through a desired set of points. Its shape can be modified by moving its control points. In general it is a smoother curve than the P-spline curve.

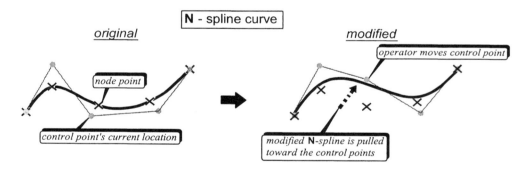

Splines are especially useful in the aircraft, automobile and shipbuilding industries.

Creating Splines

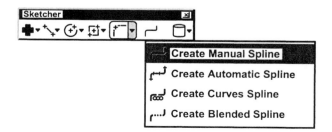

The type of spline curve to be created is selected from the System Configuration dialog box.The Spline creation type is accessed from CAD Settings

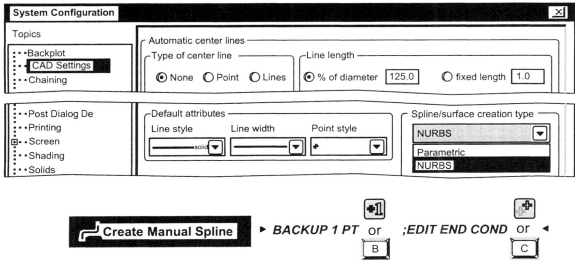

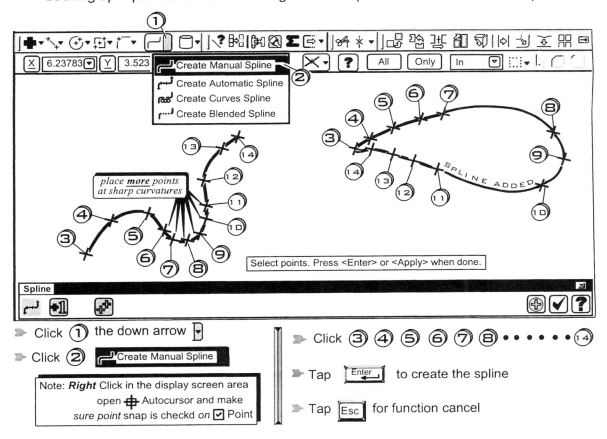

A *Parametric or NURBS spline* curve is created. *Each node* point the spline passes through *must be clicked*. After pressing Enter, the operator also has the option of backing up 1 point at a time or adding additional points to either end of the spline.

➤ Click ① the down arrow ⊡

➤ Click ② ⌐Create Manual Spline

Note: **Right** Click in the display screen area open ✛ Autocursor and make *sure point* snap is checkd *on* ☑ Point

➤ Click ③ ④ ⑤ ⑥ ⑦ ⑧ • • • • • • ⑭

➤ Tap [Enter] to create the spline

➤ Tap [Esc] for function cancel

A Parametric or NURBS spline curve is created. The operator must click *each node point* the curve is to pass through. The operator then edits the spline by adjusting its *first and last* endpoint conditions. In this case the *first* endpoint condition specifies a *tangency* to a *line* and the *last* endpoint condition a *tangency* to an *arc*.

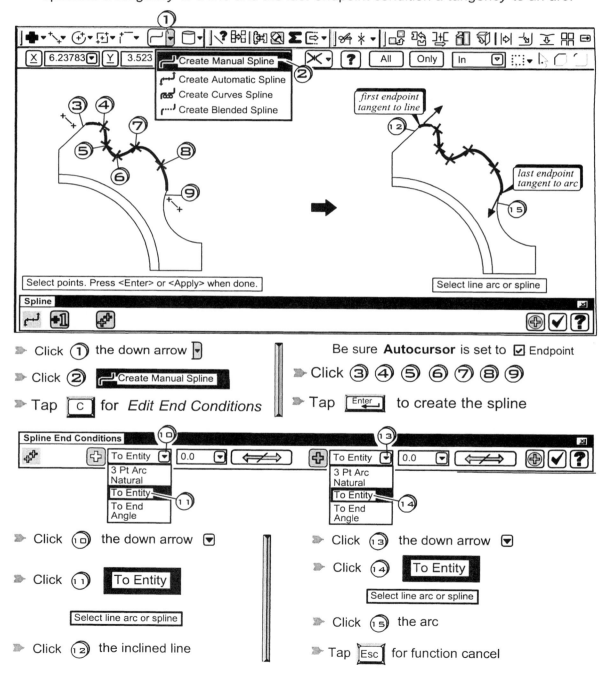

➤ Click ① the down arrow ▾   Be sure **Autocursor** is set to ☑ Endpoint

➤ Click ② ⌐ Create Manual Spline   ➤ Click ③ ④ ⑤ ⑥ ⑦ ⑧ ⑨

➤ Tap ⬚C for *Edit End Conditions*   ➤ Tap ⏎Enter to create the spline

➤ Click ①⓪ the down arrow ▾   ➤ Click ①③ the down arrow ▾

➤ Click ①① To Entity   ➤ Click ①④ To Entity

  Select line arc or spline   Select line arc or spline

➤ Click ①② the inclined line   ➤ Click ①⑤ the arc

                                                   ➤ Tap Esc for function cancel

Create Automatic Spline

A *Parametric or NURBS spline* curve is created. The operator must click the *first, second* and *last* points the curve is to *pass through.* The system will then *select any other* existing points to pass the curve through such that it *remains within the system's curve tolerance* value.

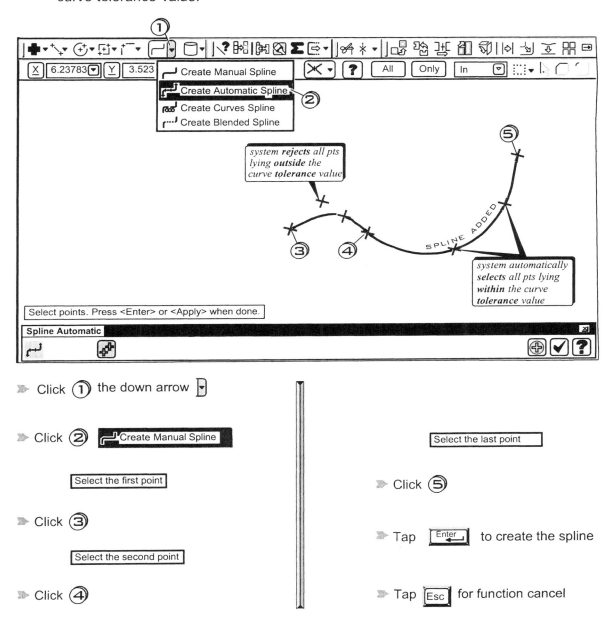

>> Click ① the down arrow ⊡

>> Click ② [Create Manual Spline]

[Select the first point]

>> Click ③

[Select the second point]

>> Click ④

[Select the last point]

>> Click ⑤

>> Tap [Enter] to create the spline

>> Tap [Esc] for function cancel

Create Curves Spline ▸ *DEVIATION* or *KEEP,BLANK,DELETE,MOVE* or ◂

A Parametric or *NURBS spline* curve is *fitted to an existing geometric pattern made by line, arc, ellipse or spline entities. Mastercam* creates a separate spline for each pattern chained. The operator can also specify the *maximum distance* a fitted *Parametric spline* can deviate from an existing geometric pattern that consists of arcs and /or NURBS splines.

The original *entities* can be kept, blanked, deleted,or moved to a different level.

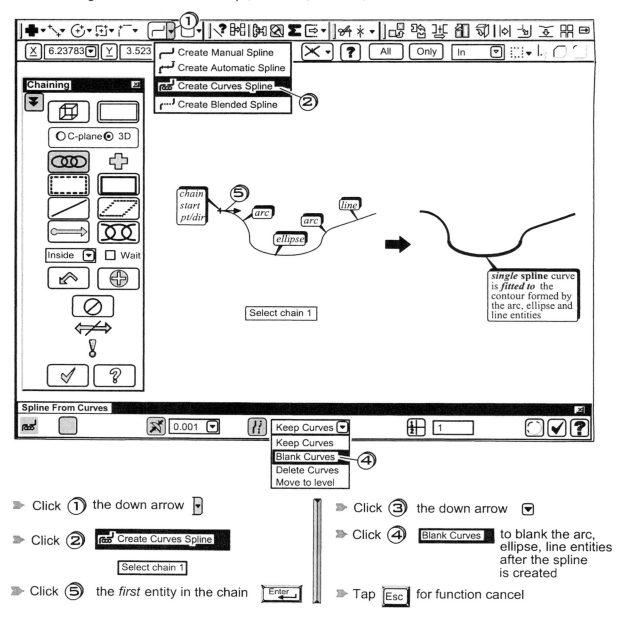

single **spline** curve is *fitted to* the contour formed by the arc, ellipse and line entities

▶ Click ① the down arrow

▶ Click ② **Create Curves Spline**

  Select chain 1

▶ Click ⑤ the *first* entity in the chain Enter

▶ Click ③ the down arrow

▶ Click ④ **Blank Curves** to blank the arc, ellipse, line entities after the spline is created

▶ Tap Esc for function cancel

A *Parametric* or *NURBS spline* curve is *fitted tangent to two existing line, arc, ellipse or spline entities.* The operator clicks the first entity, and then is prompted to slide the tangent arrow to the desired tangent location. The same is repeated for the second entity. The magnitude of the tangent spline can be specified. The way Mastercam trims the entities at their points of tangency to the spline can also be selected. The operator can *edit* any of the conditions controlling create blended spline **before** the *Apply* ☑ button is clicked

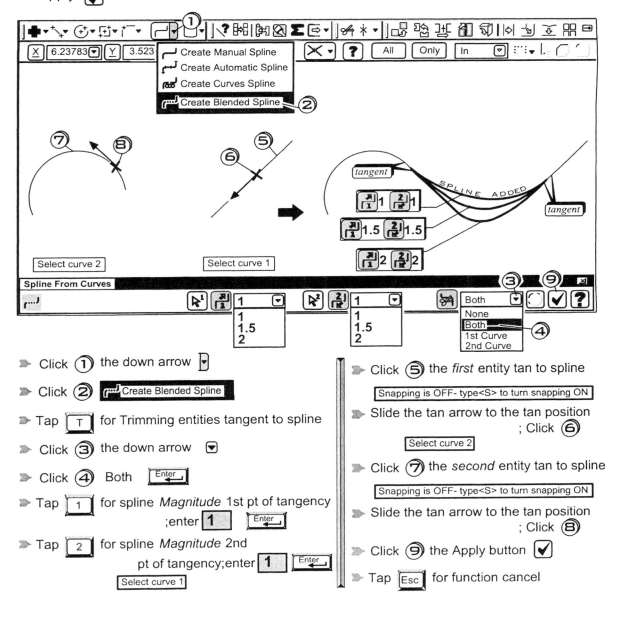

⇒ Click ① the down arrow ▾

⇒ Click ② ◢ Create Blended Spline

⇒ Tap T for Trimming entities tangent to spline

⇒ Click ③ the down arrow ▾

⇒ Click ④ Both Enter

⇒ Tap 1 for spline *Magnitude* 1st pt of tangency
;enter 1 Enter

⇒ Tap 2 for spline *Magnitude* 2nd
pt of tangency;enter 1 Enter
Select curve 1

⇒ Click ⑤ the *first* entity tan to spline

Snapping is OFF- type<S> to turn snapping ON

⇒ Slide the tan arrow to the tan position
; Click ⑥
Select curve 2

⇒ Click ⑦ the *second* entity tan to spline

Snapping is OFF- type<S> to turn snapping ON

⇒ Slide the tan arrow to the tan position
; Click ⑧

⇒ Click ⑨ the Apply button ☑

⇒ Tap Esc for function cancel

2-13 Zooming Graphics Window Displays

Mastercam X2 provides the user with the ability to control the size of the graphics window geometry displayed. Zooming can be executed while the operator is currently in an active function such as Sketcher. The various zoom functions provided are described in his section.

Zooming Graphics Window Displays

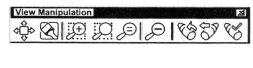

Zoom Window   or

Magnifies the geometry contained within a *rectangular* window clicked by the operator.

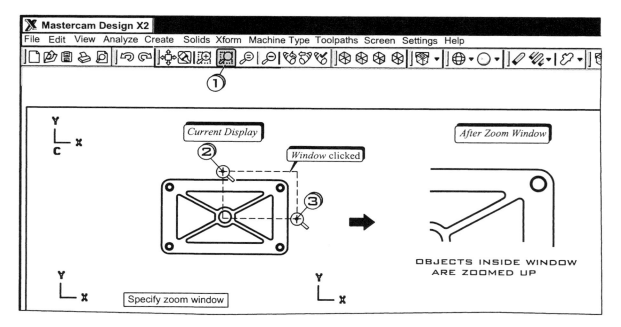

➤ Click ① the Zoom window button 🔍 or tap the F1 key

       Specify zoom window

➤ Click ② ③ corners of the zoom window

➤ Press Esc to cancel the function

Displays the geometry at the *scale of the previous zoom*. If *no* previous zoom *exists* the system will *reduce* the geometric display *by half its original size*. Unzoom can be used a *maximum* of *eight times* to reduce the current geometric display.

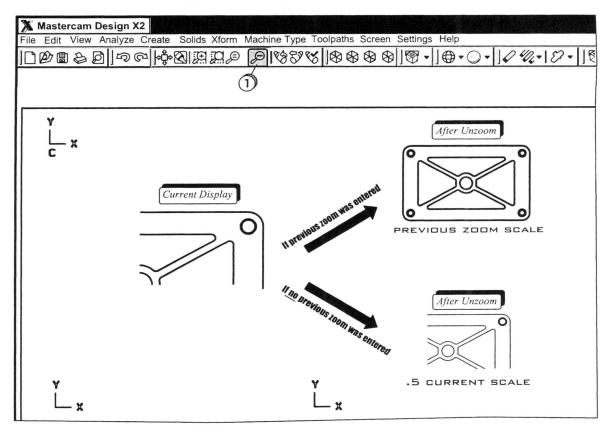

≫ Click ① the Unzoom button [🔍] or tap the [F2] key

Zoom Target

Magnifies the geometry contained within a *rectangular* window or target boundary clicked by the operator. The operator is prompted *first* to click the target point where the *center* of the window is to be placed. Next, the operator is prompted to drag the mouse cursor and click the location of the *corner* of the window to define its size.

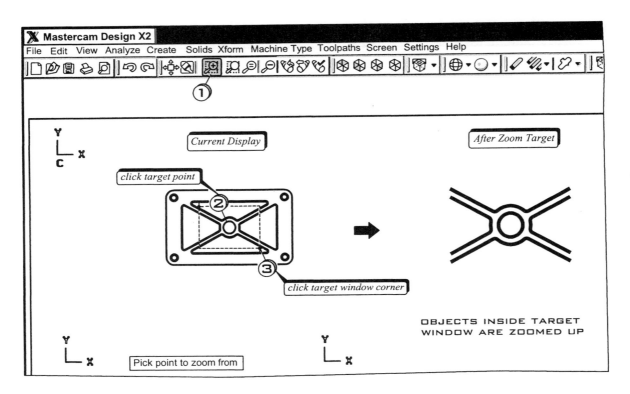

➤ Click ① the Zoom Target button

Pick point to zoom from

➤ Click ② near the center of the circle

Choose a second corner of your zooming box

➤ Click ③ the location of the corner of the target window

➤ Press [Esc] to cancel the function

Zoom Selected

The operator first *identifies the entities* to be considered *in the scaling set*. When Zoom Selected is clicked Mastercam *scales the set* such that it *fits* the graphics window display.

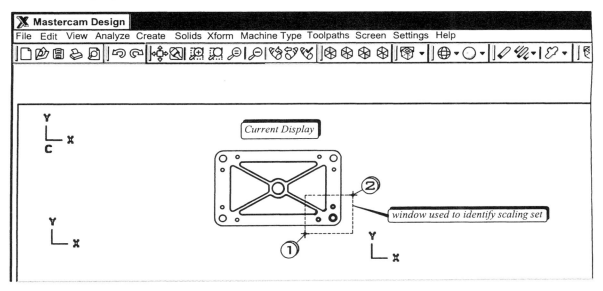

➤ Click ② ③ corners of the zoom window to *identify* the *entities in the scaling set*

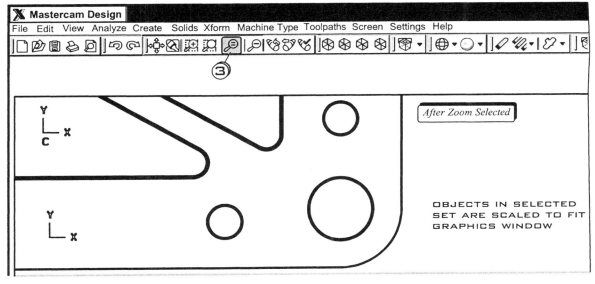

➤ Click ③ the Zoom Selected button

➤ Press Esc to cancel the function

Enables the operator to *dynamically zoom in or out* from a point selected. After clicking the focal point the function features any one of the following methods for dynamically zooming:

▸ Moving the mouse up(*zoom up*) or down(*zoom down*)

▸ Spin the mouse wheel up(*zoom up*) or down(*zoom down*)

▸ Tap Pg Up key(*zoom up*) or tap Pg Dn key(*zoom down*)

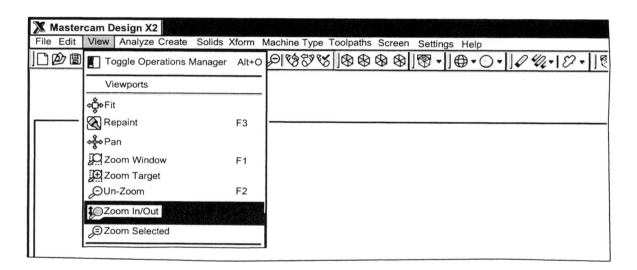

Note:
 The operator can use the mouse wheel and the Pg Up   Pg Dn keys independent of the Zoom in/out function. They can be applied when any function such as Sketcher is currently active.

2-14 Panning Screen displays

The operator can *pan* the geometric screen display in any of *four directions*: *up, down, left* or *right*. This is accomplished by one of two methods.

▶ Tapping the directional arrow keys on the keypad [↑] up [↓] down [←] left [→] right

▶ Press the [Alt] key and ***keeping it depressed*** ;

press the mouse wheel and drag the mouse in the desired pan direction

Cursor display indicates mouse is in pan mode

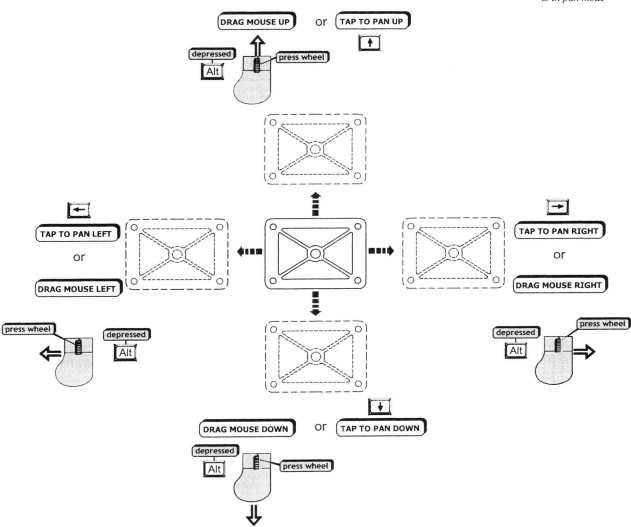

2-15 Fitting the Existing Geometry to the Screen Display Area

Scales the *shape* formed by all the *visible* graphic entities such that it *fits* the *graphics window display* area.

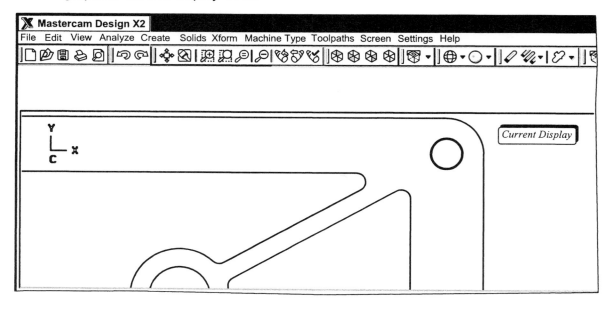

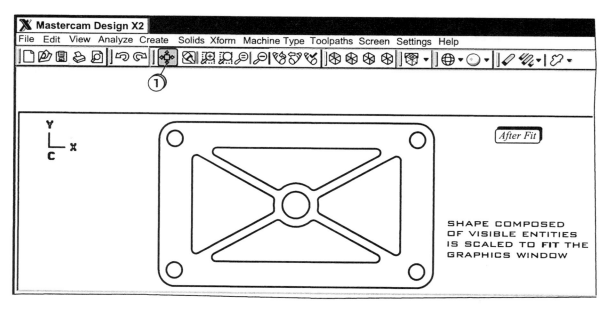

➤ Click ① the Fit button 🔹 or tap the [Home] key.

2-16 Repainting the Screen

Repaint Screen

Ocassionally, *Mastercam* may display incomplete or distorted images of the entities in the graphics window. This depends upon the graphic capabilities of your PC, the size the part file and ammount of available memory. Repaint is the *first* remedy the operator uses to correct these display problems in the graphics window.

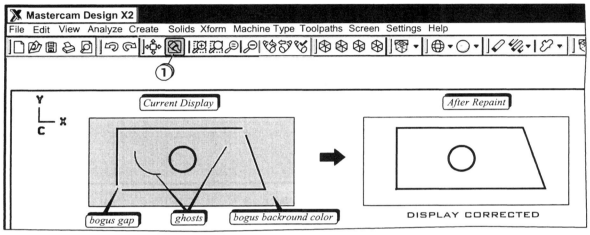

➤ Click ① the Zoom Target button

2-17 Regenerating the Screen

Regenerate Screen ⬛ or Shift Ctrl R

Regenerates all the graphic entities and *displays all* the *current existing geometry*. This command is used if Repaint *fails* to restore all the existing graphics.

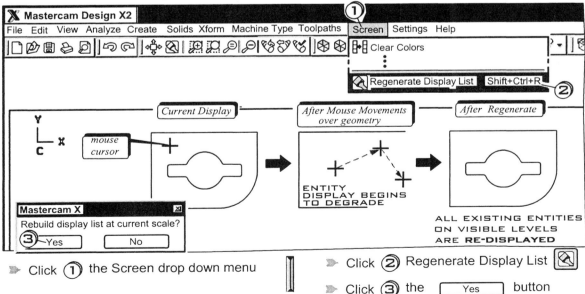

➤ Click ① the Screen drop down menu

➤ Click ② Regenerate Display List

➤ Click ③ the [Yes] button

2-18 The Undo/Redo Functions

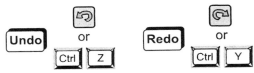

These functions are used to *undo or redo one or more events that have just been executed in sequential order in the current w*orking file. Any *function -based* operation is defined as an *event*. A single line created with the line function is an event. Similarily, the Xform function that makes 50 copies of a line is also considered as an event. When applied to Xform, the Undo and redo functions will undo and redo all 50 copies at once since *Xform* is treated as a *single event*. The default setting enables Mastercam to save up to 2 billion undo/redo events. This is restricted only by the availability of ramdom access memory(RAM) in your PC. The System Configuration Dialog box can be used to direct *Mastercam* to store only a specific number of events and allocate a maximum ammount of RAM to the undo/redo functions.

Note:
- ▸ Opening a part file or creating a new file **clears** the undo/redo list
- ▸ Undo/redo **cannot** be applied to **toolpath** or **CAM** related functions
- ▸ Undo/redo **can** only be applied to **CAD** related functions including:
  - ● entity creation/edit
  - ● drafting entities(dimensions)
  - ● file annotations
  - ● modify entity attributes

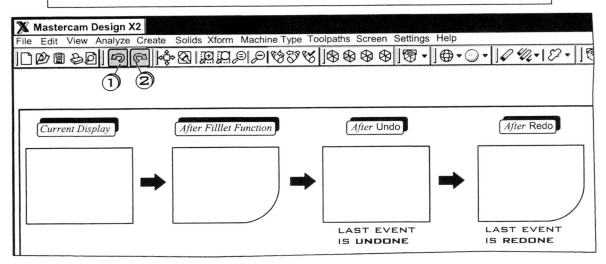

➠ Click ① the Undo button 🔙 to **undo** the last event(fillet creation)

➠ Click ② the Redo button 🔜 to **redo** the last event(fillet creation)

EXERCISES

In each exercise create the *wireframe* part geometry using *Mastercam X2's* Design package.

2-1)   YOUR INITIALS

File Name: **EX2-1JV**

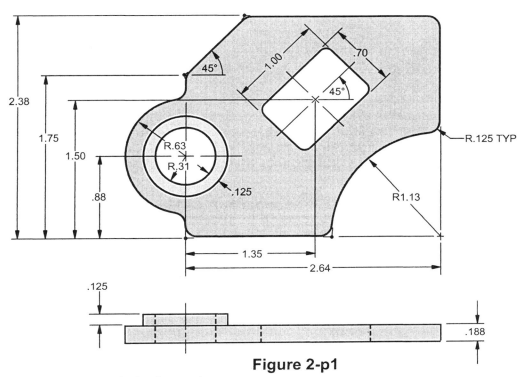

Figure 2-p1

Enter the *MastercamX2* Design package.

➤ Click ① the *Mastercam X2* icon

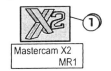

Mastercam X2
MR1

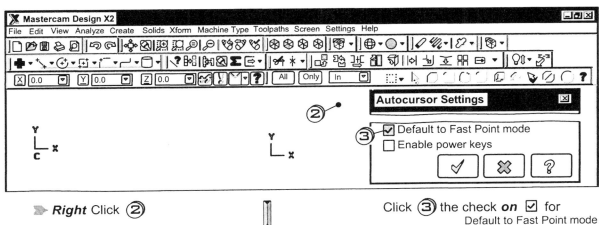

➤ ***Right* Click ②**

Click ③ the check **on** ☑ for
Default to Fast Point mode

a) Create the R.63 and R1.13 arc entities.

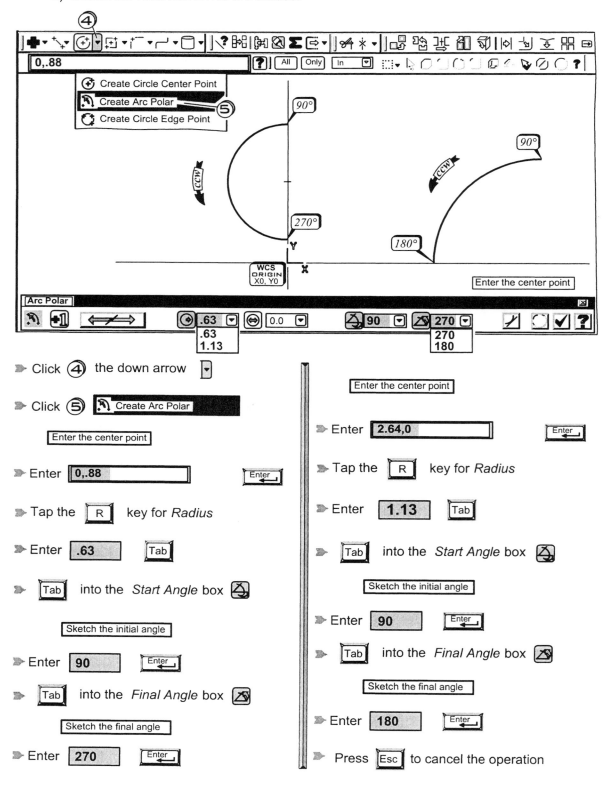

➤ Click ④ the down arrow ▾

➤ Click ⑤ ▣ Create Arc Polar

  Enter the center point

➤ Enter 0,.88   Enter↵

➤ Tap the R key for *Radius*

➤ Enter .63   Tab

➤ Tab into the *Start Angle* box ▣

  Sketch the initial angle

➤ Enter 90   Enter↵

➤ Tab into the *Final Angle* box ▣

  Sketch the final angle

➤ Enter 270   Enter↵

  Enter the center point

➤ Enter 2.64,0   Enter↵

➤ Tap the R key for *Radius*

➤ Enter 1.13   Tab

➤ Tab into the *Start Angle* box ▣

  Sketch the initial angle

➤ Enter 90   Enter↵

➤ Tab into the *Final Angle* box ▣

  Sketch the final angle

➤ Enter 180   Enter↵

➤ Press Esc to cancel the operation

b) Add the horizontal and vertical outer boundary lines .

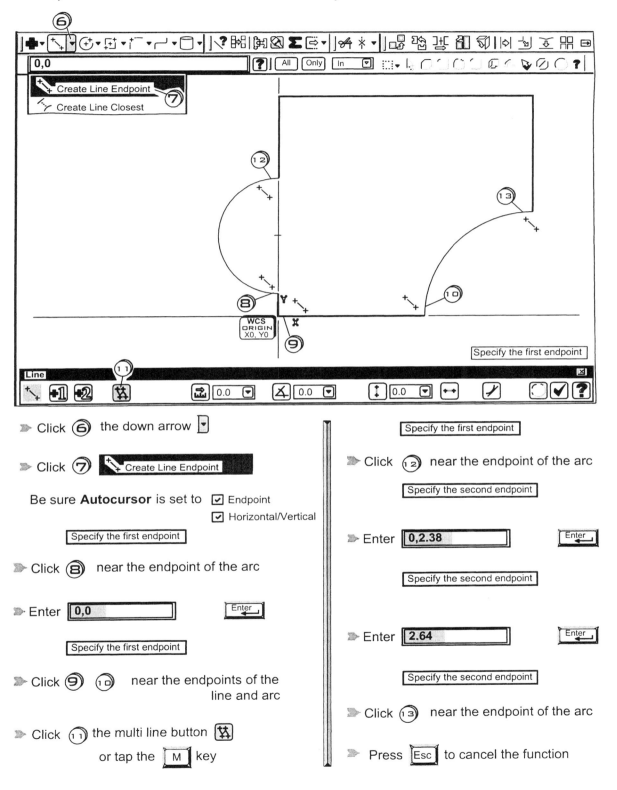

➤ Click ⑥ the down arrow ▾

➤ Click ⑦ ✛ Create Line Endpoint

  Be sure **Autocursor** is set to ☑ Endpoint
                                     ☑ Horizontal/Vertical

  Specify the first endpoint

➤ Click ⑧ near the endpoint of the arc

➤ Enter 0,0   Enter↵

  Specify the first endpoint

➤ Click ⑨ ⑩ near the endpoints of the
                         line and arc

➤ Click ⑪ the multi line button 🔀
    or tap the M key

  Specify the first endpoint

➤ Click ⑫ near the endpoint of the arc

  Specify the second endpoint

➤ Enter 0,2.38   Enter↵

  Specify the second endpoint

➤ Enter 2.64   Enter↵

  Specify the second endpoint

➤ Click ⑬ near the endpoint of the arc

➤ Press Esc to cancel the function

c) Generate the 45° chamfer.

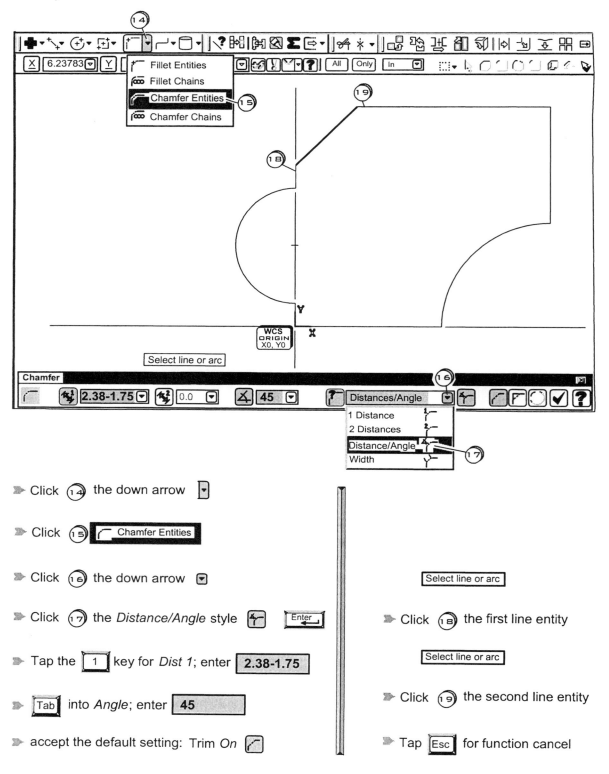

Select line or arc

➤ Click (14) the down arrow ▾

➤ Click (15) ⌐ Chamfer Entities

➤ Click (16) the down arrow ▾

➤ Click (17) the *Distance/Angle* style ⅟   Enter ⏎

➤ Tap the 1 key for *Dist 1*; enter **2.38-1.75**

➤ Tab into *Angle*; enter **45**

➤ accept the default setting: Trim *On* ⌐

Select line or arc

➤ Click (18) the first line entity

Select line or arc

➤ Click (19) the second line entity

➤ Tap Esc for function cancel

d) Create R.125 fillets and rounds on the outside contour.

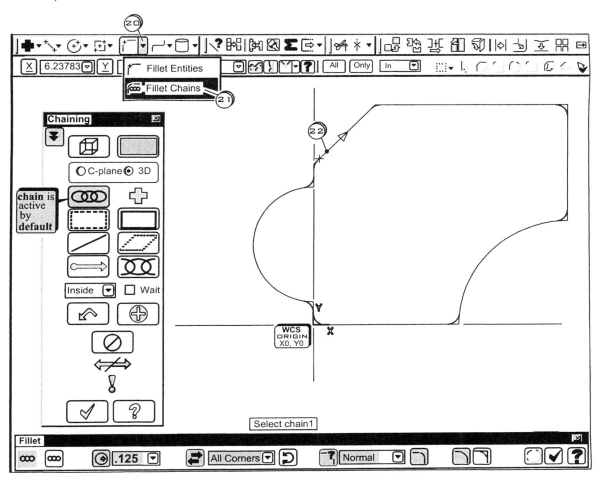

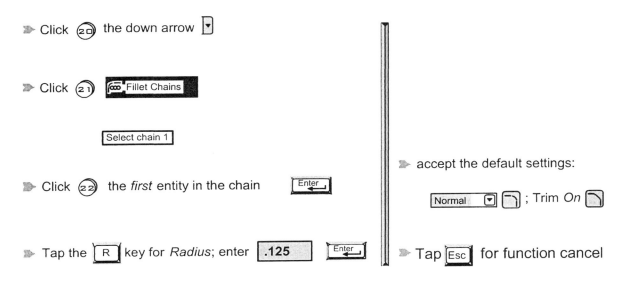

Click ⟨20⟩ the down arrow ▾

Click ⟨21⟩ 📇 Fillet Chains

Select chain 1

Click ⟨22⟩ the *first* entity in the chain   Enter ⏎

Tap the R key for *Radius*; enter .125   Enter ⏎

accept the default settings:

Normal ▾ ▢ ; Trim *On* ▢

Tap Esc for function cancel

e) Create R.31 circle and the circle offset by .125

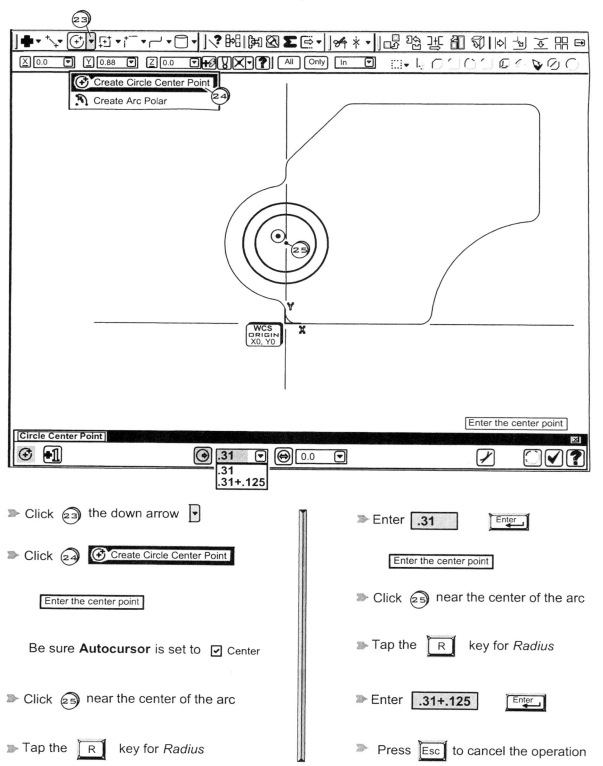

➤ Click ㉓ the down arrow ▾

➤ Click ㉔ ⊕ Create Circle Center Point

Enter the center point

Be sure **Autocursor** is set to ☑ Center

➤ Click ㉕ near the center of the arc

➤ Tap the R key for *Radius*

➤ Enter .31   Enter ↵

Enter the center point

➤ Click ㉕ near the center of the arc

➤ Tap the R key for *Radius*

➤ Enter .31+.125   Enter ↵

➤ Press Esc to cancel the operation

f) Comlete the CAD model by constructing the .7 x 1 rectangle at 45° having R.125 inside fillets.

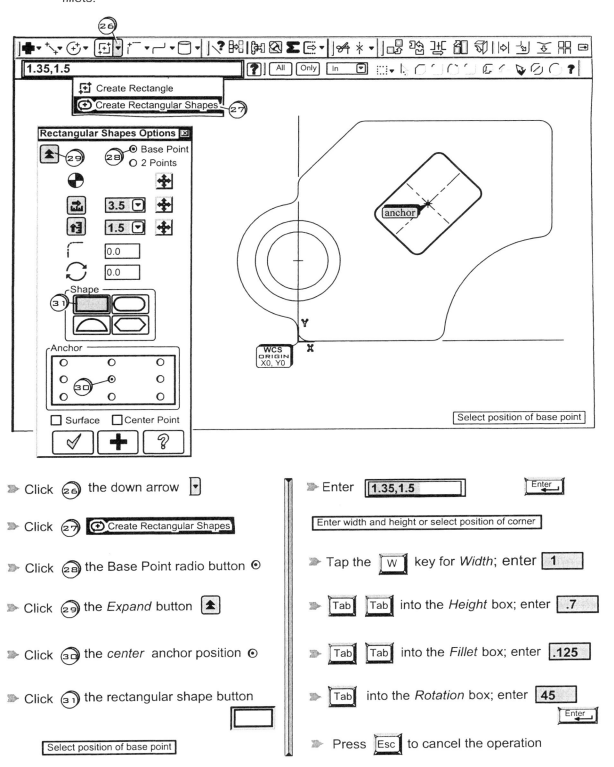

➤ Click ㉖ the down arrow ▾

➤ Click ㉗ ⊕ Create Rectangular Shapes

➤ Click ㉘ the Base Point radio button ⦿

➤ Click ㉙ the *Expand* button ⬆

➤ Click ㉚ the *center* anchor position ⦿

➤ Click ㉛ the rectangular shape button

Select position of base point

➤ Enter 1.35,1.5   Enter

Enter width and height or select position of corner

➤ Tap the W key for *Width*; enter 1

➤ Tab Tab into the *Height* box; enter .7

➤ Tab Tab into the *Fillet* box; enter .125

➤ Tab into the *Rotation* box; enter 45   Enter

➤ Press Esc to cancel the operation

g) Save the file

> Click ③② the File pull down menu

> Click ③③   [icon] Save As

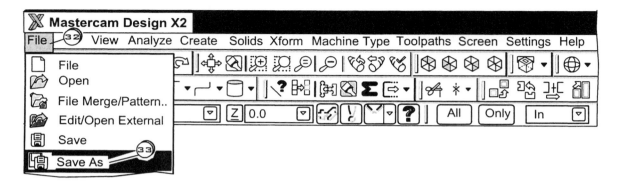

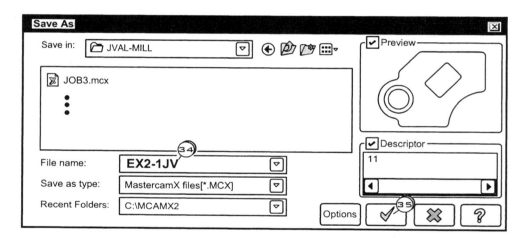

> Click ③④ in the file name box and enter the name of the **.MCX** file to be saved,
for example file, **EX2-1JV**

Note:
The descriptor 11= *Mastercam X2*

> Click ③⑤   the OK button   [✓]

For the following exercises ONLY THE FRONT VIEW is to be created for the 2D CAD model
All other views are given as reference .

2-2)   File Name: **EX2-2JV** ← YOUR INITIALS

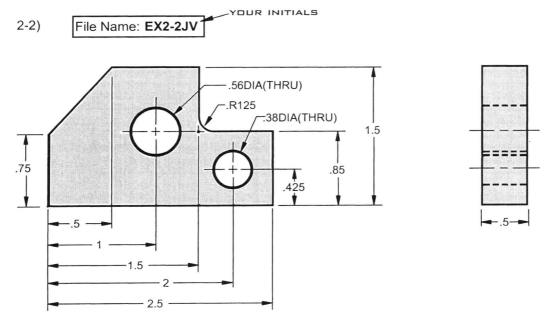

Figure 2-p2

2-3)   File Name: **EX2-3JV** ← YOUR INITIALS

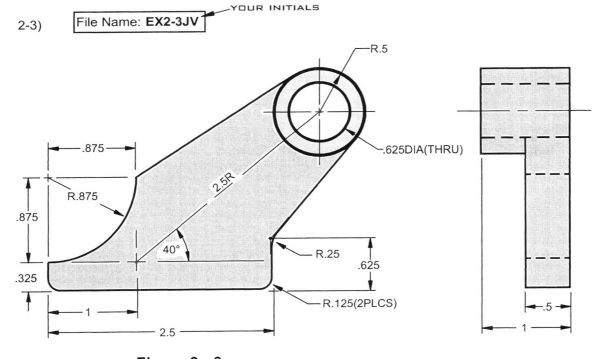

Figure 2-p3

2-4)   File Name: **EX2-4JV**   ← YOUR INITIALS

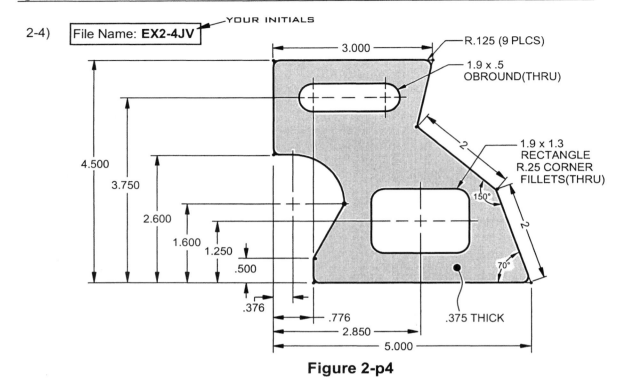

Figure 2-p4

2-5)   File Name: **EX2-5JV**   ← YOUR INITIALS

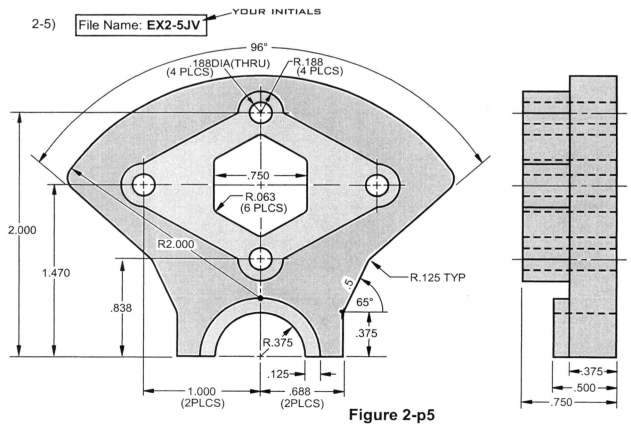

Figure 2-p5

2-6)   File Name: **EX2-6JV**   YOUR INITIALS

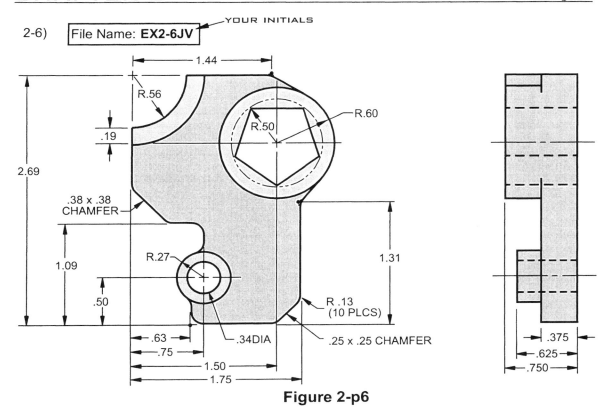

Figure 2-p6

2-7)   File Name: **EX2-7JV**   YOUR INITIALS

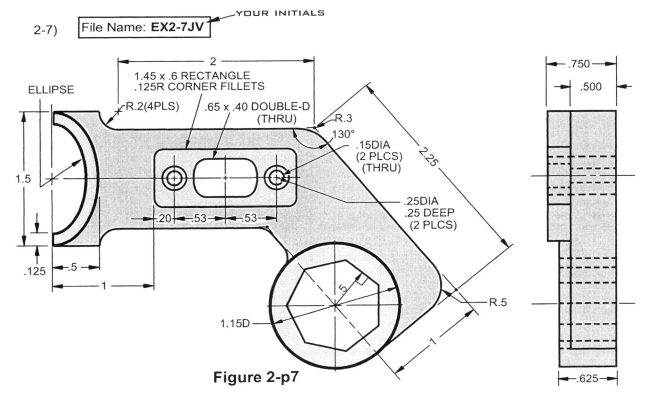

Figure 2-p7

2-8) | File Name: **EX2-8JV** | ⟵ YOUR INITIALS

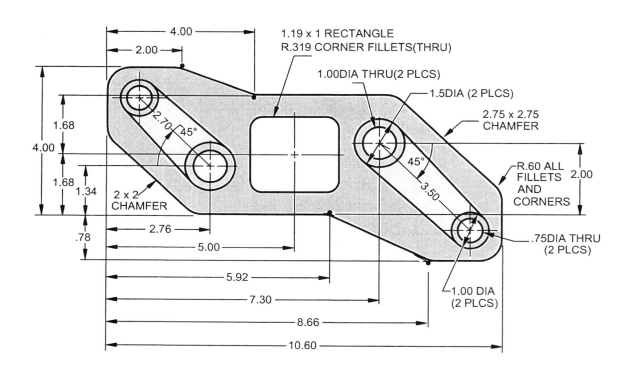

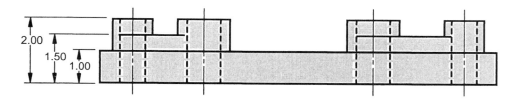

Figure 2-p8

2-9)　File Name: **EX2-9JV** — YOUR INITIALS

Material: 1030 Steel

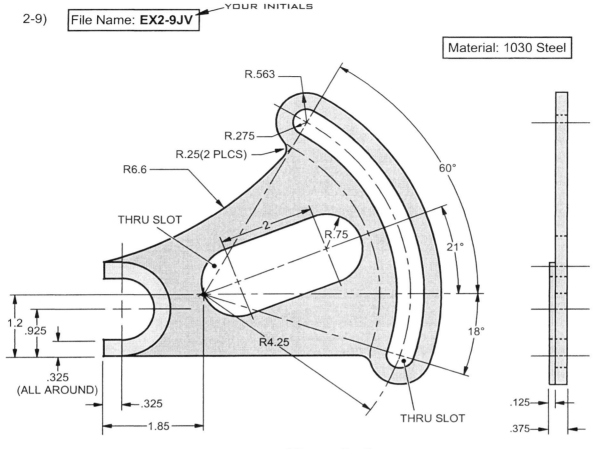

Figure 2-p9

a) Use the *Arc*, *Polar* command to create the R.6, R.925, R3.687, R3.975, R4.525 and R4.813 arcs

b) Use the *Arc*, *Endpoints* command to create the R.275 and R.563 arcs

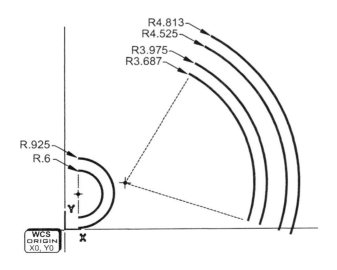

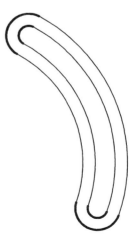

c) Add the .325 lines and the horizontal line

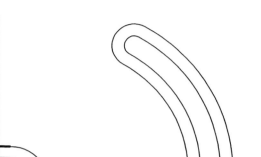

d) Use the *Arc, Endpoints* command to create the R6.6 arc

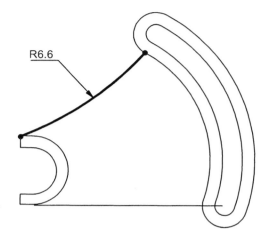

e) Erase the R3.687 arc and add the R.25 fillets

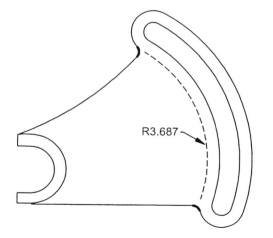

f) Create the 3.5 x 1.5 Obround at 21°

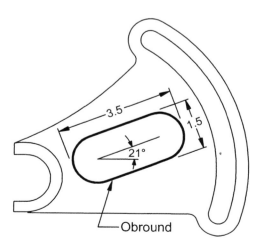

2-10)   File Name: **EX2-10JV** ← YOUR INITIALS

Generate the CAD entities and the geometric letters as shown in Figure 2p-10

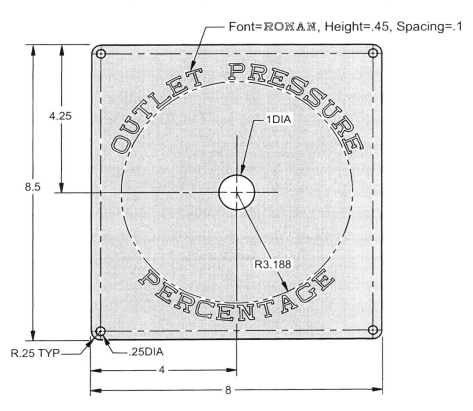

Font=ROMAN, Height=.45, Spacing=.1

Figure 2-p10

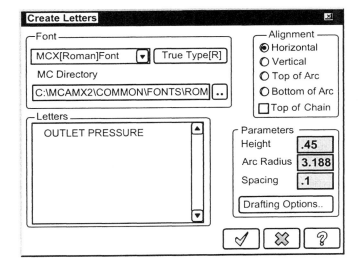

CHAPTER - 3

EDITING 2D GEOMETRY

3-1 Chapter Objectives

After completing this chapter you will be able to:

1. Explain how to delete and undelete existing geometrc entities.

2. Know how to shorten or extend an existing geometric entity to its intersection (s) with another entity or entities via the **Trim** functions

3. State the use of the **Join** function for joining separate entities into a single entity.

3. Explain how to break an existing geometric entity into separate entities via **Break.**

5. State the uses of the **Xform** function for translating, mirroring , rotating ,scaling, offseting, arraying and dragging existing geometric entities.

5. State the uses of the **Xform** function for creating copies of existing entities by translating, mirroring , rotating ,scaling, offseting, arraying and dragging.

3-2 Deleting/Undeleting Entities

The Delete/Undelete Functions

The Delete function enables the operator to *permanently* remove selected entities from the graphics window display and the *Mastercam X2* part file.
The General Selection ribbon bar contains several functions to make the job of *identifying* the entities to be deleted quick and efficient.

The Undelete function is used to *restore* entities that have *just been deleted accidentally* via Delete. The Undo function can also be used for this purpose.

The Delete Duplicates function directs Mastercam to *automatically* find and delete *duplicate entities in the current* file. Duplicate entities should be deleted since they unnecessarily increase the size of your part file and lead to *errors when chaining is executed.* Duplicates can be deleted based on their XYZ position and entity type. If the entity type mask function is used prior to using Delete Duplicates only those duplicate entity types selected in the mask will be deleted.

∞ Chain

All entities are *deleted* that are connected *end to end* in a *continuous chain*

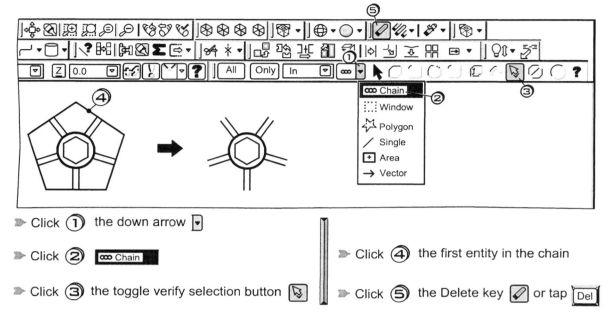

➤ Click ① the down arrow ▾

➤ Click ② ∞ Chain

➤ Click ③ the toggle verify selection button

➤ Click ④ the first entity in the chain

➤ Click ⑤ the Delete key or tap Del

Window   **in**

All entities contained *inside* a *rectangular window* clicked are deleted

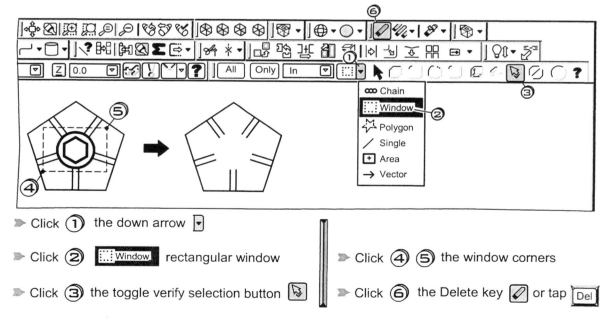

➤ Click ① the down arrow ▾

➤ Click ② Window rectangular window

➤ Click ③ the toggle verify selection button

➤ Click ④ ⑤ the window corners

➤ Click ⑥ the Delete key or tap Del

All entities contained *inside and crossing* a *rectangular window* clicked are deleted

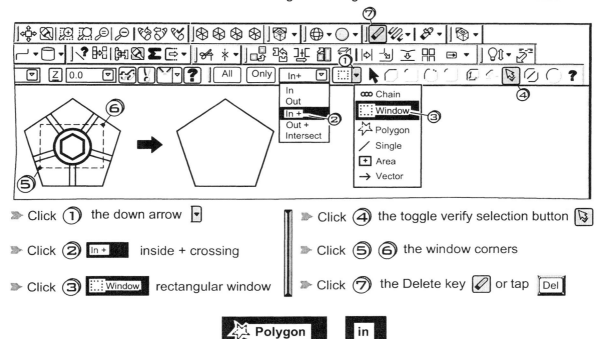

≫ Click ① the down arrow ▾

≫ Click ② **In +** inside + crossing

≫ Click ③ **Window** rectangular window

≫ Click ④ the toggle verify selection button

≫ Click ⑤ ⑥ the window corners

≫ Click ⑦ the Delete key or tap Del

All entities contained *inside* a *polygon window clicked* are *deleted*

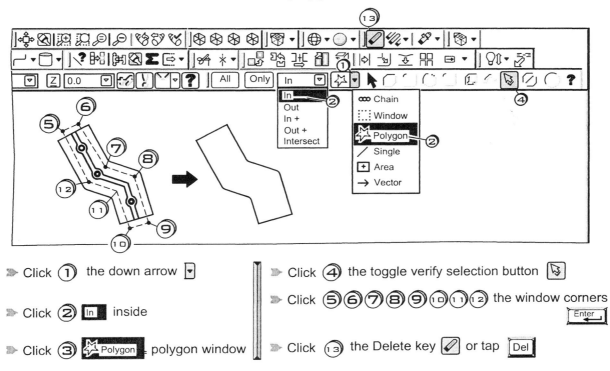

≫ Click ① the down arrow ▾

≫ Click ② **In** inside

≫ Click ③ **Polygon** polygon window

≫ Click ④ the toggle verify selection button

≫ Click ⑤⑥⑦⑧⑨⑩⑪⑫ the window corners
Enter

≫ Click ⑬ the Delete key or tap Del

Single

The operator *clicks on each* entity to be deleted

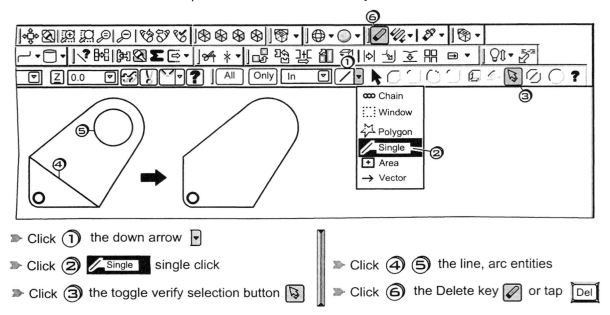

➤ Click ① the down arrow ▾

➤ Click ② **Single** single click

➤ Click ③ the toggle verify selection button 🔍

➤ Click ④ ⑤ the line, arc entities

➤ Click ⑥ the Delete key 🖊 or tap Del

Area

All *closed* boundaries in the *area that surrounds* the *point clicked* are deleted.

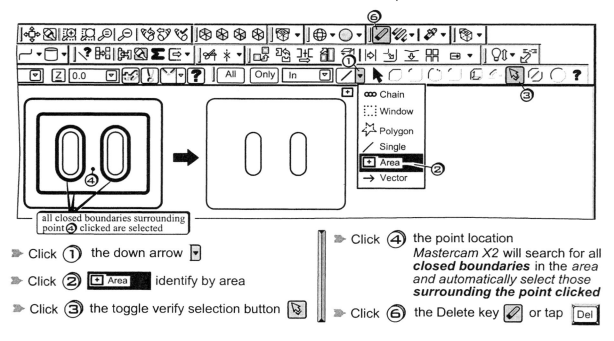

all closed boundaries surrounding point ④ clicked are selected

➤ Click ① the down arrow ▾

➤ Click ② **Area** identify by area

➤ Click ③ the toggle verify selection button 🔍

➤ Click ④ the point location
Mastercam X2 will search for all **closed boundaries** in the *area and automatically select those surrounding the point clicked*

➤ Click ⑥ the Delete key 🖊 or tap Del

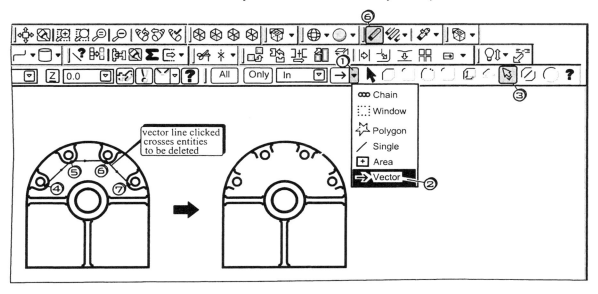

Vector

All *entities* that are *intersected* by *a vector line drawn by the operator* are deleted.

➤ Click ① the down arrow ▾

➤ Click ② 🡢Vector identify by vector line

➤ Click ③ the toggle verify selection button

➤ Click ④ ⑤ ⑥ ⑦ the endpoints of the vector line. *Mastercam X2* will select ***all the entities*** that ***cross the vector line*** Enter

➤ Click ⑥ the Delete key or tap Del

All

The operator uses Select All dialog box to define a selection mask. The mask can include any combination of entity types, colors, line widths, diameters/lengths. An arc of specific diameter and a line of specific length can also be set. When the OK button is clicked *all the entities that match the criteria in the mask are automatically* selected.

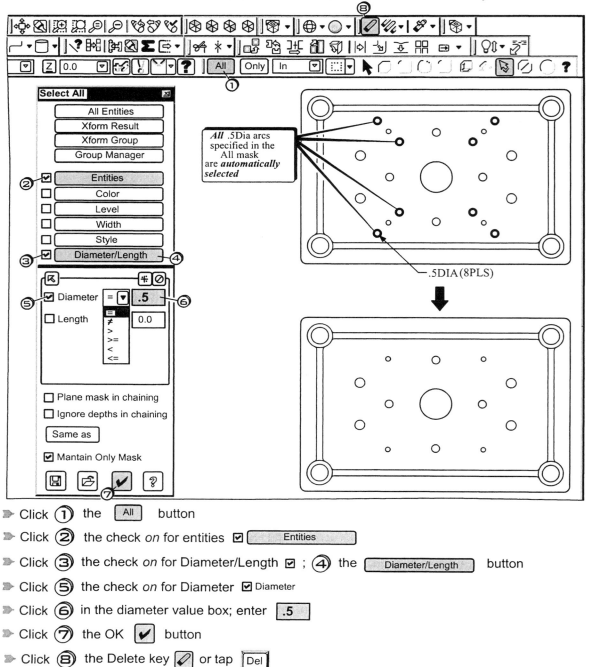

> *All* .5Dia arcs specified in the All mask are *automatically selected*

.5DIA (8PLS)

Select All

All Entities
Xform Result
Xform Group
Group Manager

② ☑ Entities
☐ Color
☐ Level
☐ Width
☐ Style
③ ☑ Diameter/Length ④

⑤ ☑ Diameter = ▼ **.5** ⑥
☐ Length　0.0
=
≠
>
>=
<
<=

☐ Plane mask in chaining
☐ Ignore depths in chaining
Same as
☑ Mantain Only Mask

➤ Click ① the All button
➤ Click ② the check *on* for entities ☑ Entities
➤ Click ③ the check *on* for Diameter/Length ☑ ; ④ the Diameter/Length button
➤ Click ⑤ the check *on* for Diameter ☑ Diameter
➤ Click ⑥ in the diameter value box; enter .5
➤ Click ⑦ the OK ✔ button
➤ Click ⑧ the Delete key or tap Del

Only

The operator uses Select Only dialog box to define a selection mask. The mask can include any combination of entity types, colors, line widths, diameters/lengths. An arc of specific diameter and a line of specific length can also be set. When the OK button is clicked *the operator must **manually select** the entities in the graphics window that meet the mask criteria.*

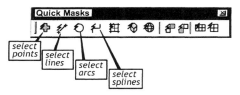

Quick Masks are new time saver functions added to *Mastercam X2*. These functions enable the operator to select existing entities by type with a single mouse click. **Left** clicking specifies masking is set to include ***all entities of the specified type***. **Right** clicking specifies masking is set for the operator to ***individually select each entity of the specified type***. Quick Masks are limited in that they *cannot be set to select arcs of specific radius or lines of specific color or length.* The All or Only dialogs must still be used to set and execute specific masking attributes. The Quick Masks toolbars can be found docked in the MRU(Most Recently Used) function bar area.of the main interface window.

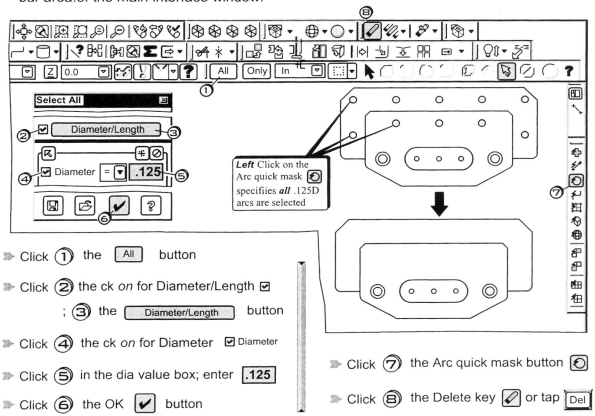

➤ Click ① the All button

➤ Click ② the ck *on* for Diameter/Length ☑

; ③ the Diameter/Length button

➤ Click ④ the ck *on* for Diameter ☑ Diameter

➤ Click ⑤ in the dia value box; enter .125

➤ Click ⑥ the OK ✔ button

➤ Click ⑦ the Arc quick mask button

➤ Click ⑧ the Delete key or tap Del

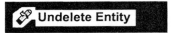

Undelete Entity

Restores deleted entities *within the **currently active .mcx part file.***
Repeatedly clicking the Undelete icon restores the entities one at a time. If
the operator deletes entities, then exits the part file, they *cannot* be restored
when the file is opened.

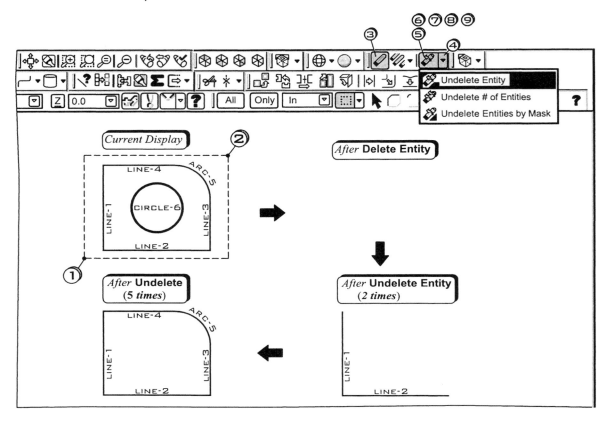

➤ Click ① ② the window corners

➤ Click ③ the Delete key 🖉 or tap [Del]

➤ Click ④ the down arrow ▾

➤ Click ⑤ ⑥ the Undelete button 🖉 *twice* to restore LINES-1,2

➤ Click ⑦ ⑧ ⑨ to restore LINES-3, 4; ARC-5

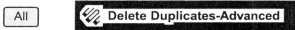

Unwanted duplicate entities can be created when translation, joining or copying functions are improperly executed. The advanced dialog box enables the operator to define the meaning of a duplicate based on attributes, in addition to its XYZ position. The All Mask specifies that *all* duplicates in the current file are to be deleted.

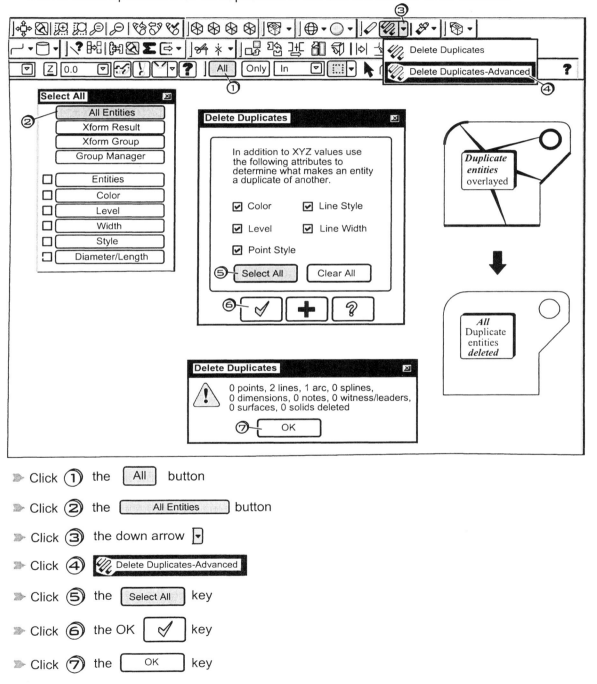

➤ Click ① the **All** button

➤ Click ② the **All Entities** button

➤ Click ③ the down arrow ▾

➤ Click ④ **Delete Duplicates-Advanced**

➤ Click ⑤ the **Select All** key

➤ Click ⑥ the OK ✓ key

➤ Click ⑦ the **OK** key

3-3 Modifying Existing 2D Geometry

The various *Mastercam X2* functions for modifying entities displayed in the graphics window are presented in this section.

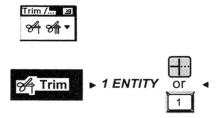

Trims(*shortens or extends*) an existing line, arc, fillet, ellipse or spline entity to its *intersection* with another entity. The operator *first* clicks the entity *to be trimmed* on the **portion to remain.** The entity to be *trimmed to* is *then* clicked.

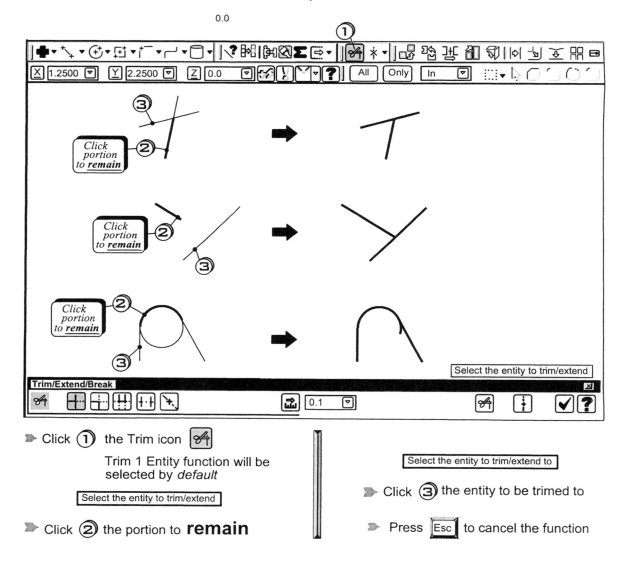

➤ Click ① the Trim icon

Trim 1 Entity function will be selected by *default*

Select the entity to trim/extend

➤ Click ② the portion to **remain**

Select the entity to trim/extend to

➤ Click ③ the entity to be trimed to

➤ Press Esc to cancel the function

Trims(shortens or extends) existing line, arc, fillet, ellipse and spline entities to their *intersection point*. The operator *first* clicks the entity *to be trimmed* on the **portion to remain.** The entity to be *trimmed to is then* clicked on the **portion to remain**.

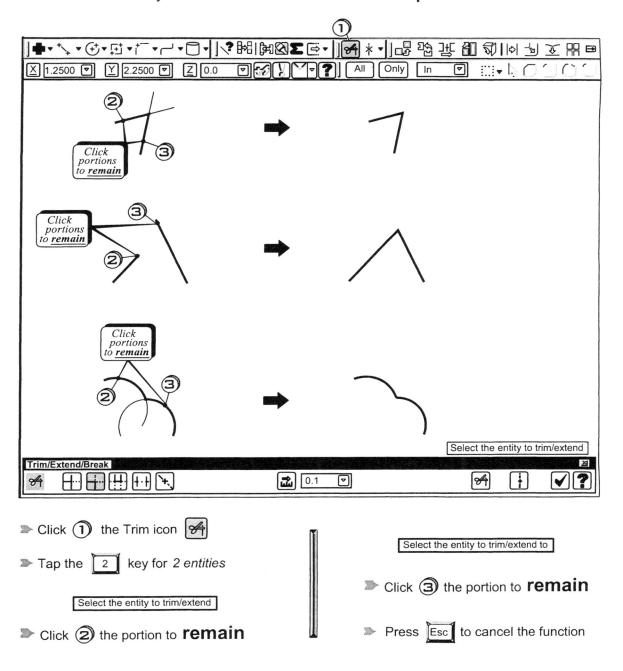

➤ Click ① the Trim icon

➤ Tap the 2 key for *2 entities*

Select the entity to trim/extend

➤ Click ② the portion to **remain**

Select the entity to trim/extend to

➤ Click ③ the portion to **remain**

➤ Press Esc to cancel the function

Trims(shortens or extends) an existing line, arc, fillet, ellipse and spline entitiy to its *intersection* with *two other* entities. The operator *first* clicks the *two trimming* entities on the **portions to remain.** The entity *to be trimmed* on the **portion to remain.** is then clicked.

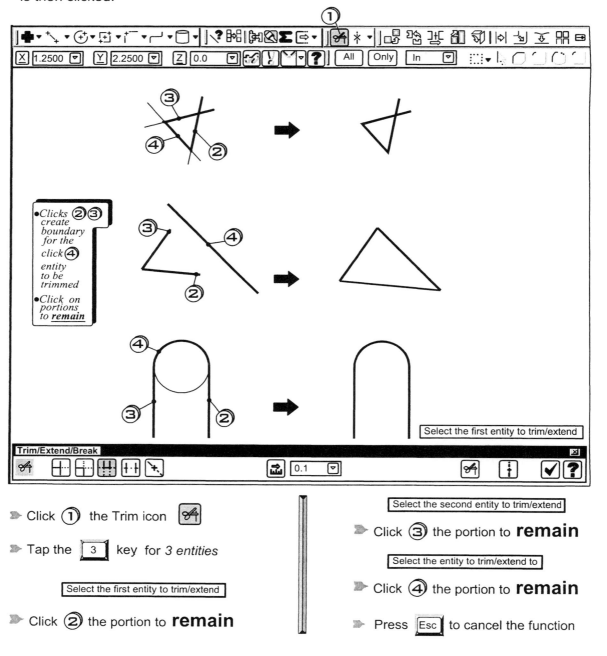

➤ Click ① the Trim icon ✂️

➤ Tap the ⬛3 key for *3 entities*

[Select the first entity to trim/extend]

➤ Click ② the portion to **remain**

[Select the second entity to trim/extend]

➤ Click ③ the portion to **remain**

[Select the entity to trim/extend to]

➤ Click ④ the portion to **remain**

➤ Press [Esc] to cancel the function

Trims an existing line, arc, fillet, ellipse or spline entitiy by *dividing it at its intersection* with *two* other *nearest* entities. The operator first clicks the **portion of the entitiy to be removed.** *Mastercam automatically* finds the nearest two intersections at each end and divides the entity between them.

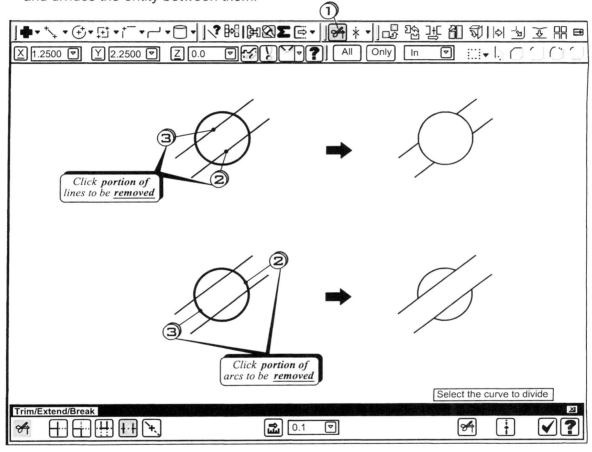

Click portion of lines to be **removed**

Click portion of arcs to be **removed**

Select the curve to divide

Trim/Extend/Break

≫ Click ① the Trim icon

≫ Tap the **D** key for *Divide entities*

Select the curve to divide

≫ Click ② ③ the portions to be **removed**

≫ Press **Esc** to cancel the function

Trims/Extends an existing line, arc, fillet, ellipse and spline entity to an apparent *intersection. Mastercam* trims or extends the entity such that it *intersects with a perpendicular computed by the software from the point clicked.*

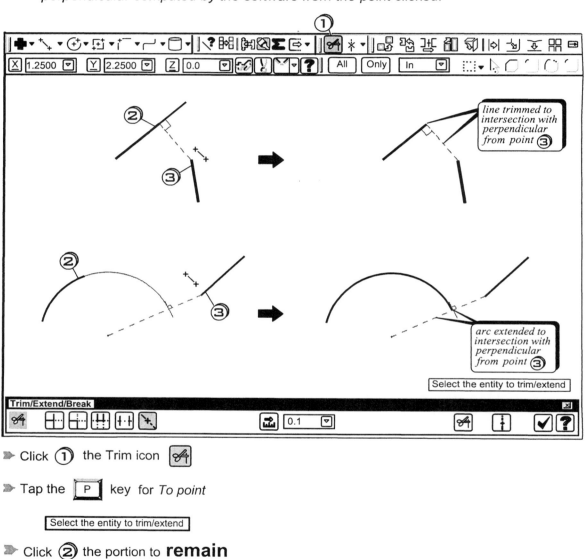

➤ Click ① the Trim icon ✁

➤ Tap the [P] key for *To point*

Select the entity to trim/extend

➤ Click ② the portion to **remain**

Indicate the trim/extend location

➤ Click ③ near the end of the line

➤ Press [Esc] to cancel the function

Trim ▸ *TO LENGTH* or ◂

Trims/Extends an existing line, arc, fillet, ellipse and spline entity to an inputted length.
A *positive* length value *extends* the entity by the ammount specified.
A *negative* length value *trims* the entity by the ammount specified. The operator
must click which end of the entity is to be trimed or extended..

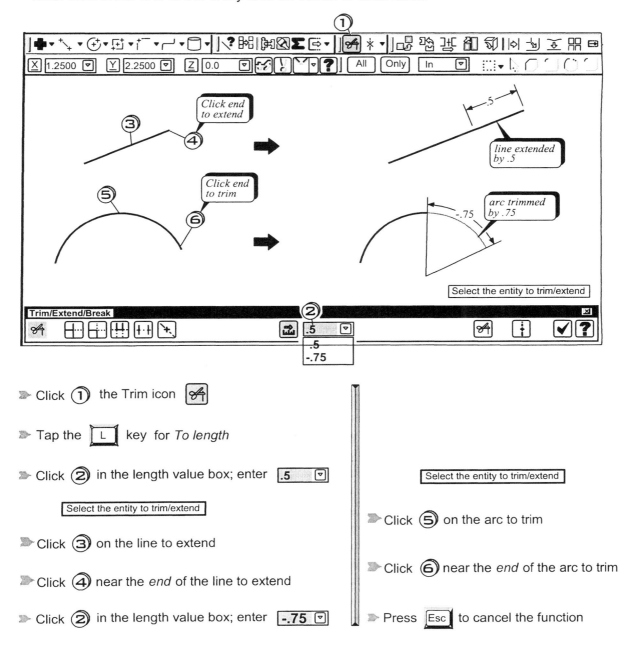

▸ Click ① the Trim icon ✂

▸ Tap the ⌊L⌋ key for *To length*

▸ Click ② in the length value box; enter .5 ▽

Select the entity to trim/extend

▸ Click ③ on the line to extend

▸ Click ④ near the *end* of the line to extend

▸ Click ② in the length value box; enter -.75 ▽

Select the entity to trim/extend

▸ Click ⑤ on the arc to trim

▸ Click ⑥ near the *end* of the arc to trim

▸ Press ⌊Esc⌋ to cancel the function

When used from the Trim/Extend/Break ribbon bar, Break either *extends or trims* an existing entity to its intersection with another entity clicked. The existing entity is **broken in two pieces**. The break point is located at the **end from which the existing entity is trimmed or extended.**

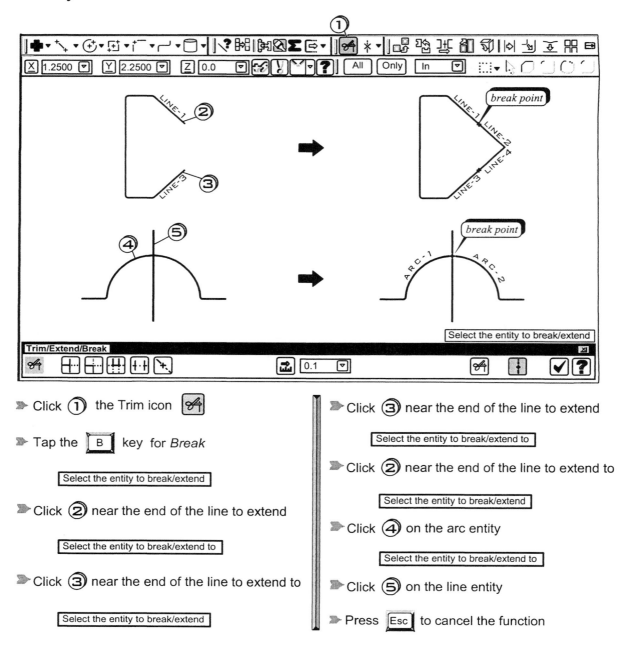

➤ Click ① the Trim icon 🖋

➤ Tap the [B] key for *Break*

    | Select the entity to break/extend |

➤ Click ② near the end of the line to extend

    | Select the entity to break/extend to |

➤ Click ③ near the end of the line to extend to

    | Select the entity to break/extend |

➤ Click ③ near the end of the line to extend

    | Select the entity to break/extend to |

➤ Click ② near the end of the line to extend to

    | Select the entity to break/extend |

➤ Click ④ on the arc entity

    | Select the entity to break/extend to |

➤ Click ⑤ on the line entity

➤ Press [Esc] to cancel the function

Break ▸ *TWO PIECES*

Breaks an existing entity into *two* pieces at any point specified.

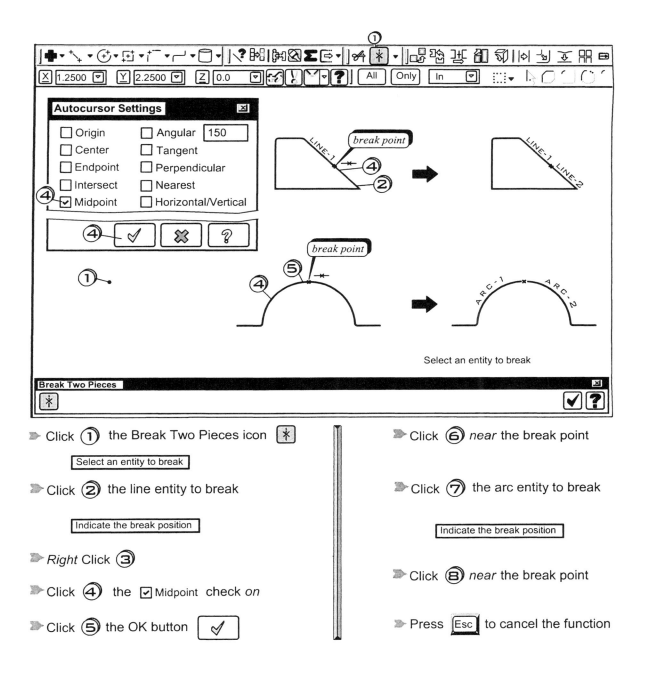

▸ Click ① the Break Two Pieces icon ✳

    Select an entity to break

▸ Click ② the line entity to break

    Indicate the break position

▸ *Right* Click ③

▸ Click ④ the ☑ Midpoint check *on*

▸ Click ⑤ the OK button ✓

▸ Click ⑥ *near* the break point

▸ Click ⑦ the arc entity to break

    Indicate the break position

▸ Click ⑧ *near* the break point

▸ Press Esc to cancel the function

Trims many existing line, arc, fillet, ellipse and spline entities to an intersection entity. The operator first clicks the entities *to be trimmed* then the entity to be trimmed to and finally clicks the **side on which portions of the entities are to remain**.

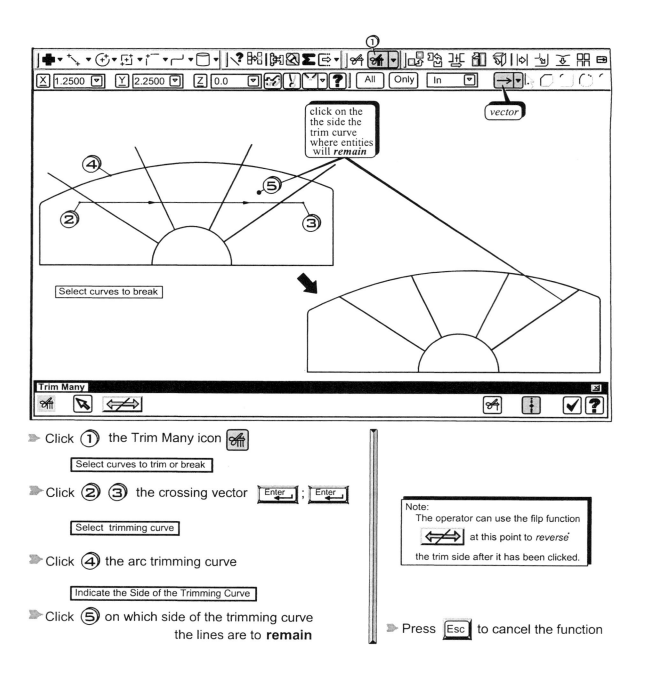

➤ Click ① the Trim Many icon

| Select curves to trim or break |

➤ Click ② ③ the crossing vector Enter ; Enter

| Select trimming curve |

➤ Click ④ the arc trimming curve

| Indicate the Side of the Trimming Curve |

➤ Click ⑤ on which side of the trimming curve the lines are to **remain**

Note:
The operator can use the filp function at this point to *reverse* the trim side after it has been clicked.

➤ Press Esc to cancel the function

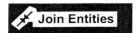 **Join Entities**

Joins existing *separate colinear lines*, *concentric arcs* having the same radii or splines into a *single* line, arc or spline entity.

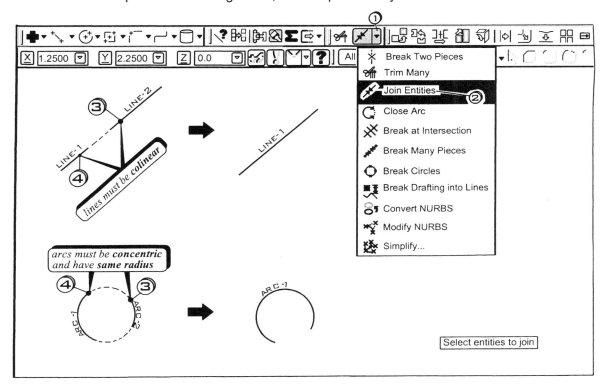

Click ① the down arrow ▾ ▾

Click ② 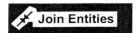 Join Entities

Select entities to join

Click ③ ④ near the ends of the entities to be joined; Enter↵

Press Esc to cancel the function

Close Arc

Extends an arc clicked such that it *closes to a full circle of 360°.*

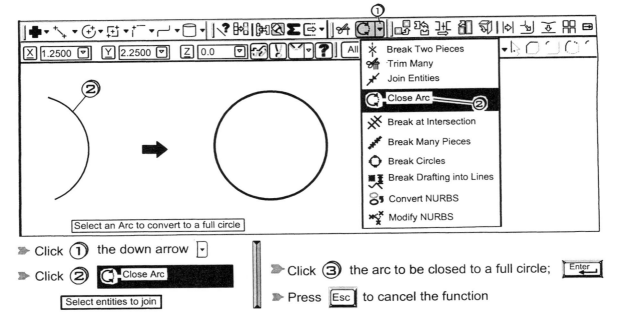

➤ Click ① the down arrow [▾]

➤ Click ② 🔵 Close Arc

Select entities to join

➤ Click ③ the arc to be closed to a full circle; [Enter]

➤ Press [Esc] to cancel the function

Break at Intersection

Breaks existing line, arc, fillet, ellipse or spline entities into a *separate length entities* at their *points of intersection*

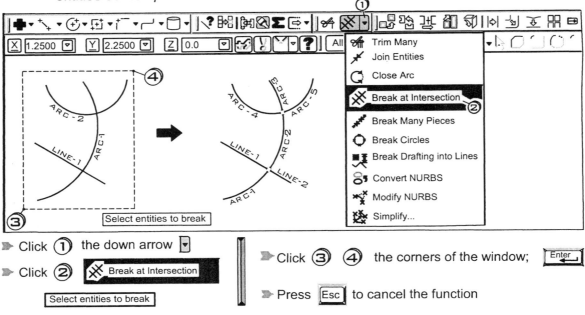

➤ Click ① the down arrow [▾]

➤ Click ② ✳ Break at Intersection

Select entities to break

➤ Click ③ ④ the corners of the window; [Enter]

➤ Press [Esc] to cancel the function

Break Many Pieces

Breaks an existing line, arc, fillet, ellipse or spline entitiy into a specified *number of equal and separate* entities. The operator has the option of specifying whether to delete, keep or blank(hide) the underlying geometry. If splines are to be broken the operator can either enter the number of equal segments or specify a tolerance based on chord height.

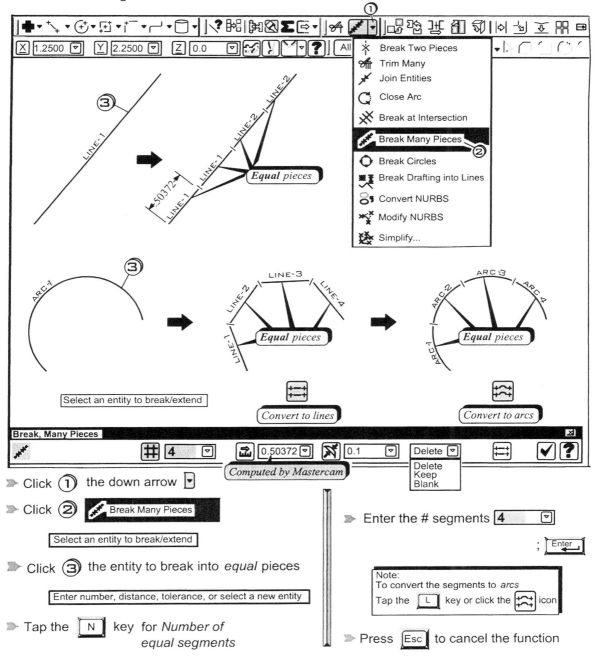

➢ Click ① the down arrow ▾

➢ Click ② [Break Many Pieces]

[Select an entity to break/extend]

➢ Click ③ the entity to break into *equal* pieces

[Enter number, distance, tolerance, or select a new entity]

➢ Tap the [N] key for *Number of equal segments*

➢ Enter the # segments [4 ▾]

; [Enter ↵]

Note:
To convert the segments to *arcs*
Tap the [L] key or click the [icon] icon

➢ Press [Esc] to cancel the function

Break Circles

Breaks an existing *full circle or circles* into any *number separate arc segments.* The operator specifies the number of arc segments each circle is to be broken into.

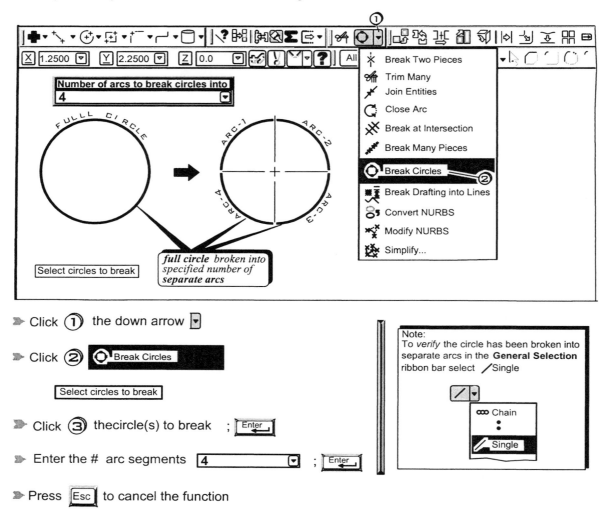

➤ Click ① the down arrow ▼

➤ Click ② Break Circles

   Select circles to break

➤ Click ③ the circle(s) to break ; Enter

➤ Enter the # arc segments [4] ; Enter

➤ Press Esc to cancel the function

Note:
To *verify* the circle has been broken into separate arcs in the **General Selection** ribbon bar select /Single

∞ Chain
⋮
/ Single

Break Drafting into Lines

Breaks existing drafting dimension entities, notes, labels witness lines as well as leader lines into geometric lines, arcs and NURBS spline curves. Breaks the lines in a crosshatch pattern into geometric line entities having the same line style as the. crosshatch pattern. Breaks copious data entities into separate geometric point and line entities depending upon the form of the data.

Convert NURBS

Changes line, arc, ellipse, P-spline entities to *NURBS spline* curves.

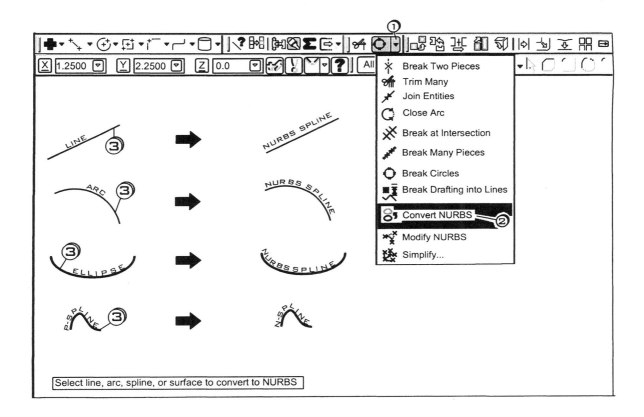

➤ Click ① the down arrow ▾

➤ Click ② Convert NURBS

> Select line, arc, spline, or surface to convert to NURBS

➤ Click ③ the entities to convert to a NURBS spline ; Enter

➤ Press Esc to cancel the function

Note:
Tap the F4 key to activate the **Analyze** function
to confirm an entity has been converted to a NURBS spline

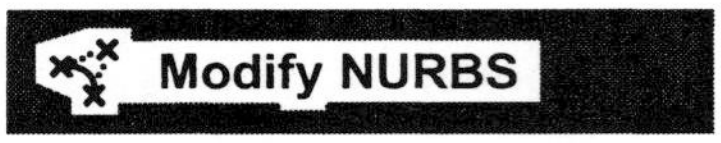

Directs Mastercam to *create arcs or lines from the existing P-spline or N-spline entities* selected. This is especially useful to more accurately locate the *center* of geometry with arcs modeled by splines. A CAD model represented by lines arcs rather than splines is more suitable for *dimensioning*. As a result of converting a CAD part file from another software package to *Mastercam*, original arcs and lines may get transformed into splines. The Simplify function can be used to change the splines back to the original arcs and lines.

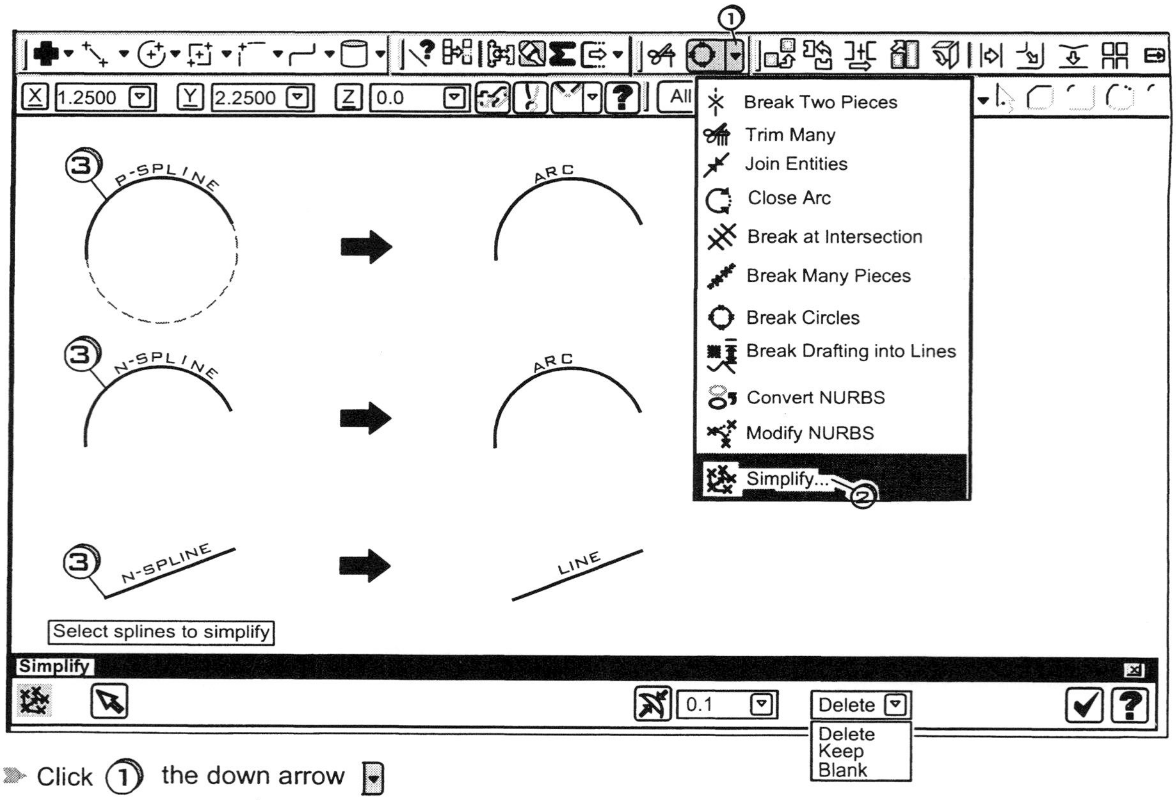

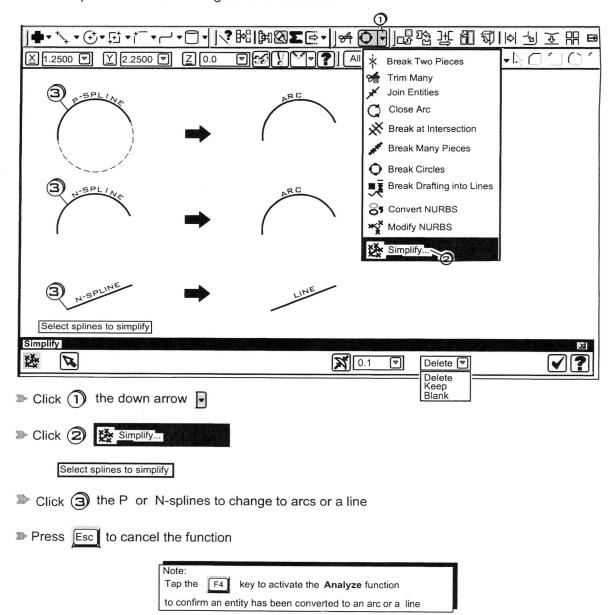

➤ Click ① the down arrow ▾

➤ Click ② 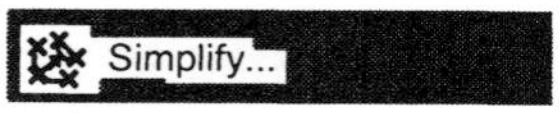

Select splines to simplify

➤ Click ③ the P or N-splines to change to arcs or a line

➤ Press Esc to cancel the function

Note:
Tap the F4 key to activate the **Analyze** function
to confirm an entity has been converted to an arc or a line

3-4 Using the Xform Functions to Transform Entities

The **XFORM** toolbar contains a set of often used functions for transforming existing entities in the graphics window by: translatiing or translating copies, rotating or rotating copies, mirroring or mirroring copies, scaling, offseting or offseting copies, arraying, rolling and dragging.The most important features of the Xform functions with respect to 2D design are considered in this section.

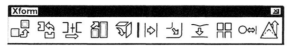

Xform Translate ▶ *MOVE* Move ⊙ *;FROM/TO points*

Moves existing geometry from a *present* reference point clicked to a *new* reference point *clicked.*

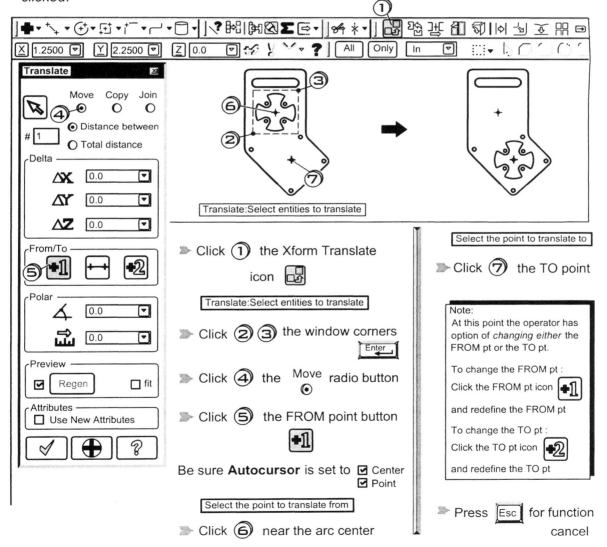

Click ① the Xform Translate icon

Translate:Select entities to translate

Click ② ③ the window corners

Enter ↵

Click ④ the Move ⊙ radio button

Click ⑤ the FROM point button

Be sure **Autocursor** is set to ☑ Center ☑ Point

Select the point to translate from

Click ⑥ near the arc center

Select the point to translate to

Click ⑦ the TO point

Note:
At this point the operator has option of *changing either* the FROM pt or the TO pt.

To change the FROM pt :
Click the FROM pt icon
and redefine the FROM pt

To change the TO pt :
Click the TO pt icon
and redefine the TO pt

Press Esc for function cancel

Creates translated copies of existing geometry. The distance between each copy is entered in terms of rectangular coordinates : ΔX , ΔY

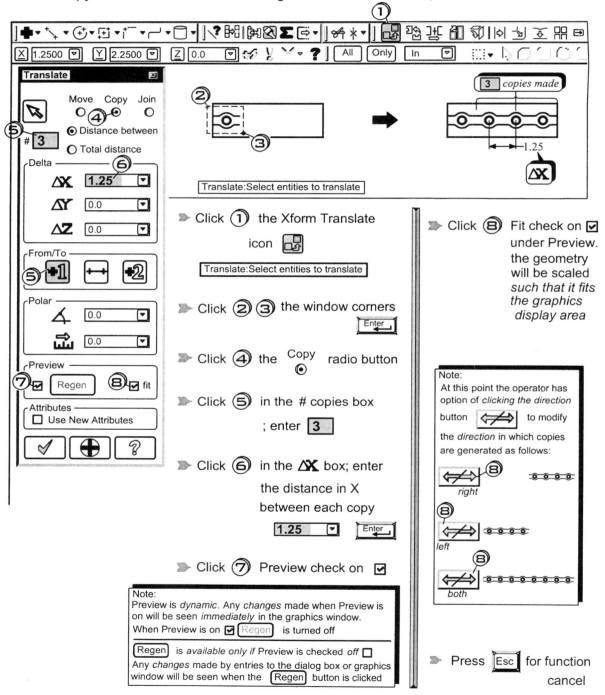

➤ Click ① the Xform Translate

icon

Translate:Select entities to translate

➤ Click ② ③ the window corners

Enter

➤ Click ④ the Copy radio button

➤ Click ⑤ in the # copies box
; enter 3

➤ Click ⑥ in the ΔX box; enter
the distance in X
between each copy

1.25   Enter

➤ Click ⑦ Preview check on ☑

Note:
Preview is dynamic. Any changes made when Preview is on will be seen immediately in the graphics window. When Preview is on ☑ Regen is turned off

Regen is available only if Preview is checked off ☐ Any changes made by entries to the dialog box or graphics window will be seen when the Regen button is clicked

➤ Click ⑧ Fit check on ☑
under Preview.
the geometry
will be scaled
such that it fits
the graphics
display area

Note:
At this point the operator has option of clicking the direction

button ⟷ to modify

the direction in which copies are generated as follows:

➤ Press Esc for function
cancel

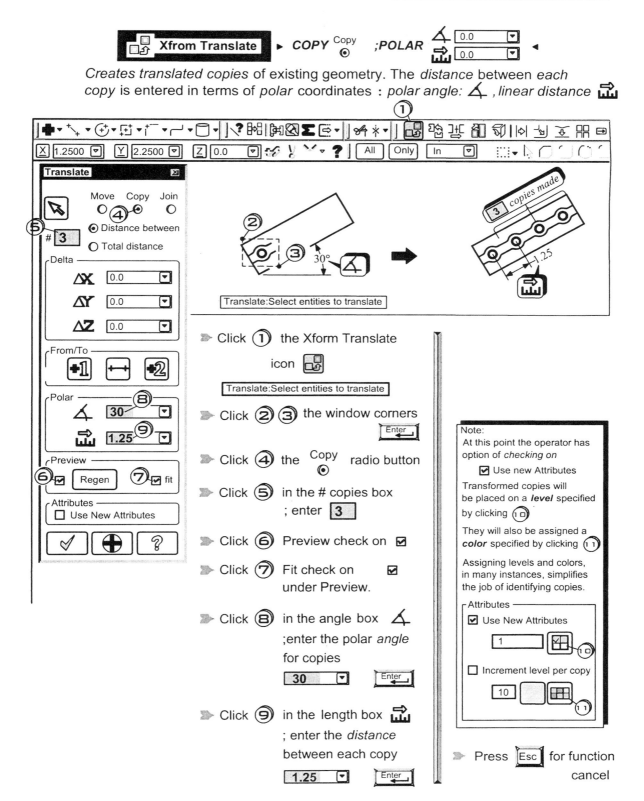

Xfrom Translate ▸ *COPY* Copy ;*POLAR* 0.0 / 0.0

Creates translated copies of existing geometry. The *distance* between *each copy* is entered in terms of *polar* coordinates : *polar angle:* ∠ , *linear distance*

Translate:Select entities to translate

➤ Click ① the Xform Translate icon

Translate:Select entities to translate

➤ Click ② ③ the window corners
Enter

➤ Click ④ the Copy radio button ⊙

➤ Click ⑤ in the # copies box ; enter 3

➤ Click ⑥ Preview check on ☑

➤ Click ⑦ Fit check on ☑ under Preview.

➤ Click ⑧ in the angle box ∠ ;enter the polar *angle* for copies
30 Enter

➤ Click ⑨ in the length box ; enter the *distance* between each copy
1.25 Enter

Note:
At this point the operator has option of *checking on*

☑ Use new Attributes

Transformed copies will be placed on a *level* specified by clicking ⑩

They will also be assigned a *color* specified by clicking ⑪

Assigning levels and colors, in many instances, simplifies the job of identifying copies.

Attributes
☑ Use New Attributes
1 ⑩
☐ Increment level per copy
10 ⑪

➤ Press Esc for function cancel

Stretches the *ends* of *existing line* entities between *two* inputted *points*
General selection *automatically assumes In+Intersect has been selected.*

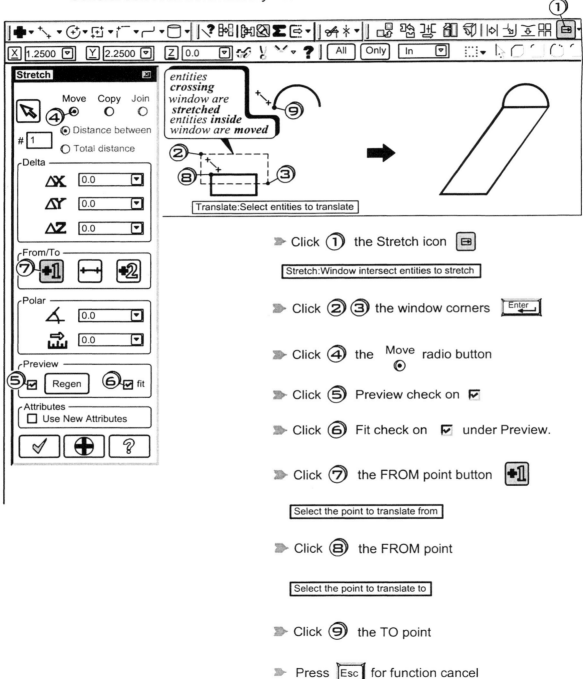

➤ Click ① the Stretch icon

Stretch:Window intersect entities to stretch

➤ Click ② ③ the window corners

➤ Click ④ the Move radio button

➤ Click ⑤ Preview check on ✔

➤ Click ⑥ Fit check on ✔ under Preview.

➤ Click ⑦ the FROM point button

Select the point to translate from

➤ Click ⑧ the FROM point

Select the point to translate to

➤ Click ⑨ the TO point

➤ Press Esc for function cancel

Xfrom Rotate ▶ *MOVE* Move ◉ *;ROTATE* ◉ Rotate ◀

Rotates existing geometry about a *specified center* and through an *inputted angle.*

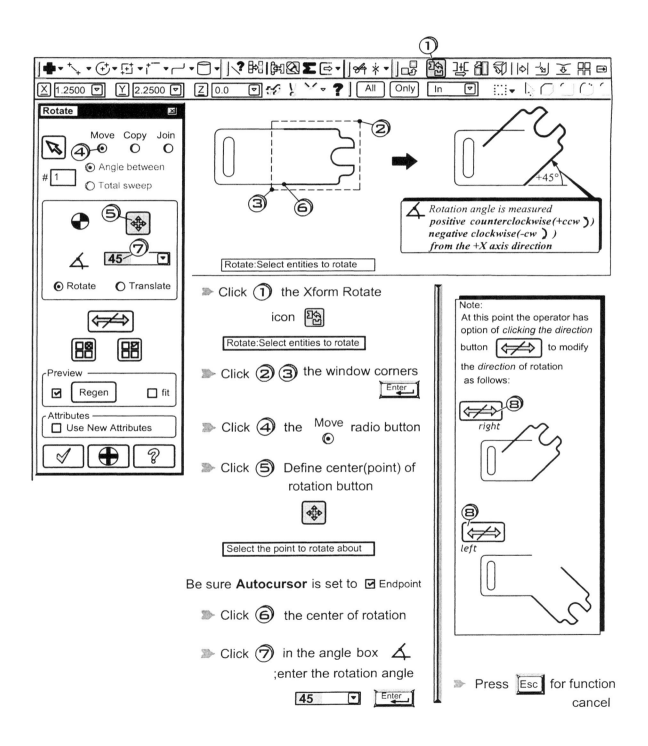

Rotation angle is measured positive counterclockwise(+ccw)) negative clockwise(-cw)) from the +X axis direction

Rotate:Select entities to rotate

➤ Click ① the Xform Rotate icon

Rotate:Select entities to rotate

➤ Click ② ③ the window corners

Enter ↵

➤ Click ④ the Move ◉ radio button

➤ Click ⑤ Define center(point) of rotation button

Select the point to rotate about

Be sure **Autocursor** is set to ☑ Endpoint

➤ Click ⑥ the center of rotation

➤ Click ⑦ in the angle box ∠ ;enter the rotation angle

45 ▼ Enter ↵

Note:
At this point the operator has option of *clicking the direction* button ⟨⇌⟩ to modify the *direction* of rotation as follows:

⟨⇌⟩ ⑧
right

⑧ ⟨⇌⟩
left

➤ Press Esc for function cancel

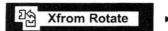

Xfrom Rotate ► *COPY* Copy ⊙ *;ROTATE* ⊙ Rotate ◄

Creates rotated copies of existing geometry about a *specified center* and through an *inputted angle*.

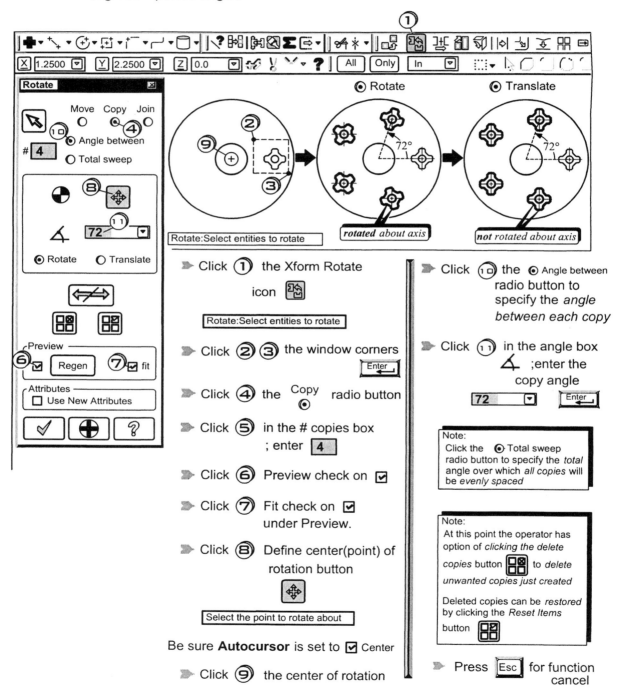

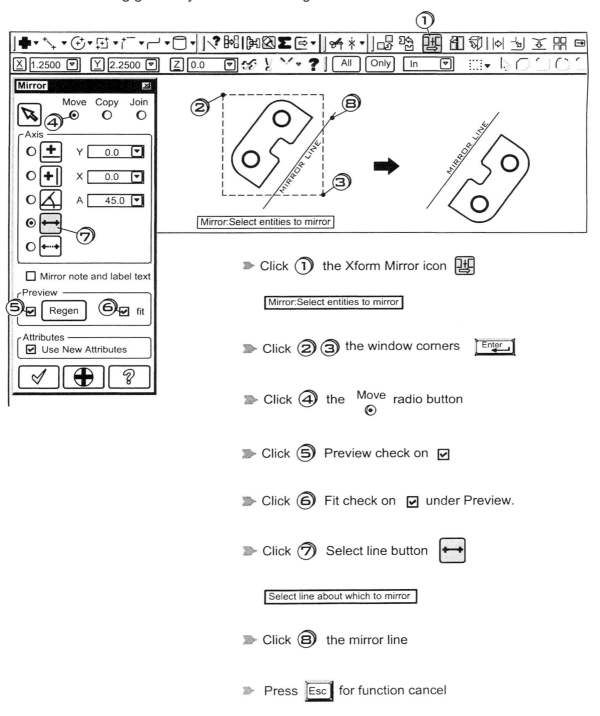

Moves existing geometry into a *mirror image* orientation about a *mirror line clicked.*

➤ Click ① the Xform Mirror icon

Mirror:Select entities to mirror

➤ Click ② ③ the window corners Enter

➤ Click ④ the Move radio button

➤ Click ⑤ Preview check on ☑

➤ Click ⑥ Fit check on ☑ under Preview.

➤ Click ⑦ Select line button

Select line about which to mirror

➤ Click ⑧ the mirror line

➤ Press Esc for function cancel

Creates a *mirror copy* of existing geometry about a *mirror line defined by clicking its two endpoints*.

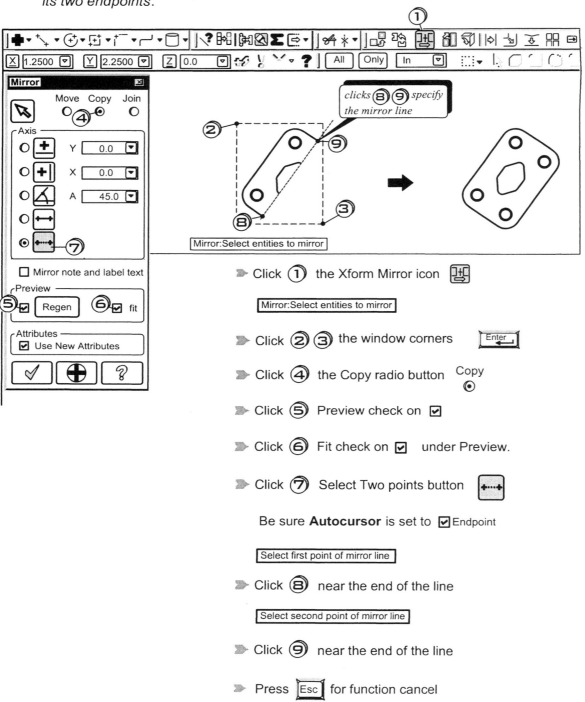

Mirror:Select entities to mirror

➣ Click ① the Xform Mirror icon 🔲

   Mirror:Select entities to mirror

➣ Click ② ③ the window corners   Enter

➣ Click ④ the Copy radio button   Copy ◉

➣ Click ⑤ Preview check on ☑

➣ Click ⑥ Fit check on ☑ under Preview.

➣ Click ⑦ Select Two points button ┝┉┥

   Be sure **Autocursor** is set to ☑ Endpoint

   Select first point of mirror line

➣ Click ⑧ near the end of the line

   Select second point of mirror line

➣ Click ⑨ near the end of the line

➣ Press Esc for function cancel

Creates a *mirror copy* of existing geometry about a selected *mirror* line parallel to either the X or Y axis of the active construction plane. The operator can input the reference point. For X-axis mirroring it specifies the ammount by which the mirror line is offset from the X- axis(Y value). For Y-axis mirroring the specification is the ammount by which the line is offset from the Y-axis(X value).

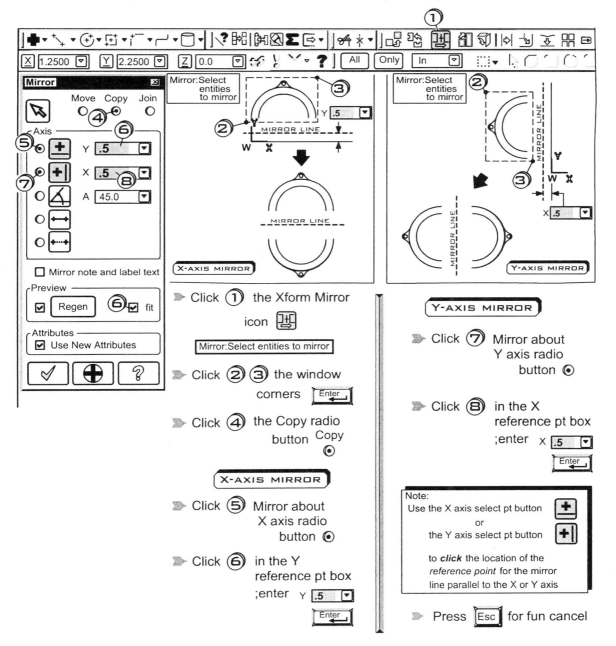

Creates a *mirror copy* of existing geometry about a selected *mirror* line defined at an angle with respect to the positive X axis of the active construction plane. The operator can input the reference point(angle 0°-360°)

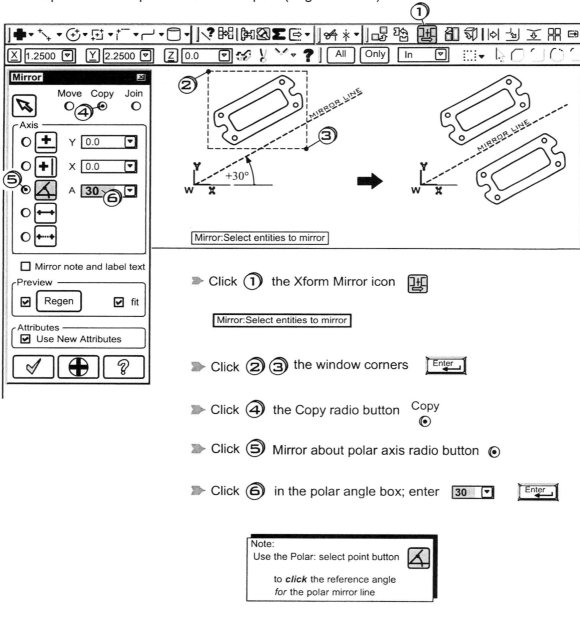

▶ Click ① the Xform Mirror icon

Mirror:Select entities to mirror

▶ Click ② ③ the window corners [Enter]

▶ Click ④ the Copy radio button   Copy ⊙

▶ Click ⑤ Mirror about polar axis radio button ⊙

▶ Click ⑥ in the polar angle box; enter [30 ▾] [Enter]

Note:
Use the Polar: select point button

to **click** the reference angle
for the polar mirror line

▶ Press [Esc] for function cancel

Xfrom Scale ▶ *MOVE* Move ⊙ *;UNIFORM* ⊙ Uniform ◀

Uniformly scales the size of existing geometry up or down by an inputter scale factor

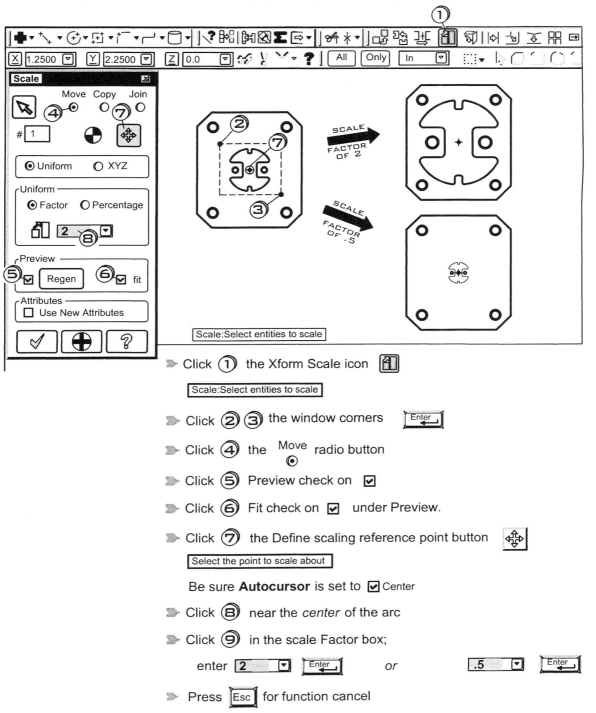

➤ Click ① the Xfrom Scale icon

Scale:Select entities to scale

➤ Click ② ③ the window corners   Enter

➤ Click ④ the Move ⊙ radio button

➤ Click ⑤ Preview check on ☑

➤ Click ⑥ Fit check on ☑ under Preview.

➤ Click ⑦ the Define scaling reference point button

Select the point to scale about

Be sure **Autocursor** is set to ☑ Center

➤ Click ⑧ near the *center* of the arc

➤ Click ⑨ in the scale Factor box;

enter 2 ▼ Enter   *or*   .5 ▼ Enter

➤ Press Esc for function cancel

Xfrom Scale ▶ *MOVE* Move ⊙ *; XYZ* ⊙ XYZ ◀

Uniformly scales the size of existing geometry *up or down by an inputter scale factor along the X, Y and Z directions of the currently active construction plane.* XYZ scaling changes the *shape* of the original geometry.

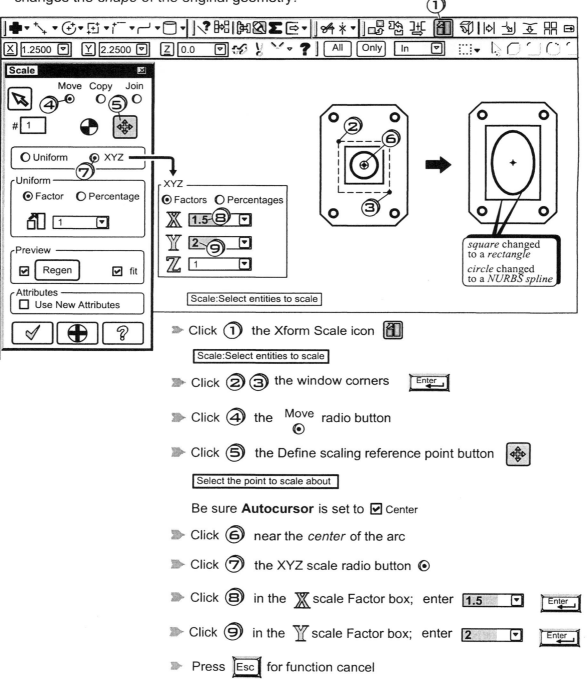

square changed to a *rectangle*

circle changed to a *NURBS spline*

▶ Click ① the Xform Scale icon

⌞Scale:Select entities to scale⌝

▶ Click ② ③ the window corners   ⌞Enter⌝

▶ Click ④ the Move ⊙ radio button

▶ Click ⑤ the Define scaling reference point button ⊕

⌞Select the point to scale about⌝

Be sure **Autocursor** is set to ☑ Center

▶ Click ⑥ near the *center* of the arc

▶ Click ⑦ the XYZ scale radio button ⊙

▶ Click ⑧ in the X scale Factor box; enter ⌞1.5⌝   ⌞Enter⌝

▶ Click ⑨ in the Y scale Factor box; enter ⌞2⌝   ⌞Enter⌝

▶ Press ⌞Esc⌝ for function cancel

| | Xfrom Offset ▸ *COPY* Copy ⊙ ; *DISTANCE* 🔛 | 0.0 ▾ | ◂

Creates an offset copy in a *perpendicular* direction to an existing line, arc, spline or curve entity. The perpendicular direction is along *every point along the entity*. This function offsets a *single* entity at a time. Use Offset Contour to offset a *chain* of entities. The Flip function is used ⇔ to *reverse the offset direction after Preview*.

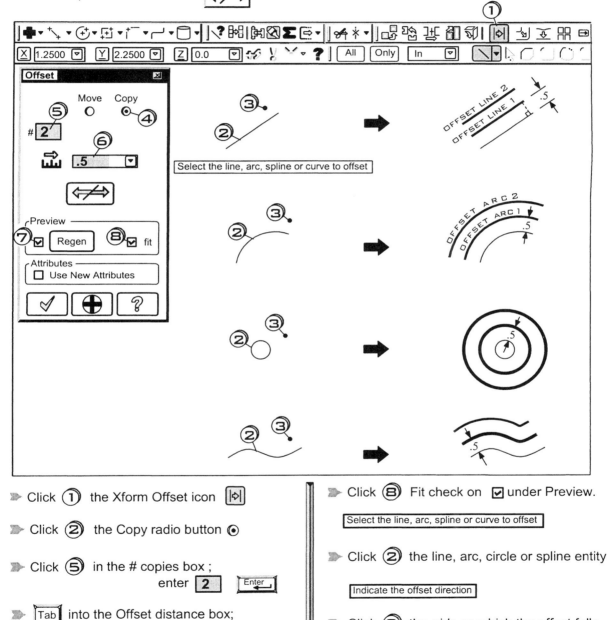

➤ Click ① the Xform Offset icon |⇼|

➤ Click ② the Copy radio button ⊙

➤ Click ⑤ in the # copies box ; enter 2 Enter

➤ Tab into the Offset distance box; enter .5 ▾ Enter

➤ Click ⑦ Preview check on ☑

➤ Click ⑧ Fit check on ☑ under Preview.

Select the line, arc, spline or curve to offset

➤ Click ② the line, arc, circle or spline entity

Indicate the offset direction

➤ Click ③ the *side* on which the offset falls

➤ Press Esc for function cancel

Creates an offset copy in a *perpendicular* direction to an existing contour. The perpendicular direction is along *every point along the contour relative to the current construction plane*. The contour may consist of connected line, arc, spline or curve entities. The Flip function is used to *reverse the direction of the offset contour after Preview.*

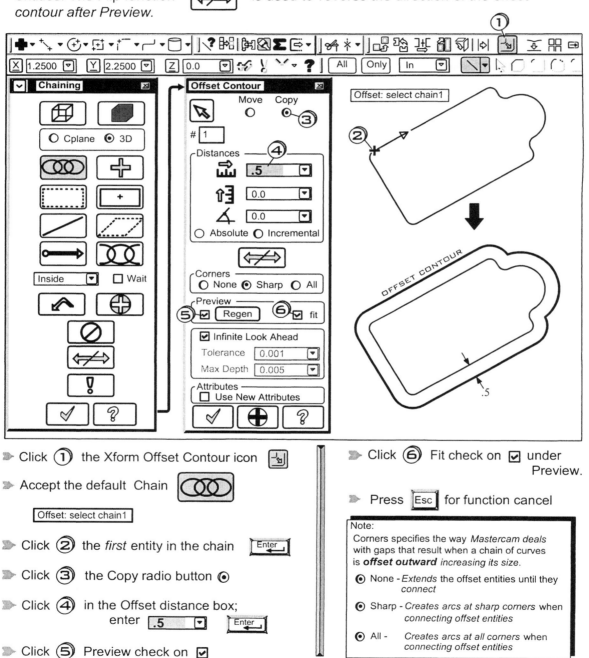

➤ Click ① the Xform Offset Contour icon

➤ Accept the default Chain

Offset: select chain1

➤ Click ② the *first* entity in the chain   Enter

➤ Click ③ the Copy radio button ◉

➤ Click ④ in the Offset distance box; enter .5   Enter

➤ Click ⑤ Preview check on ☑

➤ Click ⑥ Fit check on ☑ under Preview.

➤ Press Esc for function cancel

Note:
Corners specifies the way *Mastercam deals* with gaps that result when a chain of curves is **offset outward** increasing its size.

◉ None - *Extends* the offset entities until they *connect*

◉ Sharp - *Creates arcs at sharp corners when connecting offset entities*

◉ All - *Creates arcs at all corners when connecting offset entities*

Creates an offset copy in a *perpendicular* direction to an existing contour and optionally at a specified Z depth. The perpendicular direction is along *every point along the contour relative to the current construction plane*. The contour may consist of connected line, arc, spline or curve entities. The operator has the option of inputting *either the offset distance or the taper angle* since inputting *one determines the other*. Use the Flip function [⟷] to *reverse the direction of the offset contour after Preview.*

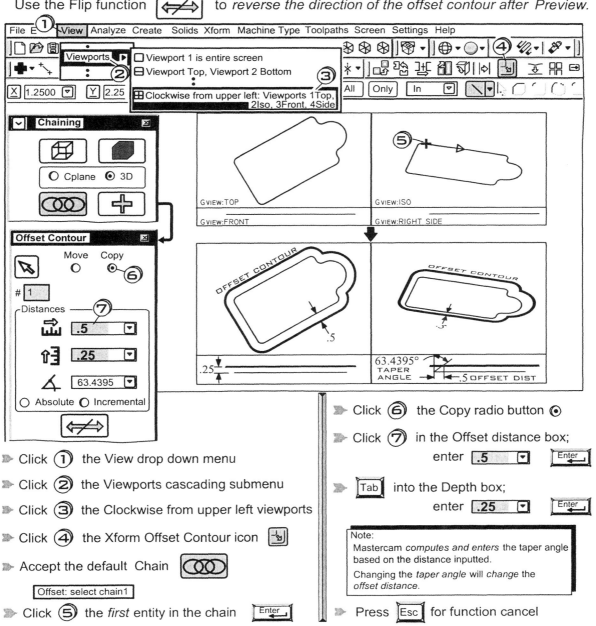

➥ Click ① the View drop down menu

➥ Click ② the Viewports cascading submenu

➥ Click ③ the Clockwise from upper left viewports

➥ Click ④ the Xform Offset Contour icon

➥ Accept the default Chain

   Offset: select chain1

➥ Click ⑤ the *first* entity in the chain [Enter]

➥ Click ⑥ the Copy radio button ⊙

➥ Click ⑦ in the Offset distance box;
   enter .5 ▾ [Enter]

➥ [Tab] into the Depth box;
   enter .25 ▾ [Enter]

Note:
Mastercam *computes and enters* the taper angle based on the distance inputted.
Changing the *taper angle* will *change* the *offset distance*.

➥ Press [Esc] for function cancel

Xfrom Rectangular Array

Creates copies of entities in an array or grid pattern. The operator can specify the number of copies to be generated in *both Direction 1(X) and Direction 2(Y)* relative to the current construction plane. Each *Direction* has assocated with it an inputted *distance* between each copy and rotation *angle* relative to the +X axis of the current construction plane.

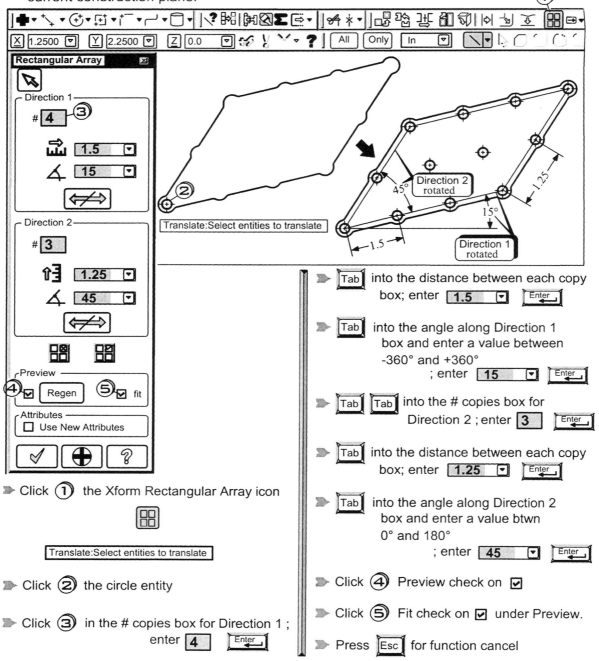

> Click ① the Xform Rectangular Array icon

Translate:Select entities to translate

> Click ② the circle entity

> Click ③ in the # copies box for Direction 1 ; enter **4** [Enter]

> [Tab] into the distance between each copy box; enter **1.5** ▼ [Enter]

> [Tab] into the angle along Direction 1 box and enter a value between -360° and +360° ; enter **15** ▼ [Enter]

> [Tab] [Tab] into the # copies box for Direction 2 ; enter **3** [Enter]

> [Tab] into the distance between each copy box; enter **1.25** ▼ [Enter]

> [Tab] into the angle along Direction 2 box and enter a value btwn 0° and 180° ; enter **45** ▼ [Enter]

> Click ④ Preview check on ☑

> Click ⑤ Fit check on ☑ under Preview.

> Press [Esc] for function cancel

Dynamically moves or copies selected entities in the graphics window as the operator moves the mouse cursor. Entities can be *translated* as well as *rotated* to their new positions. For translation, entities that have been selected using a *crossing window* (In+) will be *dynamically stretched*. Those contained *within* the window will be *translated*.

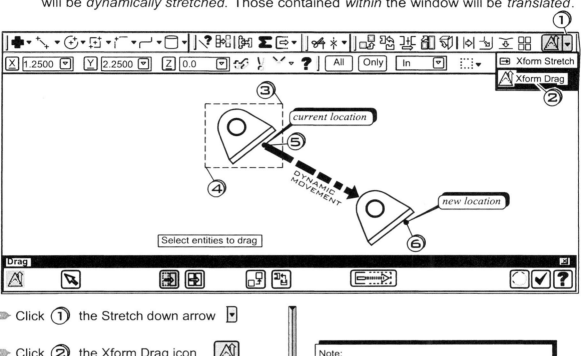

➣ Click ① the Stretch down arrow ▾

➣ Click ② the Xform Drag icon 🗻

> Select entities to drag

➣ Click ③ ④ the window corners ; Enter

➣ Tap the M key for *Move*

> Select the starting point

➣ Click ⑤ the reference *starting* pt

➣ Click ⑥ the *new* location

➣ Tap the F3 key to Repaint the screen

➣ Press Esc for function cancel

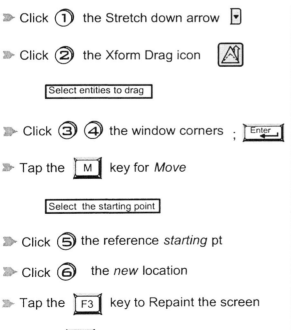

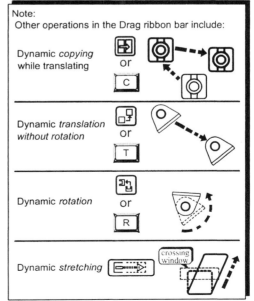

Note:
Other operations in the Drag ribbon bar include:

Dynamic *copying* while translating or C

Dynamic *translation* without rotation or T

Dynamic *rotation* or R

Dynamic *stretching*

EXERCISES

In each exercise create the *wireframe* part geometry using *MastercamX2's* Design package.

┌─────────────────────────────────┐
│ TUTORIAL PRACTICE EXERCISE   │
└─────────────────────────────────┘

3-1)   ┌──────────────────────────┐ ──── YOUR INITIALS
       │ File Name: **EX3-1JV** │
       └──────────────────────────┘

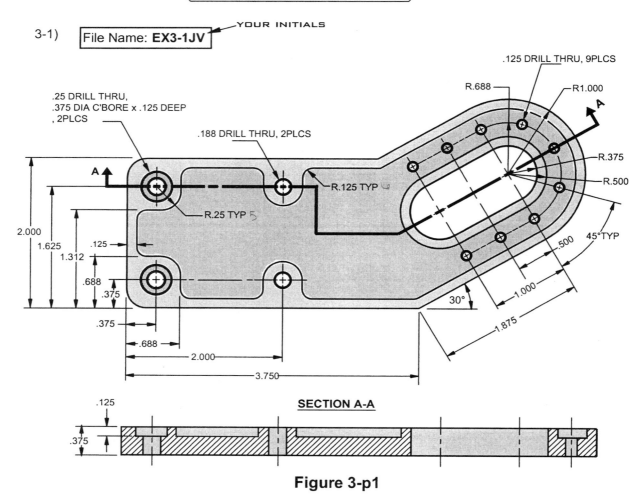

Figure 3-p1

Enter the *MastercamX2* Design package.

⫸ Click ① the *Mastercam X2* icon

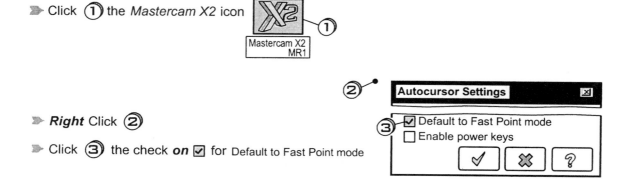

⫸ *Right* Click ②

⫸ Click ③ the check *on* ☑ for Default to Fast Point mode

a) Create the horizontal and vertical *outer boundary* lines

➤ Click ④ the down arrow ▼

➤ Click ⑤ [Create Line Endpoint]

Specify the first endpoint

➤ Click ⑥ the multi line button 🗙 or tap the M key

➤ Enter 2,2   Enter   ◀--------------------------- ⓐ

Specify the second endpoint

➤ Enter 0   Enter   Autocursor will use the *last* value inputted for Y ◀---ⓑ

Specify the second endpoint

➤ Enter ,0   Enter   Autocursor will use the *last* value inputted for X ◀---ⓒ

Specify the second endpoint

➤ Enter 3.75   Enter   ◀--------------------------- ⓓ

➤ Press Esc to cancel the operation

➤ Reposition the display by tapping the pan right → and pan up ↑ buttons on the keyboard.

b) Generate the two 1.875 long lines at an incline of 30° to the horizontal.

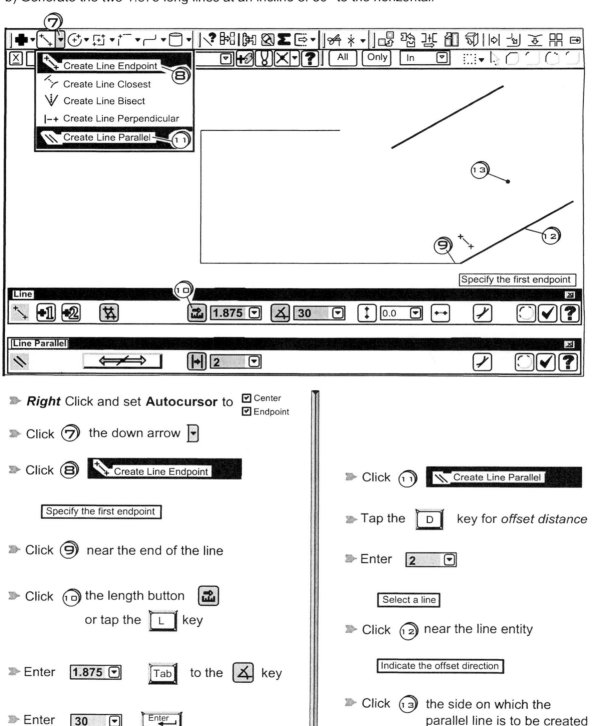

➤ **Right** Click and set **Autocursor** to ☑ Center
　　　　　　　　　　　　　　　　　　☑ Endpoint

➤ Click ⑦ the down arrow ▾

➤ Click ⑧ [➕ Create Line Endpoint]

　　[Specify the first endpoint]

➤ Click ⑨ near the end of the line

➤ Click ⑩ the length button [📊]
　　　　or tap the [L] key

➤ Enter [1.875 ▾] [Tab] to the [∠] key

➤ Enter [30 ▾] [Enter ↵]

➤ Press [Esc] to cancel the operation

➤ Click ⑪ [◸ Create Line Parallel]

➤ Tap the [D] key for *offset distance*

➤ Enter [2 ▾]

　　[Select a line]

➤ Click ⑫ near the line entity

　　[Indicate the offset direction]

➤ Click ⑬ the side on which the
　　　　　parallel line is to be created

➤ Press [Esc] to cancel the operation

c) Add the 1in radius arc tangent to the inclined lines.

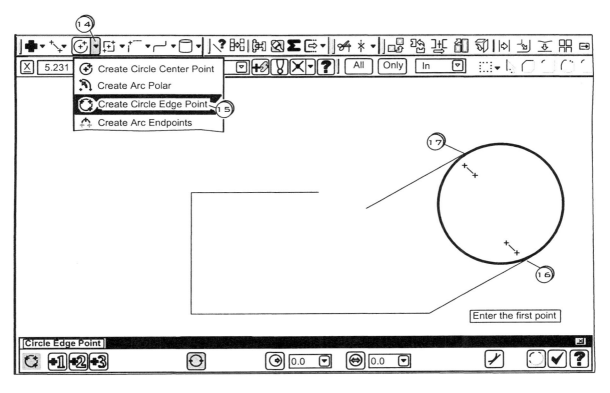

➤ Click ⑭ the down arrow 🔽

➤ Click ⑮ 🔘 Create Circle Edge Point

➤ Tap the ⎡W⎤ key for *2-pt circle* 🔘

Enter the first point

➤ Click ⑯ near the endpoint of the line

Enter the second point

➤ Click ⑰ near the endpoint of the line

➤ Press ⎡Esc⎤ to cancel the operation

d) Trim the horizontal and inclined lines to their intersection point.

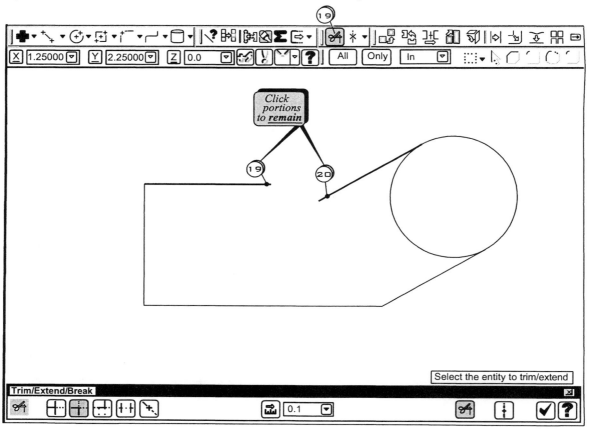

➤ Click ⑱ the Trim icon

➤ Tap the ⓶ key for *2 entities*

Select the entity to trim/extend

➤ Click ⑲ the portion to **remain**

Select the entity to trim/extend to

➤ Click ⓴ the portion to **remain**

➤ Press Esc to cancel the function

e) Use the Divide function to remove the lower left portion of the 1in radius arc.

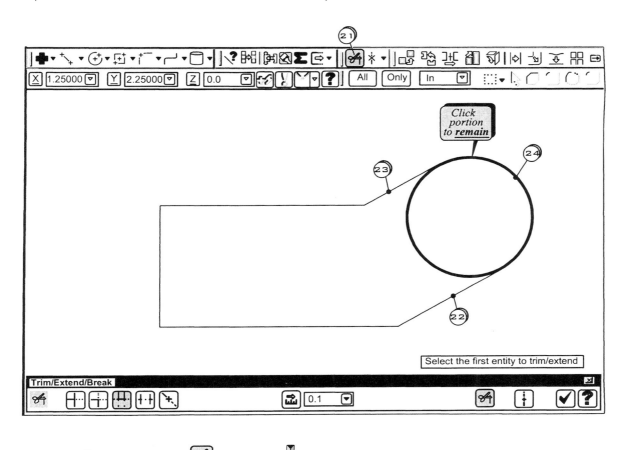

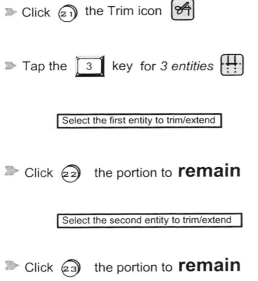

≫ Click ㉑ the Trim icon

≫ Tap the ③ key for *3 entities*

> Select the first entity to trim/extend

≫ Click ㉒ the portion to **remain**

> Select the second entity to trim/extend

≫ Click ㉓ the portion to **remain**

> Select the entity to trim/extend to

≫ Click ㉔ the portion of the arc to **remain**

≫ Press Esc to cancel the function

f) Create two R.25 circles. Generate a .188D circle, and a .25D circle centered on a .375D circle.

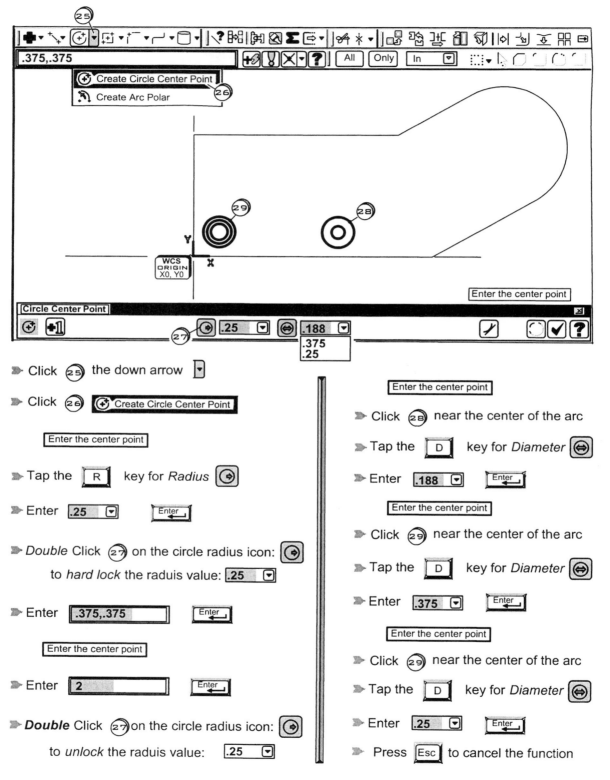

➤ Click ㉕ the down arrow ▾

➤ Click ㉖ Create Circle Center Point

Enter the center point

➤ Tap the R key for *Radius*

➤ Enter .25 ▾ Enter

➤ *Double* Click ㉗ on the circle radius icon:
     to *hard lock* the raduis value: .25 ▾

➤ Enter .375,.375 Enter

Enter the center point

➤ Enter 2 Enter

➤ *Double* Click ㉗ on the circle radius icon:
     to *unlock* the raduis value: .25 ▾

Enter the center point

➤ Click ㉘ near the center of the arc

➤ Tap the D key for *Diameter*

➤ Enter .188 ▾ Enter

Enter the center point

➤ Click ㉙ near the center of the arc

➤ Tap the D key for *Diameter*

➤ Enter .375 ▾ Enter

Enter the center point

➤ Click ㉙ near the center of the arc

➤ Tap the D key for *Diameter*

➤ Enter .25 ▾ Enter

➤ Press Esc to cancel the function

g) Generate horizontal and vertical inner boundary lines tangent to the R.25 circles.

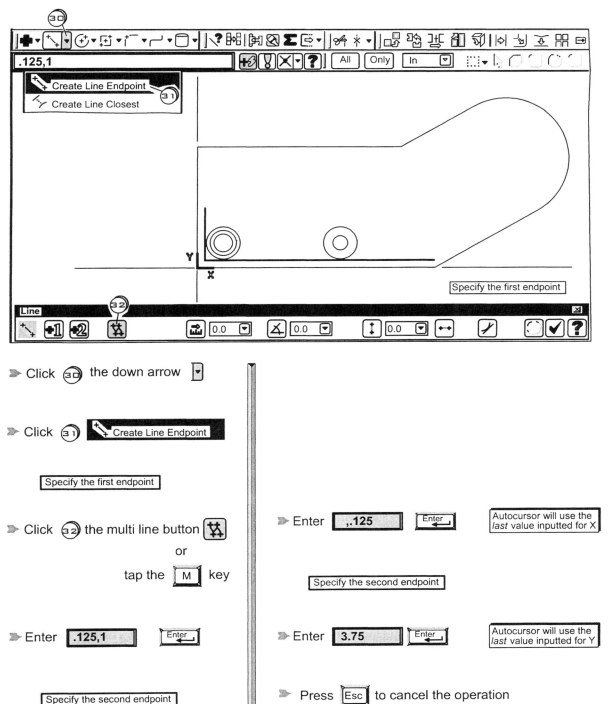

➤ Click ③⓪ the down arrow ▾

➤ Click ③① [Create Line Endpoint]

| Specify the first endpoint |

➤ Click ③② the multi line button 🔀

or

tap the [M] key

➤ Enter [.125,1] [Enter↵]

| Specify the second endpoint |

➤ Enter [,.125] [Enter] | Autocursor will use the *last* value inputted for X |

| Specify the second endpoint |

➤ Enter [3.75] [Enter↵] | Autocursor will use the *last* value inputted for Y |

➤ Press [Esc] to cancel the operation

h) Create horizontal and vertical lines *tangent* to the R.25 circles and running *perpendicular* to the inner boundary lines.

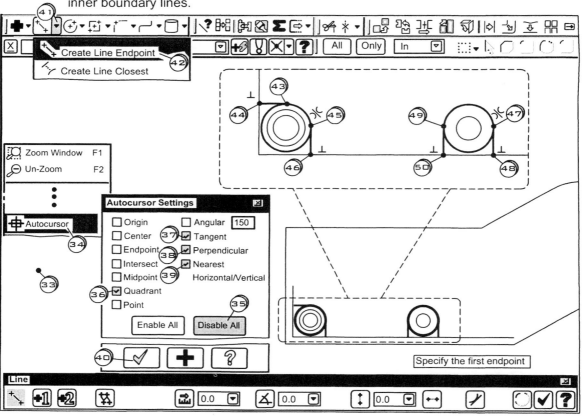

≫ **Right** Click ③③ in the display screen area

≫ Click ③④ **Autocursor ;** ③⑤ [Disable All]

≫ Click ③⑥ ③⑦ ③⑧ ③⑨ the checks on ☑for

      Quadrant, Tangent, Perpendicular and Nearest snap

≫ Click ④⓪ the OK button ☑

≫ Click ④① the down arrow ▾

≫ Click ④② [Create Line Endpoint]

    [Specify the first endpoint]

≫ Click ④③ near the *top* of the circle

    [Specify the second endpoint]

≫ Click ④④ near the *perp int pt* ; Tap [Esc] key

    [Specify the first endpoint]

≫ Click ④⑤ near the *rt side* of the circle

    [Specify the second endpoint]

≫ Click ④⑥ near the *perp int pt* ;   [Esc]

    [Specify the first endpoint]

≫ Click ④⑦ near the *rt side* of the circle

    [Specify the second endpoint]

≫ Click ④⑧ near the *perp int pt;*   [Esc]

    [Specify the first endpoint]

≫ Click ④⑨ near the *left side* of the circle

    [Specify the second endpoint]

≫ Click ⑤⓪ near the *perp int pt on*
                   the horizontal line

≫ Press [Esc] to cancel the function

i) Remove all portions of the R.25circles and the inner boundary lines such that a continuous, uninterrupted inner boundary curve is produced.

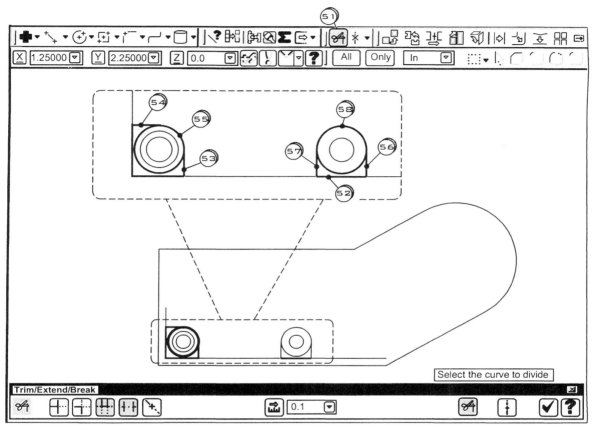

Select the curve to divide

Trim/Extend/Break

➤ Click ⑤① the Trim icon

➤ Tap the **D** key for *Divide entities*

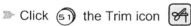

Select the curve to divide

➤ Click ⑤② the portion of the line to be **removed**

Tap the **3** key for *Trim 3 entities*

Select the first entity to trim/extend

➤ Click ⑤③ the intersecting line

Select the second entity to trim/extend

➤ Click ⑤④ the intersecting line

Select the entity to trim/extend to

➤ Click ⑤⑤ the portion of the arc to **remain**

Select the first entity to trim/extend

➤ Click ⑤⑥ the intersecting line

Select the second entity to trim/extend

➤ Click ⑤⑦ the intersecting line

Select the entity to trim/extend to

➤ Click ⑤⑧ the portion of the arc to **remain**

➤ Press **Esc** to cancel the function

j) Apply R.125 fillets to the inner boundary corners

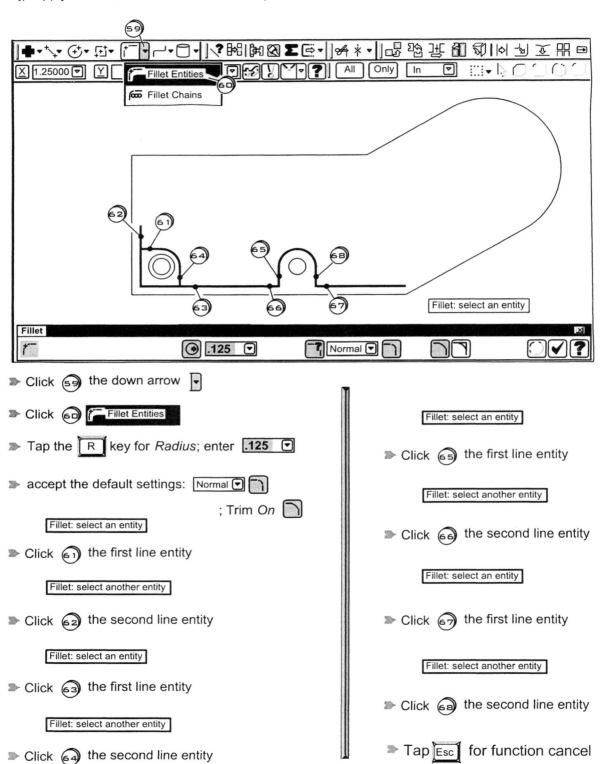

➤ Click ⑤⑨ the down arrow

➤ Click ⑥⓪ ▐ Fillet Entities ▌

➤ Tap the ▢ R key for *Radius*; enter .125 ▾

➤ accept the default settings: Normal ▾ ▢ ; Trim *On* ▢

         Fillet: select an entity

➤ Click ⑥① the first line entity

         Fillet: select another entity

➤ Click ⑥② the second line entity

         Fillet: select an entity

➤ Click ⑥③ the first line entity

         Fillet: select another entity

➤ Click ⑥④ the second line entity

         Fillet: select an entity

➤ Click ⑥⑤ the first line entity

         Fillet: select another entity

➤ Click ⑥⑥ the second line entity

         Fillet: select an entity

➤ Click ⑥⑦ the first line entity

         Fillet: select another entity

➤ Click ⑥⑧ the second line entity

➤ Tap Esc for function cancel

k) Create a copy of the inner boundary by mirroring a *copy about the horizontal* mirror line.

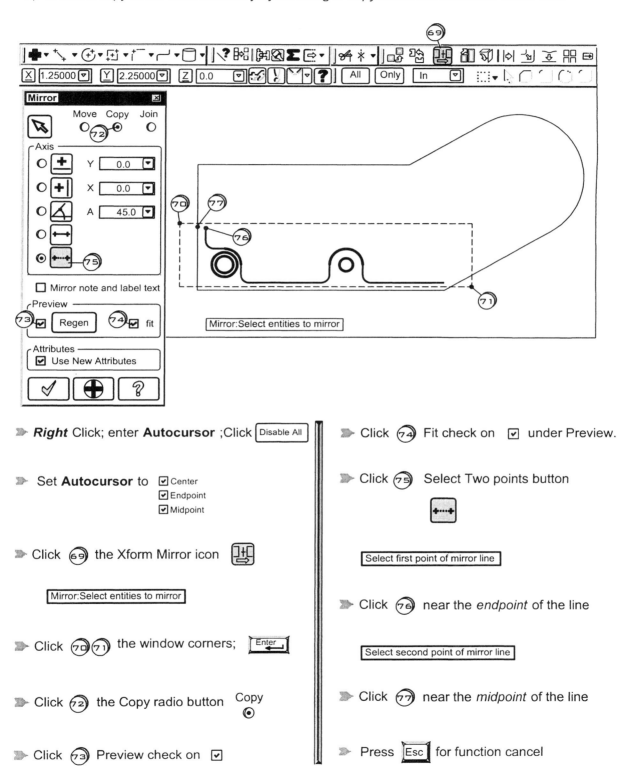

➤ **Right** Click; enter **Autocursor** ;Click Disable All

➤ Set **Autocursor** to ☑ Center
☑ Endpoint
☑ Midpoint

➤ Click ⑥⑨ the Xform Mirror icon

Mirror:Select entities to mirror

➤ Click ⑦⓪⑦① the window corners; Enter ⏎

➤ Click ⑦② the Copy radio button Copy ◉

➤ Click ⑦③ Preview check on ☑

➤ Click ⑦④ Fit check on ☑ under Preview.

➤ Click ⑦⑤ Select Two points button

➤ Select first point of mirror line

➤ Click ⑦⑥ near the *endpoint* of the line

Select second point of mirror line

➤ Click ⑦⑦ near the *midpoint* of the line

➤ Press Esc for function cancel

l) Use the Offset Contour function to generate the inclined line and arc portion of
   the inner boundary.

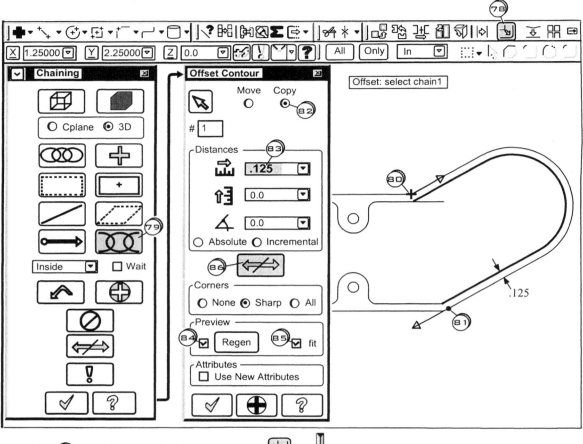

➤ Click ⑦⑧ the Xform Offset Contour icon

➤ Click ⑦⑨ the partial chain button

> Select the first entity

➤ Click ⑧⑩ the *first* entity in the chain

> Select the last entity

➤ Click ⑧① the *last* entity in the chain   Enter

➤ Click ⑧② the Copy radio button ⊙

➤ Click ⑧③ in the Offset distance box;
          enter **.125** ⊡ Enter

➤ Click ⑧④ Preview check on ☑

➤ Click ⑧⑤ Fit check on ☑ under
          Preview.

➤ Click ⑧⑥ the *left* arrow of the Flip
          direction button

          to *change* the offset
          direction from *outside to
          inside*

➤ Press Esc for function cancel

m) Trim the inner boundary lines to their points of intersection.

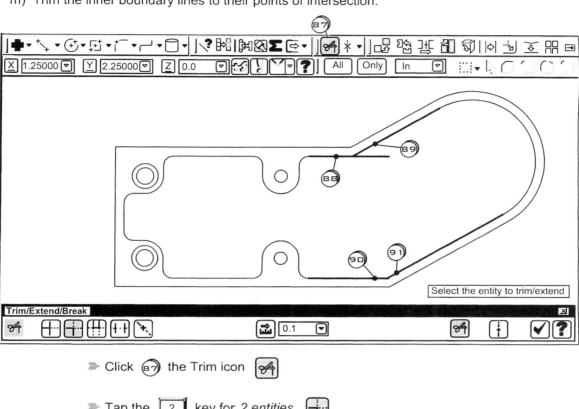

➤ Click ⑧⑦ the Trim icon

➤ Tap the ⎡2⎤ key for *2 entities*

> Select the entity to trim/extend

➤ Click ⑧⑧ the portion to **remain**

> Select the entity to trim/extend to

➤ Click ⑧⑨ the portion to **remain**

> Select the entity to trim/extend

➤ Click ⑨⑩ the portion to **remain**

> Select the entity to trim/extend to

➤ Click ⑨① the portion to **remain**

➤ Press ⎡Esc⎤ to cancel the function

n) Construct a 1in reference line to be used for *centering* a obround rectangular shape.

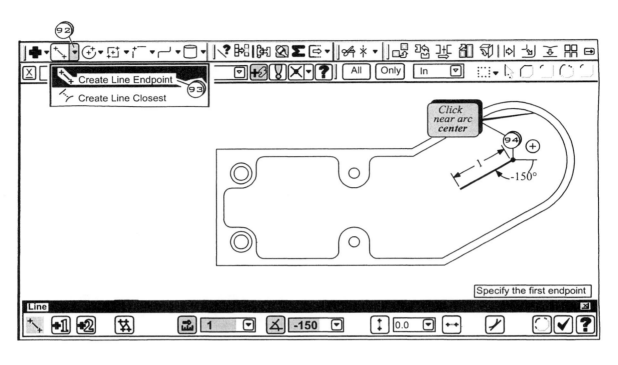

➤ Click ⑨② the down arrow ▾

➤ Click ⑨③ [Create Line Endpoint]

[Specify the first endpoint]

➤ Click ⑨④ near the *center of the arc*

➤ Tap the [L] key for line *length* 🖫

➤ Enter [1 ▾] [Tab] to the [⧊] key

➤ Enter [-150 ▾] [Enter ⏎]

➤ Press [Esc] to cancel the operation

o) Create two obround shapes centered at the 1in reference line. The first a 2in x 1in shape and the second 1.75in x .75in.

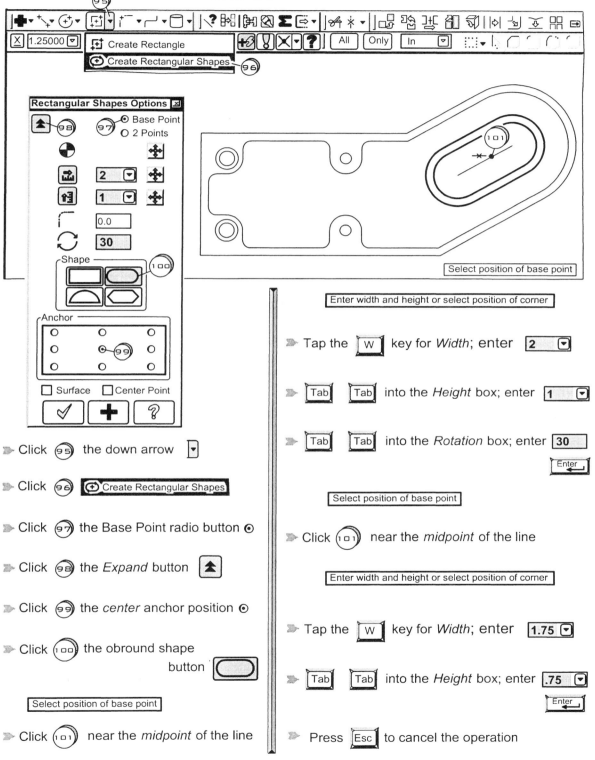

Select position of base point

Enter width and height or select position of corner

>> Tap the [W] key for *Width*; enter [2 ▼]

>> [Tab] [Tab] into the *Height* box; enter [1 ▼]

>> [Tab] [Tab] into the *Rotation* box; enter [30]

[Enter ↵]

Select position of base point

>> Click (101) near the *midpoint* of the line

Enter width and height or select position of corner

>> Tap the [W] key for *Width*; enter [1.75 ▼]

>> [Tab] [Tab] into the *Height* box; enter [.75 ▼]

[Enter ↵]

>> Press [Esc] to cancel the operation

>> Click (95) the down arrow [▼]

>> Click (96) [⊕ Create Rectangular Shapes]

>> Click (97) the Base Point radio button ⊙

>> Click (98) the *Expand* button [▲]

>> Click (99) the *center* anchor position ⊙

>> Click (100) the obround shape
button [⬭]

Select position of base point

>> Click (101) near the *midpoint* of the line

p) Place the .125DIA circle offset .688in from the reference line.

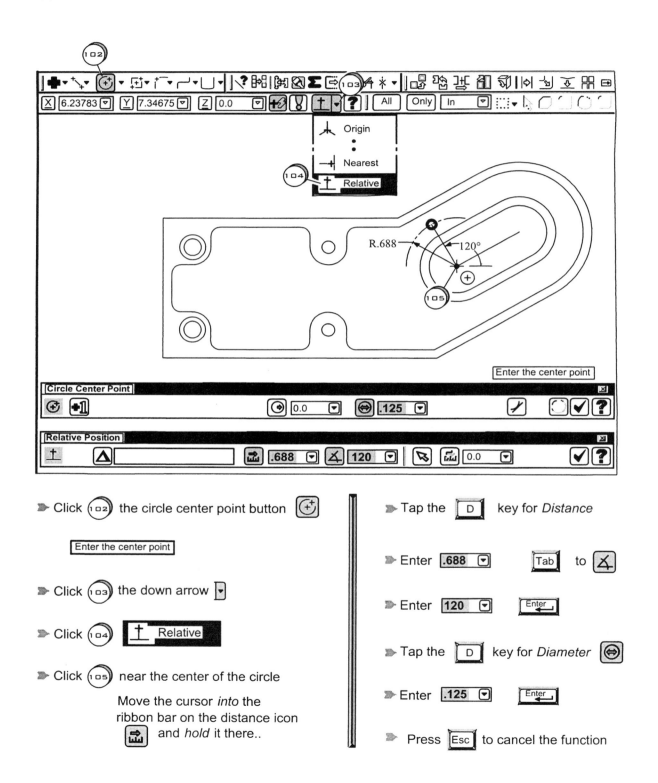

➤ Click ⟨102⟩ the circle center point button ⟨⊕⟩

Enter the center point

➤ Click ⟨103⟩ the down arrow ▾

➤ Click ⟨104⟩ ╧ Relative

➤ Click ⟨105⟩ near the center of the circle

   Move the cursor *into* the
   ribbon bar on the distance icon
   ⟨icon⟩ and *hold* it there..

➤ Tap the D key for *Distance*

➤ Enter .688 ▾   Tab to ∡

➤ Enter 120 ▾   Enter

➤ Tap the D key for *Diameter* ⊕

➤ Enter .125 ▾   Enter

➤ Press Esc to cancel the function

q) Use the Xform Translate function to create two more copies of the .125DIA circles on the 30° incline.

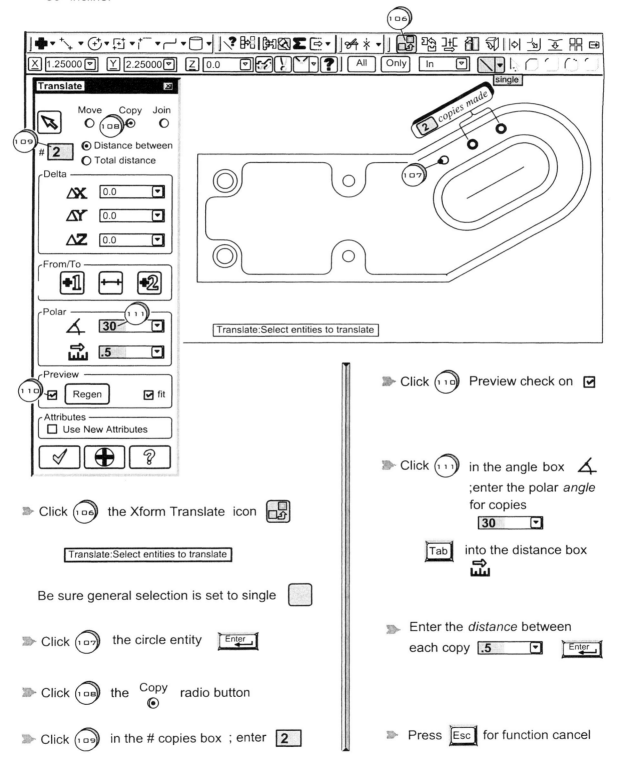

➤ Click the Xform Translate icon

Translate:Select entities to translate

Be sure general selection is set to single

➤ Click the circle entity   Enter

➤ Click the Copy radio button ⊙

➤ Click in the # copies box ; enter 2

➤ Click Preview check on ☑

➤ Click in the angle box ∡
;enter the polar *angle*
for copies
30

Tab into the distance box

➤ Enter the *distance* between
each copy .5   Enter

➤ Press Esc for function cancel

r) Create 4 copies of the .125DIA circles on the .688in radius arc by using the Xform
Rotate function.

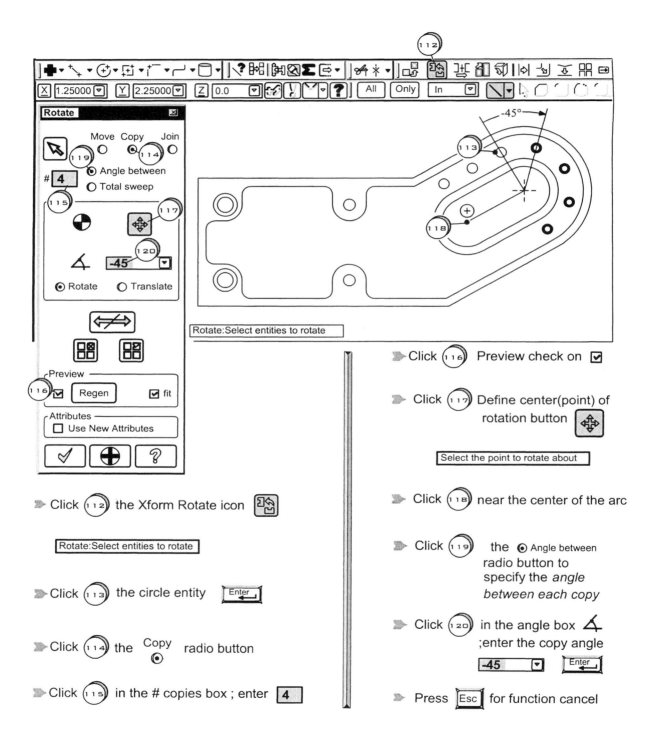

➤ Click (112) the Xform Rotate icon

Rotate:Select entities to rotate

➤ Click (113) the circle entity   Enter↵

➤ Click (114) the Copy radio button

➤ Click (115) in the # copies box ; enter 4

➤ Click (116) Preview check on ☑

➤ Click (117) Define center(point) of
rotation button

Select the point to rotate about

➤ Click (118) near the center of the arc

➤ Click (119) the ◉ Angle between
radio button to
specify the *angle
between each copy*

➤ Click (120) in the angle box ∠
;enter the copy angle
-45 ▾   Enter↵

➤ Press Esc for function cancel

s) Complete the .125DIA circle pattern by mirroring 2 copies about a 30° linclined mirror line

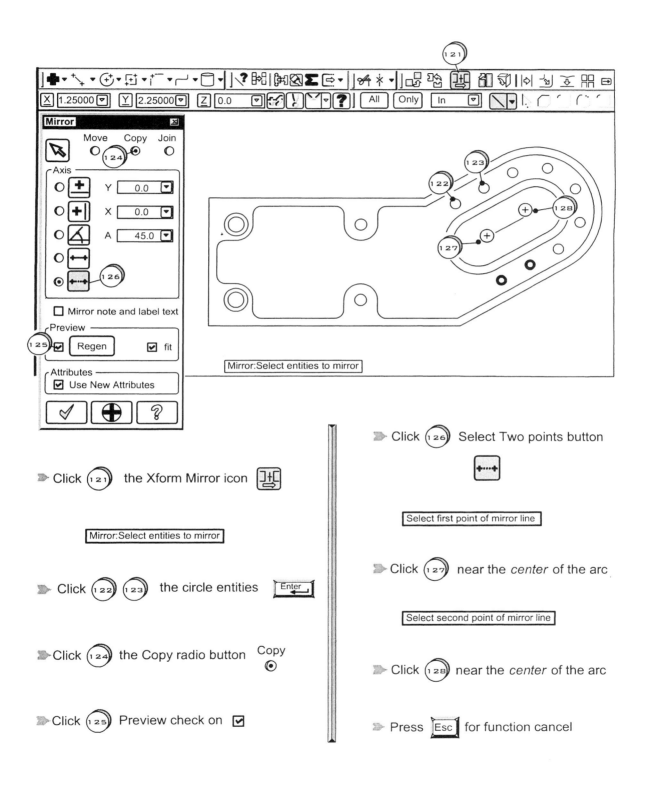

Mirror:Select entities to mirror

➤ Click (121) the Xform Mirror icon

Mirror:Select entities to mirror

➤ Click (122) (123) the circle entities   Enter

➤ Click (124) the Copy radio button   Copy ◉

➤ Click (125) Preview check on ☑

➤ Click (126) Select Two points button

Select first point of mirror line

➤ Click (127) near the *center* of the arc

Select second point of mirror line

➤ Click (128) near the *center* of the arc

➤ Press Esc for function cancel

t) Finish the CAD model by creating the R.25 fillets to the outer boundary and deleting the reference line Ⓐ

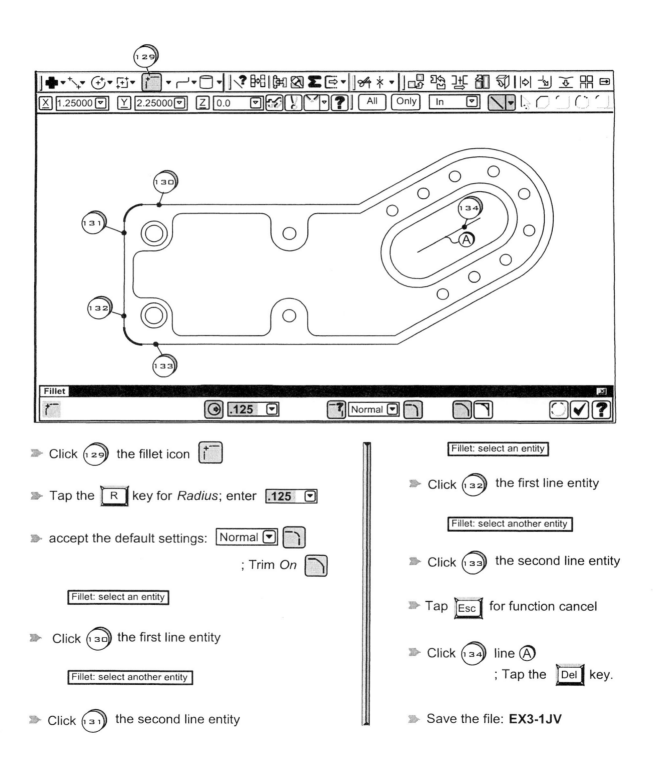

» Click ⑫⑨ the fillet icon 🖰

» Tap the [R] key for *Radius*; enter [.125 ▾]

» accept the default settings: [Normal ▾] 🖰

　　　　　　　　　　　　　　 ; Trim *On* 🖰

⌞Fillet: select an entity⌝

» Click ⑬⓪ the first line entity

⌞Fillet: select another entity⌝

» Click ⑬① the second line entity

⌞Fillet: select an entity⌝

» Click ⑬② the first line entity

⌞Fillet: select another entity⌝

» Click ⑬③ the second line entity

» Tap [Esc] for function cancel

» Click ⑬④ line Ⓐ
　　　　　　　 ; Tap the [Del] key.

» Save the file: **EX3-1JV**

For the following exercises ONLY THE FRONT VIEW is to be created for the 2D CAD model
All other views are given as reference .

3-2)

File Name: **EX3-2JV** ← YOUR INITIALS

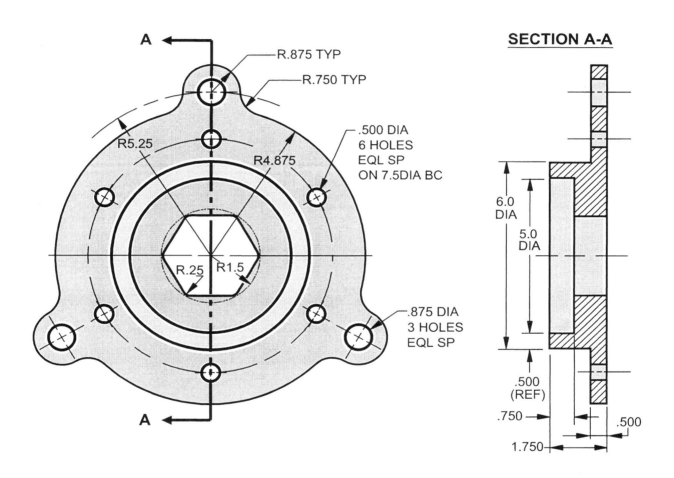

Figure 3-p2

a) Use the *Arc*, *Polar* command with Radius set to 4.875, initial angle set to 30° and final angle set to 150°

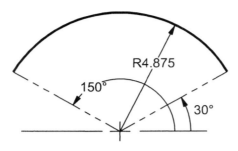

b) Add the R.875 circle and *trim*

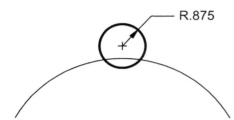

c) Add the R.75 *fillets* and the .875DIA circle

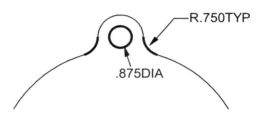

d) Use the *Xform, Rotate* command to make *two* copies. Add the .5 DIA, 5 DIA and 6DIA circles

e) Use the *Xform, Rotate* command to make *five* copies of the .5Dia circle. Use the *Polygon* command to create the 1.5R inscribed hexagon

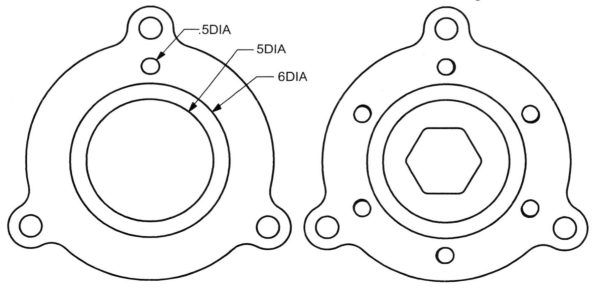

3-3)

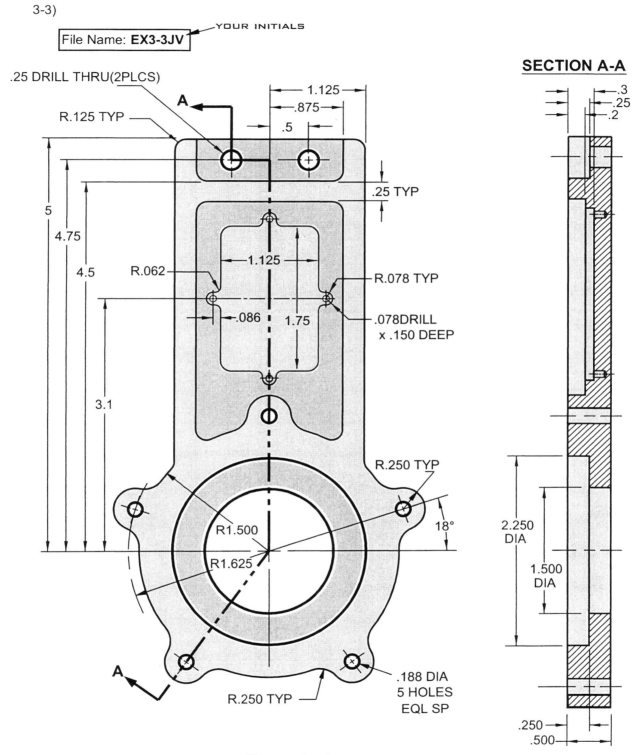

File Name: **EX3-3JV** ← YOUR INITIALS

.25 DRILL THRU(2PLCS)

SECTION A-A

R.125 TYP

A

1.125
.875
.5

.25 TYP

5

4.75

4.5

1.125

R.062

R.078 TYP

.086   1.75

.078DRILL
x .150 DEEP

3.1

R.250 TYP

18°

R1.500

R1.625

.3
.25
.2

2.250
DIA

1.500
DIA

A

R.250 TYP

.188 DIA
5 HOLES
EQL SP

.250
.500

Figure 3-p3

3-4)

File Name: **EX3-4JV**

YOUR INITIALS

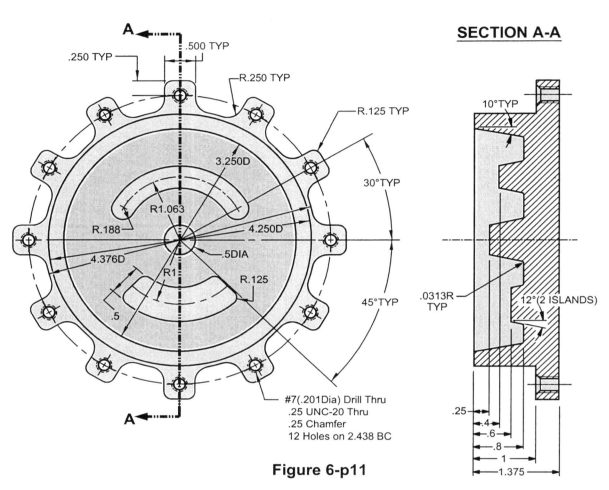

SECTION A-A

Figure 6-p11

a) Use the *Arc*, *Polar* command with Radius set to 2.188, initial angle set to 75° and final angle set to 105°

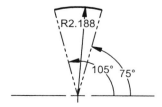

b) Complete the flange template as shown below

c) Use the *Xform, Rotate* function to make *11* copies. Add the 1.875R, and 2.125R circles.

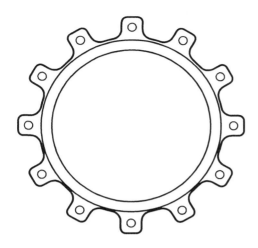

d) Create the *circle* island at a depth of -.25.

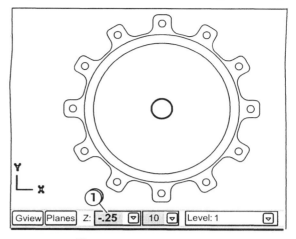

➤ Click ① in the Z depth box; enter the new depth for constructions Z: -.25

➤ Create the .5Dia circle island.

e) Create the *lower* island at a depth of -.4.

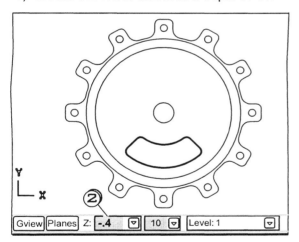

➤ Click ② in the Z depth box; enter the new depth for constructions Z: -.4

➤ Create the lower island

f) Create the *upper* island at a depth of -.6.

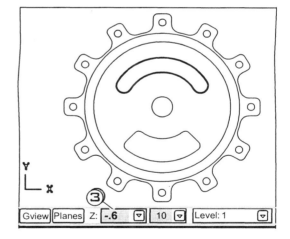

➤ Click ③ in the Z depth box; enter the new depth for constructions Z: -.6

➤ Create the upper island

➤ Click ③ in the Z depth box; enter the new depth for constructions Z: 0

3-5)

File Name: **EX3-5JV** ← YOUR INITIALS

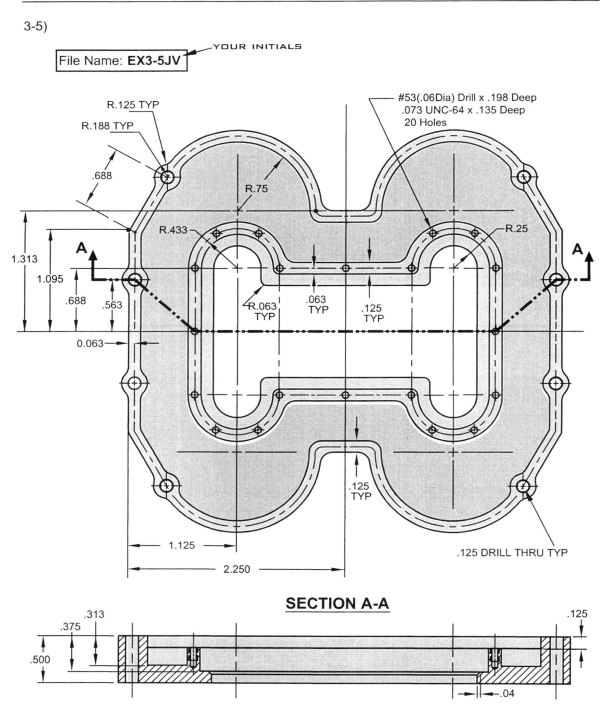

SECTION A-A

Figure 3-p5

a) Create the upper left quarter of the outside part geometry.

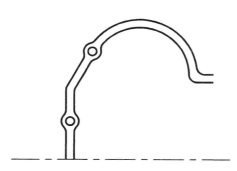

b) Create the inside geometry at a depth of -.125

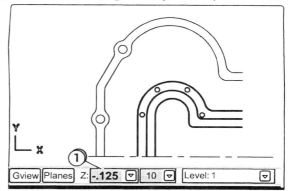

⟫ Click ① in the Z depth box; enter the new depth for constructions   Z: -.125

⟫ Create the inside geometry

⟫ Click ① in the Z depth box; enter the new depth for constructions   Z: 0

c) Use the *Xform, Mirror* command to complete the part geometry

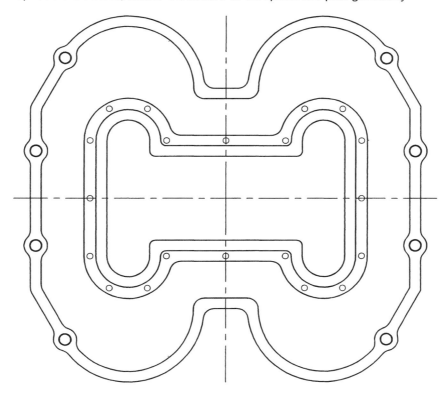

3-6)

File Name: **EX3-6JV** ←─ YOUR INITIALS

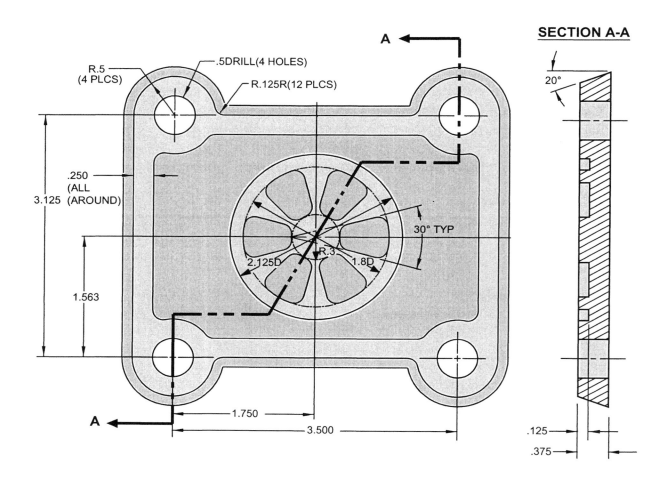

Figure 3-p6

a) Create the upper left quarter of the outside part geometry.

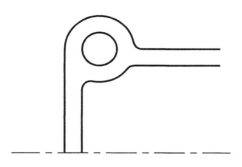

b) Use the Xform, *Mirror* function to complete the outside part geometry. add the R1.063 inner circle

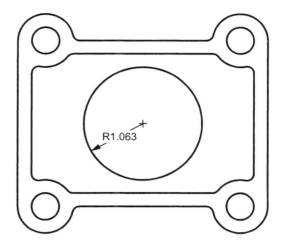

R1.063

c) Create the geometry for one inner pocket

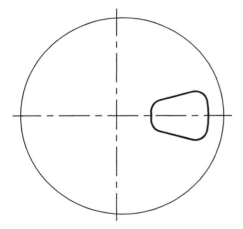

d) Use the Xform, *Rotate* function to complete the inside pocket geometry.

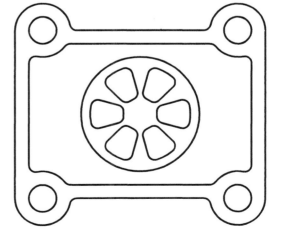

3-7)

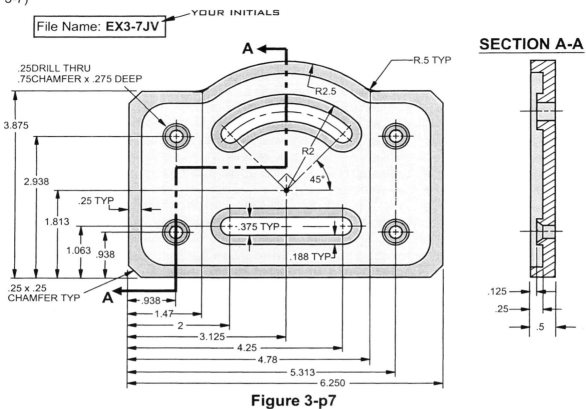

Figure 3-p7

a) Begin the creation of the outside contour

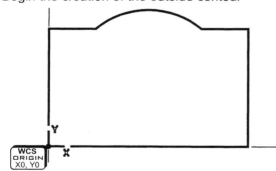

b) Add the R.5 Fillets and .25x.25 Chamfers

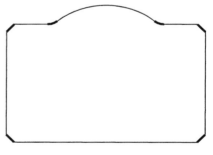

c) Use *Xform, Offset* to layout the inside contour

d) Add the R.5 Fillets to the inside geometry

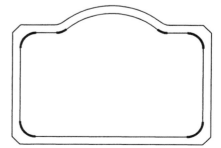

e) Create the islands at a depth of -.125.

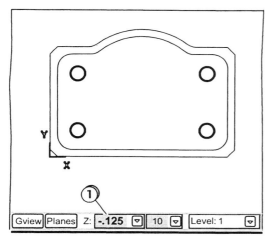

>> Click ① in the Z depth box; enter the new
depth for constructions   Z: -.125 ▽

>> Create the .5Dia circle islands.

f) Create the upper and lower islands

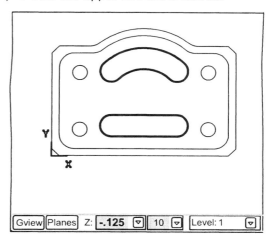

g) Use *Xform, Offset* to create the slot cutouts and add the drill holes

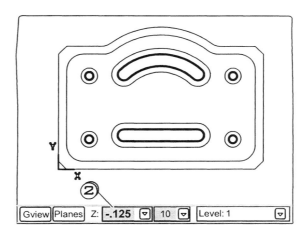

After creating slot cutouts and drill holes

>> Click ② in the Z depth box; enter the new
depth for constructions   Z: 0 ▽

3-8)

YOUR INITIALS

File Name: **EX3-8JV**

Get the file EX2-10JV completed in exercise 2-10. Add the Increase/Decrease arrows, the 2.25R circle, the line graduations and the Hartford Font text. See Figure 3p-8.
Save the file as **EX3-8JV**.

Font= **HARTFORD** , Height=.3, Width=.15

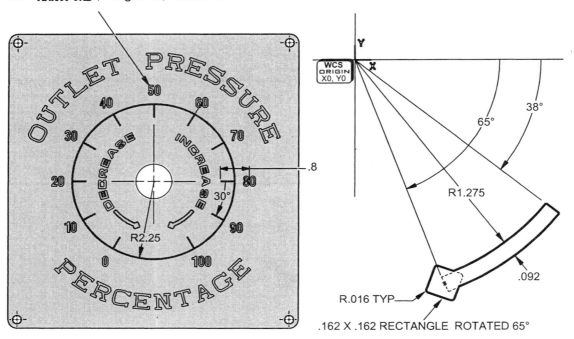

Figure 3-p8

a) Use *Create Arc* and *Rectangle* functions to begin the construction of the directional arrow.

b) Use *Delete* and *Trim* functions to complete the drectional arrow design.

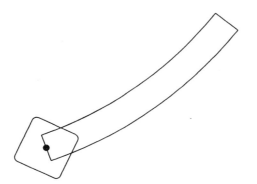

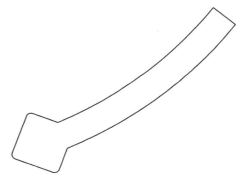

c) Use *Xform, Mirror, Copy* functions to generate the left copy of the directional arrow

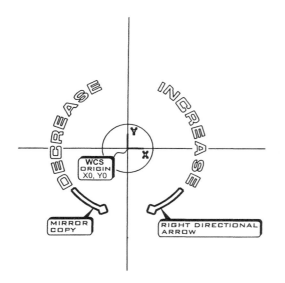

d) Create the R2.25 Circle and .8 graduation line

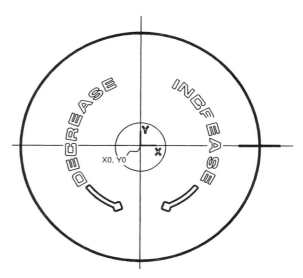

e) Use the Xform, Rotate function to rotate the graduation line down -60°

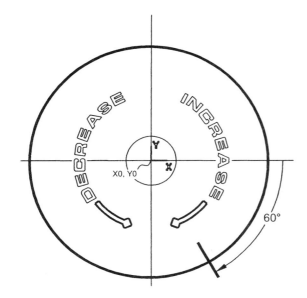

f) Use the *Xform, Rotate, Copy* function to make *nine* copies of the graduation line

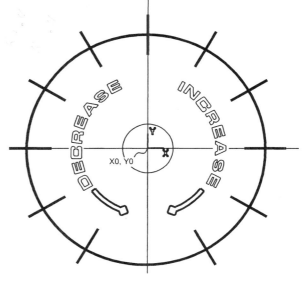

g) Insert the **HARTFORD** font geometric lettering at the *endpoint* of each graduation line.

L Create Letters

Create Letters

─ Font ─
Drafting[current]Font ▼   True Type[R]
Drafting[current]Font ①
Drafting[Dayville]Font
MC Directory
C:\MCAMX2\COMMON\FONTS\BOX ..

─ Alignment ─
⦿ Horizontal
○ Vertical
○ Top of Arc
○ Bottom of Arc
☐ Top of Chain

─ Letters ─
0 ⑪⑪

─ Parameters ─
Height   .5
Arc Radius 1.0
Spacing   0.0
② Drafting Options..

⑫ ✓   ✖   ?

DECREASE INCREASE ⑬

Enter starting location of letters

Note Text

─ Size ─
Text Height: ④ .3
Spacing ⦿ Fixed ○ Proportional
Aspect Ratio 0.5
Character Width ⑤ .15
Extra Char Spacing 0.2
Extra Line Spacing 0.2157
Factors   ☑ Use Factor

string alignment
Abc

─ Lines ─
First Line of Text
☐ Base ☐ Cap
All Lines of Text
☐ Base ☐ Over
Text Box Lines
☐ Top   ☐ Left
☐ Bottom ☐ Right

─ Path ─
⦿ Right
○ Left
○ Up
○ Down

─ Font ─
Hartford ③ ▼ Add True Type(R)

─ Alignment ─
─ Horizontal ─
⑧ ○ Left
○ Center
○ Right

─ Vertical ─
○ Top
○ Cap
⦿ Half
○ Base
⑨ ○ Bottom

─ Mirror ─
⦿ None
○ X Axis
○ Y Axis
○ X + Y

─ Angle ─ ⑥
30

─ Slant ─ ⑦
10

─ Rotation ─

⑩ ✓   ✖   ?

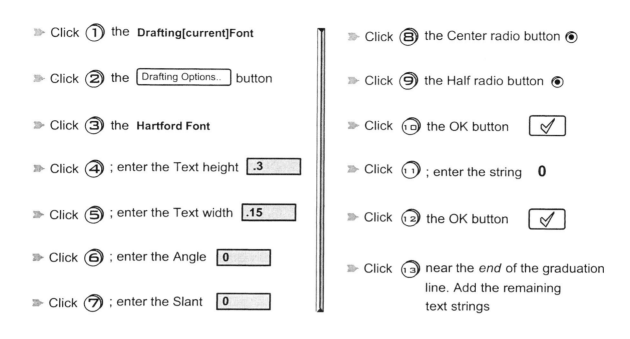

➤ Click ① the **Drafting[current]Font**

➤ Click ② the [Drafting Options..] button

➤ Click ③ the **Hartford Font**

➤ Click ④ ; enter the Text height `.3`

➤ Click ⑤ ; enter the Text width `.15`

➤ Click ⑥ ; enter the Angle `0`

➤ Click ⑦ ; enter the Slant `0`

➤ Click ⑧ the Center radio button ⊙

➤ Click ⑨ the Half radio button ⊙

➤ Click ⑩ the OK button ✓

➤ Click ⑪ ; enter the string **0**

➤ Click ⑫ the OK button ✓

➤ Click ⑬ near the *end* of the graduation
line. Add the remaining
text strings

h) Use *Modify, Trim* functions to trim back the excess portion of the graduation lines to complete
the CAD model.

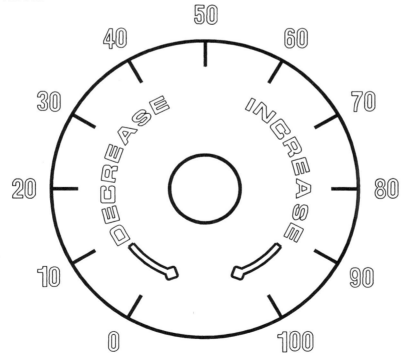

CHAPTER - 4

ADDITIONAL TOOLS FOR CAD

4-1 Chapter Objectives

After completing this chapter you will be able to:

1. Know the importance of layers and how to use them.

2. Use the Smart Dimension function to check a CAD model's dimensions

3. Use the Analyze function to check and edit CAD models

3. Explain the difference between neutral and a native file conversions

4. Execute read and write conversions between MastercamX and other files

5. Understand how to send a *MastercamX2* file as a E-mail attachement

4-2 Using Levels to Organize Information in a *Mastercam X2* Part File

A *level* can be considered as a *transparent* drawing surface. The operator can place each different type of information related to the part on its own named level. For example, the part geometry can be placed on level 1, named GEOMETRY; the dimensions on level 2, named DIMENSIONS; the fixturing on level 3, named FIXTURE and the toolpaths on level 4 named TOOLPATHS. This allows the operator to *make visible* only the information needed to complete a particular operation. MastercamX2 allows the user to create up to *2 billion* levels. The current working level is called the *main* level. All new entities are placed on the current main level. Any level can be assigned as the main level. Entities can be moved or copied from one level to another. Entities on levels that have been tagged as frozen will be *hidden* from view. MastercamX2 can also be instructed to identify several levels by a single level set name.

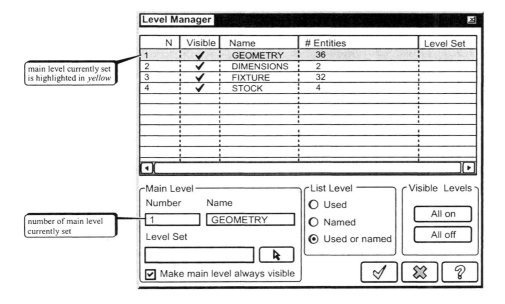

CREATING A NEW LEVEL

EXAMPLE 4-1
Create a new level named STOCK

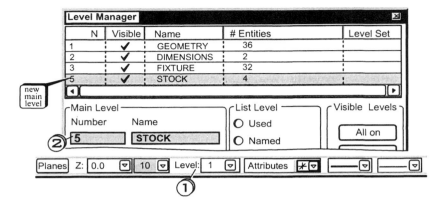

➤ Click ① the level button to display the Level Manager dialog box

➤ Click ② in the level number box; enter the new level number 5

➤ Press Tab ; enter the new level name   STOCK

➤ Press Esc to complete the operation and cancel the function

SETTING A LEVEL AS THE CURRENT MAIN LEVEL AND CONTROLLING LEVEL VISIBILITY

EXAMPLE 4-2
Set DIMENSIONS as the current *main* level. Turn the *visibility* of levels
FIXTURE, and STOCK *off*.

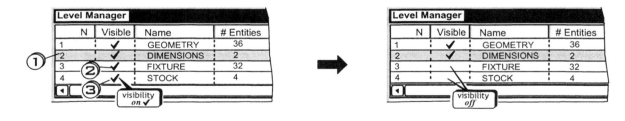

> SETTING A NEW MAIN LEVEL

➤ Click ① on the *number* of the level to be set as main   2

> CONTROLLING LEVEL VISIBILITY

➤ Click ② ③ check *off* (visibility off) for levels FIXTURE,
, and STOCK

➤ Press Esc to complete the operation and cancel the function

SETTING THE MAIN LEVEL TO THE LEVEL OF AN ENTITY CLICKED IN THE GRAPHICS WINDOW

EXAMPLE 4-3

The operator wants to work on the clamp design. Set the appropriate main level to CLAMPS simply by clicking on a clamp entity in the graphics window.

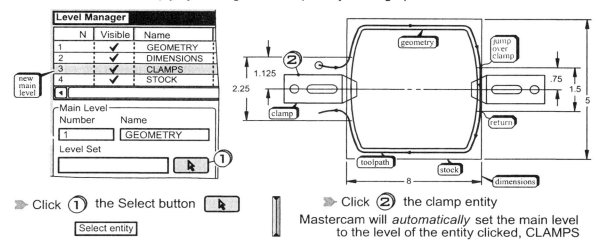

➤ Click ① the Select button

```
Select entity
```

➤ Click ② the clamp entity

Mastercam will *automatically* set the main level to the level of the entity clicked, CLAMPS

MOVING/COPYING ENTITIES FROM ONE LEVEL TO ANOTHER

EXAMPLE 4-4

The operator wants to consider two different clamping patterns for machining a part. Two levels for this purpose have been created in the Level Manager dialog box: CLAMP-SET A and CLAMP-SET B. Copy the clamps from level CLAMP-SET A to level CLAMP-SET B.

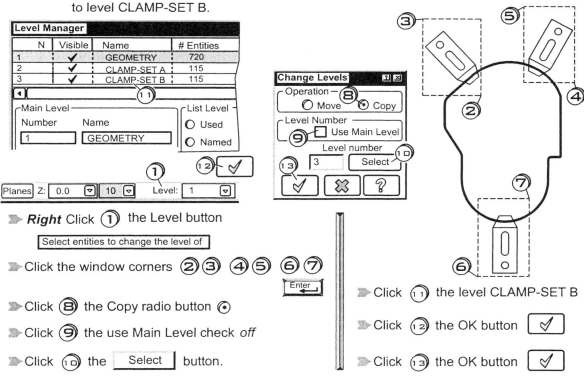

➤ *Right* Click ① the Level button

```
Select entities to change the level of
```

➤ Click the window corners ② ③ ④ ⑤ ⑥ ⑦

```
Enter
```

➤ Click ⑧ the Copy radio button ⊙

➤ Click ⑨ the use Main Level check *off*

➤ Click ⑩ the [Select] button.

➤ Click ⑪ the level CLAMP-SET B

➤ Click ⑫ the OK button ✓

➤ Click ⑬ the OK button ✓

4-3 Checking the CAD Model for Dimensional Correctness

It is highly recommended to **check** the CAD model for correctness of size and the *location* of geometric features such as lines, arcs and full circles. If any dimensional errors are found, the operator needs to quickly make corrections **before** applying machining toolpaths. *The operator can use the Drafting ribbon bar to direct Mastercam to display the current dimensions of the CAD model. .*

EXAMPLE 4-5

Verify that the existing CAD model has the dimensions as shown in Figure 4-1

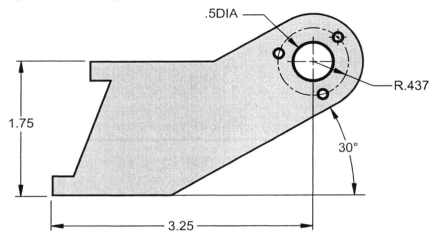

The existing CAD model to be checked

Figure 4-1

a) CREATE A NEW CURRENT MAIN LEVEL FOR DIMENSIONS

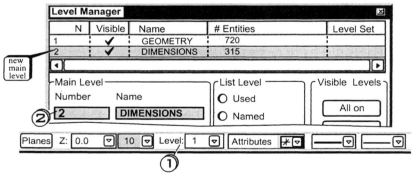

➤ Click ① the level button to display the Level Manager dialog box

➤ Click ② in the level number box; enter the new level number ❷

➤ Press [Tab] ; enter the new level name DIMENSIONS

➤ Press [Esc] to complete the operation and cancel the function

A) SET THE DESIRED GLOBAL DIMENSIONING PARAMETERS

≫ Press

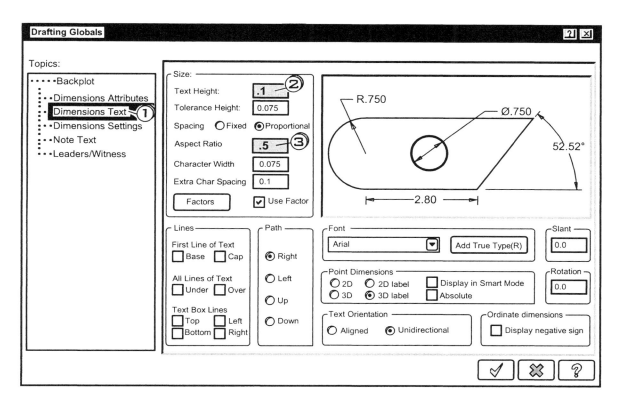

≫ Click ① the topic [Dimension Text] ,for example

≫ Click ② in the text height box and enter [.1]

≫ Click ③ in the Aspect Ratio box and enter [.5]

≫ Click ④ the OK [✓] button

B) CHECK THE APPROPRIATE DIMENSIONS

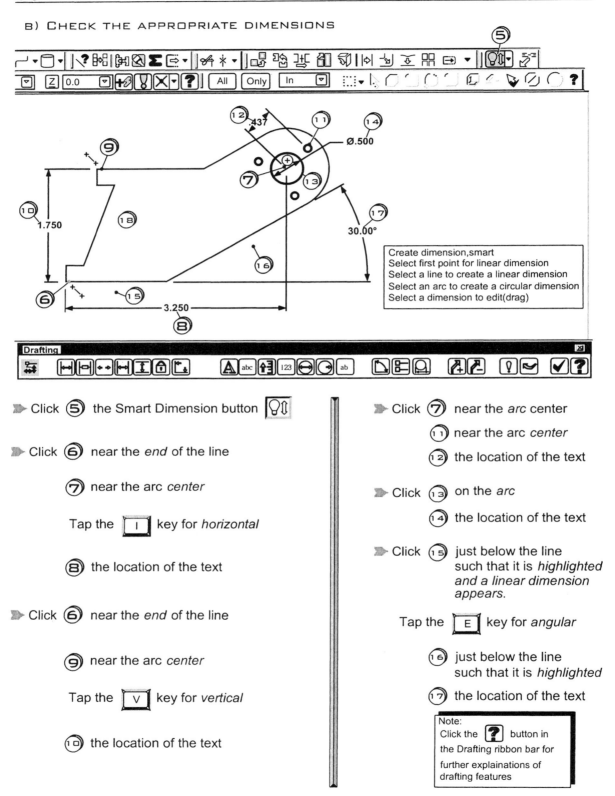

Create dimension,smart
Select first point for linear dimension
Select a line to create a linear dimension
Select an arc to create a circular dimension
Select a dimension to edit(drag)

▶ Click ⑤ the Smart Dimension button 🔆

▶ Click ⑥ near the *end* of the line

　　　　⑦ near the arc *center*

　　　　Tap the ⌷I⌷ key for *horizontal*

　　　　⑧ the location of the text

▶ Click ⑥ near the *end* of the line

　　　　⑨ near the arc *center*

　　　　Tap the ⌷V⌷ key for *vertical*

　　　　⑩ the location of the text

▶ Click ⑦ near the *arc* center

　　　　⑪ near the arc *center*

　　　　⑫ the location of the text

▶ Click ⑬ on the *arc*

　　　　⑭ the location of the text

▶ Click ⑮ just below the line
　　　　such that it is *highlighted*
　　　　and a linear dimension
　　　　appears.

　　　　Tap the ⌷E⌷ key for *angular*

　　　　⑯ just below the line
　　　　such that it is *highlighted*

　　　　⑰ the location of the text

Note:
Click the ⁇ button in
the Drafting ribbon bar for
further explainations of
drafting features

4-4 Using the Analyze Function to Check and Edit the CAD Model

The Analyze function enables the operator to spot check the size and location of selected entities. Additionally the Analyze dialog box can be used to change the parameters of existing entities such as the line lengths, angles, beginning points, end points, line widths and styles. The radius or diameter and center point location of circles and arcs can be edited. Analyze can also be applied to splines.

A) USING THE Analyze FUNCTION TO CHECK DIMENSIONAL CORRECTNESS

EXAMPLE 4-6

Verify that the existing CAD entities have the dimensions as shown in Figure 4-2

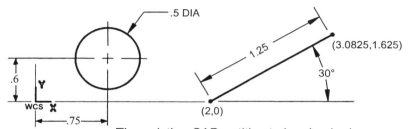

The existing CAD entities to be checked

Figure 4-2

➤ Click ① the Analyze button

➤ Click ② on the *circle* entity

Mastercam will display all the parameters of the *circle clicked*

➤ Click ③ on the *line* entity

Mastercam will display all the parameters of the *line clicked*

B) Using the Analyze function to edit the parameters of existing entities

EXAMPLE 4-7

Change the .5DIA circle to .25DIA. Change the 1.25 long line to a length of .75

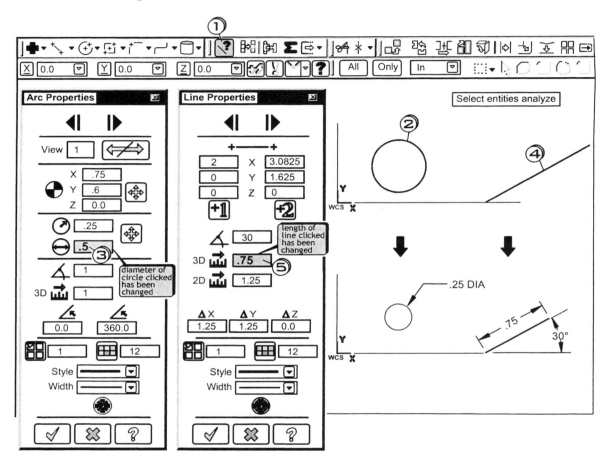

➤ Click ① the Analyze button

Select entities analyze

➤ Click ② on the *circle* entity

Mastercam will display all the parameters of the *circle clicked*

➤ Click ③ in the Diameter box; enter the *new* value .5   Enter

Select entities analyze

➤ Click ④ on the *line* entity

Mastercam will display all the parameters of the *line clicked*

➤ Click ⑤ in the 3D length box; enter the *new* value .75   Enter

➤ Press Esc to complete the operation and cancel the function

4-5 File Conversion with Other CAD/CAM Pckages

A read file conversion is the process by which a CAD file from another commonly used CAD or CAM package is converted into a *MastercamX2* **.mcx** file format. Conversely a write conversion translates a *MastercamX* file to the format of any other type of CAD or CAM package..

The trend today is to store a part, complete with its dimensions and notes, as a CAD file. A part received from a client that has been created by a particular CAD package such as AutoCAD, Solidworks, CATIA or ProE needs to be converted into a *MastercamX2* .mcx file format before it can be worked on.

MastercamX2 features two types of file conversions: *neutral* and *native*

NEUTRAL FILE CONVERSION

Each CAD/CAM software manufacturer has a different way of formatting the data in a file. A neutral translator file such as IGES, DXF or STEP is used to *convert* the manufacturer's *proprietary* CAD or CAM data format into a general industry *standard* format. *Mastercam X* can read this format and convert it into a .mcx file.

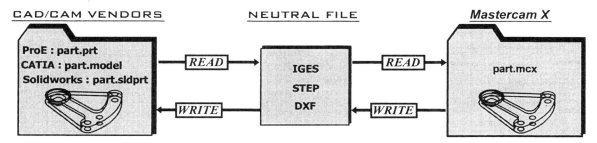

Neutral file conversion for a CAD file called **part**

NEUTRAL FILE TYPES

Neutral File	Supports	Does Not Support
IGES(Initial Graphics Exchange Standard) (**.igs**)	✓ Many types of CAD systems ✓ 2D/3D wireframe entities ✓ Surface entities, splines ✓ 2Notes, colors, levels, Analysis	✗ Solid entities ✗ IGES files tend to be *large* but compress well
DXF(Data Exchange Format) (**.dxf**)	✓ AutoCAD drawings ✓ 2D/3D wireframe entities ✓ 2Notes, colors, levels	✗ Surfaces or Solid entities Add-on pack can be purchased for solids conversions.
STEP(Standard for the Exchange of Product Model Data) (**.stp**)	✓ 2D/3D wireframe entities ✓ Solid entities Becoming the new world standard to replace IGES	✗ Other packages do not fully support STEP yet STEP is purchased as an add-on
ASCII(American Standard Code for Information Exchange) (**.txt**)	✓ Text data ✓ Bill of materials ✓ Spreadsheets ✓ CMM data and 3D digitizer data	✗ 3D wireframe entities ✗ Surface and solid entities
STL(Stereolithography) (**.stl**)	✓ CAD data for passing models to and from stereolithography machines	✗ 3D wireframe entities ✗ Surface and solid entities ✗ STL files can be very large

Neutral file conversion poses the following concerns:

- It is a *two step* process and thus more *time* consuming and *costly*.

- Conversion errors can occur due to the fact that some neutral files such as IGES come in slightly different versions or "flavors".
  STEP is intended to fix this problem since it has been created in only *one* standard *universal* version.

- The converted file size can be *large*

NATIVE FILE CONVERSION

With Native file conversion, the format of the original CAD database is converted *directly* into *MastercamX2* , .mcx format. This *one step* process *does not* require the use of a neutral file and thus *eliminates* data transfer errors. Additionally, conversion *time* and file *size* is dramatically *reduced*.

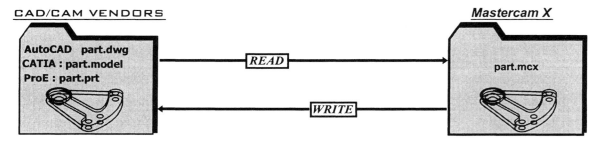

Native file conversion for a CAD file called **part**

Some native file converters are included with the basic *Mastertcam X* package others can be purchased from *Mastercam* as add-on's.

NATIVE FILE TYPES

Native File	Vendor	Standard	Add-On
AutoCAD **.dwg**	Autodesk	✔	
Parisolid **X_T**(text) **X_B**(binary)	Unigraphics,Inc	✔	
ProE **.prt**(part) **.asm**(assembly)	Parametric Technologies, Inc		✔
Solidworks **.sldprt**	Solidworks Corp	✔	
Solidedge **.par**(part) **.psm**(assembly)	Solidedge Corp	✔	
Autodesk Inventor **.ipt**(part) **.iam**(assembly)	Autodesk	✔	
CATIA V4,V5	Dassault Systems	✔	✔

EXAMPLE 4-8

Read the *neutral* IGES file, **EXAMPLE4-8.IGS**, contained on the enclosed CD into *Mastercam X2*

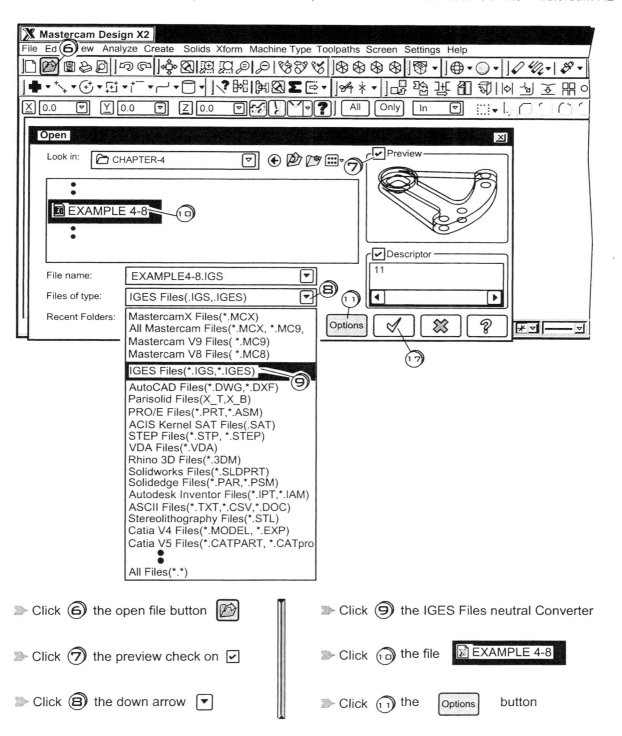

➤ Click ⑥ the open file button 📂

➤ Click ⑦ the preview check on ☑

➤ Click ⑧ the down arrow ▼

➤ Click ⑨ the IGES Files neutral Converter

➤ Click ⑩ the file 📄 EXAMPLE 4-8

➤ Click ⑪ the Options button

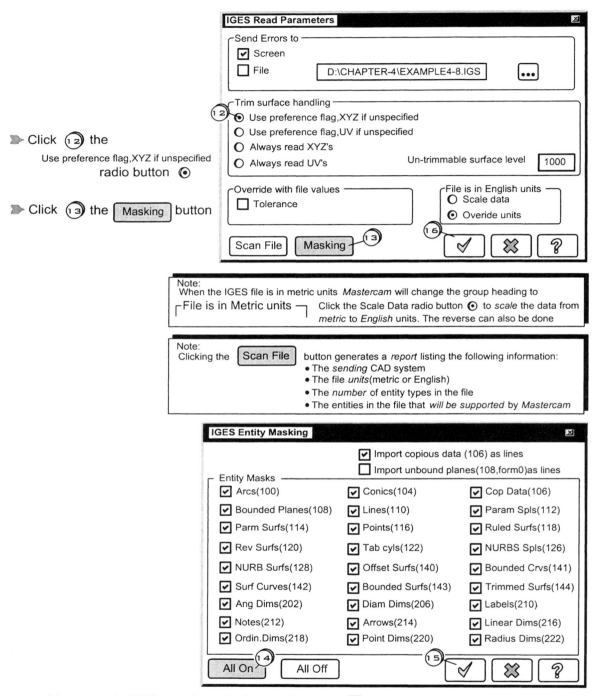

Click ⑫ the

Use preference flag,XYZ if unspecified

radio button ⊙

Click ⑬ the [Masking] button

Mastercam's IGES translator displays a check on ☑ for all the entity types that will be read. The name of the entity and its number are listed. The operator can check *on* or *off*, the types of entities to be processed

Click ⑭ the [All On] button   ‖   Click ⑮ ⑯ the OK buttons [✓]

▶ Click ⑰ the open file button [🗁] (see page 4-11) .

Mastercam X2 will then convert the file **EXAMPLE4-8.IGS** to **EXAMPLE4-8.MCX**.
and open it. All *MastercamX2* functions can be applied to process the part further.

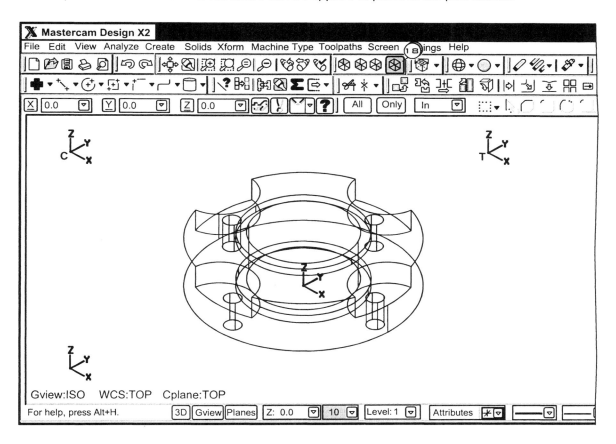

▶ Tap the [Home] key for *Zoom Fit*

▶ Click ⑱ the Iso view button [⧉]

EXAMPLE 4-9

Read the AutoCAD drawing file, **EXAMPLE4-9.DWG**, shown below *directly* to *Mastercam X2*

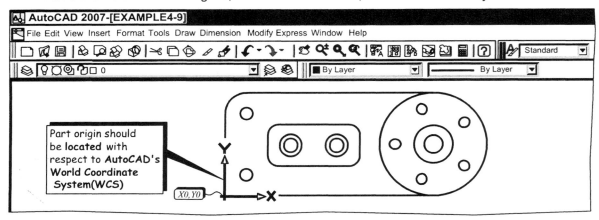

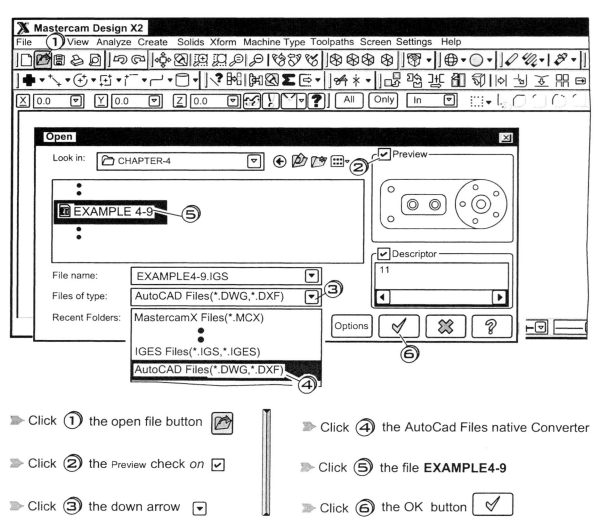

➤ Click ① the open file button

➤ Click ② the Preview check *on* ☑

➤ Click ③ the down arrow ▾

➤ Click ④ the AutoCad Files native Converter

➤ Click ⑤ the file **EXAMPLE4-9**

➤ Click ⑥ the OK button

Mastercam X2 will then convert the file **EXAMPLE4-9.DWG** to **EXAMPLE4-9.MCX**. and open it. All *MastercamX2* functions can be applied to process the part further.

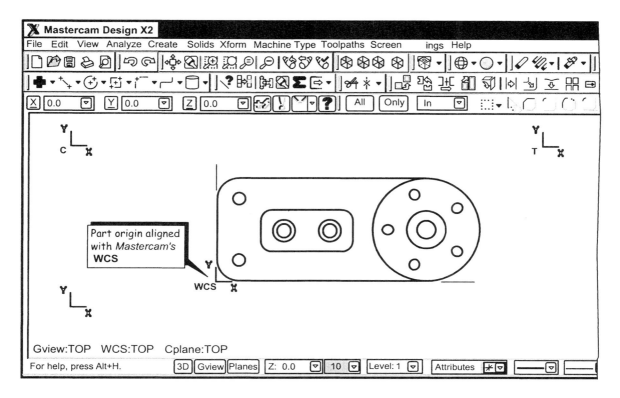

Save the file in the JVAL-MILL folder

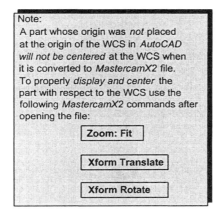

Note:
A part whose origin was *not* placed at the origin of the WCS in *AutoCAD will not be centered* at the WCS when it is converted to *MastercamX2* file. To properly *display and center* the part with respect to the WCS use the following *MastercamX2* commands after opening the file:

Zoom: Fit

Xform Translate

Xform Rotate

EXAMPLE 4-10

Open the file EXAMPLE4-10.MCX from the enclosed CD in *MastercamX*. Write it directly to the C drive as an AutoCAD file. Open the file in AutoCAD.

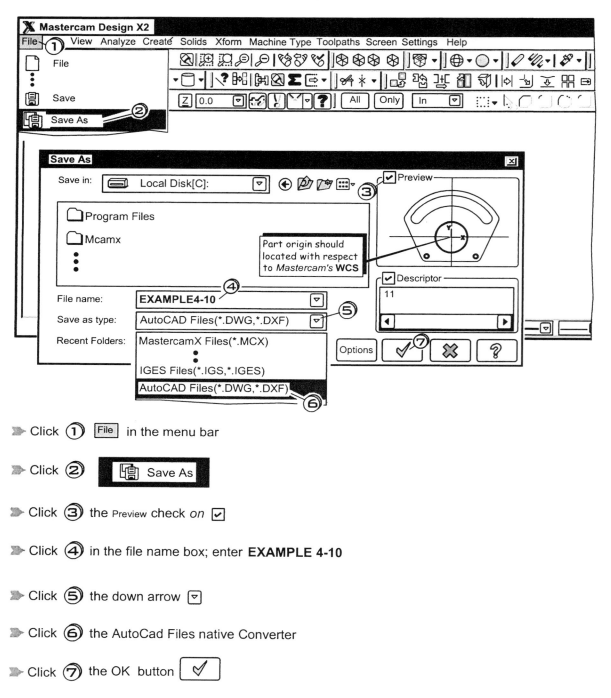

➤ Click ① [File] in the menu bar

➤ Click ② [🖫 Save As]

➤ Click ③ the Preview check *on* ☑

➤ Click ④ in the file name box; enter **EXAMPLE 4-10**

➤ Click ⑤ the down arrow ▽

➤ Click ⑥ the AutoCad Files native Converter

➤ Click ⑦ the OK button [✓]

Mastercam X2 will convert and save the file directly from **EXAMPLE4-10.MCX** to **EXAMPLE4-10.DWG.**.

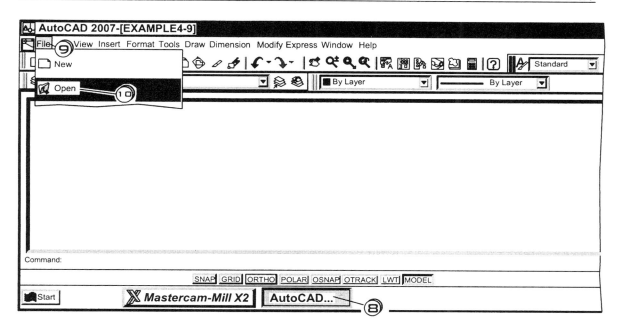

⇒ Click ⑧ **AutoCAD...** or enter the AutoCAD package from the Windows desktop

⇒ Click ⑨ the File pull down menu

⇒ Click ⑩ Open

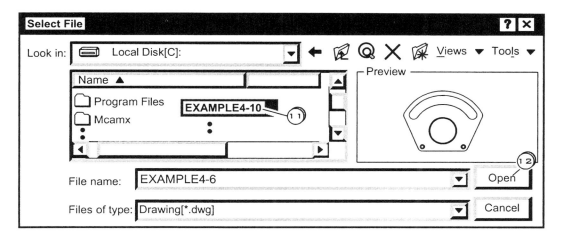

⇒ Click ⑪ on the file **EXAMPLE4-10**

⇒ Click ⑫ Open

AutoCad will then open the file **EXAMPLE4-10.DWG**. AutoCad commands can be applied
to further process the part as required..

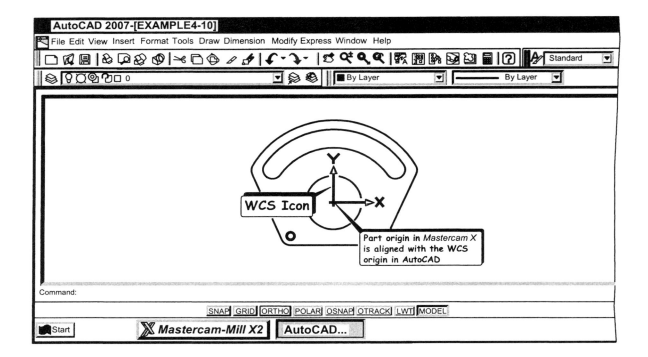

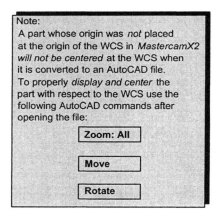

Note:
A part whose origin was *not* placed
at the origin of the WCS in *MastercamX2*
will not be centered at the WCS when
it is converted to an AutoCAD file.
To properly *display and center* the
part with respect to the WCS use the
following AutoCAD commands after
opening the file:

Zoom: All

Move

Rotate

4-6 Sending a File Over the Internet as an E-Mail Attachement

It will be assumed that the reader has an internet connection, an E-mail address and the Windows compression utility WinZip. An evaluation copy of WinZip can be downloaded from the site **www.winzip.com**. File compression software is needed to reduce the size of the attachement and improve the transmission speed. As a rule, only *modest* file sizes can be accomodated as E-mail attachements. Listed below is a table of file sizes permitted for various E-mail vendors and rates:

Vendor	Maximum File Size of Attachement
Yahoo	10MB(mail) ; 20MB(plus subscribers)
AOL	15MB
MSN-HOTMAIL	2MB(free accounts) ; 20MB(plus subscribers)

Virus protection software known as a "firewall" is also a must to guard against present and ever evolving newer infectious files that come from the internet. As a strict rule one should never download a file from an unfamilar site and never open a bogus or suspicious looking E-mail. Some good firewalls can be obtained from Mcafee,Inc, Norton Utilities, Symantec and Spyware. Some E-mail providers such as AOL already have firewalls built into their their services.

The sample exercises in this section will be done using MSN HOTMAIL, but will work for any E-mail service having the needed capacity. A free subscription to HOTMAIL can be obtained by logging on to the website: **www.hotmail.com.**

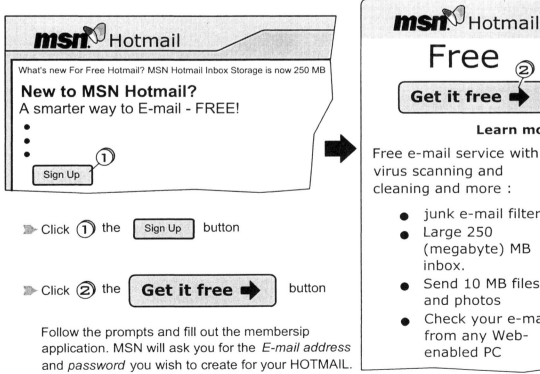

> Click ① the [Sign Up] button

> Click ② the [**Get it free ➡**] button

Follow the prompts and fill out the membersip application. MSN will ask you for the *E-mail address* and *password* you wish to create for your HOTMAIL.

EXAMPLE 4-11

a) Send the file, **EXAMPLE4-9.MCX** from EXAMPLE 4-9 as a file attachement
to the E-mail address created for your free HOTMAIL service on p4-19.
Include the message:

"Please review enclosed part design and send comments or changes via E-mail".

b) As a check, open the E-mail just sent and download the file **EXAMPLE4-9.MCX**.
Mastercam X2 should automatically open and display the part.

A) MINIMIZE *MastercamX2* RETURN TO THE WINDOWS DESKTOP AND ENTER Explorer

➤ *Right* Click ① My Computer

➤ Click ② Explore

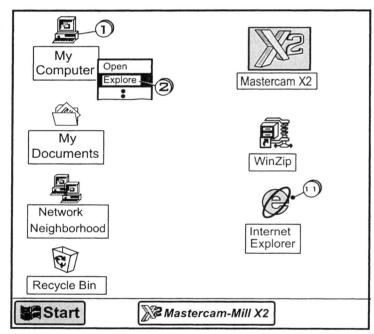

B) LOCATE AND ZIP THE FILE EXMAPLE 4-9.MCX

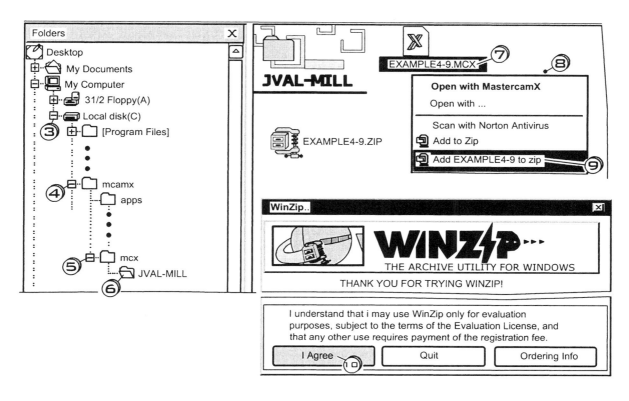

Click ③ the C drive *minus* ⊟ sign

Click ④ the mcmax folder *minus* ⊟ sign

Click ⑤ the mcx folder *minus* ⊟ sign

Click ⑥ the JVAL-MILL folder

Click ⑦ the file EXAMPLE4-9

Right Click ⑧ and move the cursor down and Click ⑨ ▸ Add EXAMPLE4-9 to zip

and Click ⑩ the ⟦ I Agree ⟧ button.

C) RETURN TO THE WINDOWS DESKTOP AND CLICK THE Internet Explorer ICON

Click ⑪ the Internet Explorer icon(See p4-19)

D) Go the the HOTMAIL WEBSITE

➣ in internet explorer go to the website **www.hotmail.com**

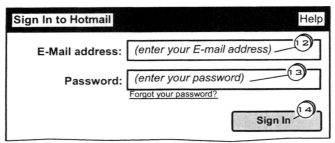

➣ Click (12) the **E-Mail address** box and enter the E-mail address you created

➣ Click (13) the **Password** box and enter the password you created

➣ Click (14) the [Sign In] button.

E) COMPOSE A NEW E-MAIL AND ATTACH THE FILE EXAMPLE 4-9 TO IT

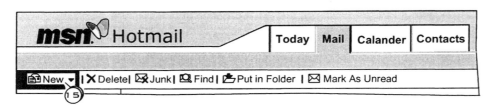

➣ Click (15) the [New ▾] drop down menu

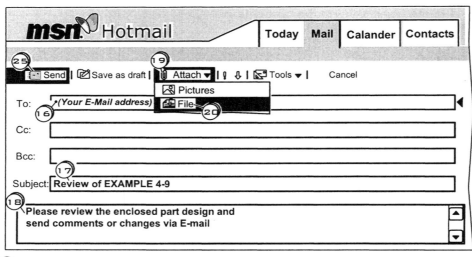

➣ Click (16) in the To box;
　　　　enter *(Your E-Mail address)*

➣ Click (17) in the Subject box; enter
　　　　Review of EXAMPLE 4-9

➣ Click (18) in the Comment box; enter
　　　　**Please review the enclosed part design and
　　　　send comments or changes via E-mail**

➣ Click (19) the [Attach ▾] drop down menu

➣ Click (20) [File]

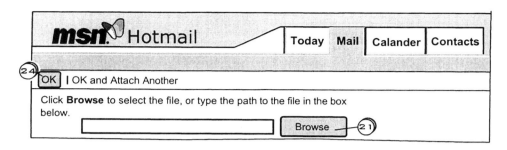

▶ Click ㉑ the [Browse] button

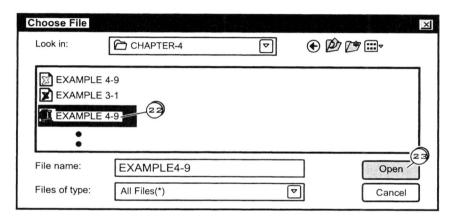

▶ Click ㉒ on the name of the zipped file: EXAMPLE4-9

▶ Click ㉓ the [Open] button.

▶ Click ㉔ the [OK] button. [OK]

▶ Click ㉕ [Send] to send the E-mail+ attached *Mastercam X* file(see page 4-22)

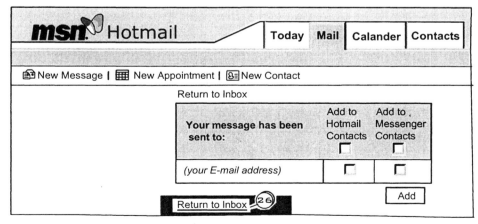

▶ Click ㉖ [Return to Inbox]

F) OPEN THE E-MAIL JUST SENT AND DOWNLOAD THE FILE EXAMPLE4-9.MCX.

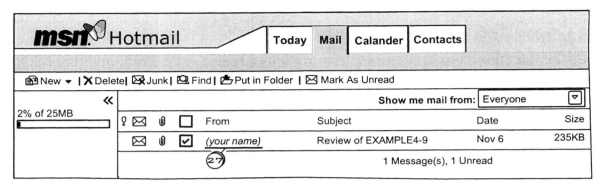

➤ Click ㉗ on the *name* of the E-mail to open

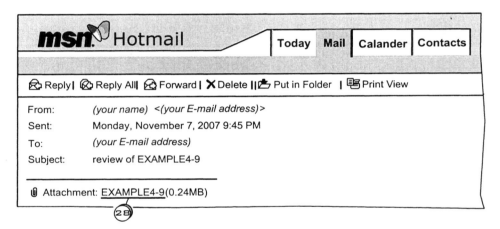

➤ Click ㉘ on the Attachment to unzip and open EXAMPLE4-9

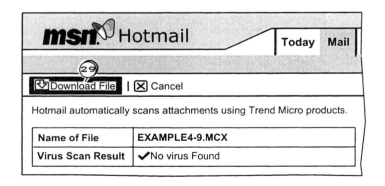

➤ Click ㉙ ⬇Download File and choose to Open it. This will automatically display the file in an active *Mastercam X2* session.

EXERCISES

4-1)   File Name: **EX4-1JV**   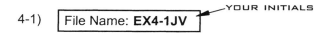 YOUR INITIALS

Create the *wireframe* CAD entities and the geometric letters for the plaque shown in Figure 4p-1

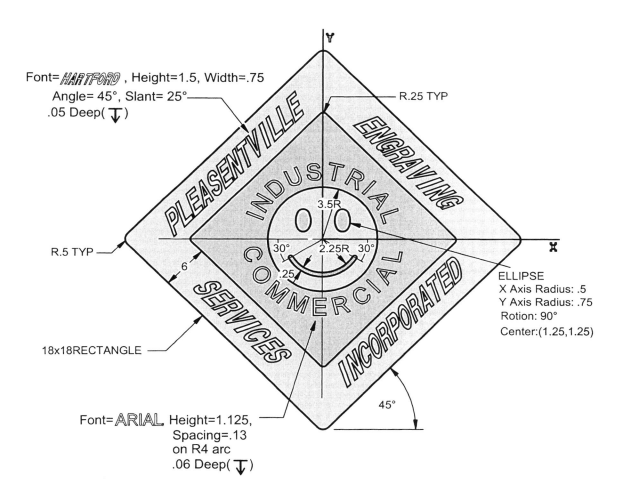

Figure 4-p1

a) Create a level and color schedule for the exercise as shown below

N	Visible	Name	# Entities	Level Set
1	✓	18x18SQ/12x12SQ	0	
2	✓	HARTFORD LET	0	
3	✓	GEO IN/ARIAL LET	0	
4	✓	15X15SQ	0	

Color
— Red
— Green
— Blue
— Magenta

b) Create the 18x18 and 12x12 rectangles on level 1

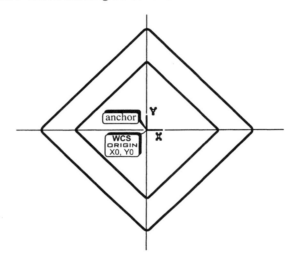

c) Create the 15x15 centerline rectangle on level 4

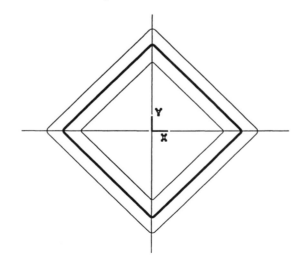

d) Insert the *HARTFORD* font geometric lettering on level 2.
   Place each string at the *midpoint* of of the sides of the 15 x 15 rectangle

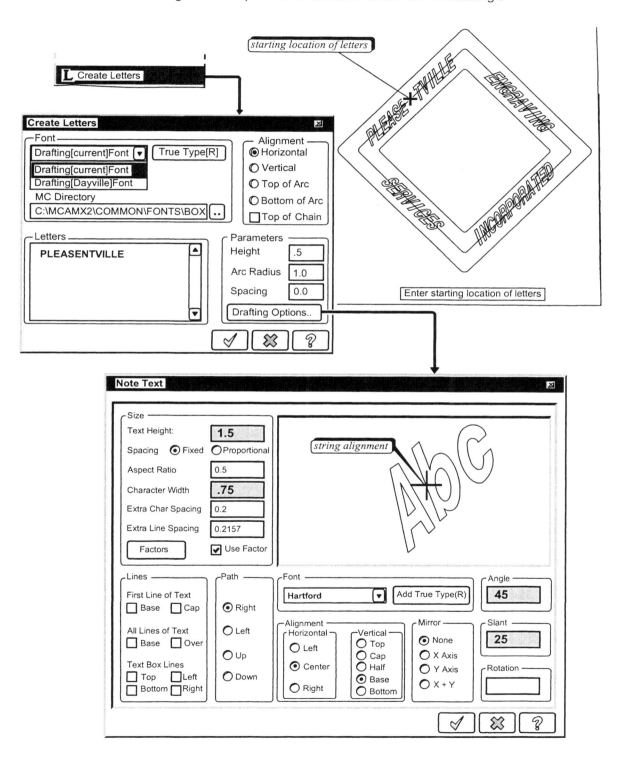

e) Insert the remaining geometry on level 3

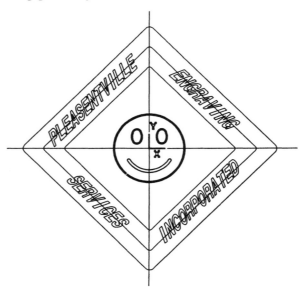

f) Insert the ARIAL font geometric lettering on level 3 . Place each string on a R4 arc.

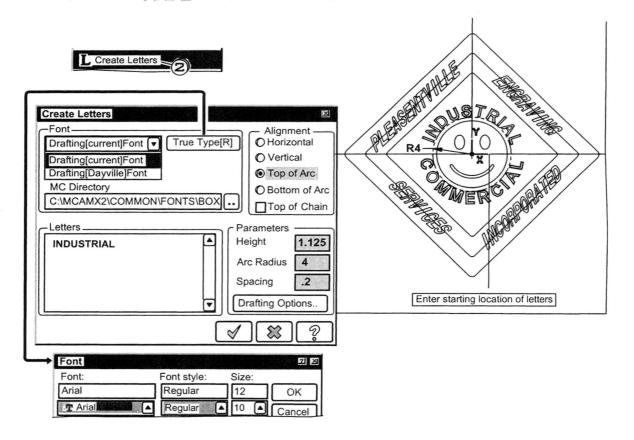

4-2) The *wireframe* CAD model shown in Figure 4p-2 is stored in the file folder
CHAPTER4 as file **EX4-2**. Open the file and save it in the folder JVAL-MILL.

a) Create a level and color schedule for the exercise as shown below

N	Visible	Name	# Entities	Level Set
1	✔	PART GEOMETRY	27	
2	✔	DIMENSIONS	0	

Level Manager

Color — Red

b) Use the **Smart Dimension** function to place check dimensions on the CAD model
as shown below

c) If necessary, edit the CAD model to conform to the dimensions in Figure 4p-2. The
Analyze and **Xform Translate** functions can be very useful for this purpose.

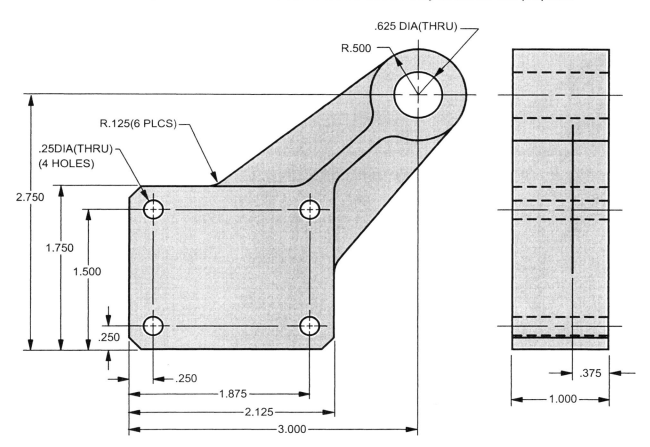

Figure 4-p2

4-3) a) Read the file **EX4-3.IGES** stored in the folder ⬜ CHAPTER4.

b) Save it in the JVAL-MILL folder

c) Use the **Zoom Fit, Xform Translate** and **Rotate** functions to align
and place the model at the origin of the WCS

 4-4) a) Read the file **EX4-4.DXF** stored in the folder ⬜ CHAPTER4.

b) Save it in the JVAL-MILL folder

c) Use the **Zoom Fit, Xform Translate** and **Rotate** functions to align
and place the model at the origin of the WCS

 4-5) a) Open the file **EX4-5.MCX** stored in the folder ⬜ CHAPTER4.

b) Zip and send it as an attachment to the E-mail address you created with
your free HOTMAIL account.
Create a subject: Review of EX4-4.
Include a comment: "Please review the enclosed part design and send comments
or changes via E-mail".

c) As a check open the E-mail just sent, download it and view the design in
an active *Mastercam X2* session.

d) Use the **Zoom Fit, Xform Translate and Rotate** functions to align and move
the part with respect to *Mastercam's* WCS

4-6)

a) What is a level?

b) List *three* uses for levels

c) What is the difference between *neutral* and *natural* file conversions

d) State *three* challenges facing neutral file conversion

e) Explain some of the advantages and shortcomings of using STEP

e) The operator is converting an IGES file and wants a report on the type
of CAD system that created the file as well as the number of entities
in the file. How can this be accomplished?

CHAPTER - 5

HOLE OPERATIONS IN 2D SPACE

5-1 Chapter Objectives

After completing this chapter you will be able to:

1. Execute commands for specifying drill points on a 2D *wireframe* CAD model.

2. Know how to use *Mastercam's* Machine Group Properties dialog box to setup stock and the Material Manager dialog box to specify stock material.

3. Explain how to select tools from *Mastercam's* tool library and assign tool speeds and feeds

4. Understand how to use the drill function to specify important drilling parameters at a drill point.

5 Explain how to check toolpaths via the backplotter.

6. Know how to use the verifier to view real time solid model animation of the entire machining process.

7. Explain how to pan and zoom and section the machined part within the machining verifier.

8. State how to machine a circle in stock using the circle mill module.

9. Understand the process of machining holes at different depths and retract heights.

10. Know how to edit existing drill tool paths.

11. Explain the steps taken in generating a word address part program to run on a CNC machine tool(postprocessing).

12. Know how to use *Mastercam's* Machine Definition Manager to select different machine tools and posts.

The Mastercam X2 Mill package must be *active before any of the drill functions can be used.*

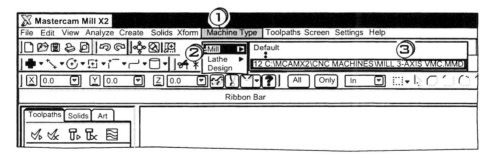

➤ Click ① the `Machine Type` drop down menu

➤ Click ② the `Mill` package

➤ Click ③ the `MILL 3-AXIS VMC.MMD` general vertical milling machine with a Fanuc type controller.

5-2 Specifying Drilling Locations on the CAD Model of a Part

The several methods of specifying the drilling points on a *wireframe* CAD model for a part are presented in this section.

Specifying Drilling Points

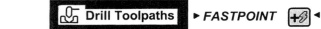

The operator specifies the *X and Y coordinates* of each drilling point ①

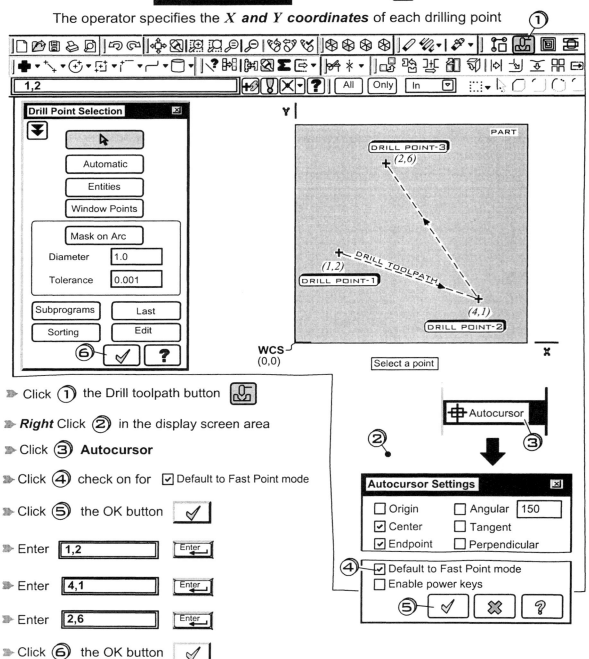

➤ Click ① the Drill toolpath button

➤ **Right** Click ② in the display screen area

➤ Click ③ **Autocursor**

➤ Click ④ check on for ☑ Default to Fast Point mode

➤ Click ⑤ the OK button

➤ Enter 1,2   Enter

➤ Enter 4,1   Enter

➤ Enter 2,6   Enter

➤ Click ⑥ the OK button

The operator specifies the drilling point by *manually snapping* to its location via one of *autocursor's snap settings, endpoint, center, midpoint, intersection, quandrant, etc* .

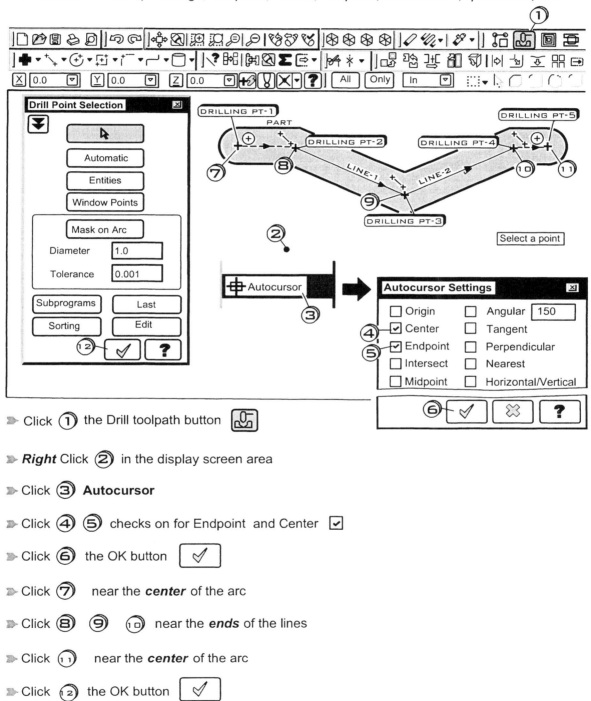

➤ Click ① the Drill toolpath button

➤ *Right* Click ② in the display screen area

➤ Click ③ **Autocursor**

➤ Click ④ ⑤ checks on for Endpoint and Center ☑

➤ Click ⑥ the OK button ✓

➤ Click ⑦ near the *center* of the arc

➤ Click ⑧ ⑨ ⑩ near the *ends* of the lines

➤ Click ⑪ near the *center* of the arc

➤ Click ⑫ the OK button ✓

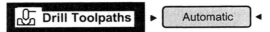

The operator automatically specifies a *set of drilling points from an existing array of points:* ● *first* point clicked indicates the *first hole to be drilled*,
 ● *second* point clicked indicates the *tool path direction*
 ● *third* point clicked indicates the *last hole to be drilled.*
Mastercam determines the point set by finding the *next closest* point in sequence.

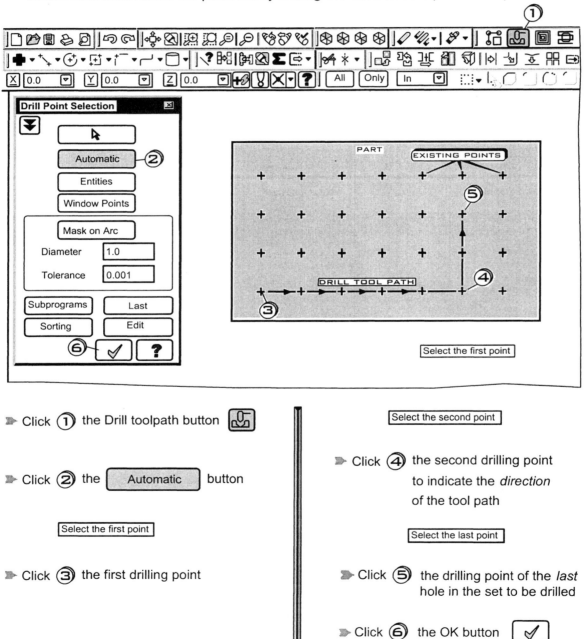

➤ Click ① the Drill toolpath button

➤ Click ② the [Automatic] button

[Select the first point]

➤ Click ③ the first drilling point

[Select the second point]

➤ Click ④ the second drilling point
to indicate the *direction*
of the tool path

[Select the last point]

➤ Click ⑤ the drilling point of the *last*
hole in the set to be drilled

➤ Click ⑥ the OK button

Mastercam places the drilling points at the **ends** of existing entities clicked. If *closed arcs* are clicked the the drill points will be placed at their **centers**. Click the [Sorting] button to control the order in which the points are sorted.

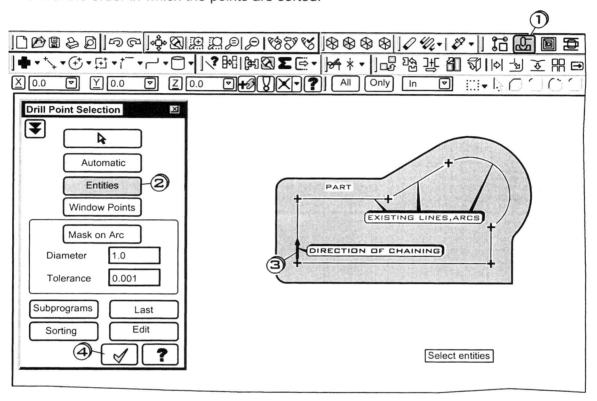

➤ Click ① the Drill toolpath button [ ]

➤ Click ② the [Entities] button

[Select entities]

➤ Click ③ near the end of the entity where the chainining is to start

➤ Click ④ the OK button [✓]

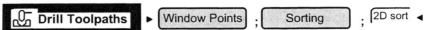

The operator *first selects* a point *sorting pattern* to indicate the *type of tool path pattern* the drill tool is to follow when machining at the point locations in the set. In this case it is selected from the 2D zig-zag set of patterns. The operator then clicks a *rectangular window* to enclose the existing set of points. Drilling will occur at *all the points enclosed* in the rectangular window.

This function will *only work* with *point* entities.

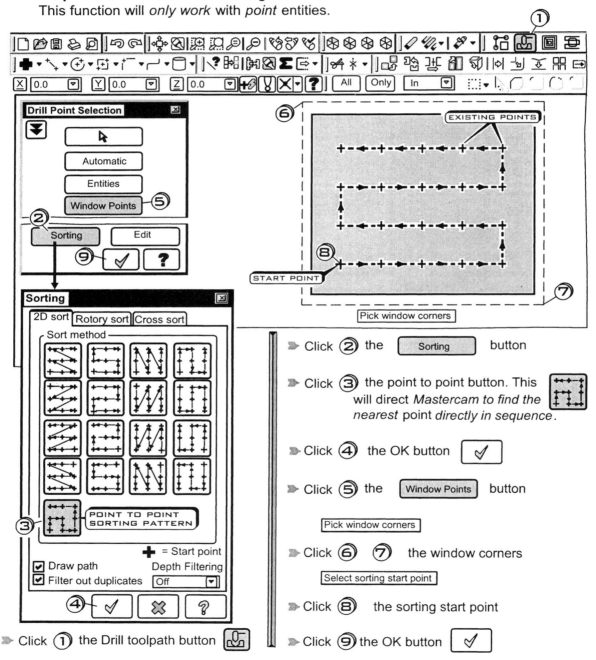

➤ Click ① the Drill toolpath button 🔧

➤ Click ② the [Sorting] button

➤ Click ③ the point to point button. This will direct *Mastercam to find the nearest* point *directly in sequence*.

➤ Click ④ the OK button ✓

➤ Click ⑤ the [Window Points] button

[Pick window corners]

➤ Click ⑥ ⑦ the window corners

[Select sorting start point]

➤ Click ⑧ the sorting start point

➤ Click ⑨ the OK button ✓

Directs Mastercam to *automatically find and select* the *center* of all **holes** **that match the diameter** of an **initial** hole clicked. This is a very useful method of specifing drill toolpaths. Be sure a sorting pattern has been selected.

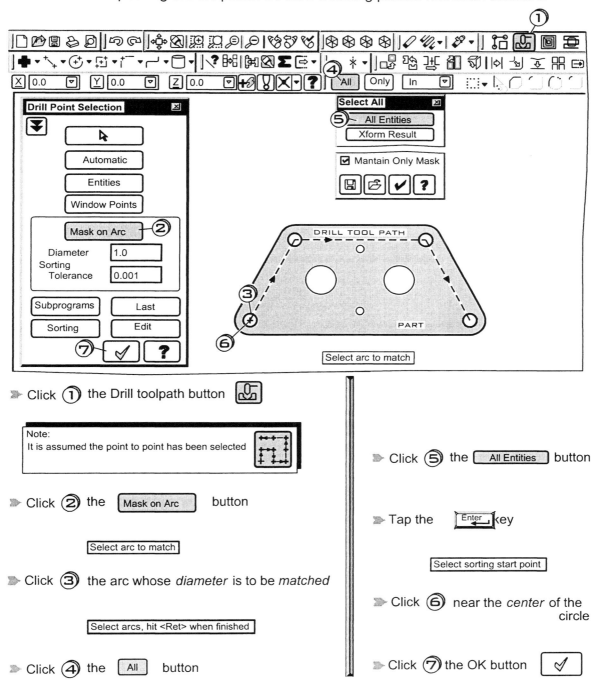

➤ Click ① the Drill toolpath button

> Note:
> It is assumed the point to point has been selected

➤ Click ② the [Mask on Arc] button

[Select arc to match]

➤ Click ③ the arc whose *diameter* is to be *matched*

[Select arcs, hit <Ret> when finished]

➤ Click ④ the [All] button

➤ Click ⑤ the [All Entities] button

➤ Tap the [Enter] key

[Select sorting start point]

➤ Click ⑥ near the *center* of the circle

➤ Click ⑦ the OK button

Directs Mastercam to **automatically create** a set of drill points according to a **rectangular grid** *pattern.* The operator enters the location of the pattern, number of points to be created and the distance between points in both X and Y directions.

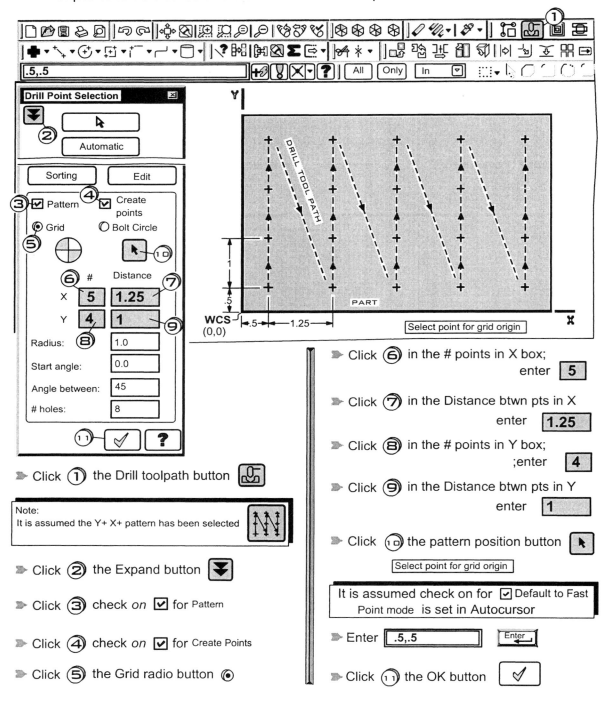

➤ Click ① the Drill toolpath button

Note:
It is assumed the Y+ X+ pattern has been selected

➤ Click ② the Expand button

➤ Click ③ check *on* ☑ for Pattern

➤ Click ④ check *on* ☑ for Create Points

➤ Click ⑤ the Grid radio button ◉

➤ Click ⑥ in the # points in X box; enter 5

➤ Click ⑦ in the Distance btwn pts in X enter 1.25

➤ Click ⑧ in the # points in Y box; ;enter 4

➤ Click ⑨ in the Distance btwn pts in Y enter 1

➤ Click ⑩ the pattern position button

Select point for grid origin

It is assumed check on for ☑ Default to Fast Point mode is set in Autocursor

➤ Enter .5,.5   Enter

➤ Click ⑪ the OK button

Directs Mastercam to **automatically create** a set of drill points according to a **bolt circle** pattern. The operator enters the location of the pattern, number of points to be created, start angle for the first point and the angle between each subsequent point.

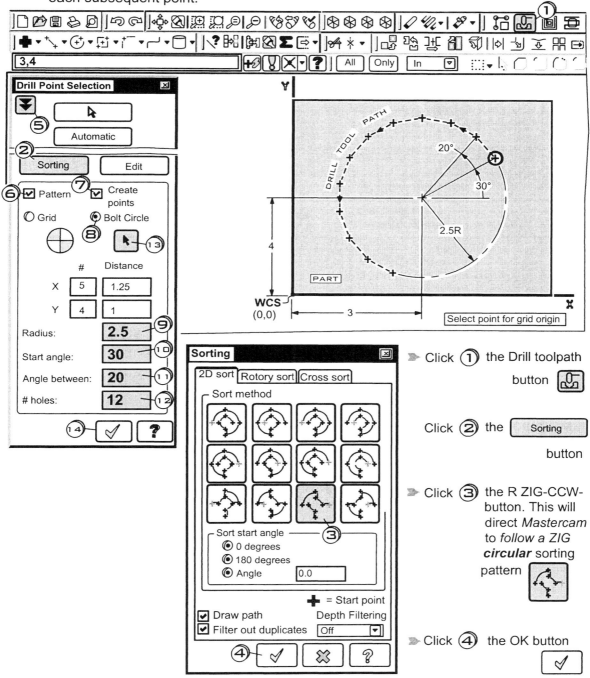

➤ Click ① the Drill toolpath button 🔲

➤ Click ② the Sorting button

➤ Click ③ the R ZIG-CCW-button. This will direct *Mastercam* to *follow a ZIG circular* sorting pattern

➤ Click ④ the OK button ✓

➤ Click ⑤ the Expand button ⬇

➤ Click ⑥ check *on* ☑ for Pattern

➤ Click ⑦ check *on* ☑ for Create Points

➤ Click ⑧ the Bolt Circle radio button ◉

➤ Click ⑨ in theRadius box; enter ‖ 2.5 ‖

➤ Click ⑩ in the Start angle ; enter ‖ 30 ‖

➤ Click ⑪ in the Angle between box; enter ‖ 20 ‖

➤ Click ⑫ in the # holes box; enter ‖ 12 ‖

➤ Click ⑬ the pattern position button ◥

Select point for grid origin

It is assumed check on for ☑ Default to Fast
Point mode is set in Autocursor

➤ Enter ‖ 3,4 ‖   Enter ↵

➤ Click ⑭ the OK button ✓

The operator specifies that *Mastercam* is to **copy the last set** of drilling point locations and sorting order just entered and **use these as the new drill point set and order** . Useful for drilling a set of *pilot* holes followed by *drilling* and *tapping* the same set.

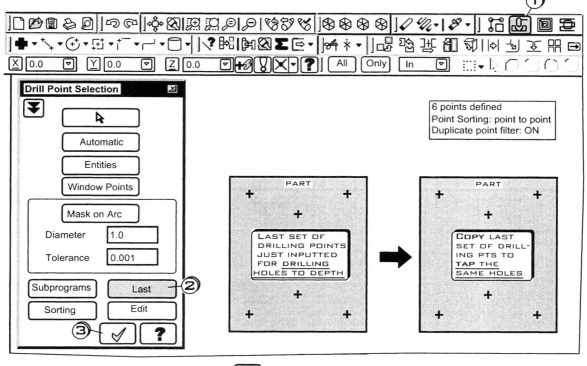

➤ Click ① the Drill toolpath button

➤ Click ② the [Last] button

6 points defined
Point Sorting: point to point
Duplicate point filter: ON

➤ Click ③ the OK button

5-3 Specifying Drill Cycle Parameters

Drilling operations are carried out with the use of *Mastercam's* drill cycle dialog box. It contains many cycles including drill/couunterbore, peck drill, chip break, tap, and bore. Special features include automatic determination of the drill depth for *thru* holes and a depth calculator for computing the proper drill depth when the finish diameter is different from the drill diameter. Important parameters in the drill cycle dialog box and their meanings are discussed in this section.

Z- DEPTH PARAMETERS

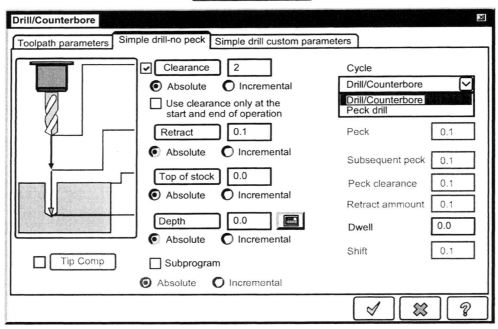

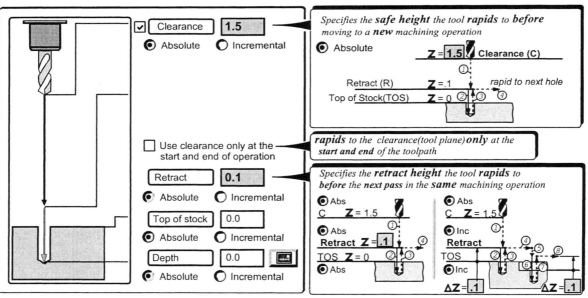

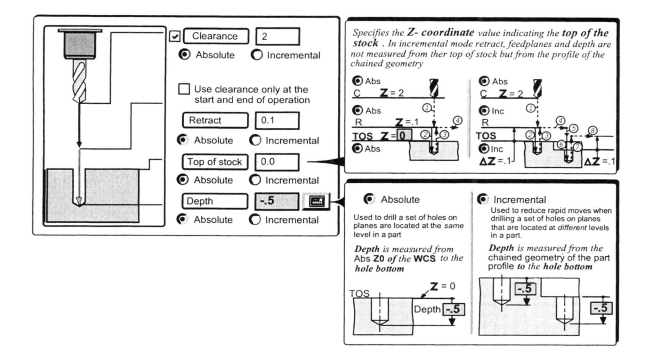

CYCLE PARAMETERS

A total of 7 pre-defined or canned drilling cycles and 13 custom cycles are provided by *Mastercam X2*.

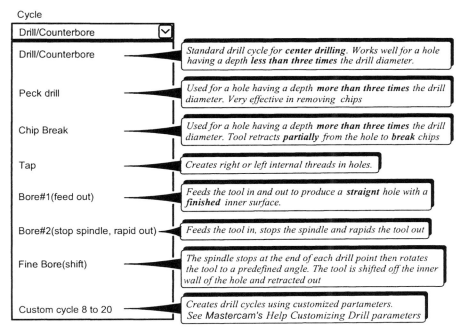

CYCLE DEPENDENT PARAMETERS

Certain drill parameters become *active only when* a certain cycle is selected for use. The Drill/Counterbore cycle activates the Dwell parameter, the Peck drill and Chip break cycles the Peck parameter and the Bore the Dwell parameter.

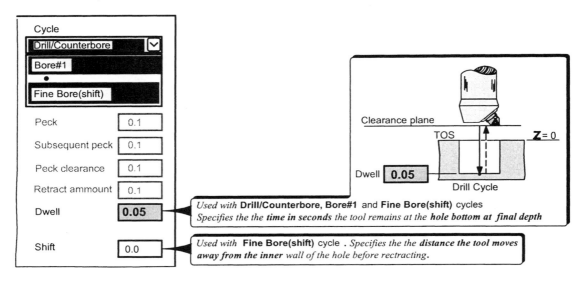

Cycle

Drill/Counterbore

Bore#1

Fine Bore(shift)

Peck	0.1
Subsequent peck	0.1
Peck clearance	0.1
Retract ammount	0.1
Dwell	0.05
Shift	0.0

Used with **Drill/Counterbore, Bore#1** and **Fine Bore(shift)** cycles
*Specifies the the **time in seconds** the tool remains at the **hole bottom** at **final depth***

Used with **Fine Bore(shift)** cycle . *Specifies the the **distance the tool moves away from the inner** wall of the hole before rectracting.*

Clearance plane
TOS   **Z** = 0
Dwell 0.05
Drill Cycle

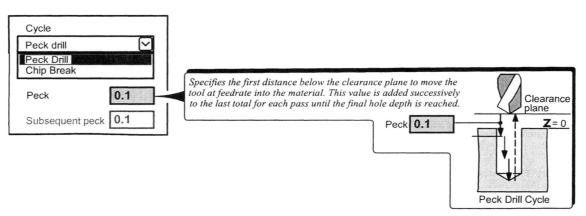

Cycle

Peck drill

Peck Drill
Chip Break

| Peck | 0.1 |
| Subsequent peck | 0.1 |

Specifies the first distance below the clearance plane to move the tool at feedrate into the material. This value is added successively to the last total for each pass until the final hole depth is reached.

Clearance plane
Peck 0.1   **Z** = 0
Peck Drill Cycle

The remaining cycle dependent parameters become active when the Custom cycles are selected. Refer to *Mastercam* Help for further definitions of these parameters.

TIP COMP PARAMETERS

The tip comp parameters are clicked on to produce a *thru* hole for a material of specified thickness.

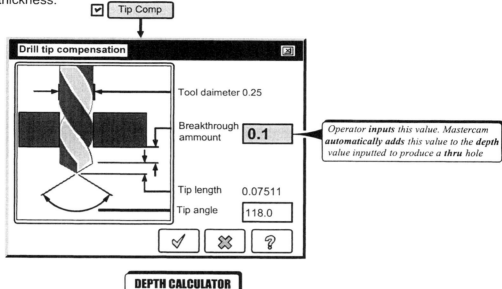

DEPTH CALCULATOR

The depth calculator function is used to compute the depth of a countersunk hole.

EXAMPLE 5-1

A .75Dia Chamfer mill tool with a 90° taper angle and a .06 flat on tip diameter has been selected to countersink a hole to .44Dia. See Fig 5-1. Direct Mastercam to automatically compute and enter the appropriate countersinking depth in the depth box.

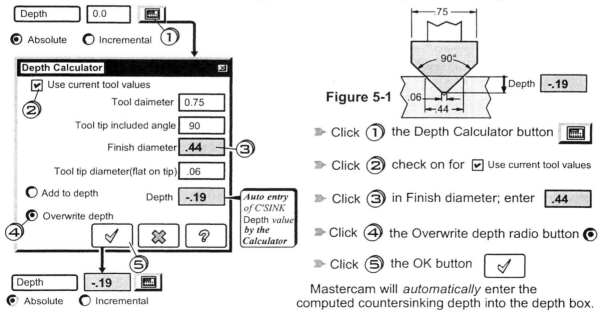

Figure 5-1

➤ Click ① the Depth Calculator button

➤ Click ② check on for ☑ Use current tool values

➤ Click ③ in Finish diameter; enter .44

➤ Click ④ the Overwrite depth radio button ⦿

➤ Click ⑤ the OK button ✓

Mastercam will *automatically* enter the computed countersinking depth into the depth box.

EXAMPLE 5-2
Assume a CAD model of the part face shown in Figure 5-2 has been created in *Mastercam*.

● Input the stock and material for the job

● Specify drill locations and assign the tools

● Select the hole operations; enter the drilling parameters listed in PROCESS PLAN 5-1

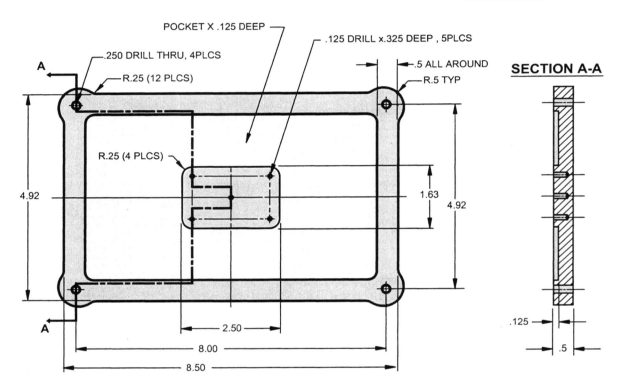

Figure 5-2

PROCESS PLAN 5-2

No.	Operation	Tooling
1	SPOT DRILL X .166 DEEP (4 HOLES)	1/8 SPOT DRILLL
2	PECK DRILL THRU(4 HOLES)	1/4 DRILL
3	SPOT DRILL X .125 DEEP (5 HOLES)	1/32 SPOT DRILL(.31-FLUTE LENGTH;1.5-OAL)
4	PECK DRILL X .325 DEEP	1/8 DRILL

A) ENTER *Mastercam's* MACHINE GROUP PROPERTIES DIALOG BOX

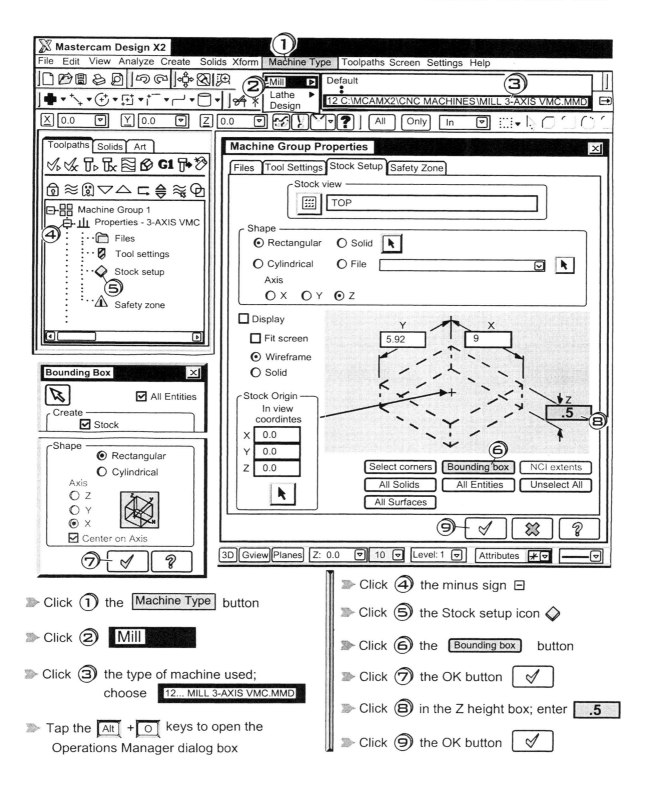

➤ Click ① the Machine Type button

➤ Click ② Mill

➤ Click ③ the type of machine used;
choose 12... MILL 3-AXIS VMC.MMD

➤ Tap the Alt + O keys to open the
Operations Manager dialog box

➤ Click ④ the minus sign ⊟

➤ Click ⑤ the Stock setup icon ◇

➤ Click ⑥ the Bounding box button

➤ Click ⑦ the OK button ✓

➤ Click ⑧ in the Z height box; enter .5

➤ Click ⑨ the OK button ✓

B) ENTER *Mastercam's* MATERIAL MANAGER DIALOG BOX

◆ SPECIFY THE STOCK MATERIAL

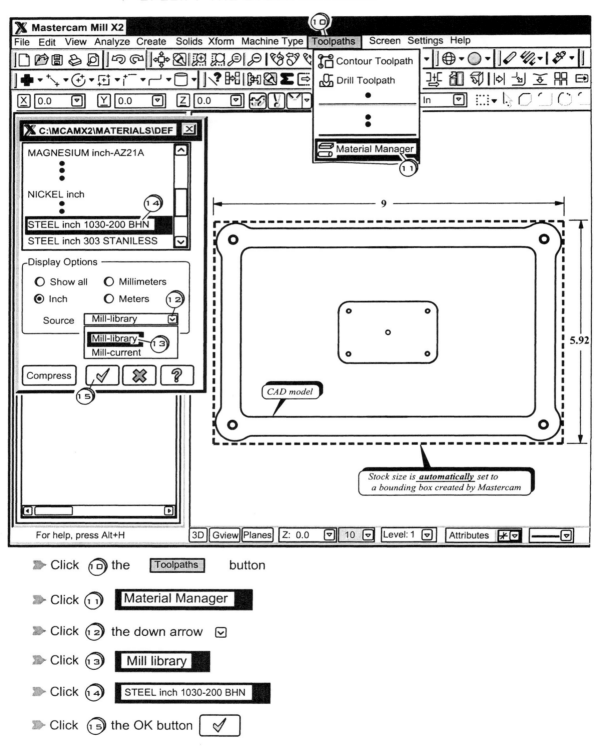

➤ Click ⑩ the [Toolpaths] button

➤ Click ⑪ [Material Manager]

➤ Click ⑫ the down arrow ▽

➤ Click ⑬ [Mill library]

➤ Click ⑭ [STEEL inch 1030-200 BHN]

➤ Click ⑮ the OK button ✓

B) DRILL THE .125DIA PILOT HOLES X .166 DEEP

◆ CREATE THE PILOT HOLE TOOLPATH

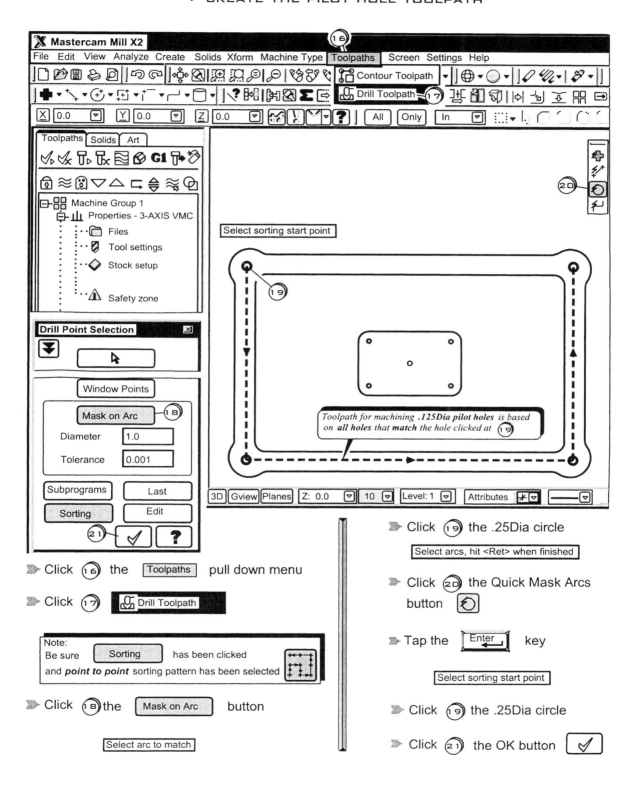

➤ Click ⑯ the [Toolpaths] pull down menu

➤ Click ⑰ 🔩 Drill Toolpath

```
Note:
Be sure [ Sorting ] has been clicked
and point to point sorting pattern has been selected
```

➤ Click ⑱ the [Mask on Arc] button

[Select arc to match]

➤ Click ⑲ the .25Dia circle

[Select arcs, hit <Ret> when finished]

➤ Click ⑳ the Quick Mask Arcs
button 🔄

➤ Tap the [Enter] key

[Select sorting start point]

➤ Click ⑲ the .25Dia circle

➤ Click ㉑ the OK button ✓

◆ OBTAIN THE NEEDED .125 DIA SPOT DRILL TOOL

Mastercam will activate and display the Tool Parameters Tab. The operator can use this tab to **select a tool, set speeds and feeds** and other general toolpath parameters.

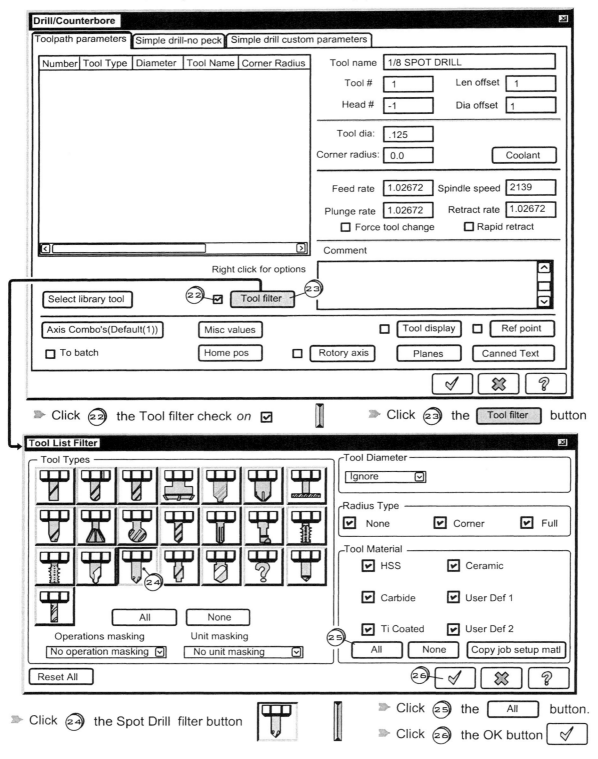

➤ Click ㉒ the Tool filter check *on* ☑   ➤ Click ㉓ the [Tool filter] button

➤ Click ㉔ the Spot Drill filter button   ➤ Click ㉕ the [All] button.

➤ Click ㉖ the OK button ☑

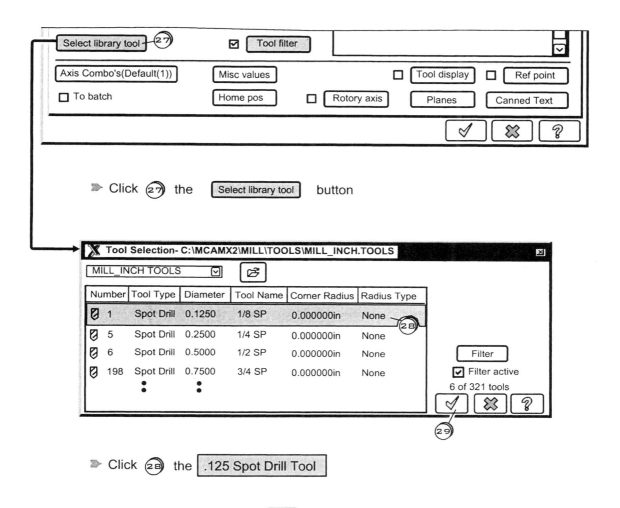

➣ Click ㉗ the [Select library tool] button

➣ Click ㉘ the [.125 Spot Drill Tool]

➣ Click ㉙ the OK button ☑ to bring the tool from the library into your part file

◆ AUTO ASSIGN SPEEDS FEEDS FOR THE TOOL

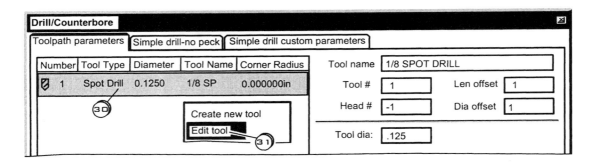

➤ Click ③⓪ on the .125 Spot Drill tool

➤ *Right* Click move the cursor down and Click ③① [Edit tool]

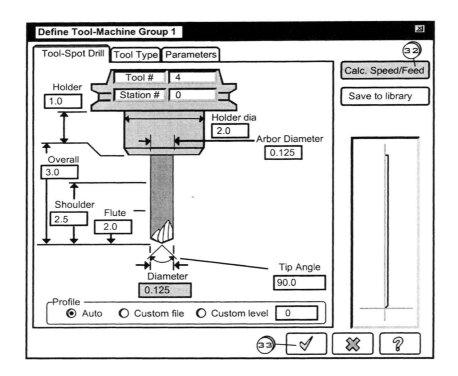

➤ Click ③② the [Calc. Speed/Feed] button. Mastercam will *automatically* assign
Speeds/Feeds for the *operation* and the *stock material* selected.

➤ Click ③③ the OK button [✓] to enter the tool parameters in the
Operations Manager under 🛡 Tool settings

◆ ENTER THE REQUIRED .125 DIA PILOT HOLE MACHINING PARAMETERS

⟫ Click ③④ ⌐Simple drill-no peck⌐ tab

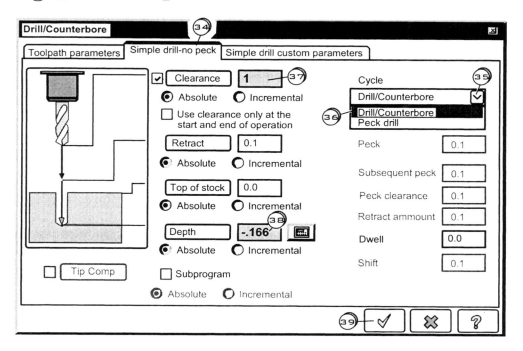

⟫ Click ③⑤ the toggle down button ▽

⟫ Click ③⑥ Drill/Counterbore

⟫ Click ③⑦ in the Clearance box and enter 1

⟫ Click ③⑧ in the Depth box and enter -.166

⟫ Click ③⑨ the OK button ✓ to *create* the operation 📂 1 - Simple drill-no peck in the Operations Manager. The accompanying Parameters file for the operation in the Operations Manager will *store* the data just entered.

C) PECK DRILL THE .25DIA HOLES THRU

- ◆ CREATE THE DRILL THRU TOOLPATH

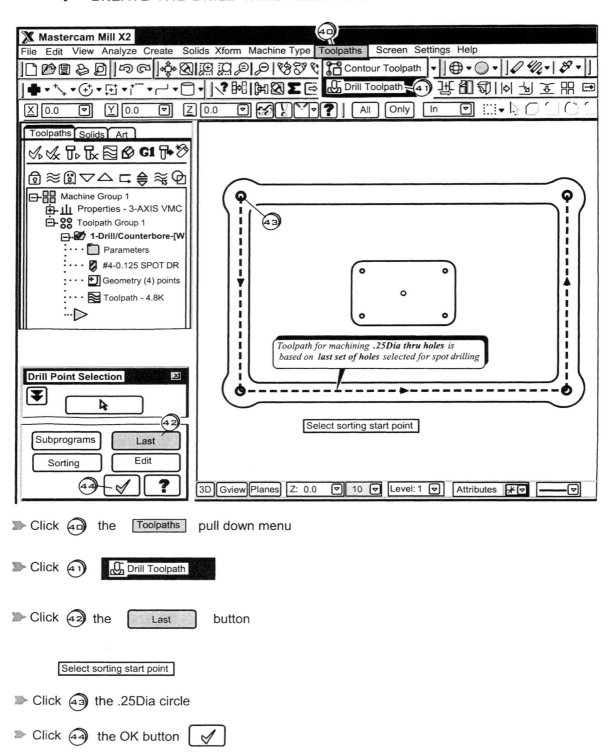

Toolpath for machining *.25Dia thru holes* is based on **last set of holes** selected for spot drilling

Select sorting start point

▶ Click ④⓪ the Toolpaths pull down menu

▶ Click ④① Drill Toolpath

▶ Click ④② the Last button

Select sorting start point

▶ Click ④③ the .25Dia circle

▶ Click ④④ the OK button ✓

◆ OBTAIN THE NEEDED .25 DIA DRILL TOOL

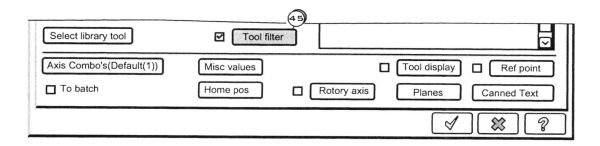

➤ Click (45) the [Tool filter] button

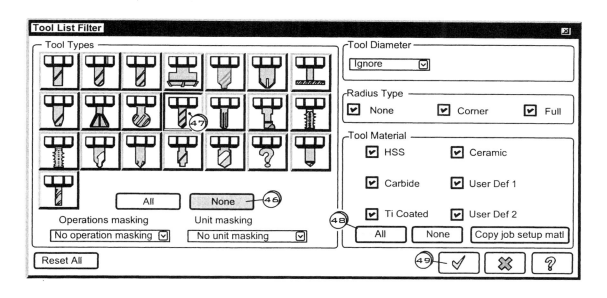

➤ Click (46) the [None] button to clear the field

➤ Click (47) the Drill filter button

➤ Click (48) the [All] button.

➤ Click (49) the OK button

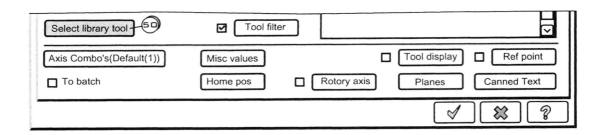

➤ Click ⑤⓪ the [Select library tool] button

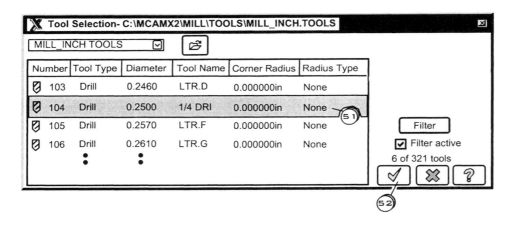

➤ Click ⑤① the [.25 Drill Tool]

➤ Click ⑤② the OK button [✓] to bring the tool from the library into the part file for the operation.

◆ AUTO ASSIGN SPEEDS FEEDS FOR THE TOOL

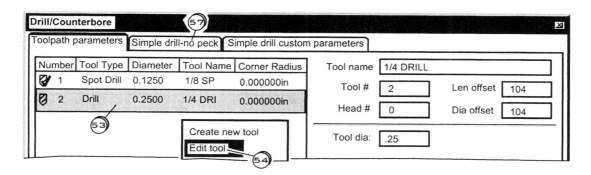

▶ Click (53) on the .25 Drill tool

▶ *Right* Click move the cursor down and Click (54) [Edit tool]

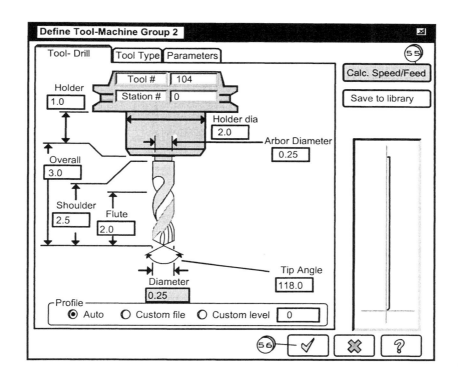

▶ Click (55) the [Calc. Speed/Feed] button. Mastercam will *automatically* assign
Speeds/Feeds for the *operation* and the *stock material* selected.

▶ Click (56) the OK button [✓] to enter the tool parameters in the
Operations Manager under ▨ Tool settings

◆ ENTER THE REQUIRED .25 DIA THRU HOLE MACHINING PARAMETERS

➤ Click (57) ⌠Simple drill-no peck⌡ tab. See page 5-27.

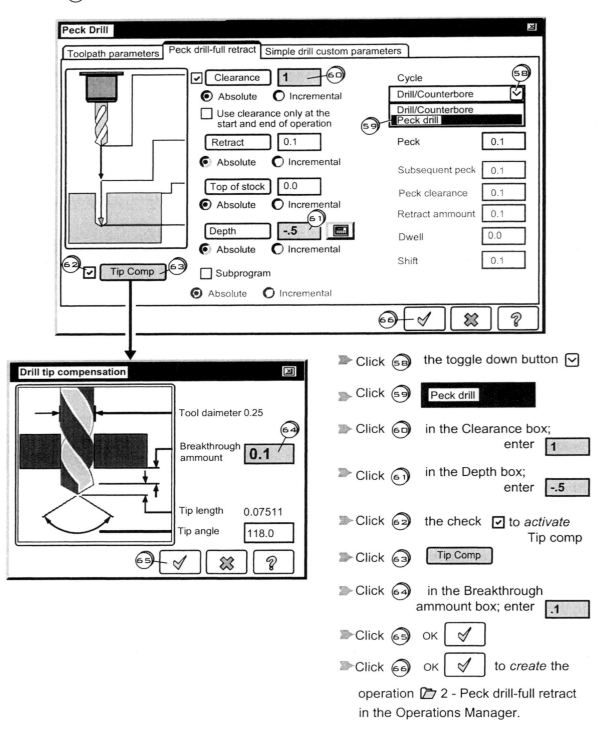

➤ Click (58) the toggle down button ☑

➤ Click (59) ▐Peck drill▌

➤ Click (60) in the Clearance box;
                          enter ▐1▌

➤ Click (61) in the Depth box;
                          enter ▐-.5▌

➤ Click (62) the check ☑ to *activate*
                                   Tip comp

➤ Click (63) ▐ Tip Comp ▌

➤ Click (64) in the Breakthrough
                     ammount box; enter ▐.1▌

➤ Click (65) OK ☑

➤ Click (66) OK ☑ to *create* the

operation 🗁 2 - Peck drill-full retract
in the Operations Manager.

D) DRILL THE .031DIA PILOT HOLES X .125 DEEP

♦ CREATE THE PILOT HOLE TOOLPATH

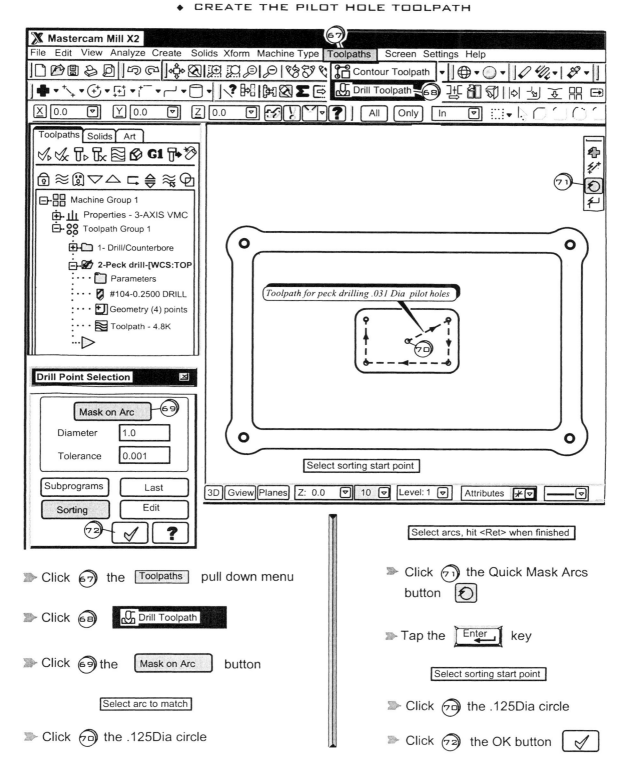

Toolpath for peck drilling .031 Dia pilot holes

>> Click (67) the [Toolpaths] pull down menu

>> Click (68) [Drill Toolpath]

>> Click (69) the [Mask on Arc] button

[Select arc to match]

>> Click (70) the .125Dia circle

>> Click (71) the Quick Mask Arcs button

>> Tap the [Enter] key

[Select sorting start point]

>> Click (70) the .125Dia circle

>> Click (72) the OK button

◆ CREATE THE NEEDED .031 DIA SPOT DRILL TOOL

The smallest spot drill currently in Mastercam's library is .125 Dia. So the tool we need to use will have to be created.

Note: The Create new tool function **cannot be run** from *Mastercam Mill X2(DEMO)*.

    If *Mastercam Mill X2(DEMO) is used* : **Select the 0.125 Spot Drill tool**

                                             Skip all the steps for creating the new .031 spot drill tool.

                                            Continue with the exercise startinng on page 5-34.

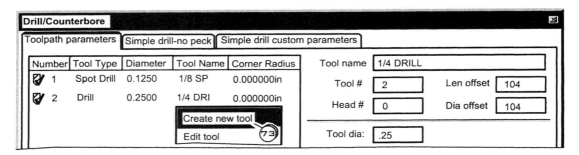

⇒ **Right** Click move the cursor down and Click (73) Create new tool

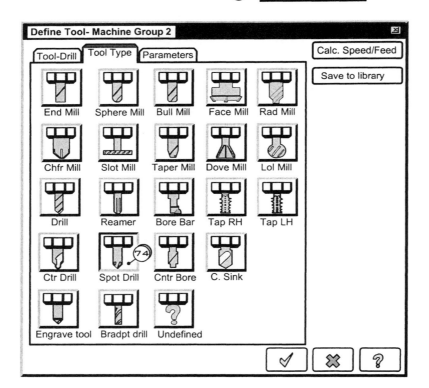

⇒ Click (74) the Spot Drill filter button

The tool dimensions are listed in the PROCESS PLAN 5-2 on page 5-16 as :

1/32 Spot Drill(.31 - Flute length; 1.5 - OAL)

Enter these values into the Define Tool dialog box :

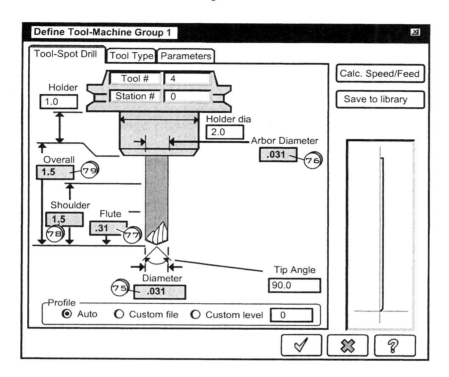

➤ Click ⑦⑤ in the Diameter box; enter [.031]

➤ Click ⑦⑥ in the Arbor Diameter box; enter [.031]

➤ Click ⑦⑦ in the Flute box; enter [.31]

➤ Click ⑦⑧ in the Shoulder box; enter [1.5]

➤ Click ⑦⑨ in the Overall box; enter [1.5]

Open the Parameters Tab and open the SPOTDRIL sub-file.
Create the name 1/32 SPOT DRILLx1.5L for the new tool.

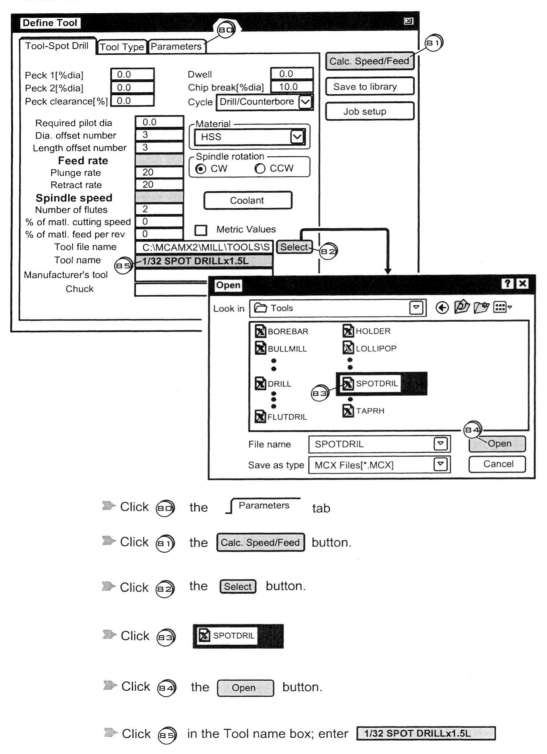

➤ Click (80) the ⌐Parameters⌐ tab

➤ Click (81) the [Calc. Speed/Feed] button.

➤ Click (82) the [Select] button.

➤ Click (83) [◢ SPOTDRIL]

➤ Click (84) the [Open] button.

➤ Click (85) in the Tool name box; enter [1/32 SPOT DRILLx1.5L]

Save the tool in the SPOTDRIL sub-file which is contained within the main tool library file MILL_INCH.Tools

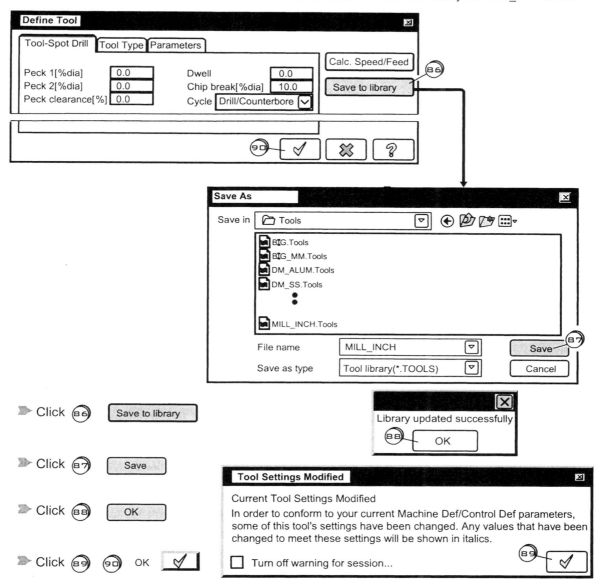

> Click ⑧⑥ [Save to library]

> Click ⑧⑦ [Save]

> Click ⑧⑧ [OK]

> Click ⑧⑨ ⑨⓪ OK ✓

After creating and storing the new tool in the tools library, *Mastercam* will then load it into the current job file.

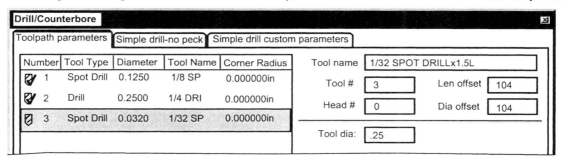

♦ ENTER THE REQUIRED .031 DIA PILOT HOLE MACHINING PARAMETERS

⟫ Click ⟨91⟩ ⌠Simple drill-no peck⌡   tab

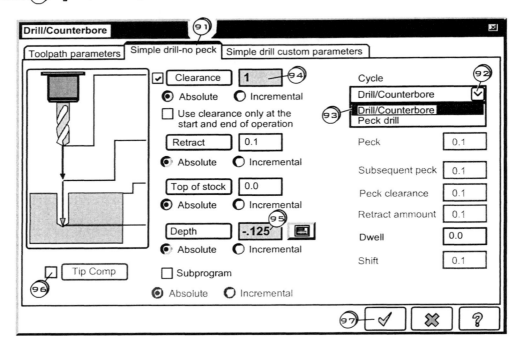

⟫ Click ⟨92⟩   the toggle down button   ☑

⟫ Click ⟨93⟩   ▐ Drill/Counterbore ▌

⟫ Click ⟨94⟩   in the Clearance box and enter   ▐ 1 ▌

⟫ Click ⟨95⟩   in the Depth box and enter   ▐ -.125 ▌

⟫ Click ⟨96⟩   such that a check ☐ *does not* appear in the Tip Comp box
                                              (Tip Comp *off* for *blind* holes)

⟫ Click ⟨97⟩   the OK button ▐ ✓ ▌ to *create* operation ⟋ 3 - Simple drill-no peck in the .
                                    Operations Manager.

E) PECK DRILL THE .125DIA HOLES x .325 DEEP

◆ CREATE THE PECK DRILL DRILL TOOLPATH

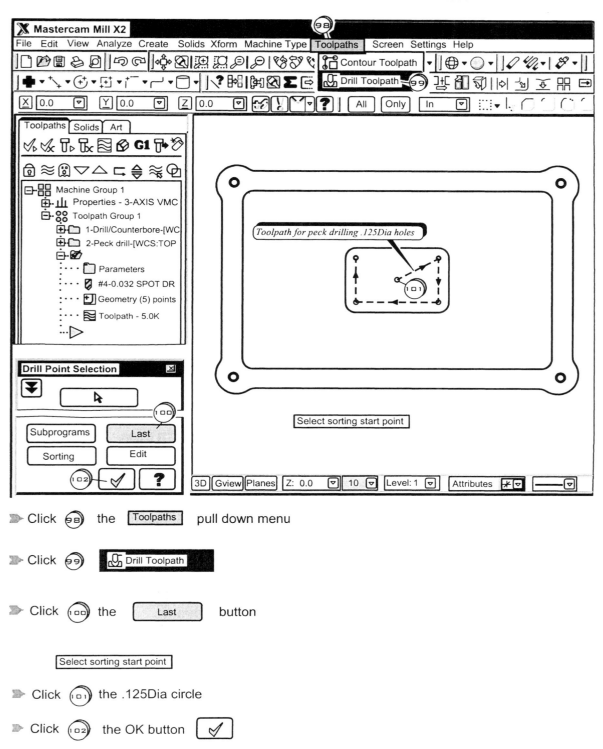

Toolpath for peck drilling .125Dia holes

Select sorting start point

➡ Click ⑨⑧ the [Toolpaths] pull down menu

➡ Click ⑨⑨ [Drill Toolpath]

➡ Click ⑩⑩ the [Last] button

[Select sorting start point]

➡ Click ⑩① the .125Dia circle

➡ Click ⑩② the OK button [✓]

◆ OBTAIN A .125 DIA DRILL TOOL FROM THE TOOL LIBRARY

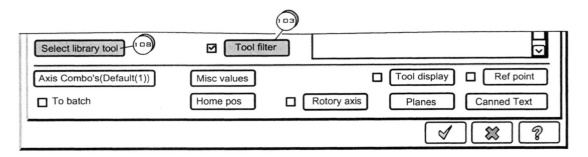

➤ Click ⓐ the [Tool filter] button

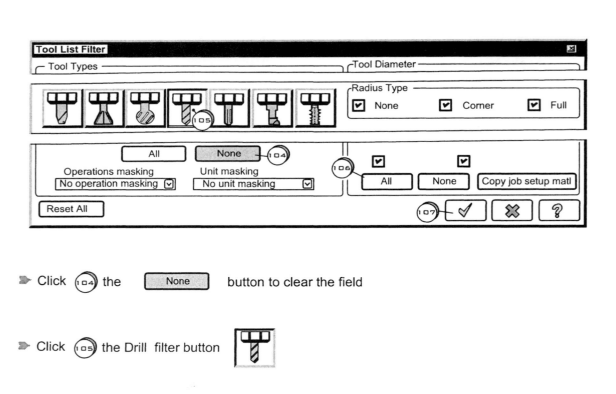

➤ Click ⓘ₀₄ the [None] button to clear the field

➤ Click ⓘ₀₅ the Drill filter button

➤ Click ⓘ₀₆ the [All] button.

➤ Click ⓘ₀₇ the OK button

➤ Click (108) the [Select library tool] button. See page 5-36.

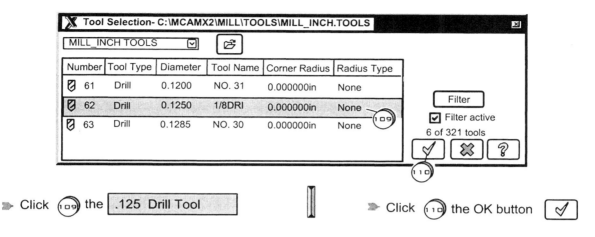

➤ Click (109) the [.125 Drill Tool]   ➤ Click (110) the OK button [✓]

◆ AUTO ASSIGN SPEEDS FEEDS FOR THE TOOL

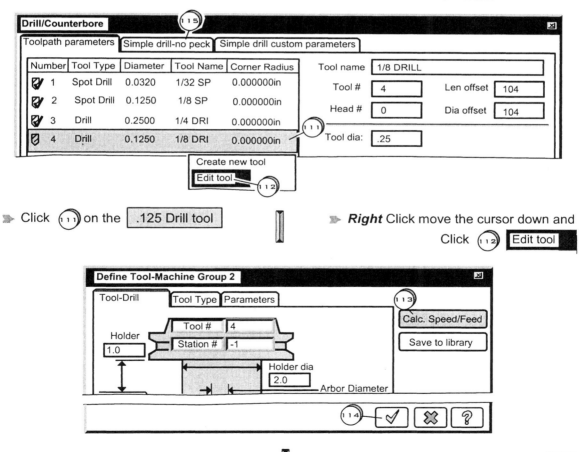

➤ Click (111) on the [.125 Drill tool]   ➤ *Right* Click move the cursor down and

Click (112) [Edit tool]

➤ Click (113) the [Calc. Speed/Feed] key   ➤ Click (114) the OK button [✓]

◆ ENTER THE REQUIRED .125 DIA PECK DRILL MACHINING PARAMETERS

▶Click ⑪⑤ ⌐Simple drill-no peck   tab. See page 5-37.

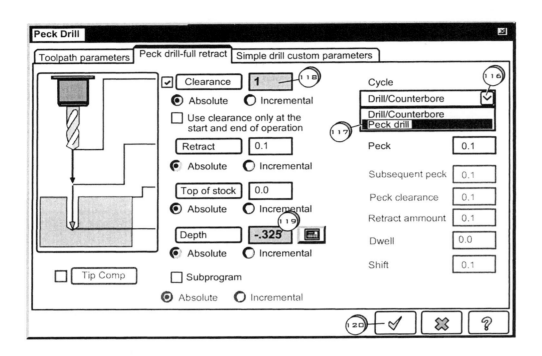

▶Click ⑪⑥ the toggle down button ☑

▶Click ⑪⑦

▶Click ⑪⑧ in the Clearance box; enter **1**

▶Click ⑪⑨ in the Depth box; enter **-.325**

▶Click ⑫⓪ OK ✓ to *create* the operation 📁 4 - Peck drill-full retract in the
          Operations Manager.

5-4 Backplotting Machining Operations

Previously generated toolpaths are contained in the job's NCI file(*.NCI) and can be   played back and visually checked for correctness by selecting the **Backplot** function from the **Operations Manager** dialog box. The Operations manager is a powerful feature that enables the user to see, backplot, verify and edit the job's machining operations.

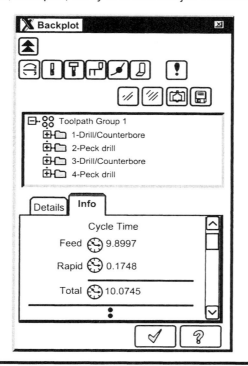

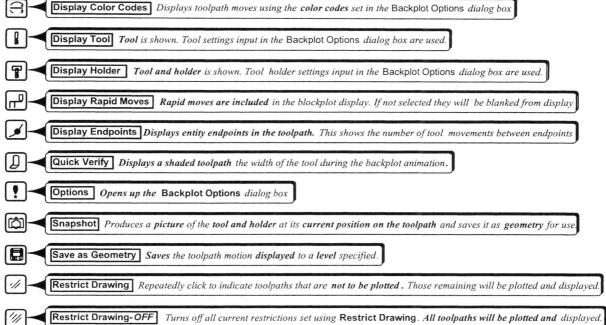

Display Color Codes *Displays toolpath moves using the **color codes** set in the* Backplot Options *dialog box*

Display Tool *Tool is shown. Tool settings input in the* Backplot Options *dialog box are used.*

Display Holder *Tool and holder is shown. Tool holder settings input in the* Backplot Options *dialog box are used.*

Display Rapid Moves *Rapid moves are included in the blockplot display. If not selected they will be blanked from display*

Display Endpoints *Displays entity endpoints in the toolpath. This shows the number of tool movements between endpoints*

Quick Verify *Displays a shaded toolpath the width of the tool during the backplot animation.*

Options *Opens up the **Backplot Options** dialog box*

Snapshot *Produces a picture of the **tool and holder** at its **current position on the toolpath** and saves it as **geometry** for use.*

Save as Geometry *Saves the toolpath motion **displayed** to a **level** specified.*

Restrict Drawing *Repeatedly click to indicate toolpaths that are **not to be plotted**. Those remaining will be plotted and displayed.*

Restrict Drawing-OFF *Turns off all current restrictions set using **Restrict Drawing**. All toolpaths will be plotted and displayed.*

EXAMPLE 5-3

Use the Backplot function in the Operations Manager dialog box to display the drilling tool paths created in Example 5-2.

A) BACKPLOTTING THE TOOLPATHS FOR ALL OPERATIONS

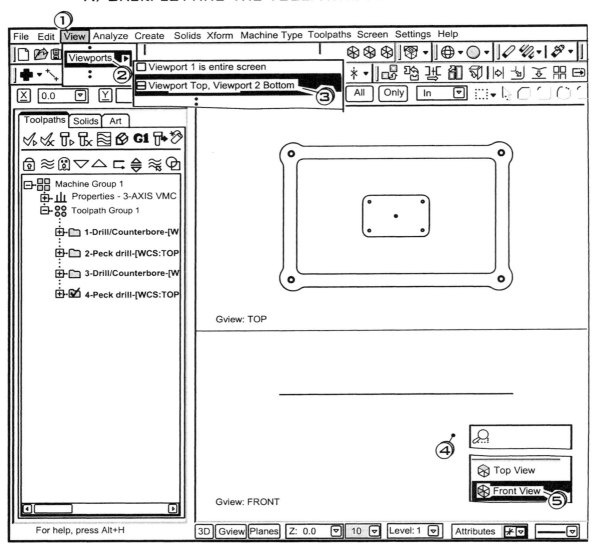

▶ Click ① the [View] pull down menu

▶ Click ② [Viewports]

▶ Click ③ [⊟ Viewport Top, Viewport 2 Bottom]

▶ **Right** Click ④ in the lower view; move the cursor down to  ;Click ⑤

Specify *all* the operations listed in the Operations Manager are to be *backplotted*

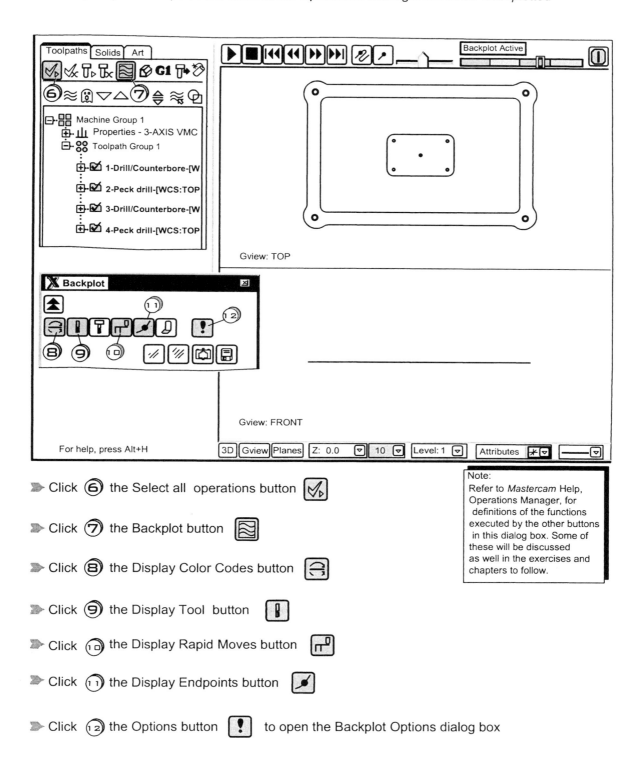

➤ Click ⑥ the Select all operations button ☑▹

➤ Click ⑦ the Backplot button ≋

➤ Click ⑧ the Display Color Codes button ⤝

➤ Click ⑨ the Display Tool button ▯

➤ Click ⑩ the Display Rapid Moves button ▱⁰

➤ Click ⑪ the Display Endpoints button ▱

➤ Click ⑫ the Options button ❗ to open the Backplot Options dialog box

Note:
Refer to *Mastercam* Help, Operations Manager, for definitions of the functions executed by the other buttons in this dialog box. Some of these will be discussed as well in the exercises and chapters to follow.

Enter backplot display options such as the tool's appearence.

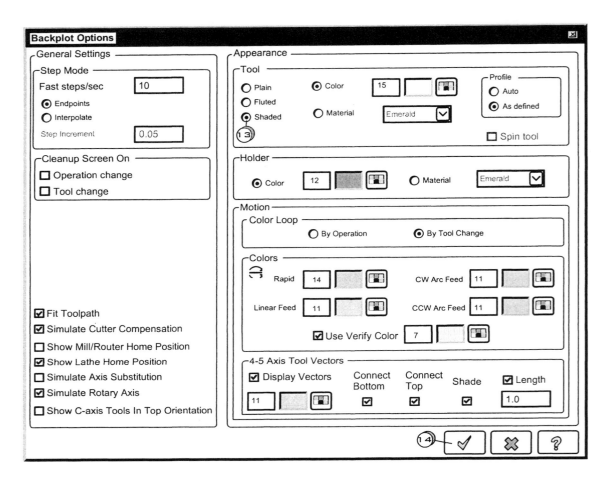

➤ Click ⑬ the Shaded radio button ⊙ .Mastercam will display the tool *shaded* in the
backplot animation.

➤ Click ⑭ the OK button [✓]

Note:
Refer to *Mastercam* Help,
Backplot Options, for
definitions of the functions
executed by the other buttons
in this dialog box.

Move the tool in *single step* mode along the toolpath

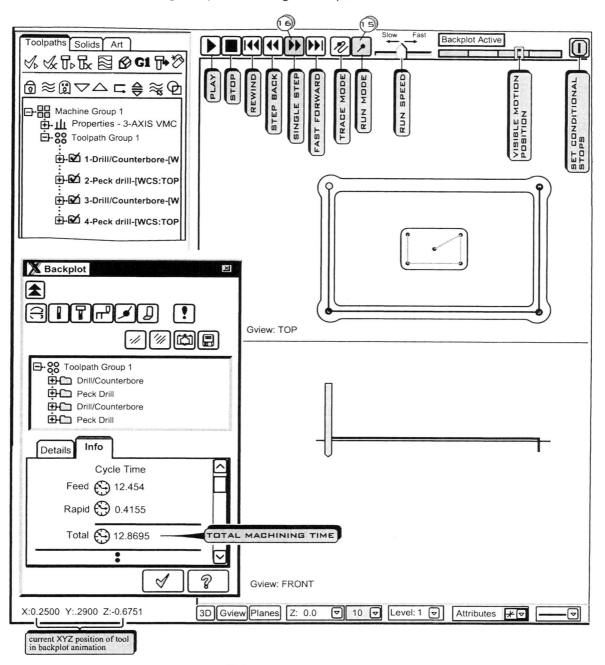

current XYZ position of tool in backplot animation

➤ Click ⑮ the Run Mode button to direct *Mastercam* to draw the toolpath as the backplot progresses

➤ *Repeatedly* Click ⑯ the Single Step button to see step by step tool movement.

B) BACKPLOTTING THE TOOLPATHS FOR SELECTED OPERATIONS

Suppose the operator wants to backplot *only* operations 3 and 4.

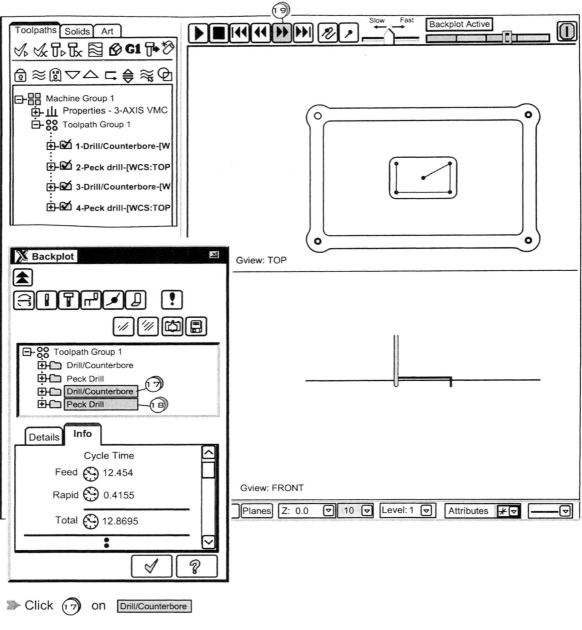

➤ Click ⟨17⟩ on Drill/Counterbore

➤ Depress the Ctrl key on the keyboard and ***keeping it depressed***

   Click ⟨18⟩ on the *next* operation Peck Drill

➤ ***Repeatedly*** Click ⟨19⟩ the Single Step button ⏩ to see step by step tool movement.

5-5 Verifying Machining Operations

Mastercam's verifier uses the NCI file to produce the most realistic animation of previously defined machining operations. The part is displayed as a solid model. The operator can see material being removed as the tool moves along a specified toolpath. Verification provides for *off line* checking or checking that *does not* cause a CNC machine to be tied up for the job proveout. This reduces possible tool and machine damage, increases the capability of machining a correct part on the first run and thus *increases the return on investment* for the job.

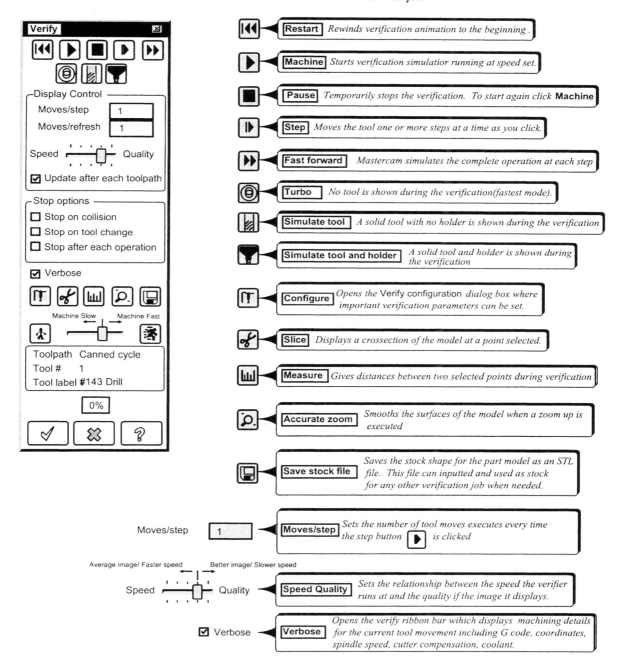

Restart *Rewinds verification animation to the beginning .*

Machine *Starts verification simulatior running at speed set.*

Pause *Temporarily stops the verification. To start again click* **Machine**

Step *Moves the tool one or more steps at a time as you click.*

Fast forward *Mastercam simulates the complete operation at each step*

Turbo *No tool is shown during the verification(fastest mode).*

Simulate tool *A solid tool with no holder is shown during the verification*

Simulate tool and holder *A solid tool and holder is shown during the verification*

Configure *Opens the* Verify *configuration dialog box where important verification parameters can be set.*

Slice *Displays a crossection of the model at a point selected.*

Measure *Gives distances between two selected points during verification*

Accurate zoom *Smooths the surfaces of the model when a zoom up is executed*

Save stock file *Saves the stock shape for the part model as an STL file. This file can inputted and used as stock for any other verification job when needed.*

Moves/step   1   **Moves/step** *Sets the number of tool moves executes every time the step button* ▶ *is clicked*

Average image/ Faster speed   Better image/ Slower speed
Speed ⊢—⊟—⊣ Quality   **Speed Quality** *Sets the relationship between the speed the verifier runs at and the quality if the image it displays.*

☑ Verbose   **Verbose** *Opens the verify ribbon bar wihich displays machining details for the current tool movement including G code, coordinates, spindle speed, cutter compensation, coolant.*

EXAMPLE 5-4

Use the Verify function in the Operations Manager dialog box to generate a
proveout run of *all* the the drilling operations created in Example 5-2.

Specify *all* the operations listed in the Operations Manager are to be *verified*

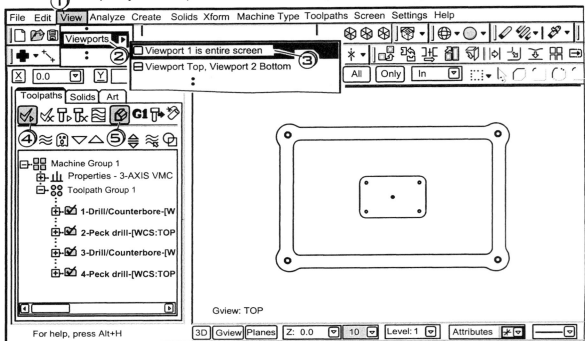

▷ Click ① the View pull down menu

▷ Click ② Viewports

▷ Click ③ ☐ Viewport is entire screen

▷ Click ④ the Select all operations button 🗹

▷ Click ⑤ the Verify button 🖉

Set the verification display options such as Remove chips and the stock's color.

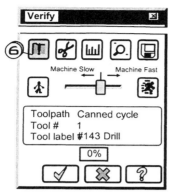

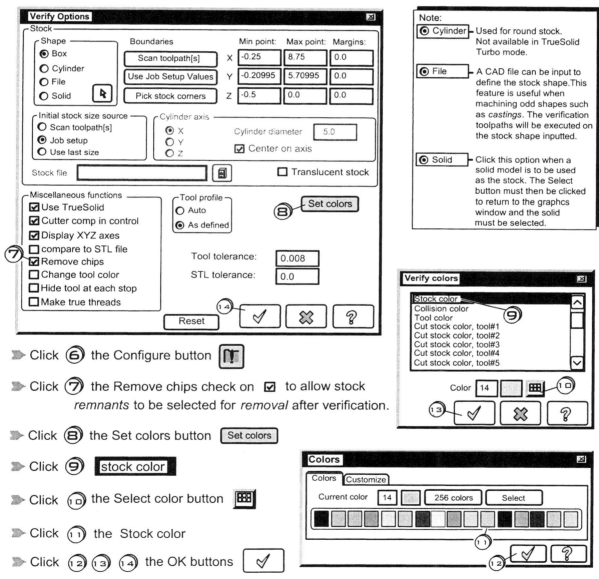

➤ Click ⑥ the Configure button 🕅

➤ Click ⑦ the Remove chips check on ☑ to allow stock
  remnants to be selected for *removal* after verification.

➤ Click ⑧ the Set colors button [Set colors]

➤ Click ⑨ [stock color]

➤ Click ⑩ the Select color button ▦

➤ Click ⑪ the Stock color

➤ Click ⑫ ⑬ ⑭ the OK buttons ✓

Move the tool in Machine mode at the Machine speed set.

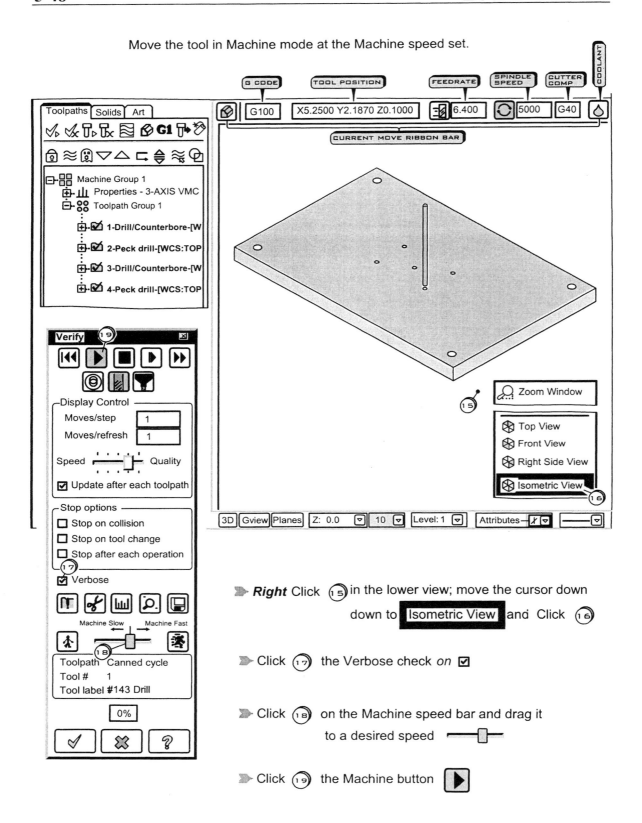

> ➤ **Right** Click (15) in the lower view; move the cursor down down to Isometric View and Click (16)

> ➤ Click (17) the Verbose check *on* ☑

> ➤ Click (18) on the Machine speed bar and drag it to a desired speed ▭

> ➤ Click (19) the Machine button ▶

5-6 Panning, Dynamic Rotation and Sectioning within the Verifier

A) Dynamically Panning the Machined Model

Use the Pan screen display methods discussed in Chapter 2 on p2-71.

► Tapping the directional arrow keys on the keypad ⬆up ⬇down ⬅left ➡right

► Press the Alt key and **keeping it depressed** ;
press the mouse wheel and drag the mouse in the desired pan direction

press wheel

Cursor display indicates mouse is in pan mode

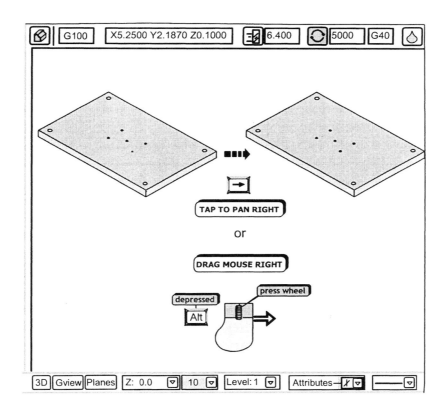

B) Dynamically Rotating the Machined Model

EXAMPLE 5-5
Dynamically rotate the machined model as shown below

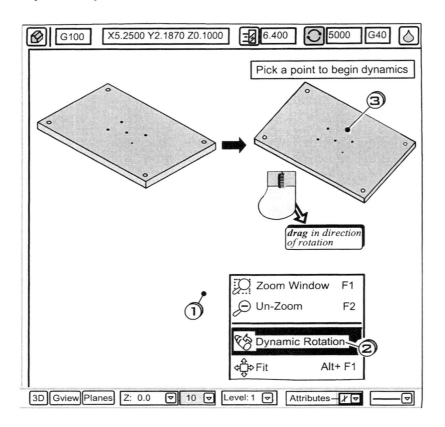

➤ **Right** Click ① in the lower view; move the cursor down

down to ⬛Dynamic Rotation⬛ and Click ②

Pick a point to begin dynamics

➤ Click ③ the reference point to rotate about.

➤ Drag the mouse cursor in the direction of desired rotation.

Click a mouse button when done

➤ Click any mouse button to exit the dynamic rotation function

C) SECTIONING THE MACHINED MODEL

EXAMPLE 5-6
Section the machined model shown below

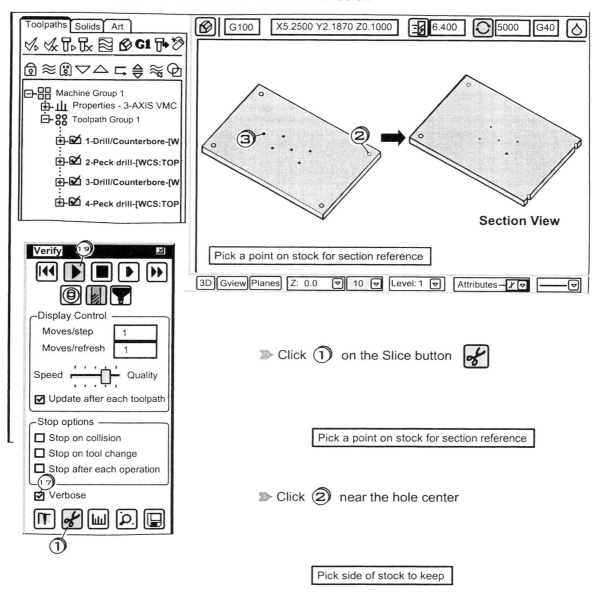

➤ Click ① on the Slice button

Pick a point on stock for section reference

➤ Click ② near the hole center

Pick side of stock to keep

➤ Click ③ the portion of the part to *remain*

5-7 Using the Circle Mill Function

The circle mill function *automatically* creates tool paths for machining a circle in stock. This is especially useful when using an endmill to produce a counterbore or a spotface.

EXAMPLE 5-7

Use a .75Dia Endmill to machine a 1.5Dia Counterbore .25 deep in the stock shown in Figure 5-3.

1.5DIA C'BORE x .25 DEEP

Figure 5-3

◆ SELECT THE CIRCLE MILL FUNCTION AND IDEBTIFT THE HOLE CENTER

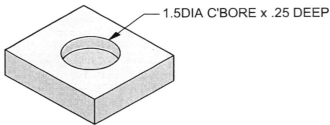

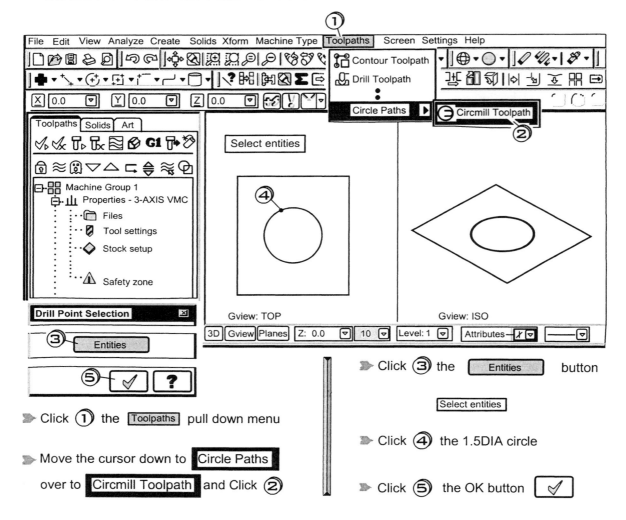

▶ Click ① the Toolpaths pull down menu

▶ Move the cursor down to Circle Paths

over to Circmill Toolpath and Click ②

▶ Click ③ the Entities button

Select entities

▶ Click ④ the 1.5DIA circle

▶ Click ⑤ the OK button ✓

◆ OBTAIN A .75 DIA FLAT END MILL TOOL FROM THE TOOL LIBRARY

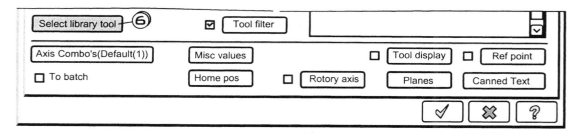

➤ Click ⑥ the [Select library tool] button

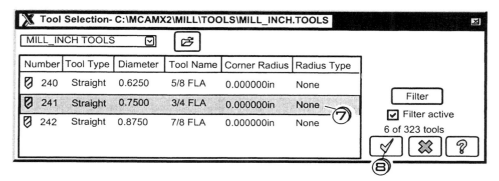

➤ Click ⑦ the | 3/4 FLA Tool |

➤ Click ⑧ the OK button [✓]

◆ AUTO ASSIGN SPEEDS FEEDS FOR THE TOOL

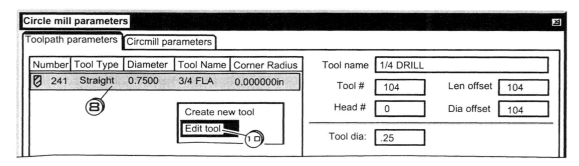

➤ Click ⑨ on the 3/4 FLA tool

➤ *Right* Click move the cursor down and Click ⑩ [Edit tool]

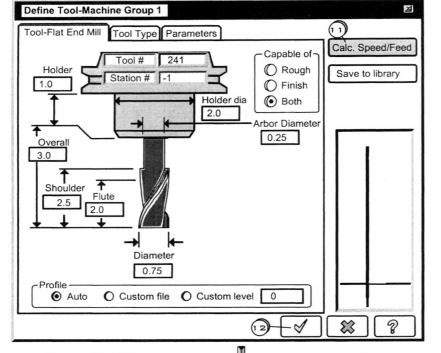

Click ⓘ **the** [Calc. Speed/Feed] **button** ‖ ⯈ **Click** ⑫ **the OK button** ✓

◆ **SPECIFY THE CIRCLE MILL TOOL PATH PARAMETERS**

Mastercam executes the circle mill function by moving the tool as shown in Figure 5-4

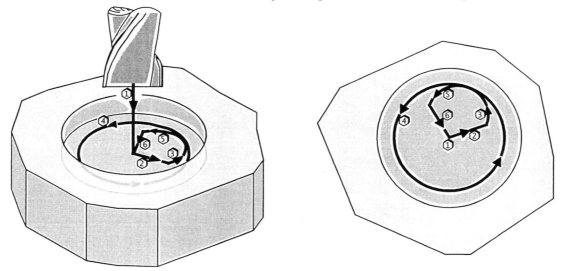

Circle mill function tool path with straight plunge entry

Figure 5-4

Note: Other options are available with the Circle Mill function including, helical entry for rough cuts, multiple passes, depth cuts, and an automatic breakthough calculator. Use *Mastercam's* Help, *Circle Mill* for further definitions.

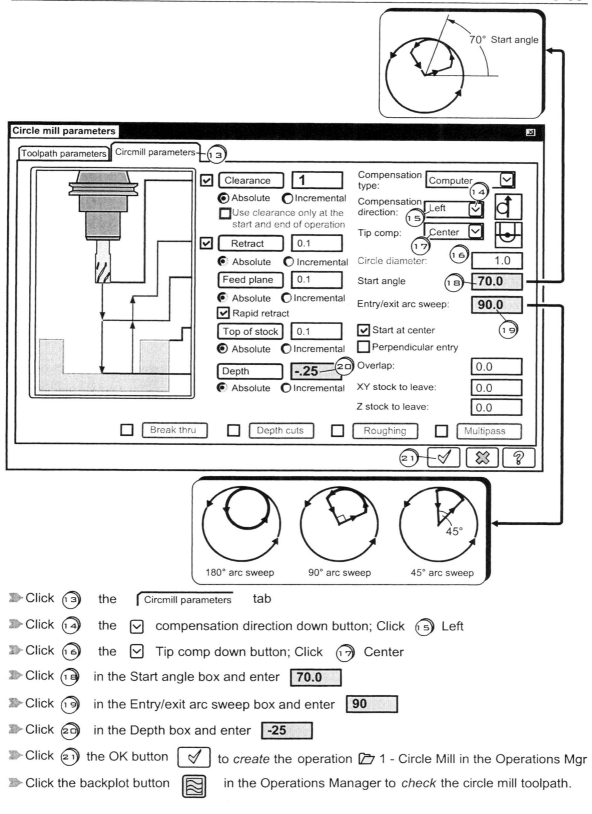

➤ Click ⑬ the [Circmill parameters] tab

➤ Click ⑭ the ☑ compensation direction down button; Click ⑮ Left

➤ Click ⑯ the ☑ Tip comp down button; Click ⑰ Center

➤ Click ⑱ in the Start angle box and enter **70.0**

➤ Click ⑲ in the Entry/exit arc sweep box and enter **90**

➤ Click ⑳ in the Depth box and enter **-25**

➤ Click ㉑ the OK button ☑ to *create* the operation 📂 1 - Circle Mill in the Operations Mgr

➤ Click the backplot button ≋ in the Operations Manager to *check* the circle mill toolpath.

5-8 Machining Holes at Different Depths and Retract Heights

For some drilling operations *rapid moves can be significantly reduced* by specifying different depths and different retract heights. This is the case for parts whose thickness changes within the drill area. The operator may also need to change the retract height in order to *avoid collisions with clamps or other workholding devices* .

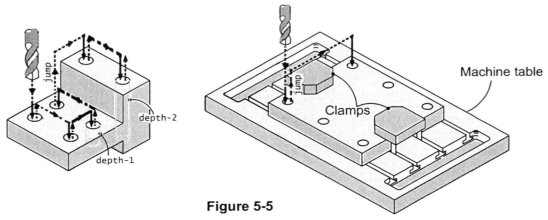

Figure 5-5

EXAMPLE 5-8

Use a .25Dia drill to machine the through holes in the part shown in Figure 5-6.

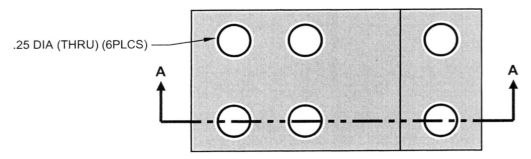

.25 DIA (THRU) (6PLCS)

SECTION A-A

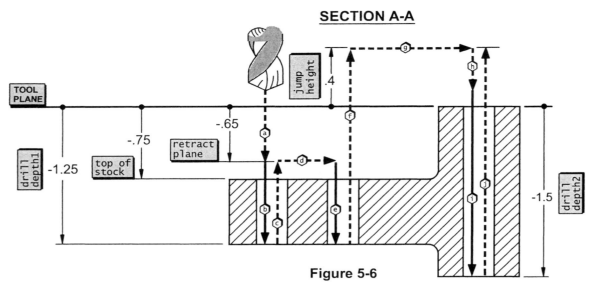

Figure 5-6

◆ CREATE THE .25DIA PECK DRILL THRU TOOLPATH

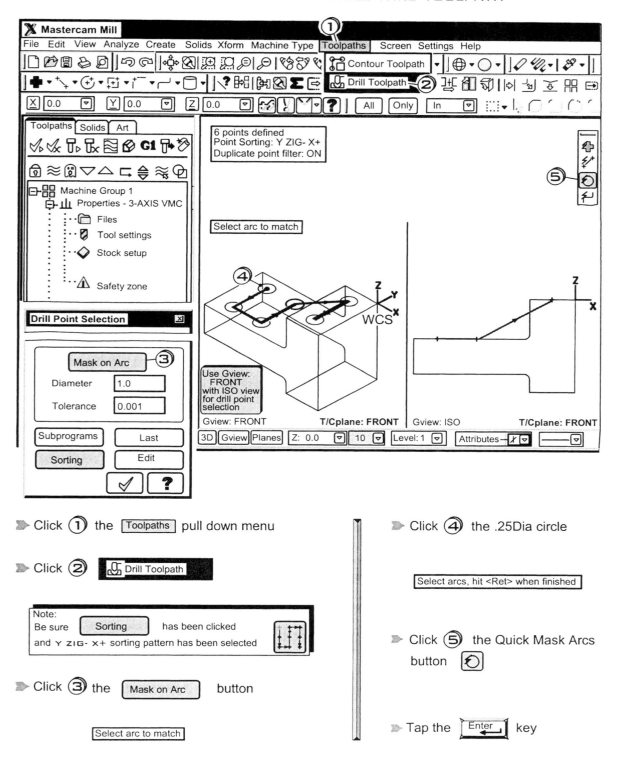

➤ Click ① the [Toolpaths] pull down menu

➤ Click ② [🔩 Drill Toolpath]

```
Note:
Be sure [ Sorting ] has been clicked
and Y ZIG- X+ sorting pattern has been selected
```

➤ Click ③ the [Mask on Arc] button

[Select arc to match]

➤ Click ④ the .25Dia circle

[Select arcs, hit <Ret> when finished]

➤ Click ⑤ the Quick Mask Arcs button [🔄]

➤ Tap the [Enter] key

◆ CREATE THE .4 JUMP IN +Z AT POINT Ⓐ AND THE [-1.5-.2]
DRILL DEPTH AT POINT Ⓑ IN THE TOOLPATH

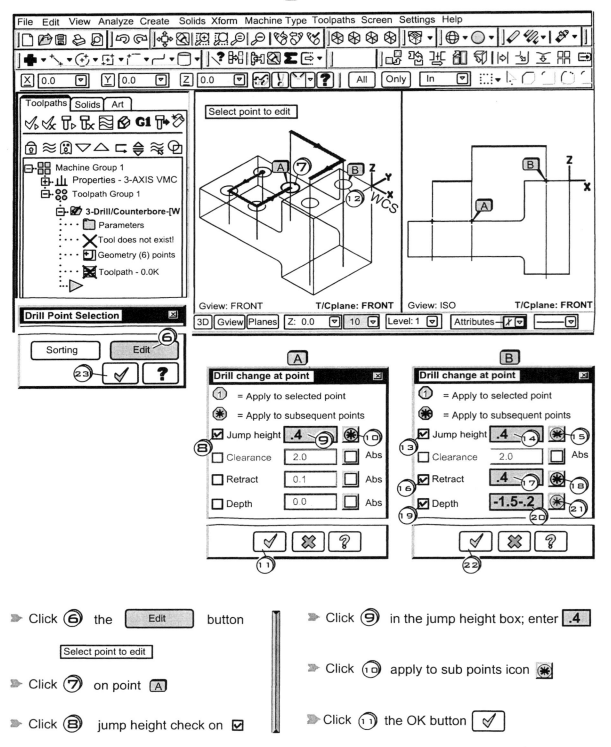

➤ Click ⑥ the [Edit] button

[Select point to edit]

➤ Click ⑦ on point Ⓐ

➤ Click ⑧ jump height check on ☑

➤ Click ⑨ in the jump height box; enter .4

➤ Click ⑩ apply to sub points icon ✳

➤ Click ⑪ the OK button ✓

➤ Click ⑥ the [Edit] button

[Select point to edit]

➤ Click ⑫ on point [B]

➤ Click ⑬ jump height check on ☑

➤ Click ⑭ in the jump height box; enter [.4]

➤ Click ⑮ apply to subsuquent points icon [✳]

➤ Click ⑯ retract check on ☑

➤ Click ⑰ in the retract box; enter [.4]

➤ Click ⑱ apply to subsuquent points icon [✳]

➤ Click ⑲ depth check on ☑

(Stock height)   (Breakthrough ammount)

➤ Click ⑳ in the depth box; enter [-1.5 - .2]

➤ Click ㉑ apply to subsuquent points icon [✳]

➤ Click ㉒ ㉓ the OK buttons [✓]

◆ OBTAIN A .25 DIA DRILL TOOL FROM THE TOOL LIBRARY

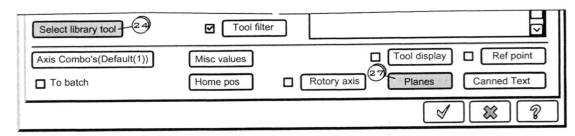

➤ Click ㉔ the [Select library tool] button

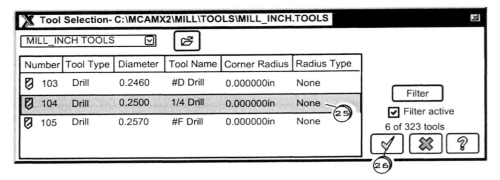

➤ Click ㉕ the **1/4 Drill Tool**   ➤ Click ㉖ the OK button ✓

Because the Gview :TOP view was *changed* to a Gview:FRONT *Mastercam* also changed the Tool and construction planes to **FRONT**. The tool plane needs to be changed back to a **TOP** orientation to drill the holes in the proper direction.

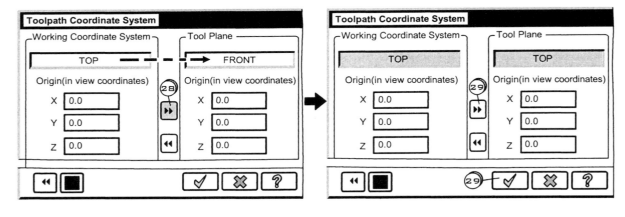

➤ Click ㉗ the [Planes] button

➤ Click ㉘ the copy entry button ▸▸

➤ Click ㉙ the OK button ✓

The tool plane will then be *changed* back to a **TOP** orientation

◆ ENTER THE REQUIRED .25 DIA THRU HOLE MACHINING PARAMETERS
FOR ALL THE HOLES PRIOR TO AND AT THE JUMP POINT

➤ Click ③① ⌠Peck drill-full retract⌡ tab to enter the .25DIA hole machining parameters.

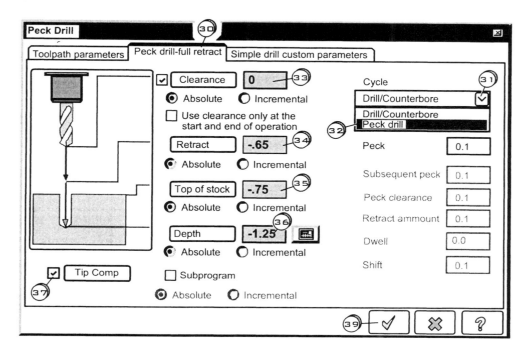

➤ Click ③① the toggle down button ☑

➤ Click ③② Peck drill

➤ Click ③③ in the Clearance box; enter 0

➤ Click ③④ in the Retract box; enter -.65

➤ Click ③⑤ in the Top of stock box ; enter -.75

➤ Click ③⑥ in the Depth box; enter -1.25

➤ Click ③⑦ the tip comp check ☑

➤ Click ③⑧ the OK button ✓

Click the Backplot button ▨ in the Operations Manager to ckeck the the drill toolpaths.

5-9 Editing Drill Tool Paths

Mastercam's Toolpath Manager is contained within the Operations Manager. It has many tools for editing existing toolpaths as discussed in Chapter 7 . It will be briefly introduced in this section to provide the operator with some *basic* editing capibilities when working with drilling toolpaths .

DELETING AN EXISTING DRILL TOOL PATH

EXAMPLE 5-9

Delete the existing drill tool path shown below

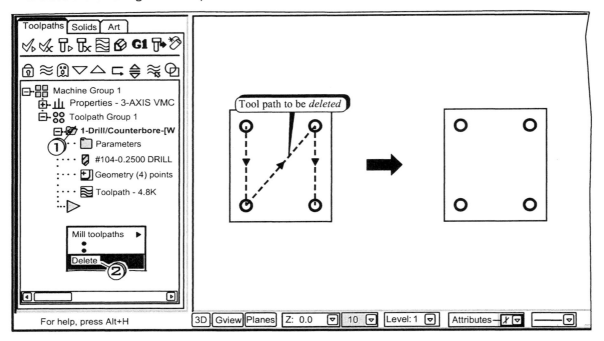

➤ Click ① on the file folder of the operation to delete such that a check ✍ appears

➤ *Right* Click move the cursor down to Delete ; Click ② to delete the drilling operation

ADDING DRILL POINTS TO AN EXISTING DRILL TOOL PATH

EXAMPLE 5-10

Add the additional drill points to the existing drill tool path shown below

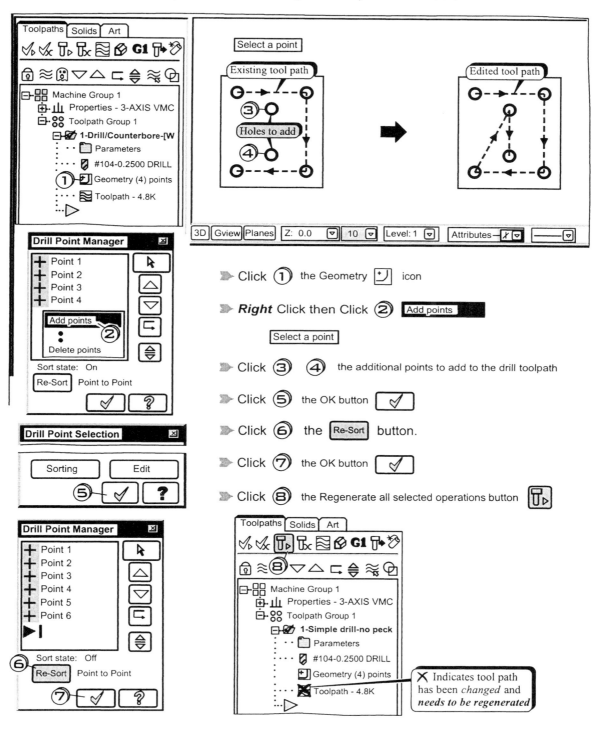

➤ Click (1) the Geometry [↵] icon

➤ **Right** Click then Click (2) [Add points]

   [Select a point]

➤ Click (3) (4) the additional points to add to the drill toolpath

➤ Click (5) the OK button [✓]

➤ Click (6) the [Re-Sort] button.

➤ Click (7) the OK button [✓]

➤ Click (8) the Regenerate all selected operations button [T▷]

X Indicates tool path has been *changed* and ***needs to be regenerated***

DELETING POINTS FROM AN EXISTING DRILL TOOL PATH

EXAMPLE 5-11

Remove the drill points from the existing drill tool path shown below

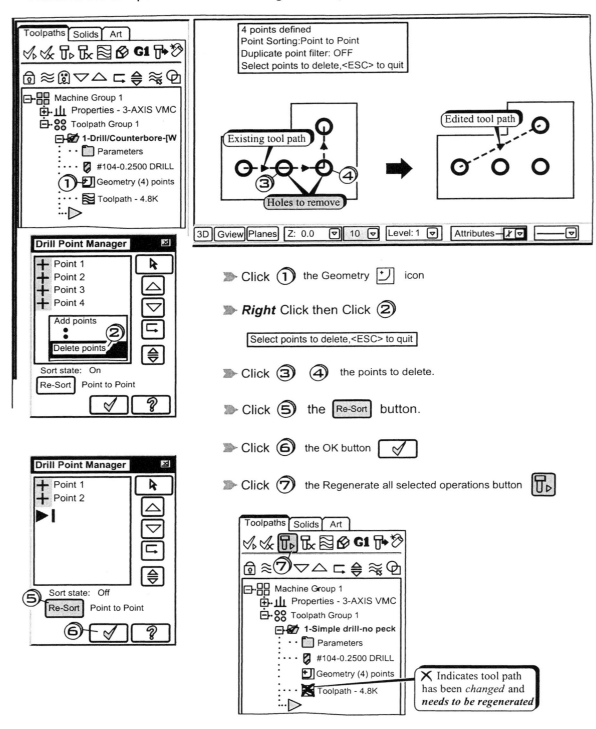

➤ Click ① the Geometry [◡] icon

➤ **Right** Click then Click ②

Select points to delete,<ESC> to quit

➤ Click ③ ④ the points to delete.

➤ Click ⑤ the Re-Sort button.

➤ Click ⑥ the OK button ✓

➤ Click ⑦ the Regenerate all selected operations button

5-10 Postprocessing

Postprocessing involves the generation of a word address part program(G-codes) for producing a part on a *particular* CNC machine tool.

The operator must select the appropriate post (FANUC, FADAL, MAZAK,etc) for the CNC machine to be used. *Mastercam* provides a library of posts for many types of CNC machines. Customized posts can also be built to satisfy particular needs.

The **.NCI** file for a job must also be accessed. Recall the . **NCI** file contains important machining information for the job such as the tool path coordinates, tool speeds and feeds, etc. *Mastercam* uses the post file **.PST** and the **.NCI** file to *automatically* generate the word address part program or .NC file.

EXAMPLE 5-12

Direct Mastercam to generate a word address part program for the machining operations as listed below.

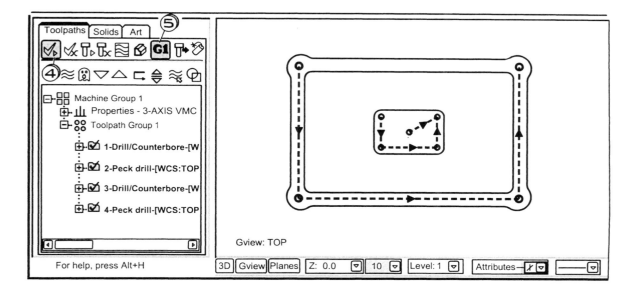

▶ Click ① the Select all operations button

▶ Click ② the Post selected operations button **G1**

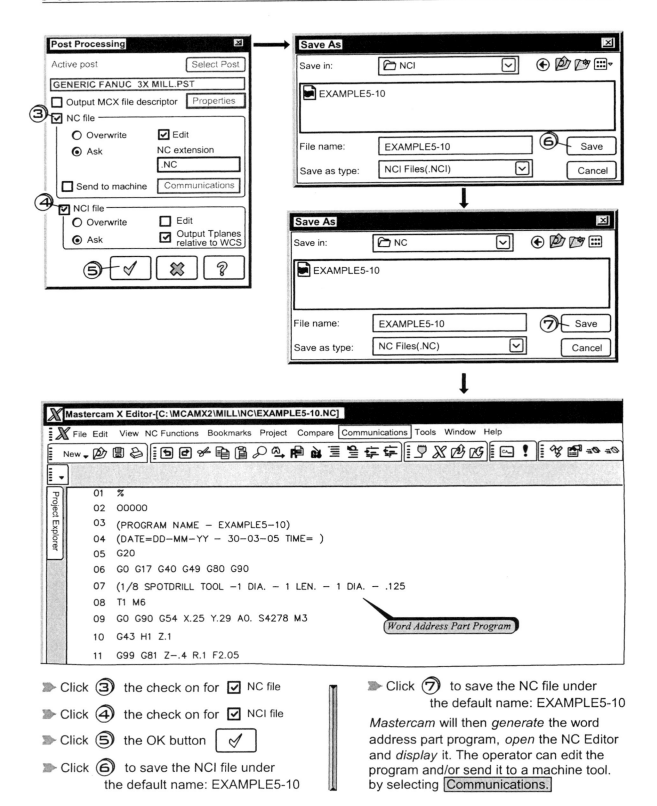

Post Processing ⊠

Active post [Select Post]

| GENERIC FANUC 3X MILL.PST |

☐ Output MCX file descriptor [Properties]

③ ☑ NC file
- ○ Overwrite ☑ Edit
- ◉ Ask NC extension
  | .NC |
- ☐ Send to machine [Communications]

④ ☑ NCI file
- ○ Overwrite ☐ Edit
- ◉ Ask ☑ Output Tplanes relative to WCS

⑤ [✓] [✗] [?]

Save As ⊠

Save in: [📁 NCI] [∨] ← 🗁 🗁 ⊞

📄 EXAMPLE5-10

File name: | EXAMPLE5-10 | ⑥ [Save]

Save as type: | NCI Files(.NCI) | [∨] [Cancel]

Save As ⊠

Save in: [📁 NC] [∨] ← 🗁 🗁 ⊞

📄 EXAMPLE5-10

File name: | EXAMPLE5-10 | ⑦ [Save]

Save as type: | NC Files(.NC) | [∨] [Cancel]

Mastercam X Editor-[C:\MCAMX2\MILL\NC\EXAMPLE5-10.NC]

File Edit View NC Functions Bookmarks Project Compare Communications Tools Window Help

New ▾

```
01   %
02   O0000
03   (PROGRAM NAME - EXAMPLE5-10)
04   (DATE=DD-MM-YY - 30-03-05 TIME= )
05   G20
06   G0 G17 G40 G49 G80 G90
07   (1/8 SPOTDRILL TOOL -1 DIA. - 1 LEN. - 1 DIA. - .125
08   T1 M6
09   G0 G90 G54 X.25 Y.29 A0. S4278 M3
10   G43 H1 Z.1
11   G99 G81 Z-.4 R.1 F2.05
```

Word Address Part Program

➤ Click ③ the check on for ☑ NC file

➤ Click ④ the check on for ☑ NCI file

➤ Click ⑤ the OK button [✓]

➤ Click ⑥ to save the NCI file under the default name: EXAMPLE5-10

➤ Click ⑦ to save the NC file under the default name: EXAMPLE5-10

Mastercam will then *generate* the word address part program, *open* the NC Editor and *display* it. The operator can edit the program and/or send it to a machine tool. by selecting [Communications.]

5-11 Choosing a Different Control and Post

Mastercam offers a wide array of control types and accompanying postprocessors to choose from.

To change controllers and posts follow the steps outlined below.

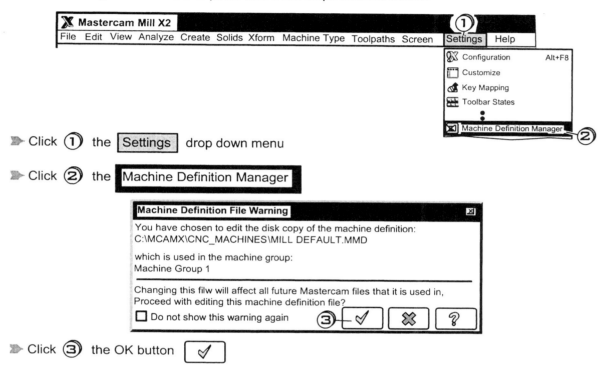

➤ Click ① the Settings drop down menu

➤ Click ② the Machine Definition Manager

➤ Click ③ the OK button

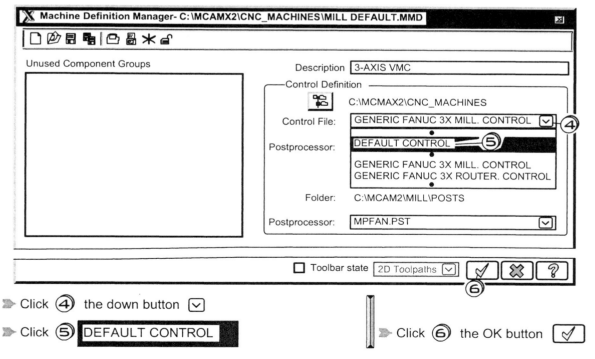

➤ Click ④ the down button

➤ Click ⑤ DEFAULT CONTROL

➤ Click ⑥ the OK button

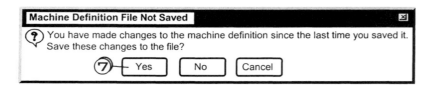

> Click ⑦ the [Yes] button

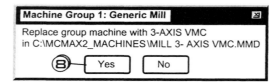

> Click ⑧ the [Yes] button

Mastercam will then use MPFAN.PST AS as the new post

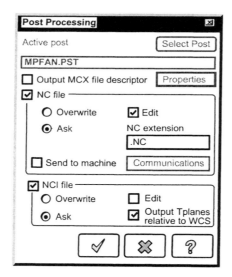

Note: In order to generate the *correct* word address NC code for a *particular* CNC machine tool and application the machine definition, control definition and .PST files must be *properly configured*. Refer to *Mastercam* Help, *About post processing*, for further information on this subject. Your authorized *Mastercam* reseller can be of great assistance in configuring or obtaining a custom post for a particular CNC machine tool.

EXERCISES

5-1) Consider the part shown in Figure 5p-1 and the operations listed PROCESS PLAN 5P-1

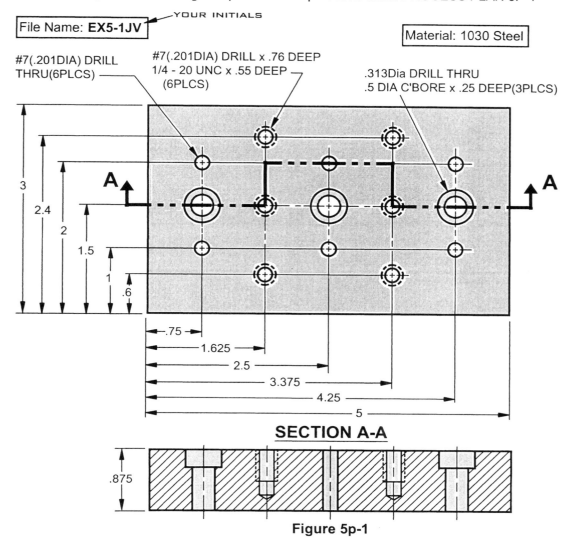

Figure 5p-1

PROCESS PLAN 5P-1

No.	Operation	Tooling
1	CENTER DRILL x .2 DEEP(ALL HOLES)	1/8 CENTER DRILL
2	PECK DRILL THRU(6PLCS)	#7(.201) DRILL
3	PECK DRILL x .76 DEEP(6PLCS)	#7(.201) DRILL
4	TAP x .55 DEEP(6PLCS)	1/4-20UNC TAP
5	PECK DRILL THRU (3PLCS)	5/16 DRILL
6	CIRCLE MILL x .25 DEEP (3PLCS)	3/8 FLAT ENDMILL

a) Use the Create Rectangle, Create Circle and the Xform Translate functions to generate the *wireframe* CAD model as shown below

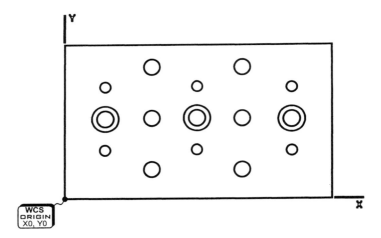

b) Setup the stock and assign the stock material

c) Generate the required drill toolpaths for executing PROCESS PLAN 5P-1

d) Backplot and verify the toolpaths

e) Direct Mastercam to generate the word address part program

5-2) a) Create the *wireframe* CAD model(top view *only*). Set up the stock and the material.
    b) Generate the required drill tool paths for executing PROCESS PLAN 5P-2
    c) Backplot and verify the tool paths
    d) Direct Mastercam to generate the word address part program

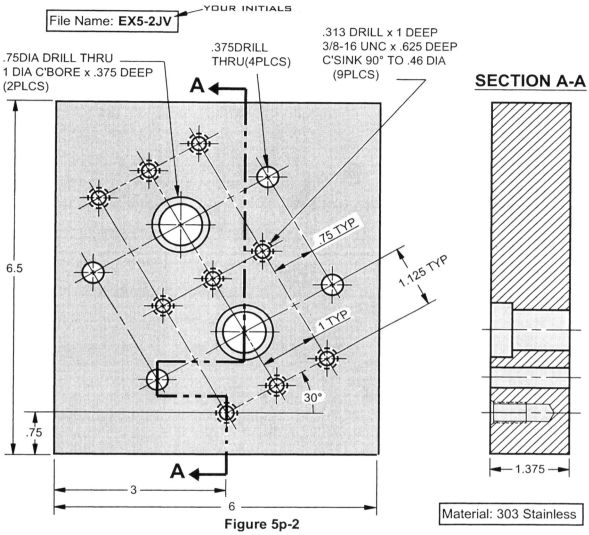

Figure 5p-2

Material: 303 Stainless

PROCESS PLAN 5P-2

No.	Operation	Tooling
1	Center Drill x .2 deep(all holes)	1/8 Center drill
2	Peck drill x 1 Deep(9PLCS)	5/16 Drill
3	Tap x .625 Deep(9 PLCS)	.375-16UNC Tap
4	C'sink 90° to .46 dia(9 plcs) [Use Depth Calculator]	3/4 Chamfer Mill
5	Peck drill thru(4PLCS)	3/8 Drill
6	Peck drill thru(2PLCS)	3/4 Drill
7	Circle Mill x .375 Deep(2PLCS)	1 Flat Endmill

5-3) a) Create the *wireframe* CAD model(top view *only*). Set up the stock and the material.
   b) Generate the required drill tool paths for executing PROCESS PLAN 5P-3
   c) Backplot and verify the tool paths
   d) Direct Mastercam to generate the word address part program

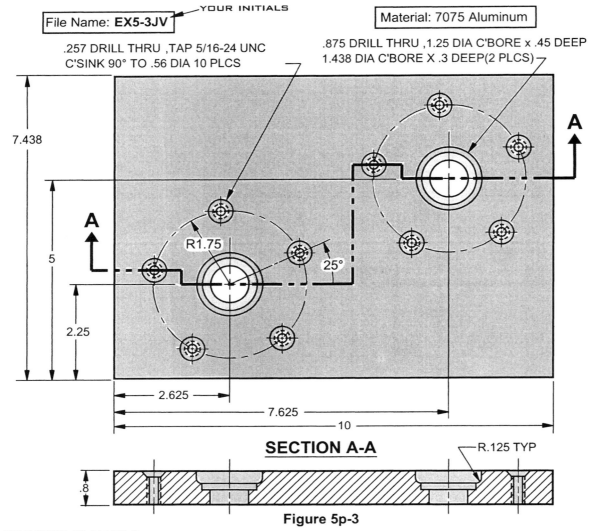

YOUR INITIALS

File Name: **EX5-3JV**

Material: 7075 Aluminum

.257 DRILL THRU ,TAP 5/16-24 UNC
C'SINK 90° TO .56 DIA 10 PLCS

.875 DRILL THRU ,1.25 DIA C'BORE x .45 DEEP
1.438 DIA C'BORE X .3 DEEP(2 PLCS)

7.438

5

2.25

R1.75

25°

2.625

7.625

10

SECTION A-A

R.125 TYP

.8

Figure 5p-3

PROCESS PLAN 5P-3

No.	Operation	Tooling
1	CENTER DRILL x .2 DEEP(ALL HOLES)	1/8 CENTER DRILL
2	PECK DRILL THRU(2PLCS)	7/8 DRILL
3	CIRCLE MILL x .45 DEEP (2PLCS)	3/4 FLAT ENDMILL
4	CIRCLE MILL x .3 DEEP (2PLCS)	3/4 BULL ENDMILL .125R
5	PECK DRILL THRU(10PLCS)	LTR 'F' DRILL
6	TAP THRU(10PLCS)	.313-24UNC TAP
7	C'SINK 90° TO .56 DIA(10 PLCS) [Use Depth Calculator]	1 CHAMFER MILL

5-4) a) Get the CAD model from the CD at the back of this text located in the folder ☐ CHAPTER5.
   b) Setup the stock and the material
   c) Generate the required drill tool paths for executing PROCESS PLAN 5P-4
   d) Backplot and verify the tool paths
   e) Direct Mastercam to generate the word address part program

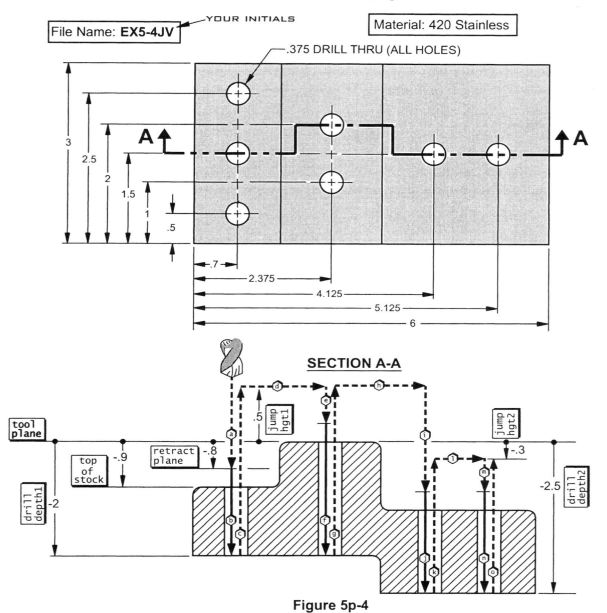

Figure 5p-4

PROCESS PLAN 5P-4

No.	Operation	Tooling
1	CENTER DRILL x .2 DEEP(ALL HOLES)	1/8 CENTER DRILL
2	PECK DRILL THRU(ALL HOLES)	3/8 DRILL

5-5) a) Create the *wireframe* CAD model(top view *only*). Set up the stock and the material.
    b) Generate the required drill tool paths for executing PROCESS PLAN 5P-5
    c) Backplot and verify the tool paths

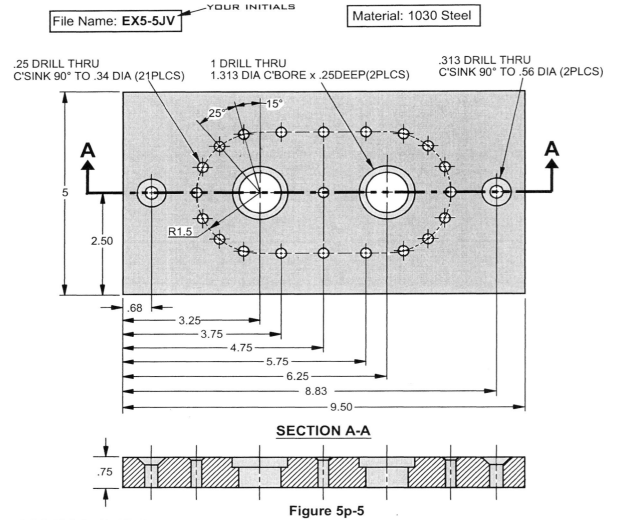

Figure 5p-5

PROCESS PLAN 5P-5

No.	Operation	Tooling
1	CENTER DRILL x .2 DEEP(ALL HOLES)	1/8 CENTER DRILL
2	PECK DRILL THRU(21PLCS)	1/4 DRILL
3	C'SINK 90° TO .34 DIA(21PLCS) [Use Depth Calculator]	1 CHAMFER MILL
4	PECK DRILL THRU(2PLCS)	1 DRILL
5	CIRCLE MILL x .25 DEEP (2PLCS)	1 FLAT ENDMILL
6	PECK DRILL THRU(2PLCS)	5/16 DRILL
7	C'SINK 90° TO .56 DIA(2PLCS) [Use Depth Calculator]	1 CHAMFER MILL

The hole design has been changed as shown in Figure 5p-6.

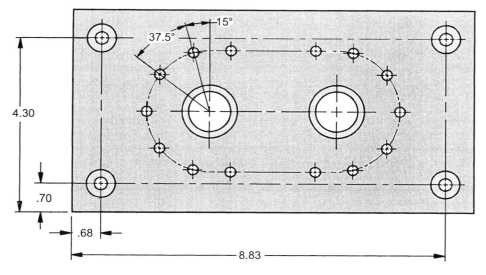

Figure 5p-6

d) Use the Geometry icon in the Operations Manager to delete the appropriate drill points.

e) Erase the required hole geometry

f) Add the new hole geometry

g) Use the Geometry icon in the Operations Manager to add the appropriate drill points.

h) Use the Regenerate all selected operations button in the Operations manager to regenerate the edited tool paths

i) Backplot and verify the edited tool paths

CHAPTER - 6

PROFILING AND POCKETING IN 2D SPACE

6-1 Chapter Objectives

After completing this chapter you will be able to:

1. Know the various methods of chaining 2D entities for contour machining

2. State the common problems in 2D chaining and the methods of correction.

3. Understand the meaning of the parameters in the Contour(2D) dialog box .

4. Know how to specify the chaining, tooling and contouring parameters required to contour machine a sample part.

5. Understand the meaning of the parameters in the Pocket(Standard) dialog .box .

6. Know how to specify the chaining, tooling and pocketing parameters required to machine a pocket in a part.

 ## 6-2 Creating a 2D Contour for Profiling

A set of points, lines, arcs and splines chained together into a *single* curve is called a machining *contour*. When *Mastercam* is directed to contour a part it does so by moving the cutting tool along the chain.

The various methods of creating a contour are described below.

The operator clicks the chain **start** point on an entity. Chaining will proceed **from the start** point in the **direction toward the midpoint** of the entity. **All** entities connected **end to end will be included** in the chain. The point at the end of the last entity in the chain is the **endpoint.** For **closed** chains the start and endpoints coincide.

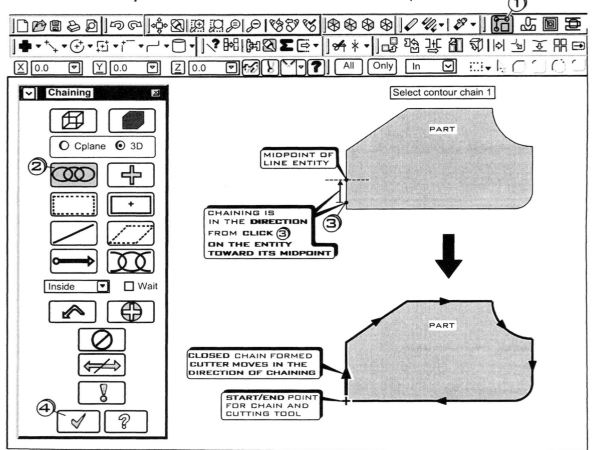

➤ Click ① the Contour Toolpaths button

➤ Click ② on the chain button

Select contour chain 1

➤ Click ③ on the entity to specify the chain **start/end** point

➤ Click ④ the OK button

Options gives the operator **control** over several important chaining parameters.
Entity mask, for example, allows the operator to selectively **ignore or include**
point, line, arc and spline entities when executing chaining.

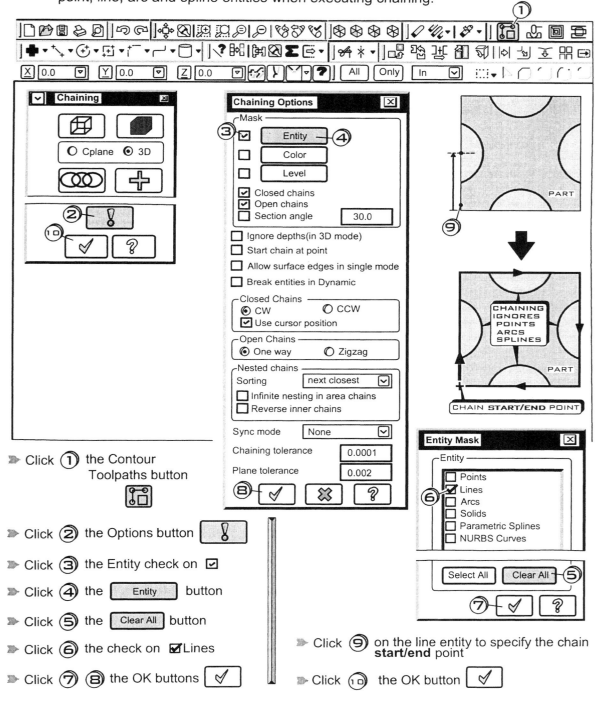

➤ Click ① the Contour
Toolpaths button

➤ Click ② the Options button

➤ Click ③ the Entity check on ☑

➤ Click ④ the [Entity] button

➤ Click ⑤ the [Clear All] button

➤ Click ⑥ the check on ☑Lines

➤ Click ⑦ ⑧ the OK buttons ✓

➤ Click ⑨ on the line entity to specify the chain
start/end point

➤ Click ⑩ the OK button ✓

 Contour Toolpath ▸ *PARTIAL* ◂

Partial creates an *open* chain with *two* mouse clicks. The *first* click specifies the *start* point of the chain. The *second* click the *endpoint* of the chain.

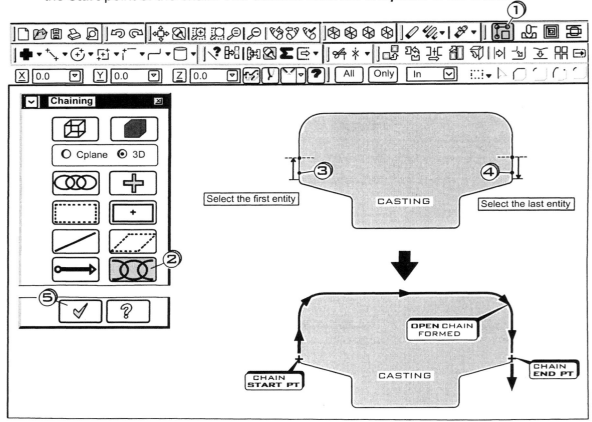

➤ Click ① the Contour Toolpaths button

➤ Click ② on the partial button

Select the first entity

➤ Click ③ to specify the chain *start point*

Select the last entity

➤ Click ④ to specify the chain *end point*

➤ Click ⑤ the OK button

Reverse gives the operator control over the *chaining direction*

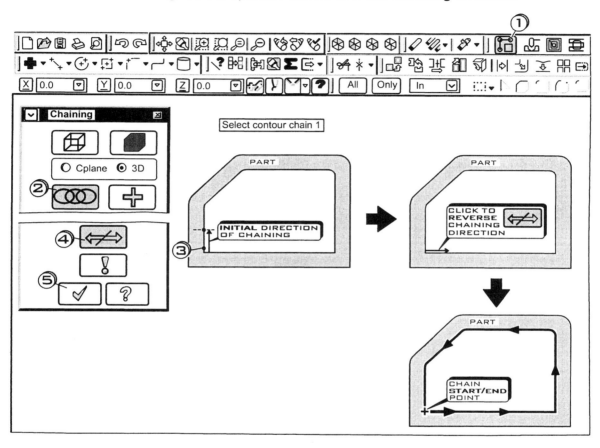

Click ① the Contour Toolpaths button

Click ② on the chain button

Select contour chain 1

Click ③ on the entity to specify the chain **start/end** point

Click ④ on the reverse button

Click ⑤ the OK button

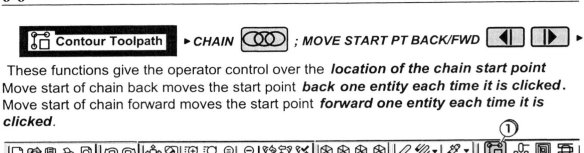

These functions give the operator control over the *location of the chain start point*
Move start of chain back moves the start point *back one entity each time it is clicked*.
Move start of chain forward moves the start point *forward one entity each time it is clicked*.

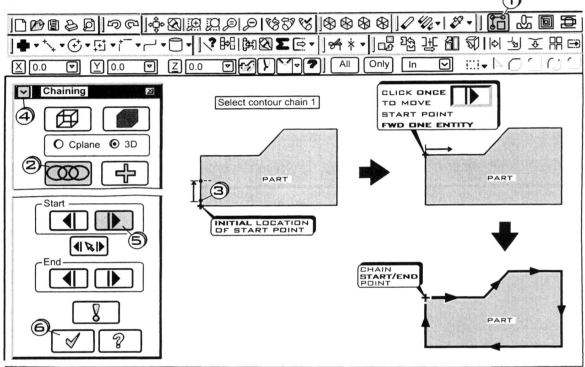

➤ Click ① the Contour Toolpaths button

➤ Click ② on the chain button

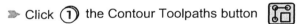

➤ Click ③ on the entity to specify the chain **start/end** point

➤ Click ④ on the Expand button

➤ Click ⑤ on the Move start of chain forward button   to move the *start point* of the chain *forward one* entity.

➤ Click ⑥ the OK button

Unselect **cancels** the last chain created. Click repeatedly to cancel single chains in sequence. Unselects all chains that were created by using a window or polygon chaining. Unselect must be used *before* the OK button is clicked.

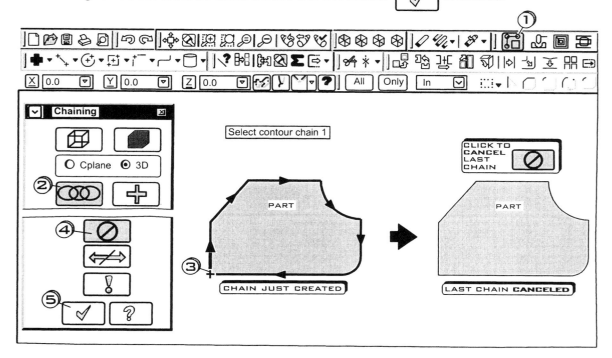

➤ Click ① the Contour Toolpaths button

➤ Click ② on the chain button

> Select contour chain 1

➤ Click ③ on the entity to specify the chain **start/end** point

➤ Click ④ on the Unselect button to cancel the *last* chain created

➤ Click ⑤ the OK button

Window creates *one or more chaines* from the entities enclosed *within* a clicked *window*. The *first corner* of the window clicked specifies the *search point* and determines the *chaining order*.

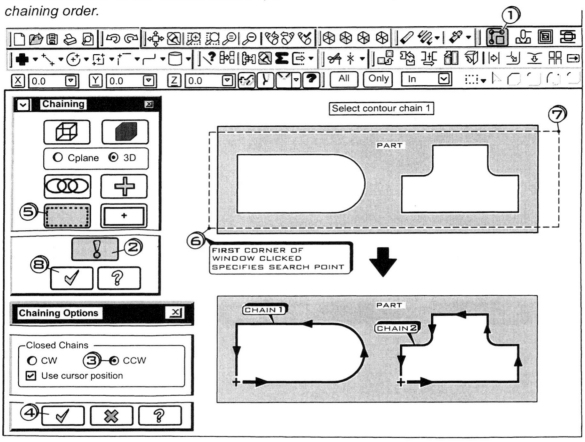

➤ Click ① the Contour Toolpaths button

➤ Click ② the Options button

➤ Click ③ the CCW radio button ⊙   to climb mill on inside cuts.

➤ Click ④ the OK button

➤ Click ⑤ the Window button

       | Select contour chain 1 |

➤ Click ⑥ the first window corner(*search point*) ; ⑦ the other window corner.

➤ Click ⑧ the OK button

Contour Toolpath ▶ *POLYGON* ◀

The operator creates a *custom* window *especially shaped* to fit around or pass through the entities forming one or more chains..

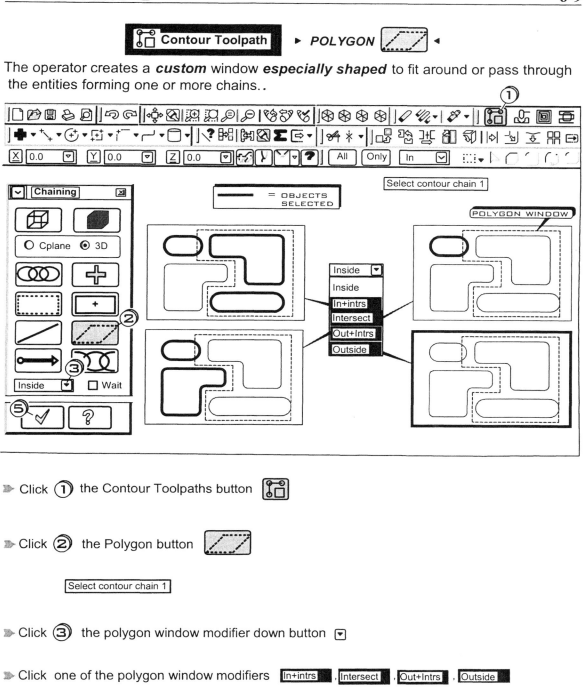

▶ Click ① the Contour Toolpaths button

▶ Click ② the Polygon button

Select contour chain 1

▶ Click ③ the polygon window modifier down button ▼

▶ Click one of the polygon window modifiers In+intrs , Intersect , Out+Intrs , Outside

▶ Click the corners of the polygon wingow

▶ Click ⑤ the OK button

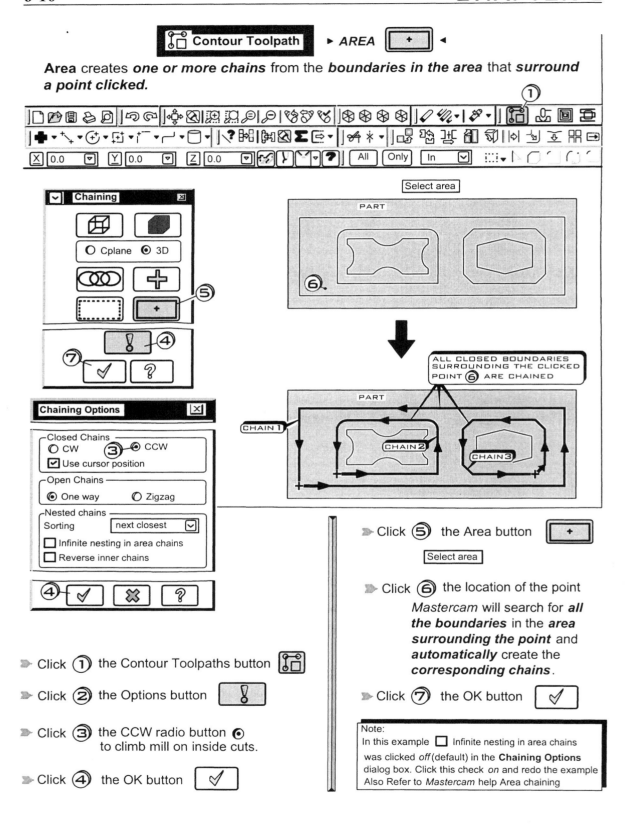

Contour Toolpath ▸ *AREA*

Area creates *one or more chains* from the *boundaries in the area* that *surround a point clicked.*

Select area

PART

ALL CLOSED BOUNDARIES SURROUNDING THE CLICKED POINT ⑥ ARE CHAINED

PART

CHAIN 1

CHAIN 2

CHAIN 3

Chaining

○ Cplane ◉ 3D

⑤

⑦

!④

Chaining Options

┌ Closed Chains ─────────
○ CW   ③ ◉ CCW
☑ Use cursor position

┌ Open Chains ─────────
◉ One way   ○ Zigzag

┌ Nested chains ─────────
Sorting   [next closest ▽]
☐ Infinite nesting in area chains
☐ Reverse inner chains

④

▸ Click ① the Contour Toolpaths button

▸ Click ② the Options button

▸ Click ③ the CCW radio button ◉
     to climb mill on inside cuts.

▸ Click ④ the OK button

▸ Click ⑤ the Area button

     Select area

▸ Click ⑥ the location of the point

     Mastercam will search for *all the boundaries* in the *area surrounding the point* and *automatically* create the *corresponding chains*.

▸ Click ⑦ the OK button

Note:
In this example ☐ Infinite nesting in area chains
was clicked *off* (default) in the **Chaining Options**
dialog box. Click this check *on* and redo the example
Also Refer to *Mastercam* help Area chaining

Enables the operator to make a **point** entity a **chain**. This chain can then be **joined** to any other chain consisting of lines, arcs or spline entities. A very useful feature when it is necessary to include a **plunge** point or a **move to** point in the chained toolpath.

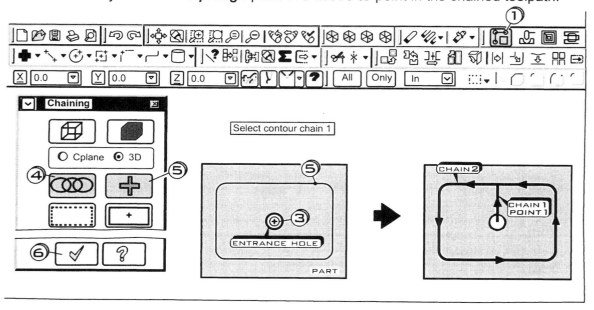

➤ Click ① the Contour Toolpaths button [image]

➤ Click ② the Point button [image]

      | Select contour chain 1 |

➤ Click ③ near the center of the circle

➤ Click ④ on the chain button [image]

      | Select contour chain 2 |

➤ Click ⑤ on the entity to specify the chain **start/end** point. *Mastercam* will then **join** chain1 consisting of point1 entity to chain2.

➤ Click ⑥ the OK button [image]

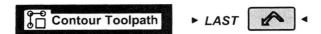

Allows the operator to **_reselect the last_** set of entities chained. This is useful for correcting a toolpath that was chained incorrectly or to execute different machining operations using the same toolpath.

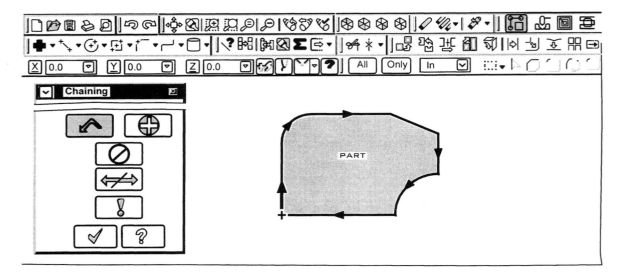

6-3 Common Problems Encountered in 2D Chaining

Below are listed the types of problems most commonly encounteted in attempting to chain 2D contours.

Problem	What's Wrong	Remedy	Result
Chaining stops at first entity of the boundary	All entities in the boundary are **not connected end to end**	Use the Trim 2 function to trim the lines together	Entire boundary is chained
Chaining stops at first entity of the boundary	All entities in the boundary are **not connected end to end. Entities that cross will cause chaining to stop**	Use the Trim 2 function to trim the lines together	Entire boundary is chained
Chaining stops at first entity of the boundary	Branch points where **more than two entities intersect will cause chaining to stop**	Use the Chain function and select Partial The operator must now click **each entity that forms the desired boundary.**	Desired boundary is chained
Chaining stops at a point on an entity	**Duplicate** entities **on top of each other** will cause **erratic chaining**	Use the Delete duplicates function	Desired boundary is chained

6-4 Specifying 2D Contouring Parameters

The Contour(2D) dialog box uses the *chained contour* and other key *contour machining parameters inputted by the operator* to generate the required tool paths. A description of These parameters and their effect on the toolpath is considered in this section.

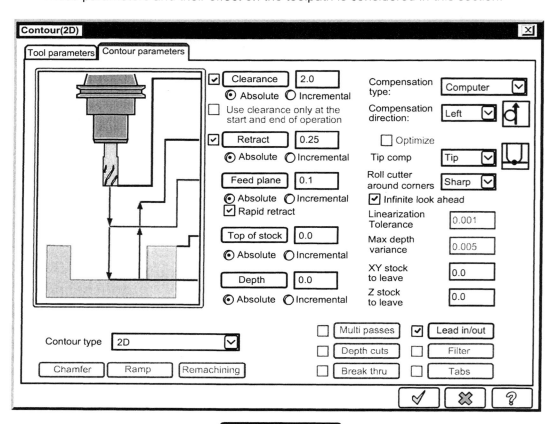

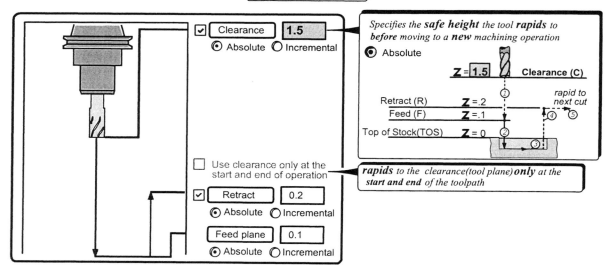

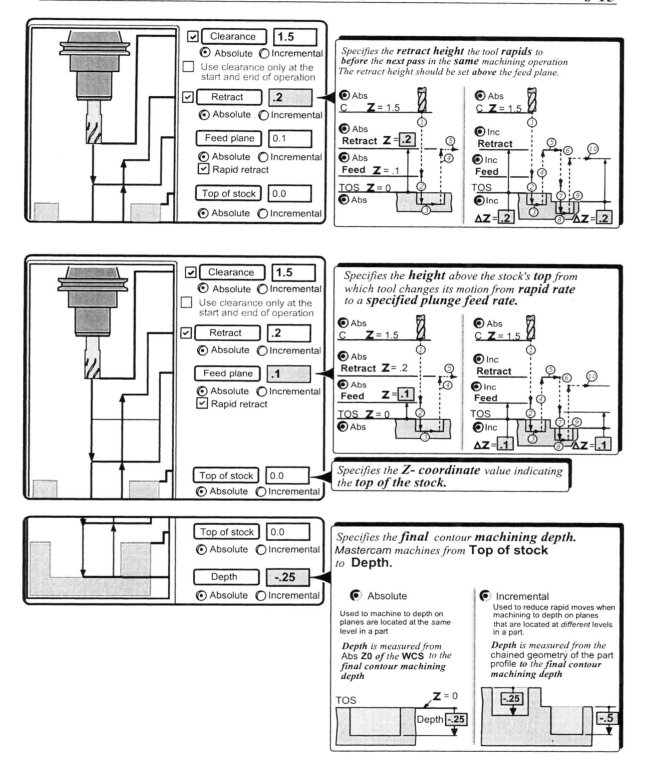

Clearance `1.5`
⊙ Absolute ○ Incremental
☐ Use clearance only at the start and end of operation

☑ **Retract** `.2`
⊙ Absolute ○ Incremental

Feed plane `0.1`
⊙ Absolute ○ Incremental
☑ Rapid retract

Top of stock `0.0`
⊙ Absolute ○ Incremental

*Specifies the **retract height** the tool **rapids** to before the **next pass** in the **same** machining operation. The retract height should be set **above** the feed plane.*

Clearance `1.5`
⊙ Absolute ○ Incremental
☐ Use clearance only at the start and end of operation

☑ **Retract** `.2`
⊙ Absolute ○ Incremental

Feed plane `.1`
⊙ Absolute ○ Incremental
☑ Rapid retract

Top of stock `0.0`
⊙ Absolute ○ Incremental

*Specifies the **height** above the stock's **top** from which tool changes its motion from **rapid rate** to a **specified plunge feed rate.***

*Specifies the **Z- coordinate** value indicating the **top of the stock.***

Top of stock `0.0`
⊙ Absolute ○ Incremental

Depth `-.25`
⊙ Absolute ○ Incremental

*Specifies the **final** contour **machining depth**. Mastercam machines from **Top of stock** to **Depth**.*

◉ Absolute

Used to machine to depth on planes are located at the *same* level in a part

***Depth** is measured from Abs **Z0** of the **WCS** to the final contour machining depth*

◉ Incremental

Used to reduce rapid moves when machining to depth on planes that are located at *different* levels in a part.

***Depth** is measured from the chained geometry of the part profile to the final contour machining depth*

DEPTH CUTS

The depth cuts parameters allow the operator to control the *number* and *depth* of *roughing* and *finishing* passes in the Z direction

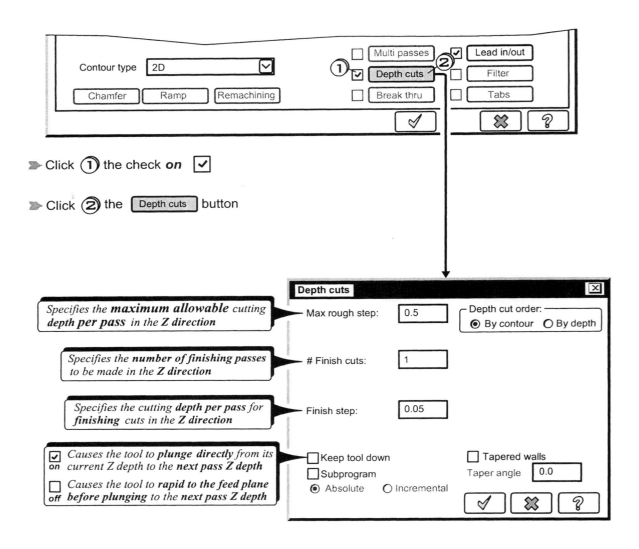

➤ Click ① the check *on* ☑

➤ Click ② the [Depth cuts] button

Specifies the **maximum allowable** cutting *depth per pass* in the Z direction ◀ — Max rough step: 0.5

Specifies the **number of finishing passes** to be made in the **Z direction** ◀ — # Finish cuts: 1

Specifies the cutting **depth per pass** for *finishing* cuts in the **Z direction** ◀ — Finish step: 0.05

☑ Causes the tool to **plunge directly from its**
on *current Z depth to the* **next pass Z depth**

☐ Causes the tool to **rapid to the feed plane**
off *before plunging to the next pass Z depth*

Depth cut order:
⦿ By contour   ○ By depth

☐ Keep tool down
☐ Subprogram
⦿ Absolute   ○ Incremental

☐ Tapered walls
Taper angle   0.0

Case-a : ***No finishing cuts are specified***. In this case *Mastercam* will ***divide the total depth (Top of stock+ Depth)*** such that an ***equal number of roughing passes*** is made. Depth per pass ***will not exceed*** current Max rough step value.

Case-a : ***No finishing cuts specified***

EXAMPLE 6-1

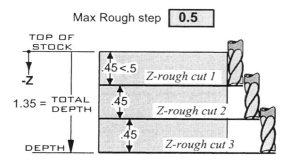

Figure 6-1

Case-b: ***Finishing cuts are specified.*** In this case *Mastercam* will ***divide the total depth (Top of stock+ Depth)*** such that an ***equal number*** of ***shallower roughing passes*** is made. This is followed by the specified finishing passes.

Case-b : ***Finishing cuts specified***

EXAMPLE 6-2

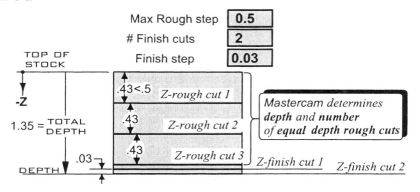

Figure 6-2

MULTI PASSES

The multi passes parameters allow the operator to control the **number** and **stepover distances** of **roughing** and **finishing** passes in the X and Y directions

Contour type	2D	③ ☑ Multi passes ④ ☑ Lead in/out
		☑ Depth cuts ☐ Filter
Chamfer Ramp Remachining		☐ Break thru ☐ Tabs

➤ Click ③ the check **on** ☑

➤ Click ④ the **Multi passes** button

Multi Passes

Roughing passes

*Specifies the **number of roughing passes** to be made in the **XY direction*** — Number `1`

*Specifies the **stepover distance per pass** for **roughing cuts** in the **XY direction*** — Spacing `0.1`

Finishing passes

*Specifies the **number of finishing passes** to be made in the **XY direction*** — Number `1`

*Specifies the **stepover distance per pass** for **finishing cuts** in the **XY direction*** — Spacing `0.05`

⦿ Final depth *Stepover finish passes are made at the **final Z-depth only***
⦿ All depths *Stepover finish passes are be made at after **every Z-depth pass***

Machine finish passes at
○ Final depth ⦿ All depths

☑ *Tool **moves directly** from its current position to the **next finish pass XY***
on   *stepover*

☐ Keep tool down

☐ *Tool **rapids to the feed plane before moving** to the next **finish pass XY***
off *stepover. This option is **not** to be used when machining **open** contours. If the tool is retracted when moving from the end of the open contour to its start point gouging may occur.*

Determination of Spacing and Number of cuts for Roughing passes

Spacing between cuts ≈ (.60 to .75) x Tool Dia

Number of cuts ≈ XY Stock to remove/ Spacing

EXAMPLE 6-3   Tool Dia =.5in ; XY Stock to remove = .65in

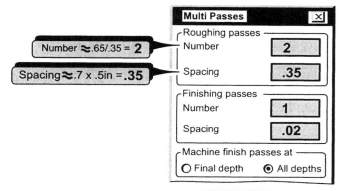

Number ≈.65/.35 = **2**

Spacing ≈.7 x .5in = **.35**

Machine finish passes at ⦿ **Final depth**

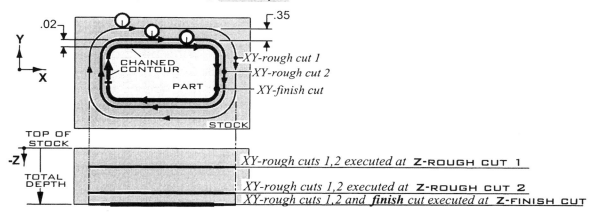

Machine finish passes at ⦿ **All depths**

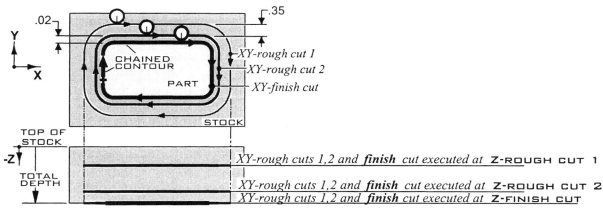

Figure 6-3

CUTTER COMPENSATION

The compensation parameters allow the operator to specify **whether or not** the cutter is positioned **offset** from the **chained contour it follows**. When compensation is set to **OFF** the cutter **center** moves **along** the contour in the direction of chaining. When compensation is set to **ON** Mastercam **automatically offsets** the cutter by its **radius** as it moves along the contour in the direction of chaining.

The parameters also control the **offset direction**(LEFT or RIGHT) with respect to the contour.

The tip compensation parameters specify whether the tool is to be moved by its **tip** or **center** point. This parameter only affects such tools as ball and bull end mills, etc.

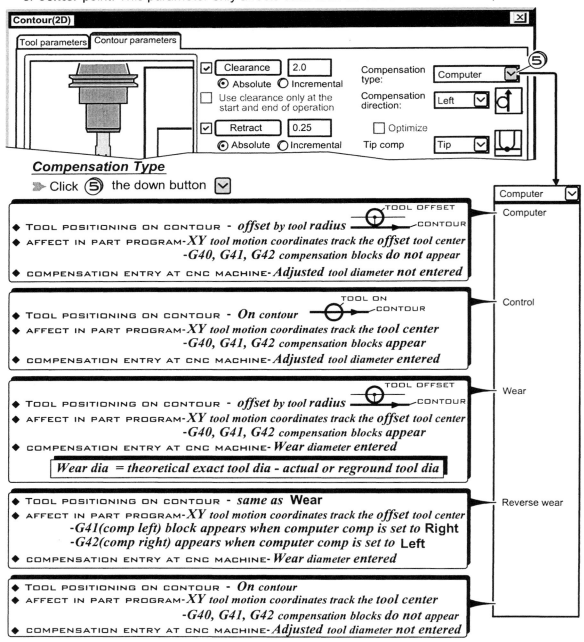

Compensation Type

➤ Click ⑤ the down button ▽

♦ TOOL POSITIONING ON CONTOUR - *offset by tool radius* ⎯ TOOL OFFSET / CONTOUR
♦ AFFECT IN PART PROGRAM-*XY tool motion coordinates track the offset tool center*
  -G40, G41, G42 compensation blocks do not appear
♦ COMPENSATION ENTRY AT CNC MACHINE-*Adjusted tool diameter not entered*

♦ TOOL POSITIONING ON CONTOUR - *On contour* ⎯ TOOL ON / CONTOUR
♦ AFFECT IN PART PROGRAM-*XY tool motion coordinates track the tool center*
  -G40, G41, G42 compensation blocks appear
♦ COMPENSATION ENTRY AT CNC MACHINE-*Adjusted tool diameter entered*

♦ TOOL POSITIONING ON CONTOUR - *offset by tool radius* ⎯ TOOL OFFSET / CONTOUR
♦ AFFECT IN PART PROGRAM-*XY tool motion coordinates track the offset tool center*
  -G40, G41, G42 compensation blocks appear
♦ COMPENSATION ENTRY AT CNC MACHINE-*Wear diameter entered*

Wear dia = theoretical exact tool dia - actual or reground tool dia

♦ TOOL POSITIONING ON CONTOUR - *same as* **Wear**
♦ AFFECT IN PART PROGRAM-*XY tool motion coordinates track the offset tool center*
  -G41(comp left) block appears when computer comp is set to **Right**
  -G42(comp right) appears when computer comp is set to **Left**
♦ COMPENSATION ENTRY AT CNC MACHINE-*Wear diameter entered*

♦ TOOL POSITIONING ON CONTOUR - *On contour*
♦ AFFECT IN PART PROGRAM-*XY tool motion coordinates track the tool center*
  -G40, G41, G42 compensation blocks do not appear
♦ COMPENSATION ENTRY AT CNC MACHINE-*Adjusted tool diameter not entered*

Note: In many cases the best option is to choose **WEAR** . This allows the setup person to input the **difference** between the theoretical **exact** cutter **diameter** and the **actual diameter** into CNC machine control unit.

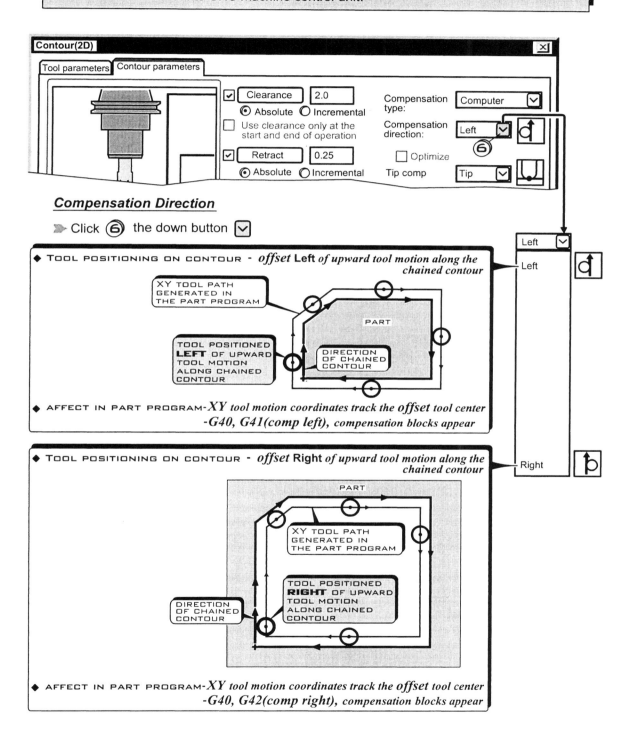

Compensation Direction

➤ Click ⑥ the down button ☑

♦ TOOL POSITIONING ON CONTOUR - *offset* **Left** *of upward tool motion along the chained contour*

XY TOOL PATH GENERATED IN THE PART PROGRAM

PART

TOOL POSITIONED **LEFT** OF UPWARD TOOL MOTION ALONG CHAINED CONTOUR

DIRECTION OF CHAINED CONTOUR

♦ AFFECT IN PART PROGRAM-*XY tool motion coordinates track the offset tool center*
-G40, G41(comp left), compensation blocks appear

♦ TOOL POSITIONING ON CONTOUR - *offset* **Right** *of upward tool motion along the chained contour*

PART

XY TOOL PATH GENERATED IN THE PART PROGRAM

TOOL POSITIONED **RIGHT** OF UPWARD TOOL MOTION ALONG CHAINED CONTOUR

DIRECTION OF CHAINED CONTOUR

♦ AFFECT IN PART PROGRAM-*XY tool motion coordinates track the offset tool center*
-G40, G42(comp right), compensation blocks appear

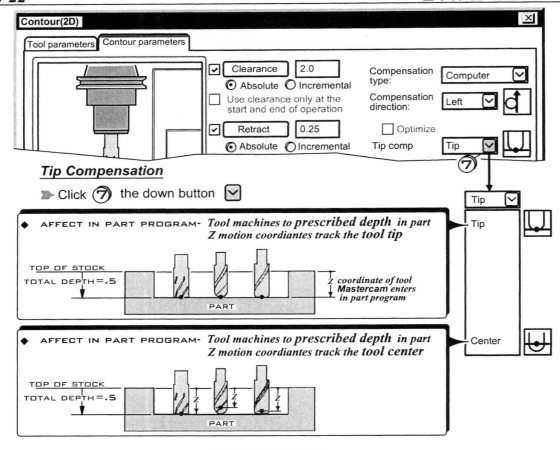

Tip Compensation

➧ Click ⑦ the down button ☑

INFINITE LOOK AHEAD

When enabled, the Infinite look ahead parameter instructs Mastercam to search the **entire chained contour** for any **self intersections** that can cause **tool gouging**. The determination is made based on the current **offset distance** and **cutter compensation**. Mastercam **automatically adjusts** the **tool path** to **prevent gouging**.

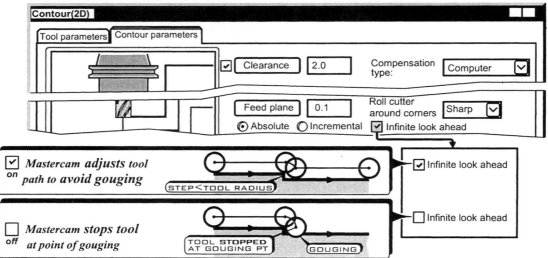

STOCK TO LEAVE

The Stock to leave parameters enable the operator to specify the ammount of **stock to be left after all roughing and finishing pases for a particular machining operation have been made**. The remaining stock can then be machined via a **new machining operation**.

In the XY direction: Stock **remains** on the **left side** if Compensation direction is set to **LEFT**

Stock **remains** on the **right side** if Compensation direction is set to **RIGHT**

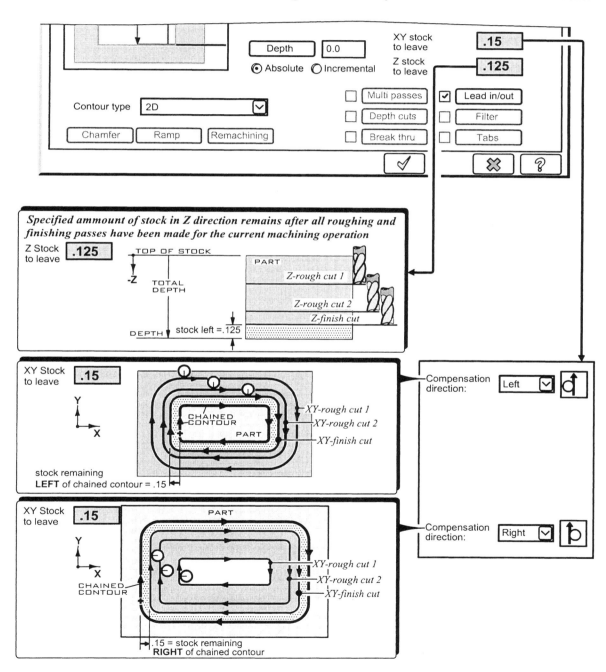

Specified ammount of stock in Z direction remains after all roughing and finishing passes have been made for the current machining operation

ROLL CUTTER AROUND CORNERS

The Roll cutter around corners parameters allows the operator to choose whether the cutter is to experience a *sudden change* in motion at a *corner* where two lines meet or have a *smoother motion* by following an *arc path(rolling)* around the corner.

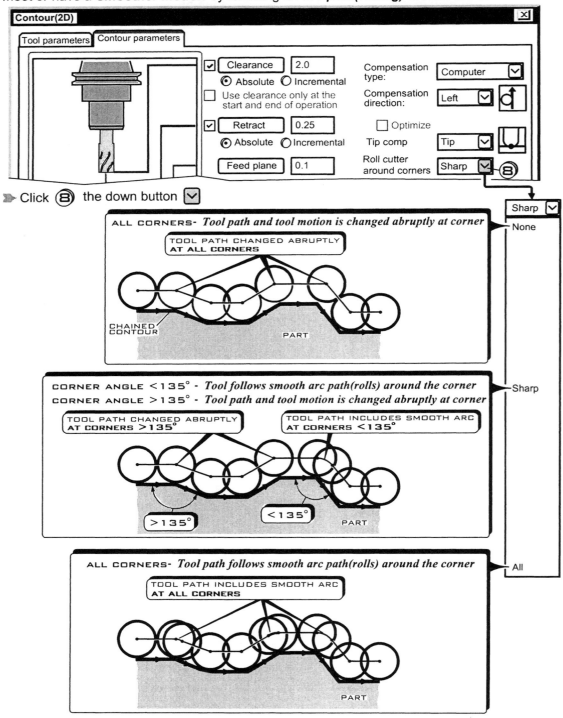

➤ Click ⑧ the down button 🔽

LEAD IN/OUT

The Lead in/out parameters enable the operator to control the ***tool's smooth entry
path into and out of the part*** . *Mastercam* features several choices for in/out tool
paths: ***straight lines, lines and arcs or arcs***.

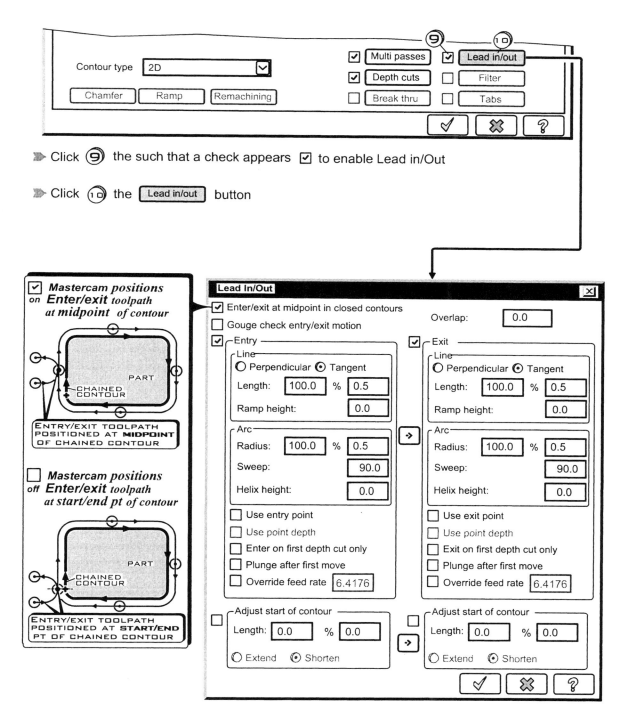

➤ Click ⑨ the such that a check appears ☑ to enable Lead in/Out

➤ Click ⑩ the [Lead in/out] button

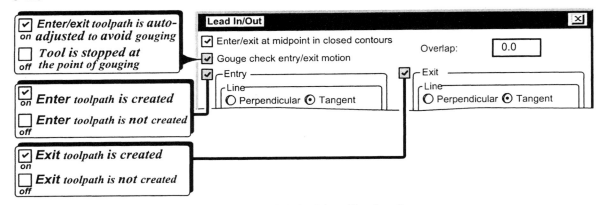

EXAMPLE 6-4

Enter: Line / Exit: Line Toolpaths

The contour shown in Figure 6-4 is to be machined using a 1/4 Flat end mill.
Direct *Mastercam* to create the required Line-Line Enter/exit tool paths.

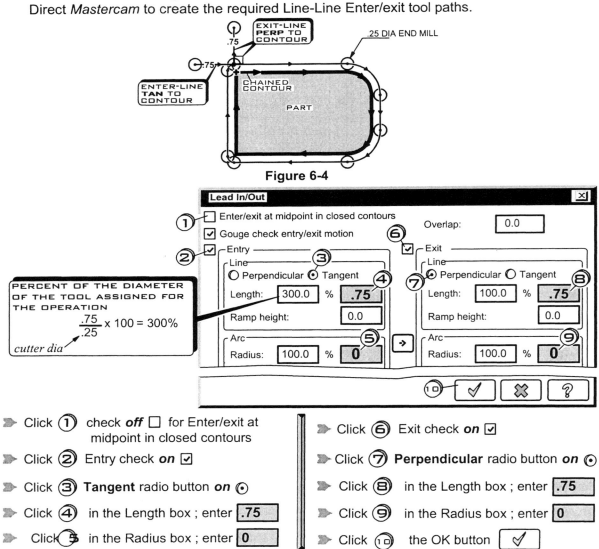

Figure 6-4

➤ Click ① check *off* ☐ for Enter/exit at midpoint in closed contours

➤ Click ② Entry check *on* ☑

➤ Click ③ **Tangent** radio button *on* ⊙

➤ Click ④ in the Length box ; enter `.75`

➤ Click ⑤ in the Radius box ; enter `0`

➤ Click ⑥ Exit check *on* ☑

➤ Click ⑦ **Perpendicular** radio button *on* ⊙

➤ Click ⑧ in the Length box ; enter `.75`

➤ Click ⑨ in the Radius box ; enter `0`

➤ Click ⑩ the OK button ✓

Enter: Line-Arc / Exit: Line-Arc Toolpaths

EXAMPLE 6-5

Direct *Mastercam* to create the required Line-Arc Enter/exit tool paths for the part shown in Figure 6-5.

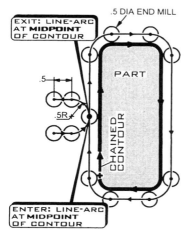

Figure 6-5

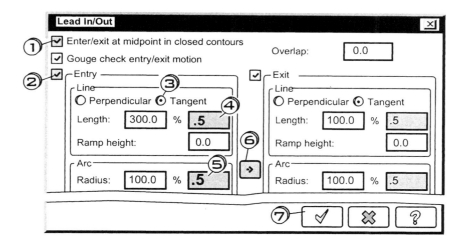

➤ Click ① check **on** ☑ for Enter/exit at midpoint in closed contours

➤ Click ② check **on** ☑

➤ Click ③ **Tangent** radio button **on** ◉

➤ Click ④ in the Length box ; input **.5**

➤ Click ⑤ in the Radius box ; input **.5**

➤ Click ⑥ the **copy Entry parameters** button [→]

➤ Click ⑦ the OK button [✓]

Enter: Line-Arc from Point/ Exit: Line-Arc to Point Toolpaths

EXAMPLE 6-6

Direct *Mastercam* to create the required Line-Arc from/to Point Enter/exit tool paths.

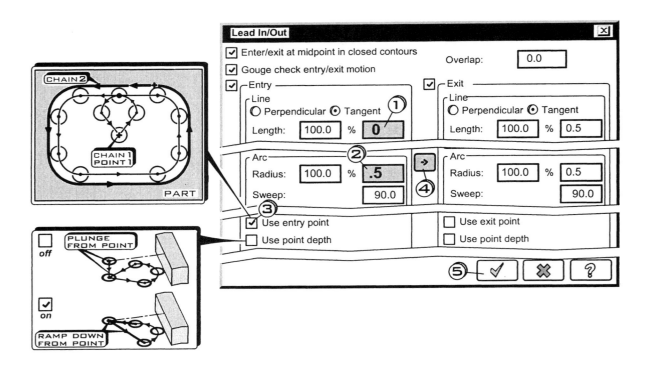

➤ Click ① in the Length box ; enter ⬛ **0**

➤ Click ② in the Radius box ; enter **.5**

➤ Click ③ the Use entry point check ☑

➤ Click ④ the *copy Entry parameters* button ➡

➤ Click ⑤ the OK button ✓

The additional parameters in the Lead In/Out dialog box are described below.

Overlap

Determines how far to shift the tool past it default lead out point on the toolpath.

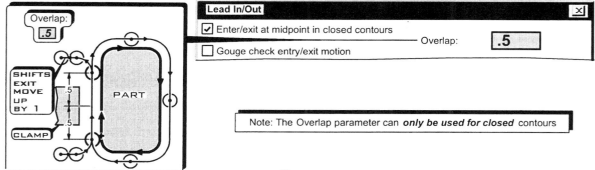

Sweep

The tool enters and exits the toolpath along a sweep angle

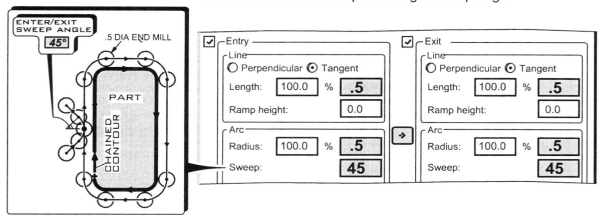

Ramp Height

The tool enters and exits the toolpath along a ramp

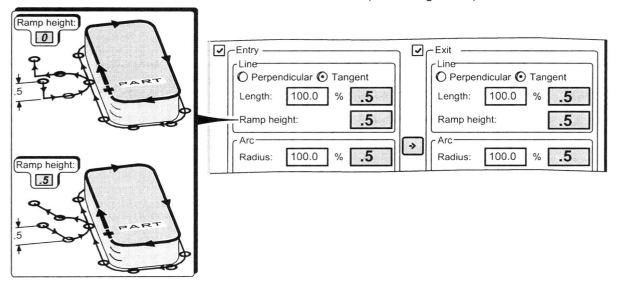

Enter on first depth cut only/Exit on last depth cut only

Applies *lead in move* at start of toolpath for *first* depth cut only. For all subsequent depth cuts tool stays on the contour with no lead in moves. *Lead out* move is made after the *last* depth cut is finished..

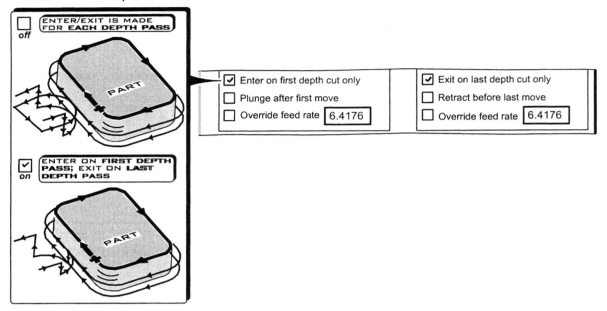

Helix height

Applies *depth to the lead in/out arc turning it into a helix.*

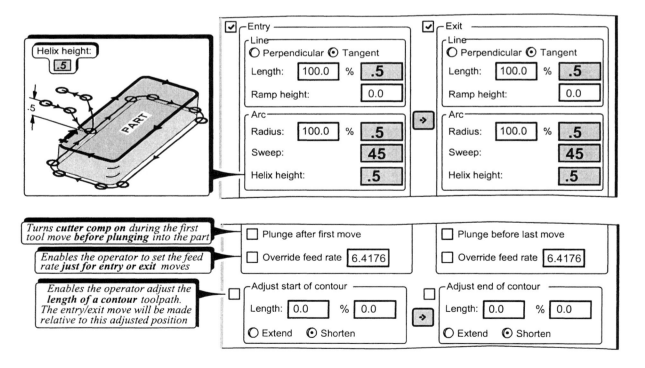

EXAMPLE 6-7

Contour machine the part given in Example 5-2

- Create a machining contour by chaining

- Assign the cutting tool given in PROCESS PLAN 6-1

- Enter the Contouring parameters as listed in PROCESS PLAN 6-1

Material
1030 Steel

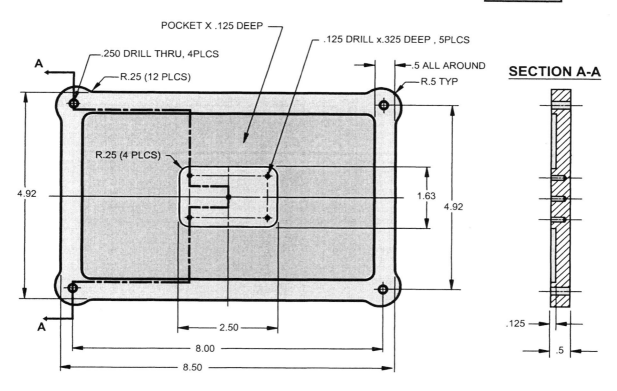

Figure 5-2

PROCESS PLAN 5-2

No.	Operation	Tooling
1	SPOT DRILL X .166 DEEP (4 HOLES)	1/8 SPOT DRILLL
2	PECK DRILL THRU(4 HOLES)	1/4 DRILL
3	SPOT DRILL X .125 DEEP (5 HOLES)	1/32 SPOT DRILL(.31-FLUTE LENGTH;1.5-OAL)
4	PECK DRILL X .325 DEEP	1/8 DRILL
5	PROFILE X .5 DEEP LEAVE .01 FOR FINISH CUTS IN XY AND Z	1/4 FLAT END MILL

A) CREATE A MACHINING CONTOUR BY CHAINING

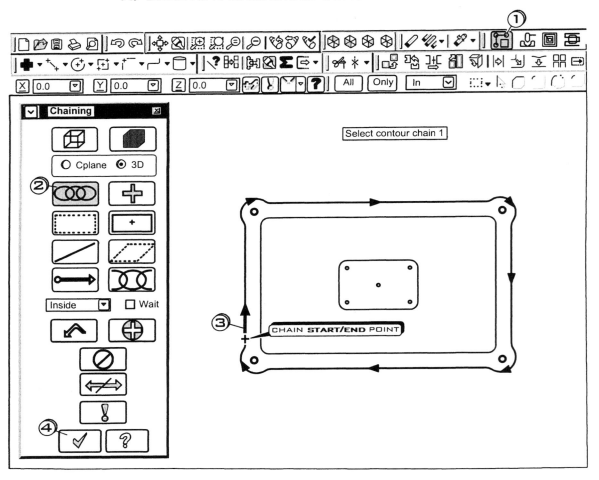

➤ Click ① the Contour Toolpaths button

➤ Click ② on the chain button

Select contour chain 1

➤ Click ③ on the entity to specify the chain **start/end** point

➤ Click ④ the OK button

B) Obtain the 1/4 End Mill tool from the tools library

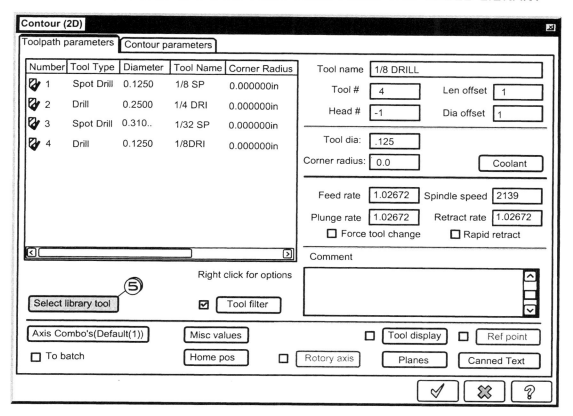

➤ Click ⑤ the [Select library tool] button

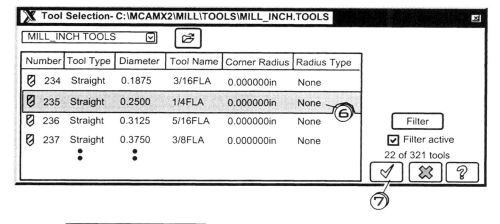

➤ Click ⑥ the [.25 Flat End Mill Tool]

➤ Click ⑦ the OK button [✓]

C) AUTO ASSIGN THE TOOL SPEED/FEED BASED ON THE STOCK MATERIAL

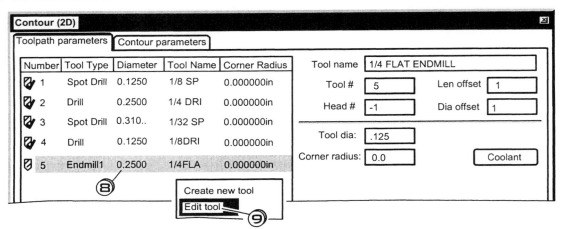

⟩ Click ⑧ on the 1/4 FLA end mill tool

⟩ *Right* Click move the cursor down and Click ⑨ Edit tool

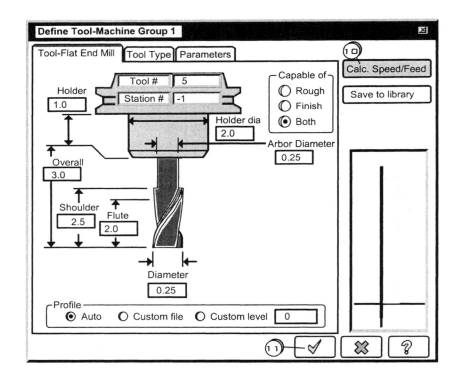

⟩ Click ⑩ the Calc. Speed/Feed button

⟩ Click ⑪ the OK button ✓

D) ENTER THE CONTOUR(2D) PARAMETERS LISTED IN PROCESS PLAN 6-1

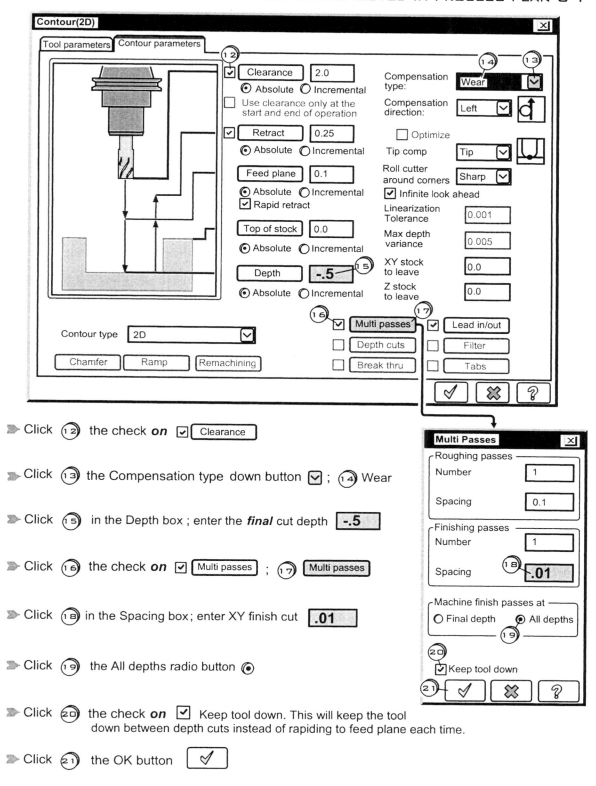

▶ Click ⑫ the check **on** ☑ Clearance

▶ Click ⑬ the Compensation type down button ☑ ; ⑭ Wear

▶ Click ⑮ in the Depth box ; enter the *final* cut depth **-.5**

▶ Click ⑯ the check **on** ☑ Multi passes ; ⑰ Multi passes

▶ Click ⑱ in the Spacing box; enter XY finish cut **.01**

▶ Click ⑲ the All depths radio button ⦿

▶ Click ⑳ the check **on** ☑ Keep tool down. This will keep the tool
down between depth cuts instead of rapiding to feed plane each time.

▶ Click ㉑ the OK button ☑

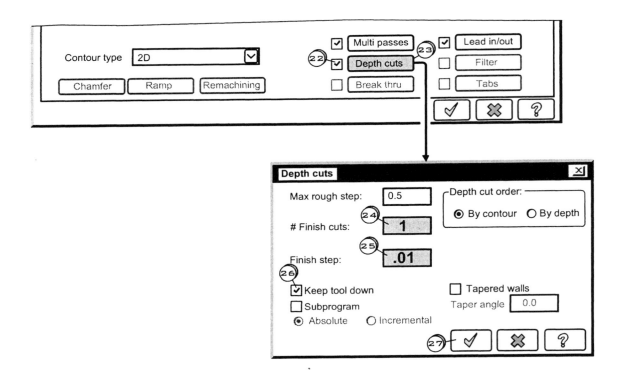

➤ Click ㉒ the check **on** ☑ Depth cuts ; ㉓ Depth cuts

➤ Click ㉔ # Finish cuts box; enter 1

➤ Click ㉕ Finish step box; enter .01

➤ Click ㉖ the check **on** ☑ Keep tool down. This will direct Mastercam to keep the cutter down
between passes instead of retracting to the clearance plane for each pass.

➤ Click ㉗ the OK button ✓

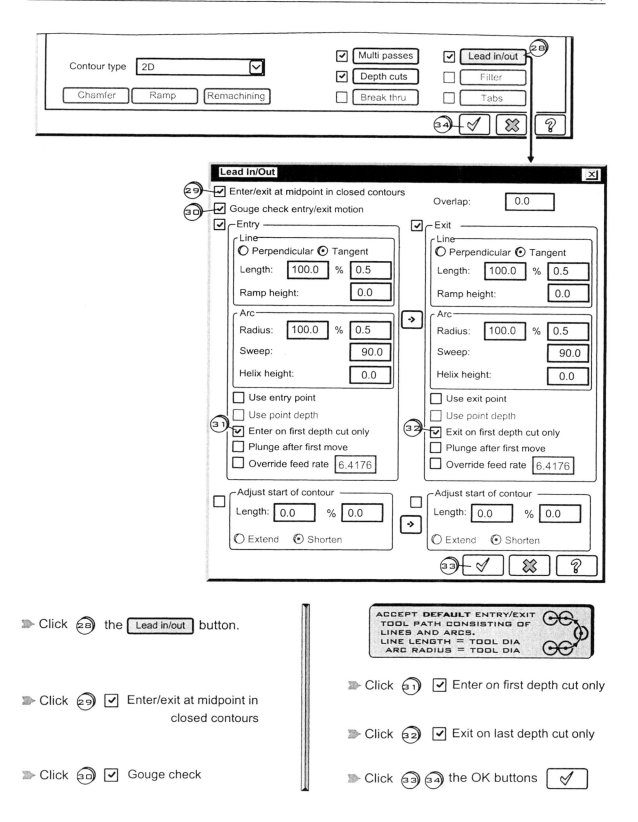

⟫ Click ㉘ the [Lead in/out] button.

⟫ Click ㉙ ☑ Enter/exit at midpoint in closed contours

⟫ Click ㉚ ☑ Gouge check

ACCEPT **DEFAULT** ENTRY/EXIT
TOOL PATH CONSISTING OF
LINES AND ARCS.
LINE LENGTH = TOOL DIA
ARC RADIUS = TOOL DIA

⟫ Click ㉛ ☑ Enter on first depth cut only

⟫ Click ㉜ ☑ Exit on last depth cut only

⟫ Click ㉝ ㉞ the OK buttons ☑

E) Backplot the Contour(2D) operation

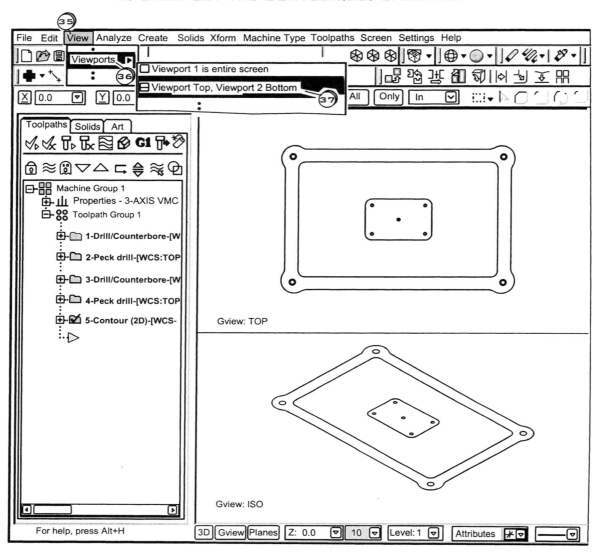

➤ Click ㉟ the View button

➤ Click ㊱  Viewports

➤ Click ㊲  ⊟ Viewport Top, Viewport 2 Bottom

The **Contour(2D)** operation is selected ☑ for backplotting in the Operations Manager dialog box

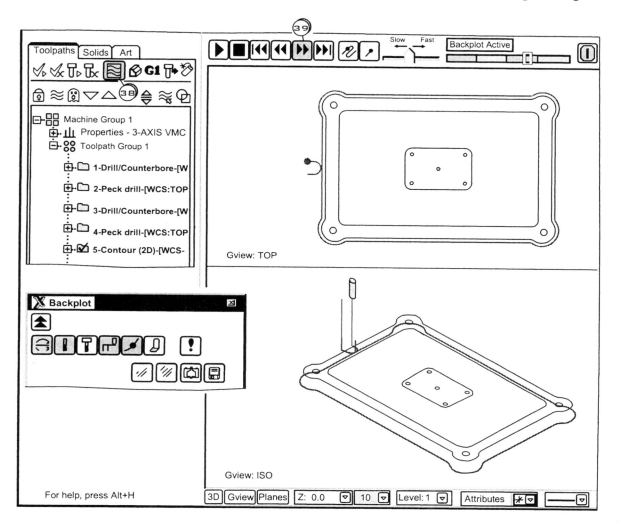

➤ Click ⟨38⟩ the Backplot button ≋

➤ *Continuously* Click ⟨39⟩ the Step forward button ▶▶ to see a *step by step* movement of the tool along the specified contour toolpath.

6-5 Specifying Pocketing Parameters

To initiate pocketing the operator clicks the Pocket Toolpath icon from the 2D Toolpaths toolbar or the Toolpaths drop down menu. The next step is to chain the entities that form the pocket's shape. After this done, *Mastercam X2* will display the **Pocket Toolpath** dialog box as shown below.

A description of the key pocketing parameters in this dialog box and their effect on the toolpath is presented in this section.

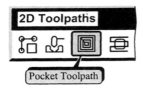

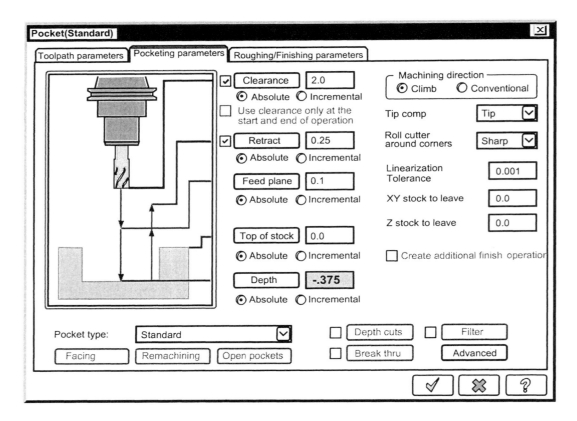

Z- DEPTH PARAMETERS

*Z depth pocketing parameters are **identical** to and carry the **same meaning** as those specified in **2D Contouring.***

NOTE: Depth   -.375   SPECIFIES THE **FINAL POCKET DEPTH**

MACHINING DIRECTION

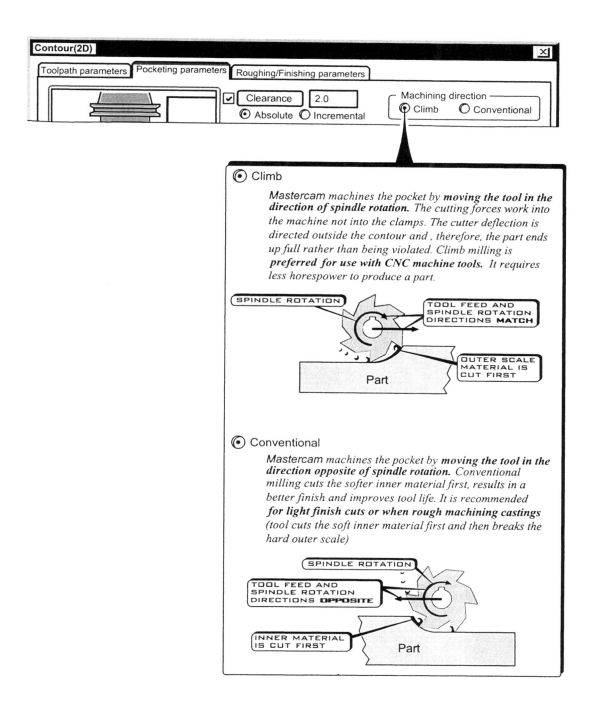

DEPTH CUTS

The depth cuts parameters allow the operator to control the **number** and **depth** of **roughing** and **finishing** passes in the Z direction

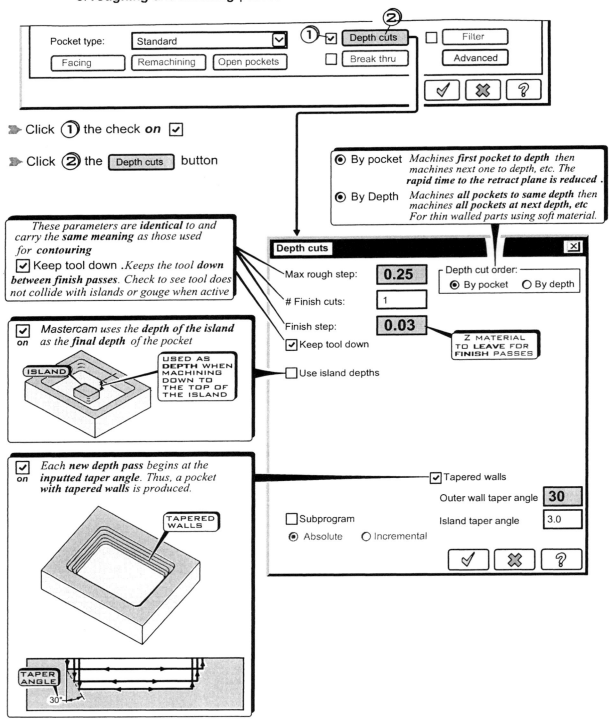

Pocket type: Standard   ①☑ Depth cuts   ②   ☐ Filter

Facing   Remachining   Open pockets   ☐ Break thru   Advanced

➤ Click ① the check **on** ☑

➤ Click ② the [Depth cuts] button

⦿ By pocket *Machines **first pocket to depth** then machines next one to depth, etc. The **rapid time to the retract plane is reduced** .*

⦿ By Depth *Machines **all pockets to same depth** then machines **all pockets at next depth**, etc For thin walled parts using soft material.*

*These parameters are **identical** to and carry the **same meaning** as those used for **contouring***
☑ *Keep tool down .Keeps the tool **down between finish passes**. Check to see tool does not collide with islands or gouge when active*

Depth cuts   ☒

Max rough step: **0.25**

Finish cuts: 1

Finish step: **0.03**

☑ Keep tool down

☐ Use island depths

Depth cut order:
⦿ By pocket   ○ By depth

Z MATERIAL TO **LEAVE** FOR FINISH PASSES

☑ on *Mastercam uses the **depth of the island** as the **final depth** of the pocket*

ISLAND   USED AS **DEPTH** WHEN MACHINING DOWN TO THE TOP OF THE ISLAND

☑ on *Each **new depth pass** begins at the **inputted taper angle**. Thus, a pocket with **tapered walls** is produced.*

☑ Tapered walls
Outer wall taper angle   **30**
Island taper angle   3.0

☐ Subprogram
⦿ Absolute   ○ Incremental

TAPERED WALLS

TAPER ANGLE
30°

POCKET TYPE

The Pocket type parameters specify whether a **Standard, Facing, Island facing**, **Remachining** or **Open** pocket is to be machined.

➤ Click ① the Pocket type down button ☑

➤ Click ② the pocket type to machine

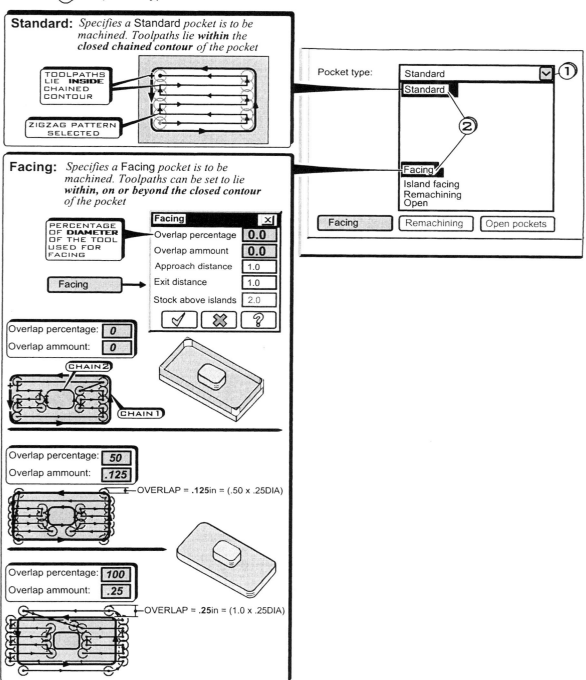

Standard: *Specifies a Standard pocket is to be machined. Toolpaths lie within the closed chained contour of the pocket*

TOOLPATHS LIE **INSIDE** CHAINED CONTOUR

ZIGZAG PATTERN SELECTED

Pocket type: Standard ☑ ①

Standard

Facing

Island facing
Remachining
Open

Facing   Remachining   Open pockets

Facing: *Specifies a Facing pocket is to be machined. Toolpaths can be set to lie within, on or beyond the closed contour of the pocket*

PERCENTAGE OF **DIAMETER** OF THE TOOL USED FOR FACING

Facing

Facing   ✕
Overlap percentage   **0.0**
Overlap ammount   **0.0**
Approach distance   1.0
Exit distance   1.0
Stock above islands   2.0
✓   ✕   ?

Overlap percentage: **0**
Overlap ammount: **0**

CHAIN 2

CHAIN 1

Overlap percentage: **50**
Overlap ammount: **.125**

OVERLAP = **.125**in = (.50 x .25DIA)

Overlap percentage: **100**
Overlap ammount: **.25**

OVERLAP = **.25**in = (1.0 x .25DIA)

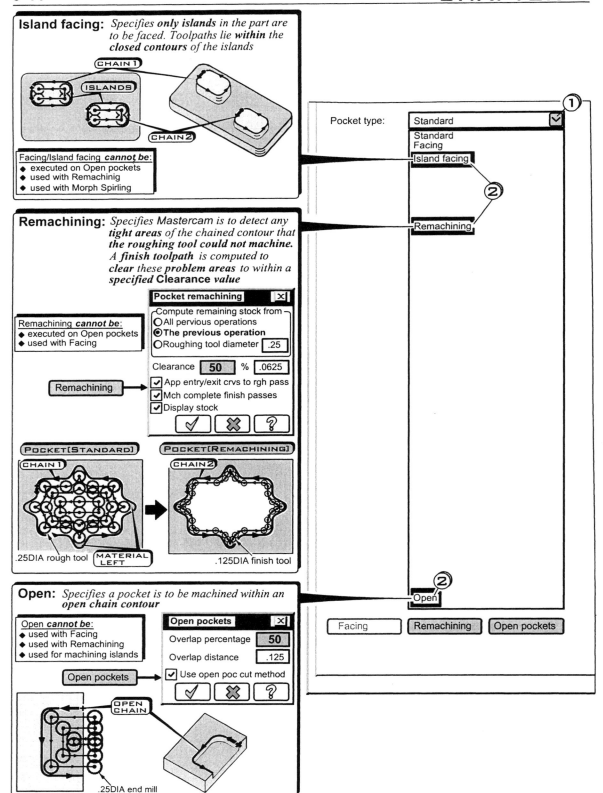

Island facing: *Specifies only islands in the part are to be faced. Toolpaths lie within the closed contours of the islands*

CHAIN 1
ISLANDS
CHAIN 2

Facing/Island facing *cannot be*:
♦ executed on Open pockets
♦ used with Remachinig
♦ used with Morph Spirling

Remachining: *Specifies Mastercam is to detect any tight areas of the chained contour that the roughing tool could not machine. A finish toolpath is computed to clear these problem areas to within a specified Clearance value*

Remachining *cannot be*:
♦ executed on Open pockets
♦ used with Facing

Remachining

Pocket remachining [X]
Compute remaining stock from
○ All pervious operations
◉ **The previous operation**
○ Roughing tool diameter [.25]

Clearance [50] % [.0625]
☑ App entry/exit crvs to rgh pass
☑ Mch complete finish passes
☑ Display stock
[✓] [✗] [?]

POCKET[STANDARD]
CHAIN 1

POCKET[REMACHINING]
CHAIN 2

.25DIA rough tool   MATERIAL LEFT
.125DIA finish tool

Open: *Specifies a pocket is to be machined within an open chain contour*

Open *cannot be*:
♦ used with Facing
♦ used with Remachining
♦ used for machining islands

Open pockets [X]
Overlap percentage [50]
Overlap distance [.125]
☑ Use open poc cut method
[✓] [✗] [?]

Open pockets

OPEN CHAIN

.25DIA end mill

Pocket type: Standard [∨]   ①
Standard
Facing
Island facing
②
Remachining
②
Open
①

Facing | Remachining | Open pockets

ADVANCED PARAMETERS

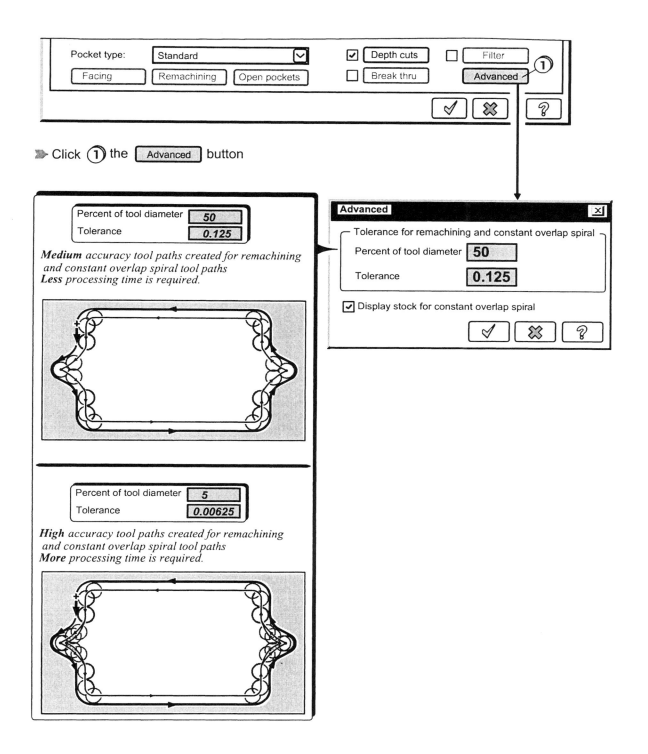

➤ Click ① the Advanced button

☑ Rough   **CUTTING METHOD**
on

*Rough pocketing is clicked **on** by default. Mastercam features **eight** types of toolpaths for*
roughing *pockets,* **Zigzag, One Way**, **Constant Overlap Spiral, Parallel Spiral,**
Parallel Spiral Clean Corners, Morph Spiral, High Speed Spiral *and* **True Spiral.**

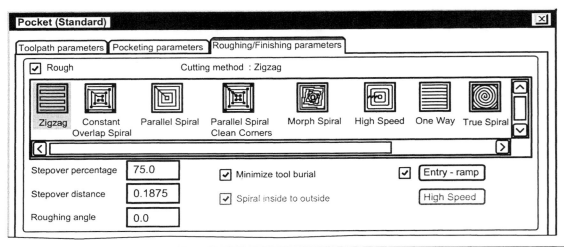

Zigzag: *Rough cuts the pocket by moving the tool back and forth along* **straight lines, switching between climb**
and conventional milling per pass. *The* **orientation** *of the pattern is set by the (0°, 90°, 180°, or 270°)*
value of the **roughing angle.**

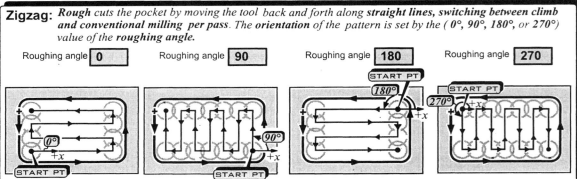

Minimize tool burial: *Used* **only** *with the* **Zigzag** *pattern to machine* **around islands** *with* **small tools**
instead of across areas where tool burial damage typically occurs.
This option may increase machining time but decrease chance of tool damage.

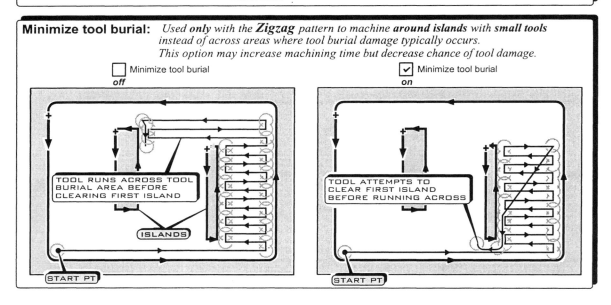

Constant Overlap Spiral: *Executes one roughing pass then determines the next pass based on the remaining area to be cut. This process is repeated until the pocket is cleared. The constant overlap spiral pattern uses more small linear moves to clear more material than the parallel spiral pattern. Machining* **direction is constant**

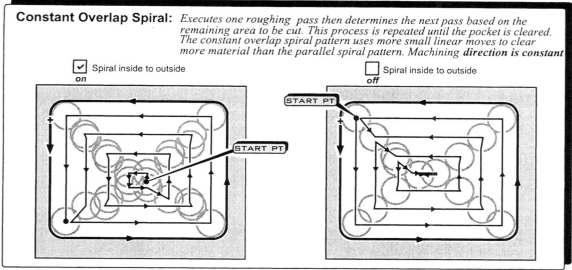

Parallel Spiral: *Roughing toolpaths follow a spiral pattern from inside to outside or outside to inside. Each new pass is* **offset from the pervious pass by the stepover value entered**. *This pattern does* **not guarantee cleanout** *of the pocket but the machining* **direction is constant.**

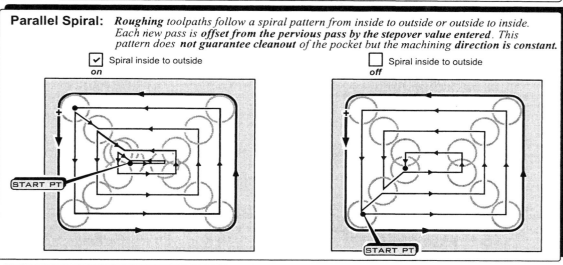

Parallel Spiral, Clean Corners: *Follows the* **same type of tool paths as the parallel spiral** *pattern except that* **corner clean out moves are added**. *This pattern increases the possibility but does not guarantee total cleanout of the pocket*

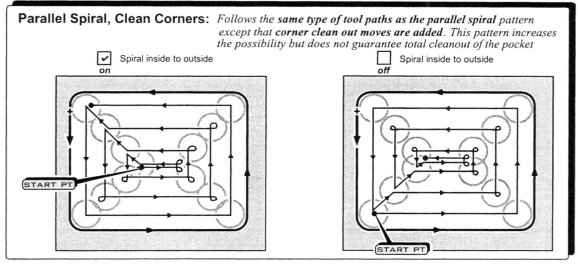

Morph Spiral: *Roughing Toolpaths are generated by gradually **interpolating** between the **outer boundary** of the pocket and the **islands.** Machining proceeds in the **same** direction.*

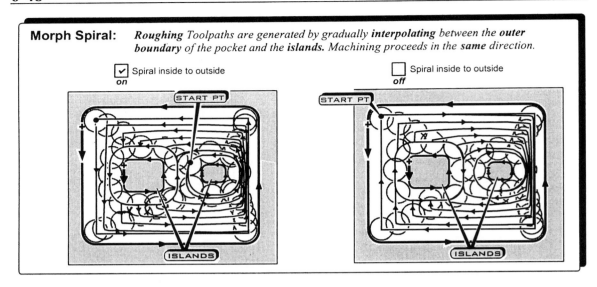

True Spiral: *Roughing toolpaths follow a spiral pattern with **tangent arc motion** from inside to outside or outside to inside in **one** direction. Tool moves **smoothly** with **minimal NC code generated** and **good pocket cleanout** resulting.*

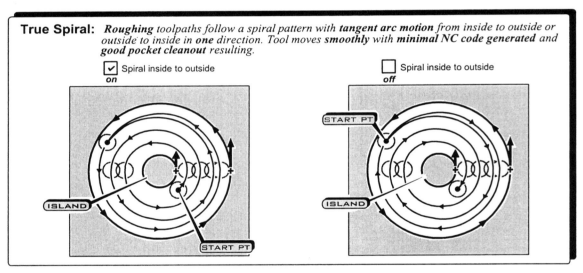

One Way: *Roughing tool paths are **linear** and cause tool to **cut** the pocket in **one direction.** Tool retracts after each pass and returns to plunge for a new pass thus increating processing time.*

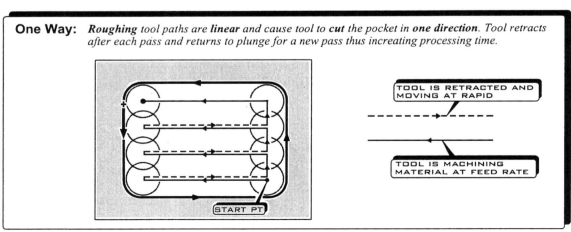

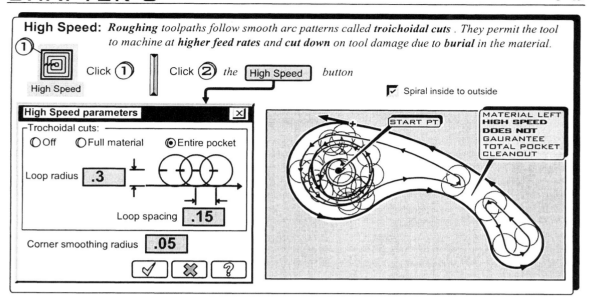

High Speed: *Roughing toolpaths follow smooth arc patterns called **troichoidal cuts**. They permit the tool to machine at **higher feed rates** and **cut down** on tool damage due to **burial** in the material.*

Click ① | Click ② *the* High Speed *button*

☑ Spiral inside to outside

High Speed parameters

Trochoidal cuts:
○ Off ○ Full material ⦿ Entire pocket

Loop radius **.3**

Loop spacing **.15**

Corner smoothing radius **.05**

START PT

MATERIAL LEFT
HIGH SPEED
DOES NOT
GAURANTEE
TOTAL POCKET
CLEANOUT

☑ Rough **STEPOVER PARAMETERS**
on

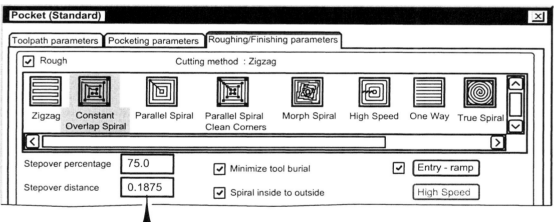

Pocket (Standard)

Toolpath parameters | Pocketing parameters | Roughing/Finishing parameters

☑ Rough   Cutting method : Zigzag

Zigzag | Constant Overlap Spiral | Parallel Spiral | Parallel Spiral Clean Corners | Morph Spiral | High Speed | One Way | True Spiral

Stepover percentage **75.0**

Stepover distance **0.1875**

☑ Minimize tool burial   ☑ Entry - ramp

☑ Spiral inside to outside   High Speed

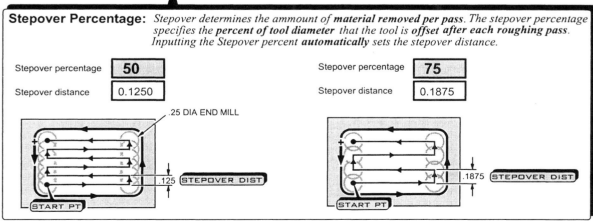

Stepover Percentage: *Stepover determines the ammount of **material removed per pass**. The stepover percentage specifies the **percent of tool diameter** that the tool is **offset after each roughing pass**. Inputting the Stepover percent **automatically** sets the stepover distance.*

Stepover percentage **50**

Stepover distance **0.1250**

Stepover percentage **75**

Stepover distance **0.1875**

.25 DIA END MILL

.125 STEPOVER DIST

START PT

.1875 STEPOVER DIST

START PT

☑ Rough   **ENTRY-HELIX/RAMP PARAMETERS**
on

The operator can select *one* of *Three* types of *entry* tool paths for *roughing a pocket : Off(straight plunge entry), Ramp(enter on zig and zag angles), Helix(enter by following a downward spiral curve).*

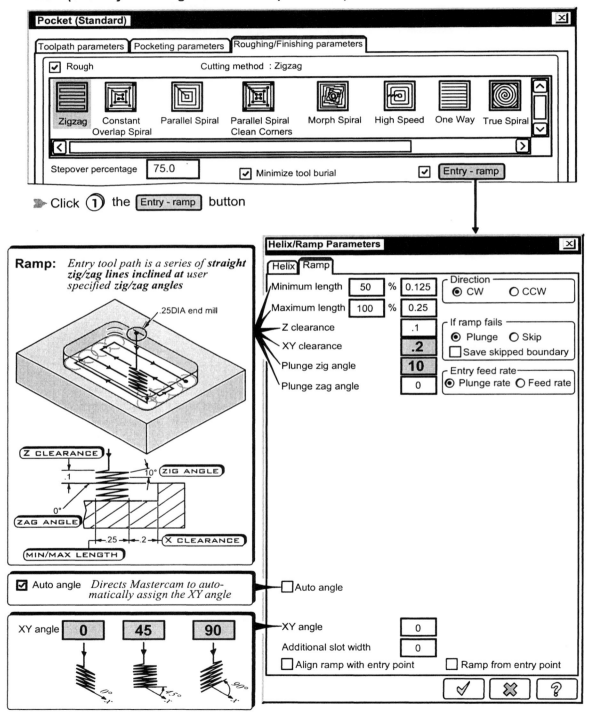

➤ Click ① the Entry - ramp button

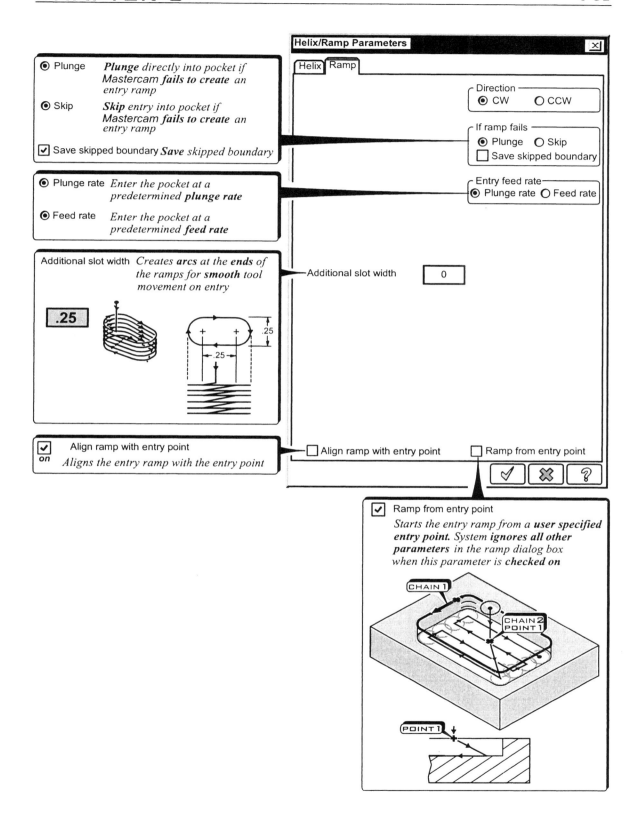

Plunge *Plunge* directly into pocket if Mastercam *fails to create* an entry ramp

Skip *Skip* entry into pocket if Mastercam *fails to create* an entry ramp

☑ Save skipped boundary *Save* skipped boundary

Plunge rate *Enter the pocket at a predetermined **plunge rate***

Feed rate *Enter the pocket at a predetermined **feed rate***

Additional slot width *Creates **arcs** at the **ends** of the ramps for **smooth** tool movement on entry*

.25

☑ Align ramp with entry point
on *Aligns the entry ramp with the entry point*

Helix/Ramp Parameters

Helix | Ramp

Direction
◉ CW ○ CCW

If ramp fails
◉ Plunge ○ Skip
☐ Save skipped boundary

Entry feed rate
◉ Plunge rate ○ Feed rate

Additional slot width 0

☐ Align ramp with entry point ☐ Ramp from entry point

☑ Ramp from entry point
*Starts the entry ramp from a **user specified entry point**. System **ignores all other parameters** in the ramp dialog box when this parameter is **checked on***

CHAIN 1

CHAIN 2 POINT 1

POINT 1

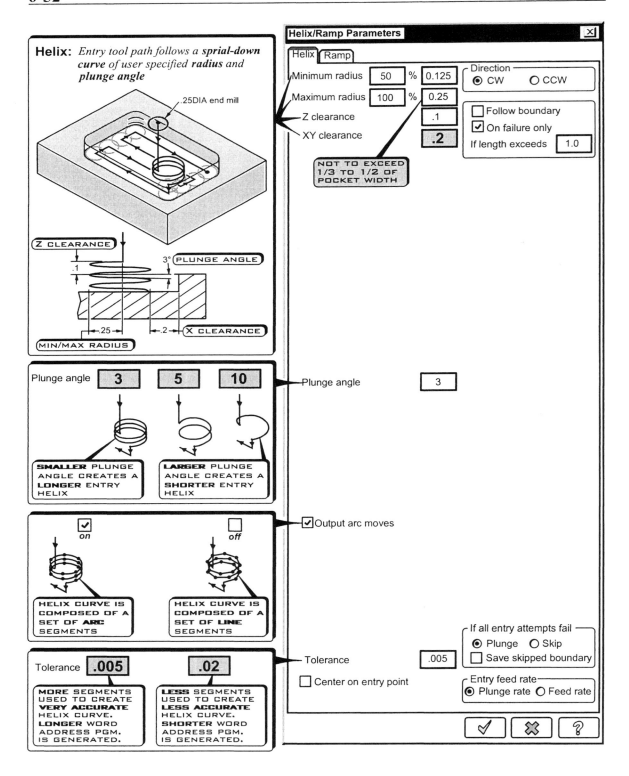

Helix: *Entry tool path follows a **sprial-down** curve of user specified **radius** and **plunge angle***

.25DIA end mill

Z CLEARANCE

.1

3° PLUNGE ANGLE

.25 .2 X CLEARANCE

MIN/MAX RADIUS

Plunge angle   **3**   **5**   **10**

SMALLER PLUNGE ANGLE CREATES A **LONGER** ENTRY HELIX

LARGER PLUNGE ANGLE CREATES A **SHORTER** ENTRY HELIX

☑ on   ☐ off

HELIX CURVE IS COMPOSED OF A SET OF **ARC** SEGMENTS

HELIX CURVE IS COMPOSED OF A SET OF **LINE** SEGMENTS

Tolerance **.005**   **.02**

MORE SEGMENTS USED TO CREATE **VERY ACCURATE** HELIX CURVE. **LONGER** WORD ADDRESS PGM. IS GENERATED.

LESS SEGMENTS USED TO CREATE **LESS ACCURATE** HELIX CURVE. **SHORTER** WORD ADDRESS PGM. IS GENERATED.

Helix/Ramp Parameters   ⊠

Helix │ Ramp

Minimum radius   50 % 0.125
Maximum radius 100 % 0.25
Z clearance   .1
XY clearance   **.2**

NOT TO EXCEED 1/3 TO 1/2 OF POCKET WIDTH

Direction
● CW   ○ CCW

☐ Follow boundary
☑ On failure only
If length exceeds   1.0

Plunge angle   3

☑ Output arc moves

If all entry attempts fail
● Plunge   ○ Skip
☐ Save skipped boundary

Tolerance   .005
☐ Center on entry point

Entry feed rate
● Plunge rate ○ Feed rate

✓   ✗   ?

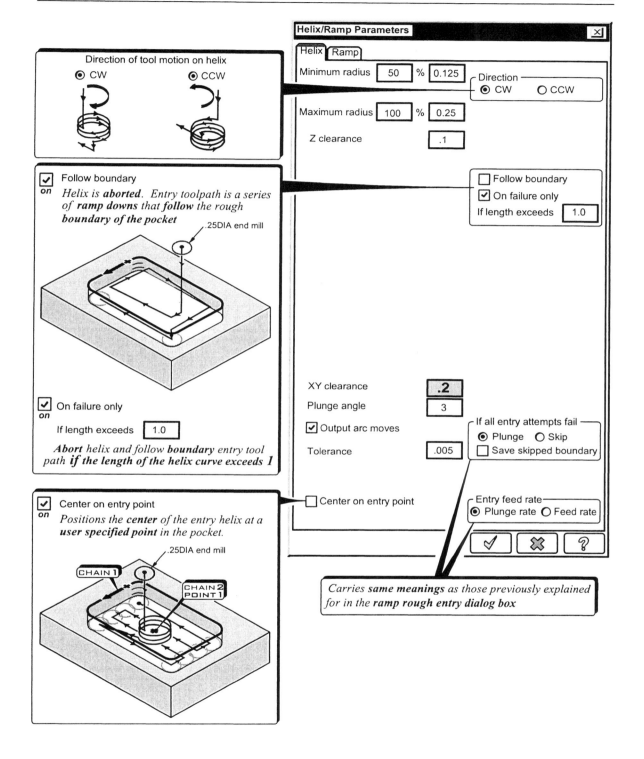

Direction of tool motion on helix

⊙ CW   ⊙ CCW

☑ **Follow boundary**
on

*Helix is **aborted**. Entry toolpath is a series of **ramp downs** that **follow** the rough **boundary** of the pocket*

.25DIA end mill

☑ **On failure only**
on

If length exceeds   1.0

Abort** helix and follow **boundary** entry tool path **if the length of the helix curve exceeds 1

☑ **Center on entry point**
on

*Positions the **center** of the entry helix at a **user specified point** in the pocket.*

.25DIA end mill

CHAIN 1   CHAIN 2 POINT 1

Helix/Ramp Parameters   ☒

Helix Ramp

Minimum radius   50 % 0.125

Direction
⊙ CW   ○ CCW

Maximum radius   100 % 0.25

Z clearance   .1

☐ Follow boundary
☑ On failure only
If length exceeds   1.0

XY clearance   **.2**
Plunge angle   3
☑ Output arc moves
Tolerance   .005

If all entry attempts fail
⊙ Plunge ○ Skip
☐ Save skipped boundary

☐ Center on entry point

Entry feed rate
⊙ Plunge rate ○ Feed rate

✓   ✖   ?

*Carries **same meanings** as those previously explained for in the **ramp rough entry** dialog box*

☑ Finish **[FINISHING PARAMETERS]**
on

Click the Finish check *on* to create finish passes in a pocket. The parameters for specifing finishing pass toolpaths are described below.

Pocket (Standard) ☒

[Toolpath parameters] [Pocketing parameters] [Roughing/Finishing parameters]

☑ Finish

┌─ Override Feed Speed ─────────

Passes   Spacing   Spring passes   Cutter compensation   ☐ Feed rate   6.4176

| 1 | | 0.01 | | 0 | | computer ▾ |

☐ Spindle speed

☑ Finish outer boundary   ☐ Optimize cutter comp in control

☐ Start finish pass at closest entity   ☐ Machine finish passes only at final depth   ☐ [Lead in/out]

☐ Keep tool down   ☐ Machine finish passes after roughing all pockets   ☐ [Thin wall]

[✓] [✗] [?]

Passes

| 1 | ⊸ *Specifies the number of finish passes to execute in the pocket toolpath*

Spacing

| 0.01 | ⊸ *Sets the **width** of stock to be removed **per finish pass***

Spring passes

| 0 | ⊸ *Executes additional finish passes along the same toolpath as the last finish pass. All spring passes follow the the same tool path with zero spacing between them. Useful for thin parts that have flexed during previous cuts.*

Cutter compensation

| computer ▾ |
| computer |
| control |
| wear |
| reverse wear |

⊸ *Carries **same meanings** as those previously explained for cutter compensation for Contour(2D)*

┌─ Override Feed Speed ─────────

☐ Feed rate   6.4176   ⊸ *Enables the operator to set a spindle feed and speed for finishing passes that is different than that used for roughing the pocket. Since less material is removed for finishing the spindle speed can be increased.*

☐ Spindle speed

☐ Finish outer boundary ── ☑ *Executes a finish pass on both the pocket's walls and its islands(default is check **on**).*
on
☐ *Ignores the pocket's outer boundary and executes a finish pass the pocket's islands only.*
off

☐ Start finish pass at closest entity ── ☑ *Begins the finish pass at the closest endpoint of the closest entity found at the end of*
on *the roughing toolpath.*
☐ *Finishes entities in the order they were selected.(default is check **off**).*
off

☐ Keep tool down ── ☑ *Tool is **not retracted** between **each** finish pass. This saves rapiding time but the operator should should*
on *verify gouging does not occur.*
☐ *Tool **retracts and rapids to the clearence plane between each** finish pass.(default is check **off**).*
off

Cutter compensation

| control | ⌄ |

☐ Optimize cutter comp in control ——— ☑ **Active only** *when cutter comp is set to* **control**. **Ignores arcs** *in the toolpath that are*
on *less than or equal to the radius of the cutting tool.* *This helps to prevent* **gouging.**

☐ Machine finish passes only at final depth ——— ☑ *Executes finish passes* **only at the final depth** *of the pocket*
on

☐ *Executes finish passes* **at all specified depth cuts** *for the pocket*
off *Useful for deep pockets.*

☐ Machine finish passes after roughing all pockets ——— ☑ *Machines roughing passes on* **all** *the pockets. Then machines*
on *finishing passes on the same set of pockets.*

☐ **Roughing and finishing** *passes are executed on the* **first** *pocket.*
off *This procedure is repeated on* **all** *the pockets.*

☑ [Lead in/out] ——— ☑ *It is recommended to add a lead in/out to the finish pass to insure that a* **dwell mark**
on **does not occur** *on the part at the end of the tool path.*

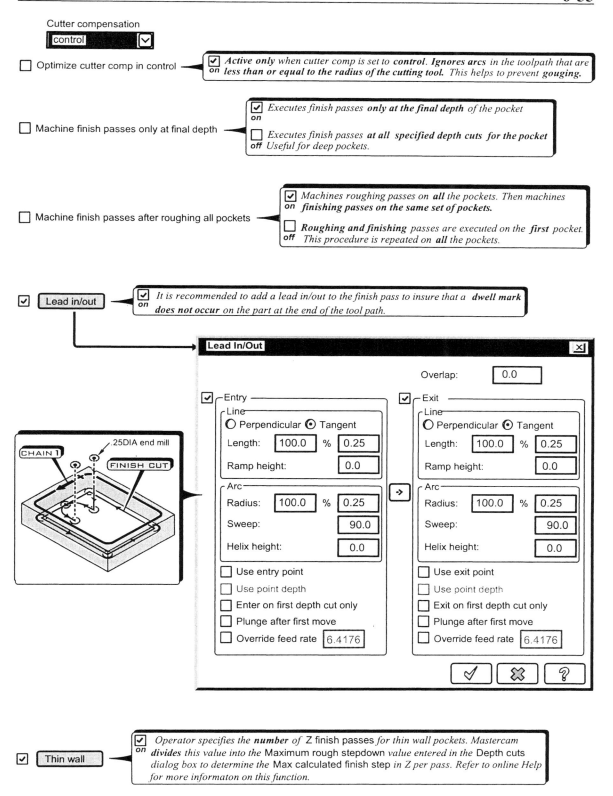

☑ [Thin wall] ——— ☑ *Operator specifies the* **number** *of* Z finish passes *for thin wall pockets. Mastercam*
on **divides** *this value into the* Maximum rough stepdown *value entered in the* Depth cuts
dialog box to determine the Max calculated finish step *in Z per pass. Refer to online Help*
for more informaton on this function.

EXAMPLE 6-8
Direct Mastercam to machine the pocket as specified in PROCESS PLAN 6-2

- Create the required pocket machining contours by chaining
- Assign the pocketing tool as listed in PROCESS PLAN 6-2
- Enter the parameters in the **Pocket(Standard)** dialog box
- Backplot and verify the pocketing toolpaths

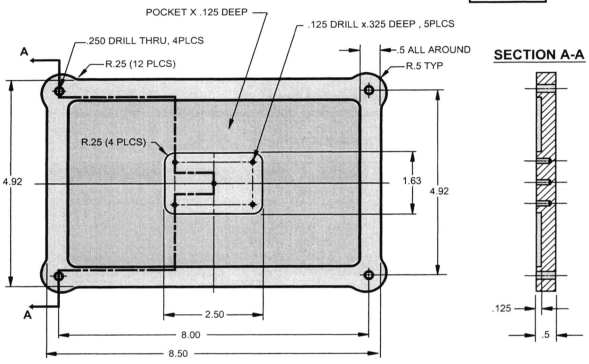

Figure 5-2

PROCESS PLAN 6-2

No.	Operation	Tooling
1	Spot drill x .166 deep (4 holes)	1/8 Spot Drilll
2	Peck drill thru(4 holes)	1/4 Drill
3	Spot drill x .125 deep (5 holes)	1/32 Spot Drill(.31-Flute length;1.5-OAL)
4	Peck drill x .325 Deep	1/8 Drill
5	Profile x .5 Deep Leave .01 for finish cuts in XY and Z	1/4 Flat End Mill
6	Pocket x .125 deep Leave .02 for finish cuts in XY and Z	1/2 Flat End Mill

A) Create the machining contours by chaining

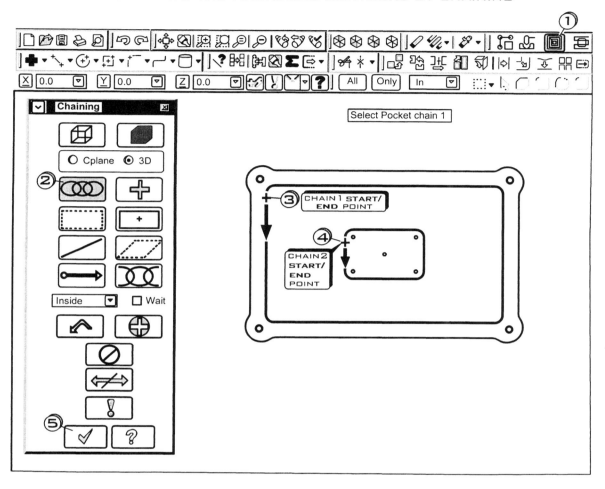

➤ Click ① the Pocket Toolpaths button

➤ Click ② on the chain button

| Select Pocket chain 1 |

➤ Click ③ on the entity to specify the chain **start/end** point

| Select Pocket chain 2 |

➤ Click ④ on the entity to specify the chain **start/end** point

| Select Pocket chain 3 |

➤ Click ⑤ the OK button

B) Obtain the 1/2 End Mill tool from the tools library

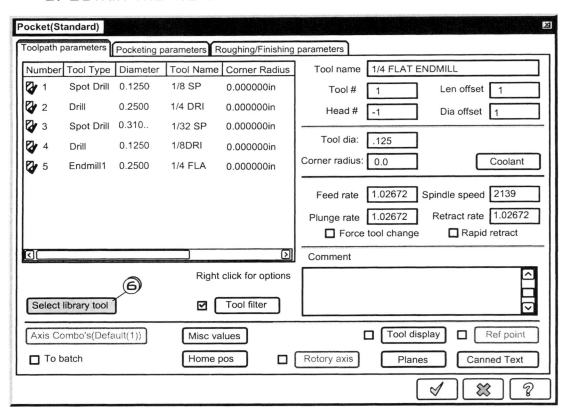

➤ Click ⑥ the [Select library tool] button

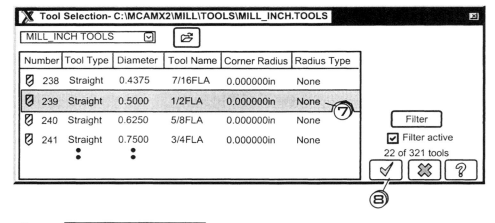

➤ Click ⑦ the [.5 Flat End Mill Tool]

➤ Click ⑧ the OK button| ✓

C) Auto assign the tool speed/feed based on the stock material

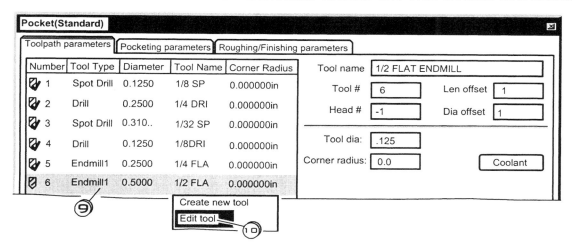

> Click ⑨ on the 1/2 FLA end mill tool

> **Right** Click move the cursor down and Click ⑩ Edit tool

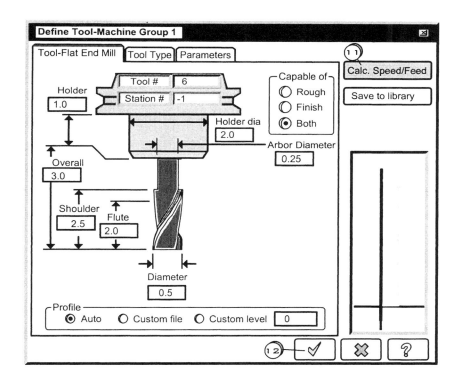

> Click ⑪ the Calc. Speed/Feed button

> Click ⑫ the OK button ✓

D) Enter the required pocket(standard) parameters listed in process plan 6-2

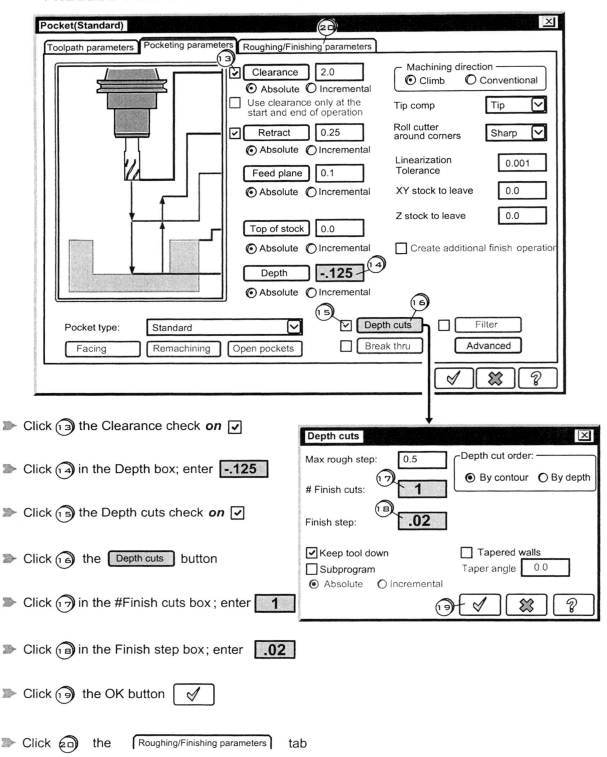

➤ Click ⑬ the Clearance check **on** ☑

➤ Click ⑭ in the Depth box; enter **-.125**

➤ Click ⑮ the Depth cuts check **on** ☑

➤ Click ⑯ the **Depth cuts** button

➤ Click ⑰ in the #Finish cuts box; enter **1**

➤ Click ⑱ in the Finish step box; enter **.02**

➤ Click ⑲ the OK button ✓

➤ Click ⑳ the **Roughing/Finishing parameters** tab

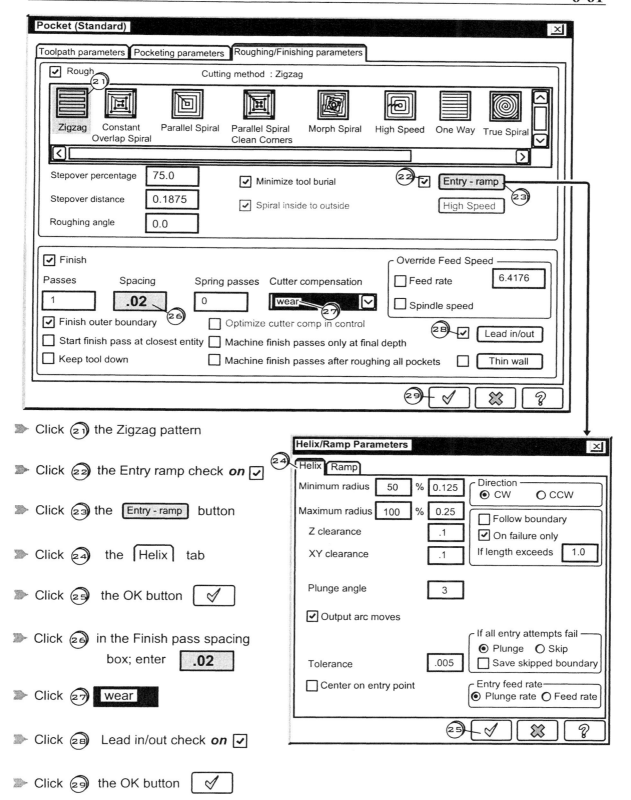

➤ Click ㉑ the Zigzag pattern

➤ Click ㉒ the Entry ramp check *on* ☑

➤ Click ㉓ the Entry - ramp button

➤ Click ㉔ the Helix tab

➤ Click ㉕ the OK button ☑

➤ Click ㉖ in the Finish pass spacing box; enter .02

➤ Click ㉗ wear

➤ Click ㉘ Lead in/out check *on* ☑

➤ Click ㉙ the OK button ☑

E) BACKPLOT THE POCKET(STANDARD) OPERATION

The Pocket(Standard) operation is selected ☑ for backplotting in the Operations Manager dialog box

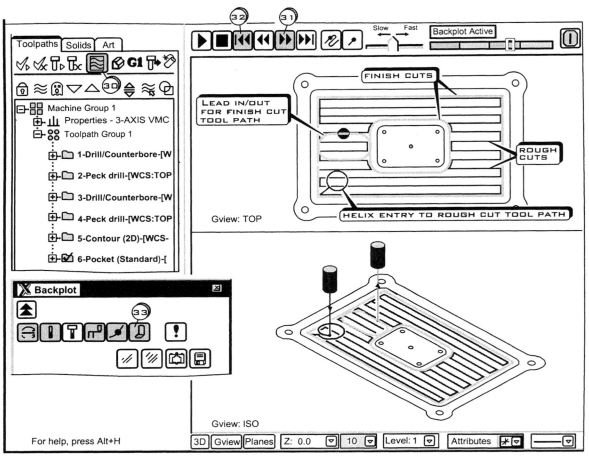

➤ Click ③⓪ the Backplot button ⧧

➤ *Continuously* Click ③① the Step forward button ⏭ to see a step by step movement of the tool along the specified pocket toolpath.

➤ Click ③② the Previous stop button ⏮

➤ Click ③③ the Quick verify button ⬛

    to get a **shaded** backplot
of the material that is
being removed from the pocket

➤ *Continuously* Click ③① the Step forward button ⏭

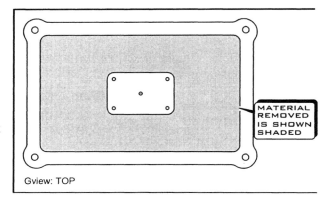

F) Verfiy the all the operations listed in process plan 6-2

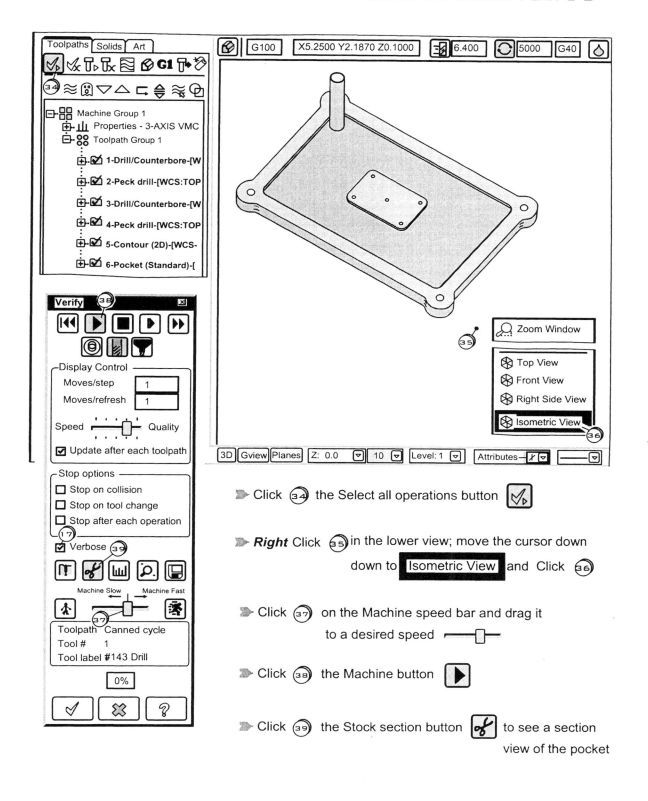

➤ Click (34) the Select all operations button

➤ **Right** Click (35) in the lower view; move the cursor down down to **Isometric View** and Click (36)

➤ Click (37) on the Machine speed bar and drag it to a desired speed

➤ Click (38) the Machine button

➤ Click (39) the Stock section button to see a section view of the pocket

EXERCISES

6-1) a) Get the *wireframe* CAD model file **EX3-1JV** :
◆ from the file generated in exercise 3-1
   or

◆ from the CD provided at the back of this text(file is located in the folder ⬜ CHAPTER6).

b) Specify the stock size and material

c) Create the required toolpaths for executing the operations listed in PROCESS PLAN 6P-1

d) Backplot and verify the toolpaths

e) Generate the word address part program

MATERIAL: 1030 STEEL

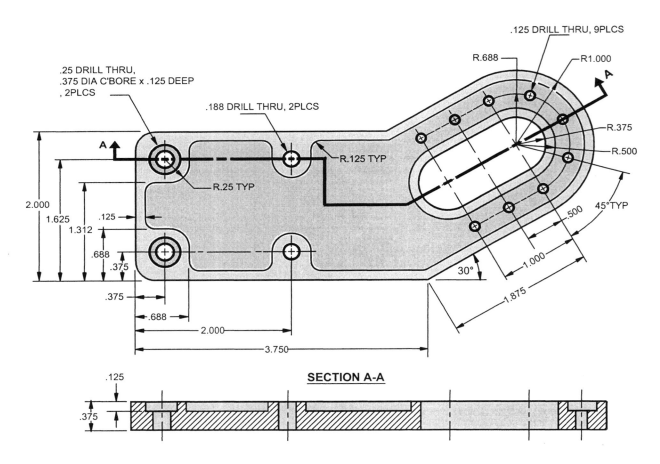

Figure 6-p1

♦ SPECIFY THE SIZE OF THE STOCK

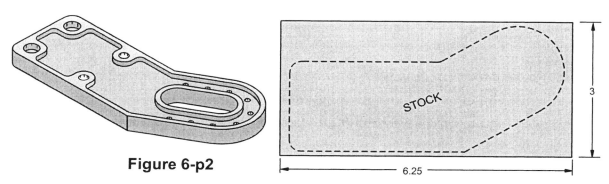

Figure 6-p2

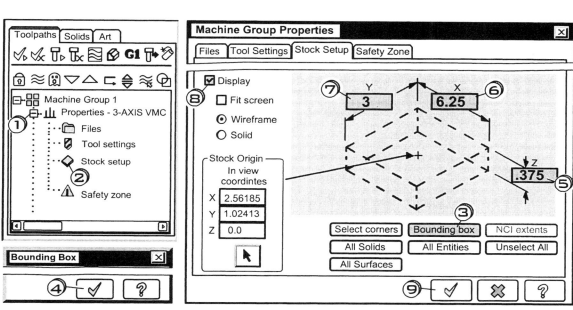

▶ Tap the [Alt] + [O] keys to open the Operations Manager dialog box

▶ Click ① the minus sign ⊟

▶ Click ② the Stock setup icon ◇

▶ Click ③ the [Bounding box] button

     This will create a *bounding box* stock shape and set the *Stock Origin* at the **center** of the box.

▶ Click ④ the OK button [✓]

▶ Click ⑤ in the Z height box; enter **.375**

▶ Click ⑥ in the X box; enter **6.25**

▶ Click ⑦ in the Y box; enter **3**

▶ Click ⑧ the check on for ☑ Display

▶ Click ⑨ the OK button [✓]

PROCESS PLAN 6P-1

No.	Operation	Tooling
1	CENTER DRILL X .166 DEEP (14 PLCS) 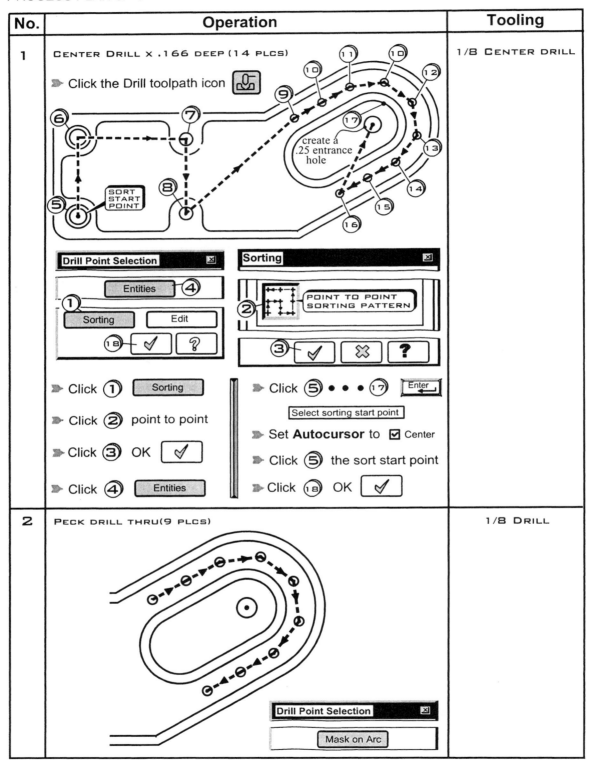	1/8 CENTER DRILL
2	PECK DRILL THRU(9 PLCS)	1/8 DRILL

PROCESS PLAN 6P-1

No.	Operation	Tooling
3	PECK DRILL THRU(2 PLCS)	3/16 DRILL
4	PECK DRILL THRU(3 PLCS)	1/4 DRILL
5	CIRCLE MILL X .125 DEEP (2 PLCS)	1/4 FLAT END MILL

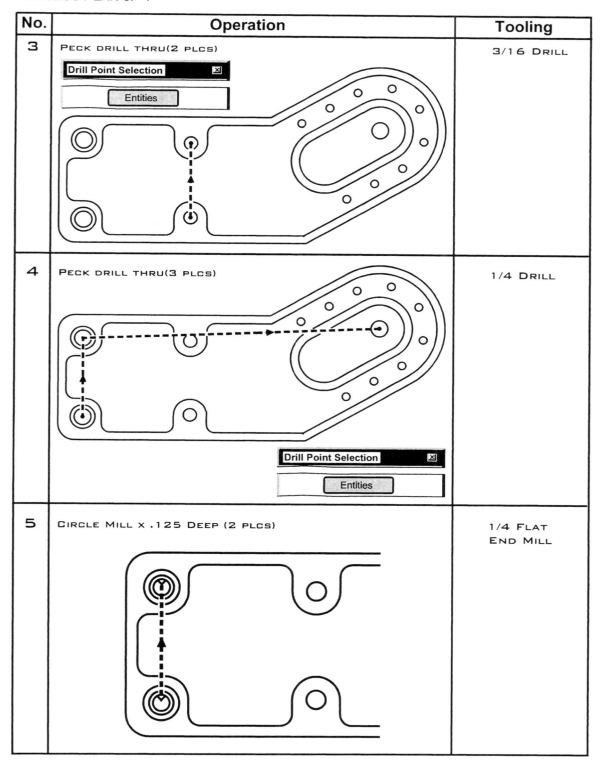

PROCESS PLAN 6P-1

No.	Operation	Tooling
6	ROUGH AND FINISH SLOT X .375 DEEP 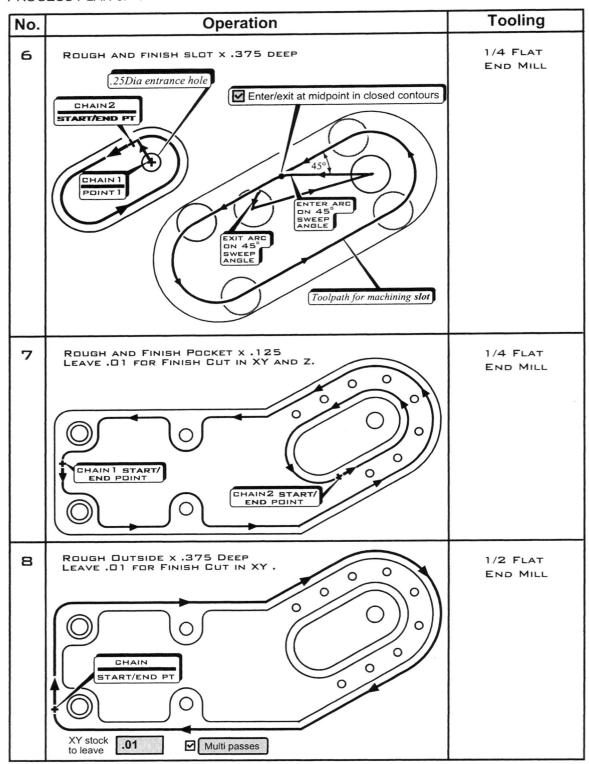	1/4 FLAT END MILL
7	ROUGH AND FINISH POCKET X .125 LEAVE .01 FOR FINISH CUT IN XY AND Z.	1/4 FLAT END MILL
8	ROUGH OUTSIDE X .375 DEEP LEAVE .01 FOR FINISH CUT IN XY .	1/2 FLAT END MILL

6-2) a) Get the *wireframe* CAD model file **EX2-1JV** generated in exercise 2-1.

    b) Specify the stock size and material

    c) Create the required toolpaths for executing the operations listed in PROCESS PLAN 6P-2

    d) Backplot and verify the toolpaths

    e) Generate the word address part program

Material: 1030 Steel

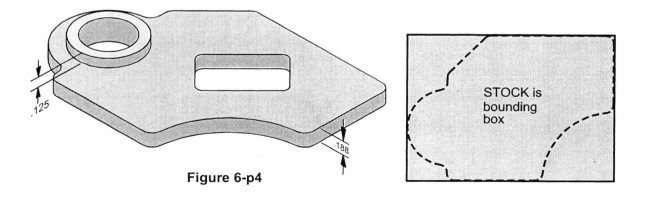

Figure 6-p3

Figure 6-p4

STOCK is bounding box

PROCESS PLAN 6P-2

No.	Operation	Tooling
1	CENTER DRILL x .166 DEEP	1/8 CENTER DRILL
	.125Drill x .166 deep (2 holes)	
2	PECK DRILL THRU	1/4 DRILL
	.25Drill (2 entrance holes)	
3	CIRCLE MILL .62 DIA THRU	1/4 END MILL

PROCESS PLAN 6P-2(*continued*)

No.	Operation	Tooling
4	ROUGH AND FINISH SLOT x .188 DEEP CHAIN2 START/ END PT CHAIN1 POINT1	1/4 END MILL
5	POCKET x .125 DEEP LEAVE .02 FOR FINISH CUT IN XY AND Z Pocket type [Facing ▼] Standard Facing ⋮ Facing Facing [×] Overlap percentage **100** Overlap ammount **.75** CHAIN2 START/ END PT CHAIN1 START/ END PT Cutting method: True Spiral	1/2 END MILL
6	ROUGH OUTSIDE x .188 DEEP LEAVE .02 FOR FINISH CUT IN XY XY stock to leave **.02** ☑ Multi passes CHAIN1 START/ END PT	1/4 END MILL

6-3) a) Get the *wireframe* CAD model file **EX2-7JV** generated in exercise 2-7.

b) Specify the stock size and material

c) Create the required toolpaths for executing the operations listed in PROCESS PLAN 6P-3

d) Backplot and verify the toolpaths

e) Generate the word address part program

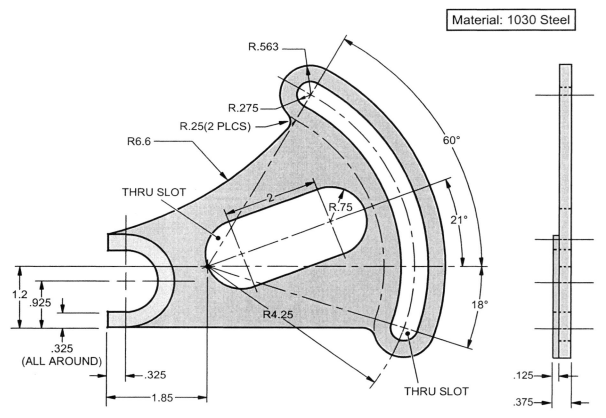

Material: 1030 Steel

Figure 6-p5

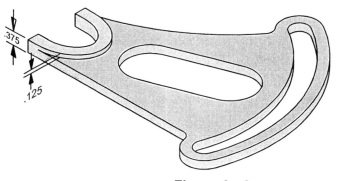

Figure 6-p6

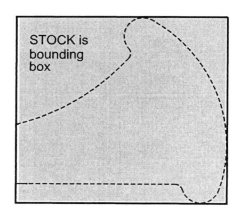

STOCK is bounding box

PROCESS PLAN 6P-3

No.	Operation	Tooling
1	CENTER DRILL X .166 DEEP(2 HOLES) .125Drill x .166 deep (2 holes)	#4 1/8 CENTER DRILL
2	PECK DRILL THRU(2 HOLES) .25Drill (2 entrance holes)	1/4 DRILL
3	ROUGH AND FINISH .75R SLOT X .375 DEEP CHAIN2 START/END PT CHAIN1 POINT1	1/4 END MILL
4	ROUGH AND FINISH .275R SLOT X .375 DEEP CHAIN1 POINT1 CHAIN2 START/END PT	1/4 END MILL

PROCESS PLAN 6P-3(*continued*)

No.	Operation	Tooling
5	POCKET X .125 DEEP. LEAVE .02 FOR FINISH CUT IN XY AND Z .25 Typ OFFSET LINES/ARCS CHAIN 1 START/ END PT Use **XFORM** ⟶ **OFFSET**	1/4 END MILL
6	ROUGH OUTSIDE X .375 DEEP LEAVE .03 FOR FINISH CUT IN XY XY stock to leave **.03** ☑ Multi passes CHAIN 1 START/ END PT	1/4 END MILL

6-4) a) Get the *wireframe* CAD model file **EX3-2JV** generated in exercise 3-2.

b) Specify the stock size and material

c) Create the required toolpaths for executing the operations listed in PROCESS PLAN 6P-4

d) Backplot and verify the toolpaths

e) Generate the word address part program

Material: 1030 Steel

SECTION A-A

A

R.875 TYP

R.750 TYP

.500 DIA
6 HOLES
EQL SP
ON 7.5DIA BC

R5.25

R4.875

R.25   R1.5

.875 DIA
3 HOLES
EQL SP

A

6.0
DIA

5.0
DIA

.500
(REF)

.750

.500

1.750

Figure 6-p7

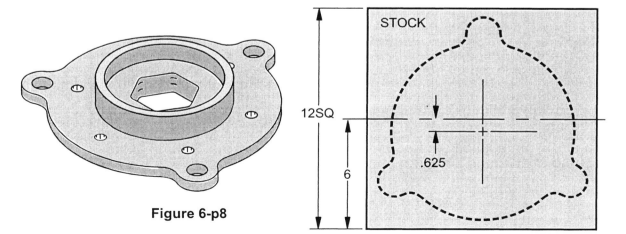

STOCK

12SQ

6

.625

Figure 6-p8

PROCESS PLAN 6P-4

No.	Operation	Tooling
1	**POCKET X 1.25 DEEP** **LEAVE .02 FOR FINISH CUT IN XY AND Z** Pocket type: Facing — Standard / Facing Facing → Facing: Overlap percentage **100**, Overlap ammount **.75** Cutting method: True Spiral 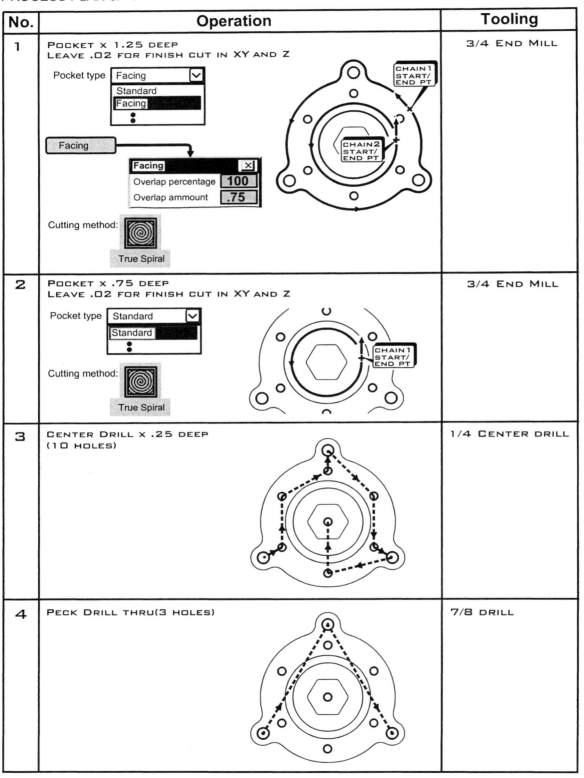	3/4 END MILL
2	**POCKET X .75 DEEP** **LEAVE .02 FOR FINISH CUT IN XY AND Z** Pocket type: Standard — Standard Cutting method: True Spiral	3/4 END MILL
3	**CENTER DRILL X .25 DEEP** **(10 HOLES)**	1/4 CENTER DRILL
4	**PECK DRILL THRU(3 HOLES)**	7/8 DRILL

PROCESS PLAN 6P-4(*continued*)

No.	Operation	Tooling
5	PECK DRILL THRU (7 HOLES)	1/2 DRILL
6	ROUGH AND FINISH HEX SLOT X 1 DEEP	1/4 END MILL
7	ROUGH OUTSIDE X .5 DEEP LEAVE .01 FOR FINISH CUT IN XY XY stock to leave .03 ☑ Multi passes	3/4 END MILL

6-5) a) Get the *wireframe* CAD model file **EX3-3JV** generated in exercise 3-3.

b) Specify the stock size and material

c) Create the required toolpaths for executing the operations listed in PROCESS PLAN 6P-5

d) Backplot and verify the toolpaths

e) Generate the word address part program

Material: 1030 Steel

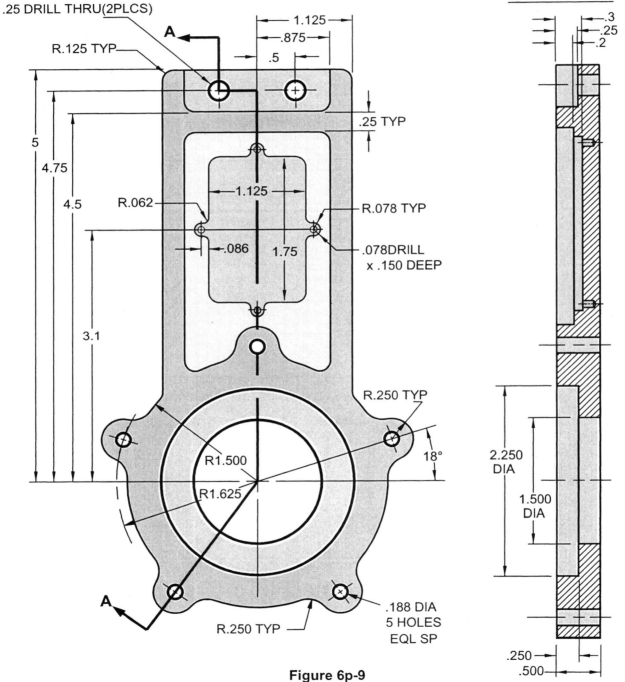

Figure 6p-9

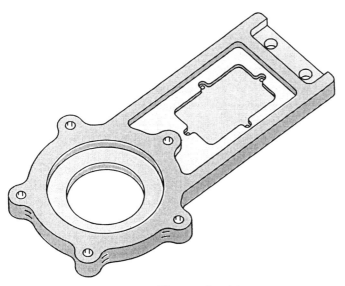

Figure 6p-10

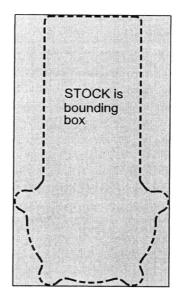

PROCESS PLAN 6P-5

No.	Operation	Tooling
1	CENTER DRILL x .166 DEEP(7 HOLES)	1/8 CENTER DRILL
2	PECK DRILL THRU(5 HOLES)	3/16 DRILL
3	PECK DRILL THRU(2 HOLES)	1/4 DRILL

PROCESS PLAN 6P-5(*continued*)

No.	Operation	Tooling
4	CIRCLE MILL 1.5DIA THRU	1/2 END MILL
5	CIRCLE MILL 2.25DIA X .25 DEEP	1/2 END MILL
6	POCKET X .2DEEP LEAVE .01 FOR FINISH CUT IN XY AND Z	1/4 END MILL
7	POCKET X .3DEEP LEAVE .01 FOR FINISH CUT IN XY AND Z	1/4 END MILL
8	POCKET[REMACHINING] X .3DEEP LEAVE .01 FOR FINISH CUT IN XY AND Z Pocket type [Remachining ▾] Standard ⋮ Remachining Open	5/32 END MILL
9	PECK DRILL X .450DEEP	5/64 DRILL

PROCESS PLAN 6P-5(*continued*)

No.	Operation	Tooling
10	POCKET X .25DEEP LEAVE .01 FOR FINISH CUT IN XY AND Z 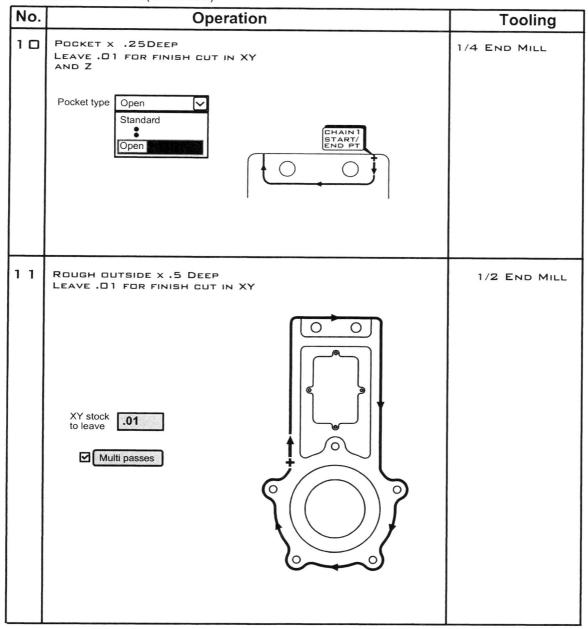	1/4 END MILL
11	ROUGH OUTSIDE X .5 DEEP LEAVE .01 FOR FINISH CUT IN XY	1/2 END MILL

6-6) a) Get the *wireframe* CAD model file **EX3-4JV** generated in exercise 3-4.

b) Specify the stock size and material

c) Create the required toolpaths for executing the operations listed in PROCESS PLAN 6P-6

d) Backplot and verify the toolpaths

e) Generate the word address part program

Material: 1030 Steel

SECTION A-A

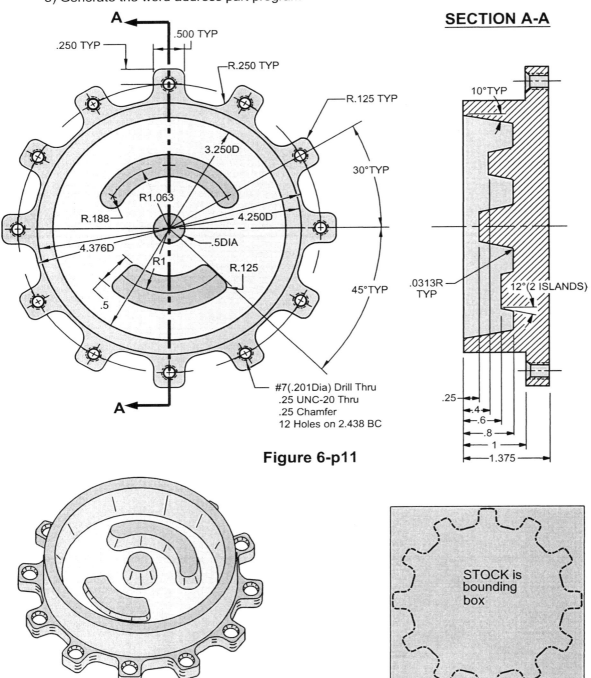

Figure 6-p11

Figure 6-p12

STOCK is bounding box

PROCESS PLAN 6P-6(*continued*)

No.	Operation	Tooling
1	**POCKET X 1 DEEP** **LEAVE .02 FOR FINISH CUT IN XY AND Z**	1/2 END MILL
2	**POCKET[ISLAND FACING]** **LEAVE .01 FOR FINISH CUT IN XY AND Z**	1/4 BULL END MILL .0313R

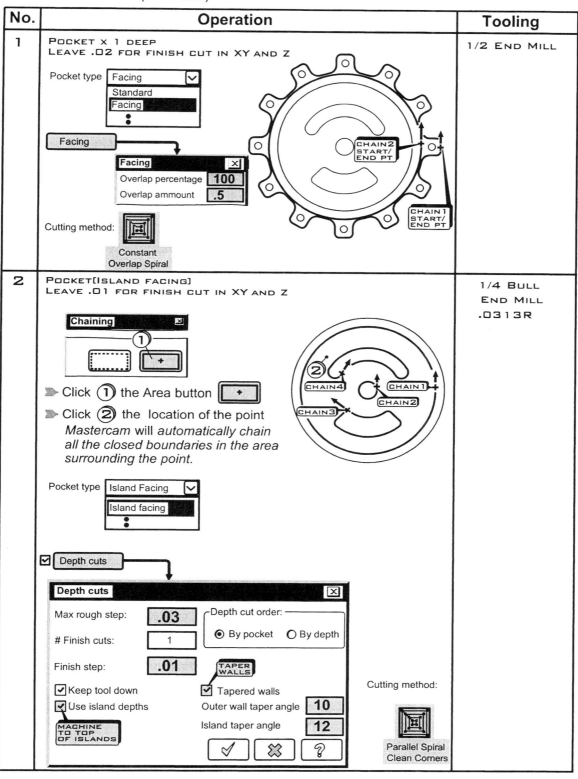

Operation 1 details:

Pocket type Facing

Standard
Facing
⋮

Facing

Facing [×]
Overlap percentage **100**
Overlap ammount **.5**

Cutting method: Constant Overlap Spiral

CHAIN2 START/END PT

CHAIN1 START/END PT

Operation 2 details:

Chaining [×]
①

▶ Click ① the Area button [+]
▶ Click ② the location of the point
Mastercam will *automatically chain all the closed boundaries in the area surrounding the point.*

CHAIN4 CHAIN1 CHAIN2 CHAIN3 ②

Pocket type Island Facing
Island facing
⋮

☑ Depth cuts

Depth cuts [×]
Max rough step: **.03**
Finish cuts: 1
Finish step: **.01**

Depth cut order:
◉ By pocket   ○ By depth

TAPER WALLS
☑ Keep tool down   ☑ Tapered walls
☑ Use island depths   Outer wall taper angle **10**
MACHINE TO TOP OF ISLANDS   Island taper angle **12**

✓ ✗ ?

Cutting method: Parallel Spiral Clean Corners

PROCESS PLAN 6P-5(*continued*)

No.	Operation	Tooling
3	CENTER DRILL X .166 DEEP(12 HOLES)	1/8 CTR DRILL
4	PECK DRILL THRU(12 HOLES)	#7(.201) DRILL
5	TAP THRU(12 HOLES)	1/4-20 TAPRH

PROCESS PLAN 6P-5(*continued*)

No.	Operation	Tooling
6	CHAMFER X 1.15 DEEP (12 HOLES)	1/2 CHAMFER MILL
7	ROUGH OUTSIDE X 1.375 DEEP LEAVE .01 FOR FINISH CUT IN XY CHAIN 1 START/END PT XY stock to leave .01 ☑ Multi passes	1/2 END MILL

6-7) a) Get the *wireframe* CAD model file **EX3-5JV** generated in exercise 3-5.

      b) Specify the stock size and material

      c) Create the required toolpaths for executing the operations listed in PROCESS PLAN 6P-7

      d) Backplot and verify the toolpaths

      e) Generate the word address part program

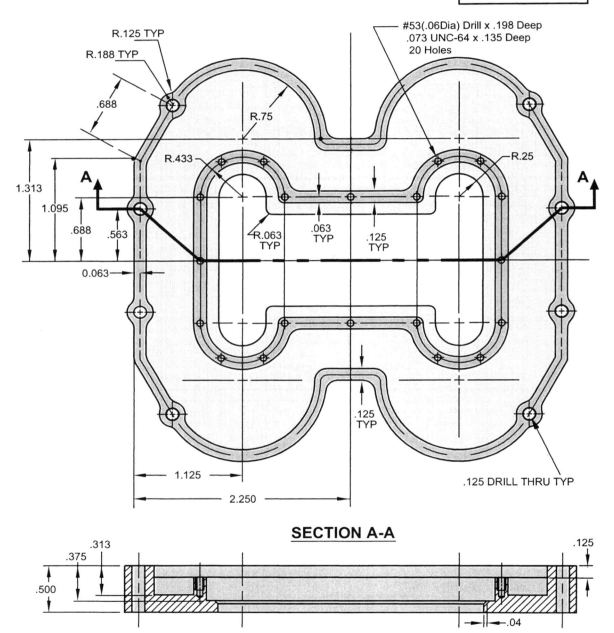

Figure 6-p13

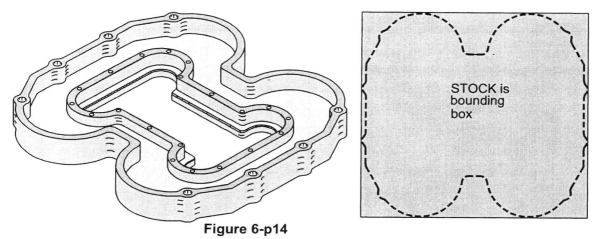

Figure 6-p14

STOCK is bounding box

PROCESS PLAN 6P-7

No.	Operation	Tooling
1	PECK DRILL X .198 DEEP(20 HOLES)	#53(.06 DIA) DRILL
2	TAP X .135 DEEP(20 HOLES)	.073-64-TAP
3	CENTER DRILL X .166 DEEP(9 HOLES) Create .4375 circle	1/8 CTR DRILL

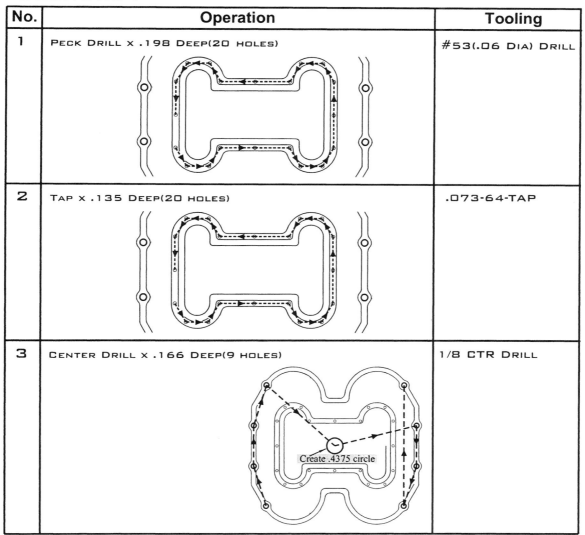

PROCESS PLAN 6P-7(*continued*)

No.	Operation	Tooling
4	PECK DRILL THRU (8 HOLES)	1/8 DRILL
5	PECK DRILL THRU	7/16 DRILL
6	POCKET[ISLAND FACING] LEAVE .01 FOR FINISH CUT IN XY AND Z	1/4 END MILL

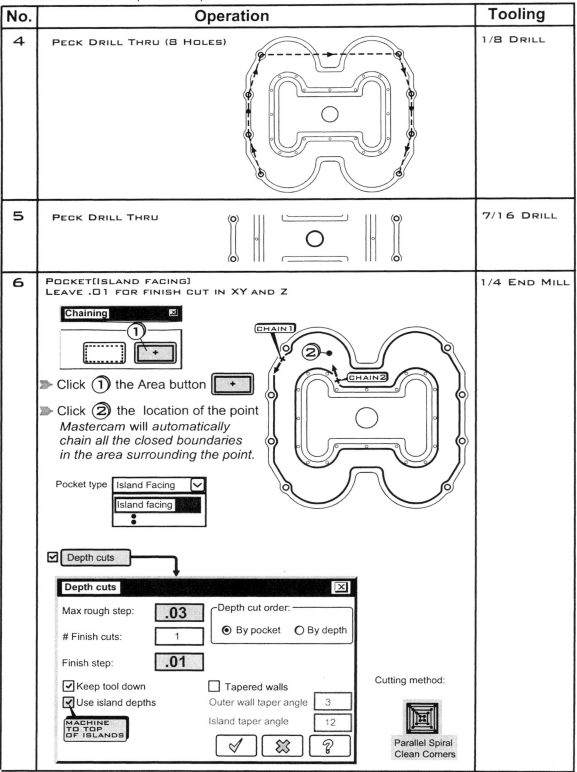

Chaining ①

➤ Click ① the Area button

➤ Click ② the location of the point
Mastercam will automatically chain all the closed boundaries in the area surrounding the point.

Pocket type   Island Facing

Island facing

☑ Depth cuts

Depth cuts

Max rough step: **.03**

Finish cuts: 1

Finish step: **.01**

Depth cut order:
⦿ By pocket   ◯ By depth

☑ Keep tool down   ☐ Tapered walls
☑ Use island depths   Outer wall taper angle   3
MACHINE TO TOP OF ISLANDS   Island taper angle   12

Cutting method:

Parallel Spiral Clean Corners

PROCESS PLAN 6P-7(*continued*)

No.	Operation	Tooling
7	ROUGH AND FINISH SLOT x .5 DEEP LEAVE .01 FOR FINISH CUT IN XY	7/16 END MILL
8	ROUGH AND FINISH INSIDE CONTOUR x .375 DEEP LEAVE .01 FOR FINISH CUT IN XY	7/16 END MILL
9	CHAMFER INSIDE CONTOUR	1/2 CHAMFER MILL
10	ROUGH OUTSIDE x .5 DEEP LEAVE .01 FOR FINISH CUT IN XY	1/4 END MILL

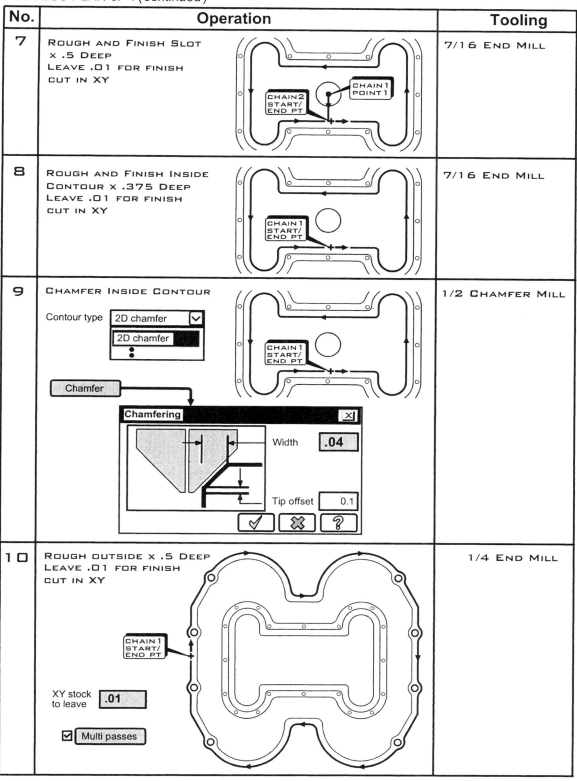

6-8) a) Get the *wireframe* CAD model file **EX3-6JV** generated in exercise 3-6.

b) Specify the stock size and material

c) Create the required toolpaths for executing the operations listed in PROCESS PLAN 6P-8

d) Backplot and verify the toolpaths

e) Generate the word address part program

Material: 1030 Steel

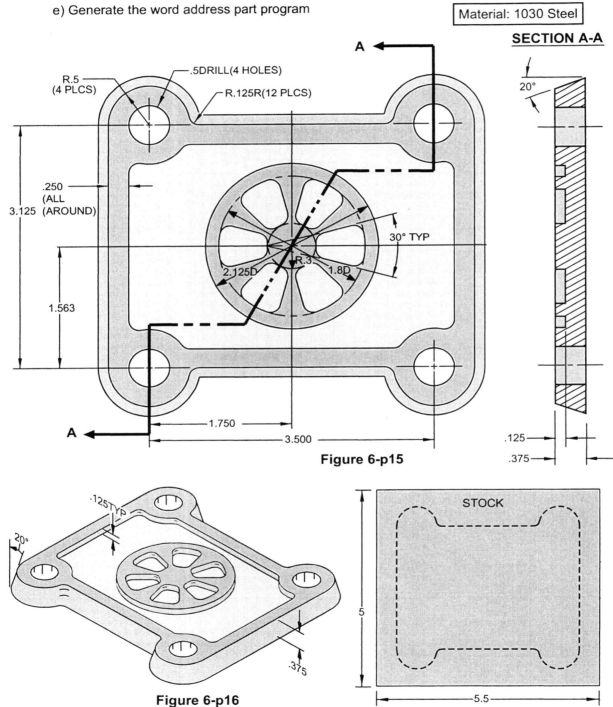

SECTION A-A

R.5 (4 PLCS)

.5DRILL(4 HOLES)

R.125R(12 PLCS)

.250 (ALL (AROUND)

3.125

30° TYP

2.125D R.3 1.8D

1.563

1.750

3.500

20°

.125

.375

Figure 6-p15

.125TYP

20°

.375

STOCK

5

5.5

Figure 6-p16

PROCESS PLAN 6P-8

No.	Operation	Tooling
1	CENTER DRILL x .166 DEEP(4 HOLES) 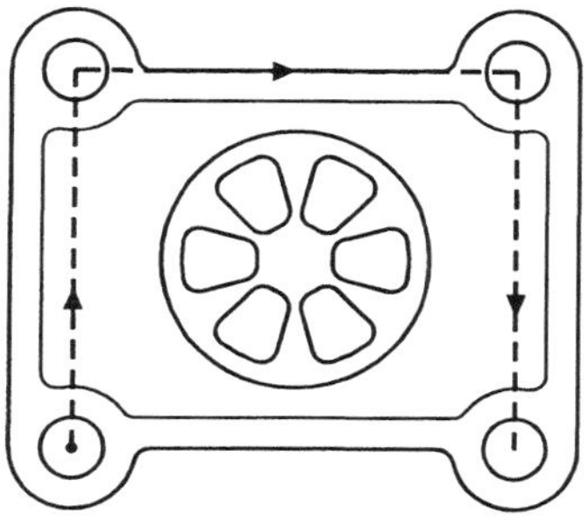	1/4 CTR DRILL
2	PECK DRILL x .375 DEEP(4 HOLES) 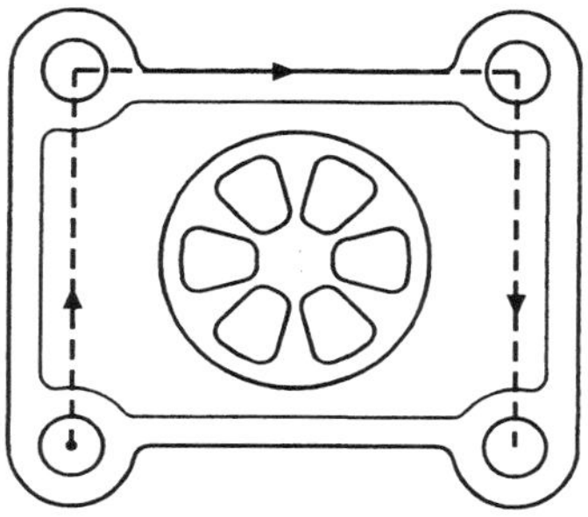	1/2 DRILL
3	 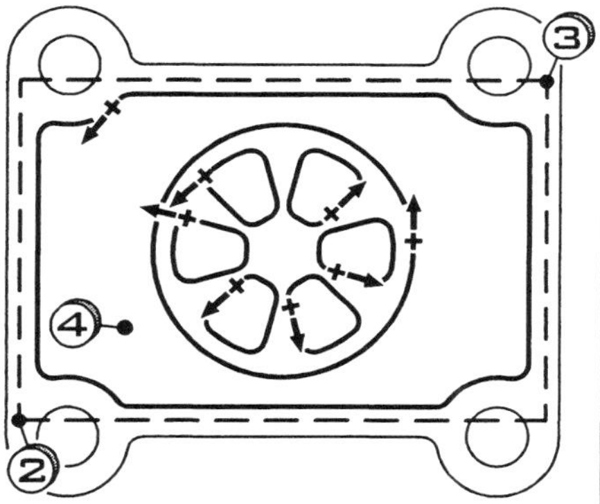 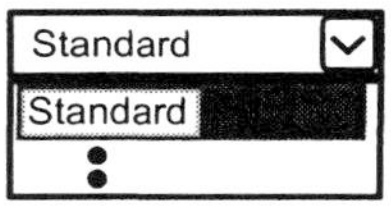 POCKET x .125 DEEP LEAVE .01 FOR FINISH CUT IN XY AND Z 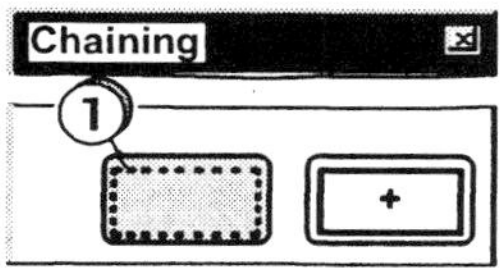 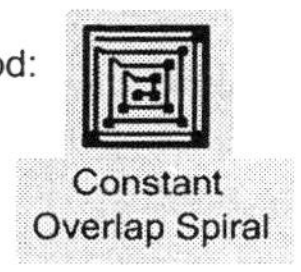	1/4 END MILL

Chaining

➤ Click ① the Window button

➤ Click ② ③ the win corners

➤ Click ④ the search point location.

Mastercam will *automatically chain all the closed boundaries contained in the window.*

Pocket type Standard
  Standard

Cutting method: Constant Overlap Spiral

PROCESS PLAN 6P-8(*continued*)

No.	Operation	Tooling
4	TAPER CUT OUTSIDE 20° x .375 DEEP LEAVE .01 FOR FINISH CUT IN XY 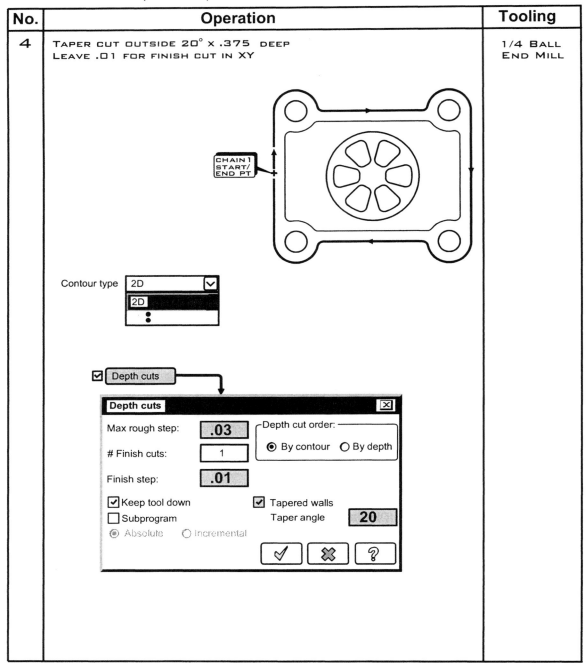	1/4 BALL END MILL

6-9) a) Get the *wireframe* CAD model file **EX3-8JV** completed in exercise 3-8.
   b) Specify the stock size and material
   c) Create the required toolpaths for executing the operations listed in PROCESS PLAN 6P-9
   d) Backplot and verify the toolpaths
   e) Generate the word address part program

Material: 2024 Aluminum

Font= HARTFORD
Height=.3, Width=.15

.25 DRILL THRU
(4 HOLES)

Font=ROMAN
Height=.45
Spacing=.1

Font=BLOCK
Height=.2,
Spacing=.04

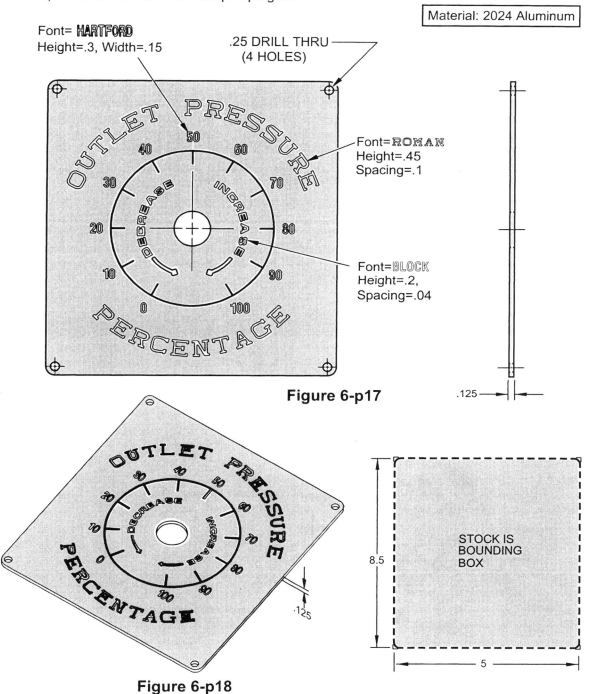

Figure 6-p17

.125

Figure 6-p18

STOCK IS
BOUNDING
BOX

8.5

5

.125

PROCESS PLAN 6P-9

No.	Operation	Tooling
1	CONTOUR X .05 DEEP **Chaining** ➤ Click ① the Window button ➤ Click ② ③ the win corners ➤ Click ④ the search point location. *Mastercam automatically chains all the closed boundaries found in the window.* Note: Click the Lead in/out Check **off** ☐ Lead in/out	1/32 BALL ENDMILL
2	DRILL THRU (4 HOLES)	1/4 DRILL
3	CIRCLE MILL 1 DIA THRU	1/2 ENDMILL

PROCESS PLAN 6P-9(*continued*)

No.	Operation	Tooling
4	ROUGH OUTSIDE x .125 DEEP LEAVE .01 FOR FINISH CUT IN XY 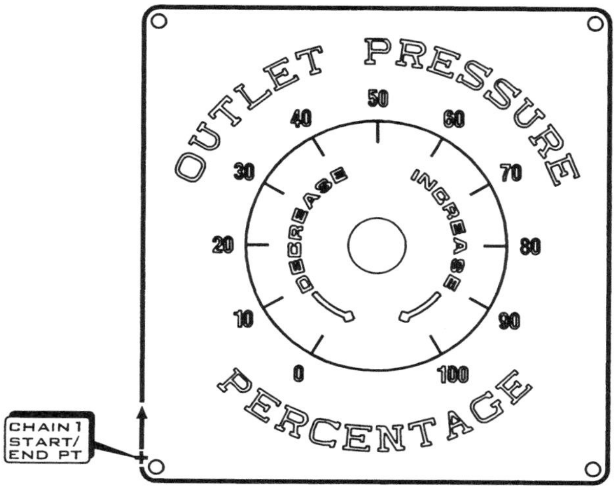 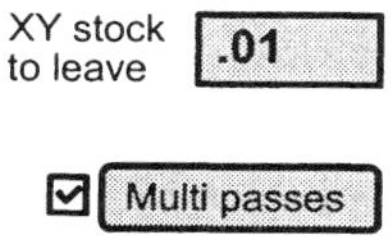	1/2 ENDMILL

CHAIN 1
START/
END PT

OUTLET PRESSURE
PERCENTAGE
DECREASE   INCREASE

XY stock
to leave   .01

☑ Multi passes

6-10) a) Get the *wireframe* CAD model file **EX4-1JV** created in exercise 4-1.
   b) Specify the stock size and material
   c) Create the required toolpaths for executing the operations listed in PROCESS PLAN 6P-10
   d) Backplot and verify the toolpaths
   e) Generate the word address part program

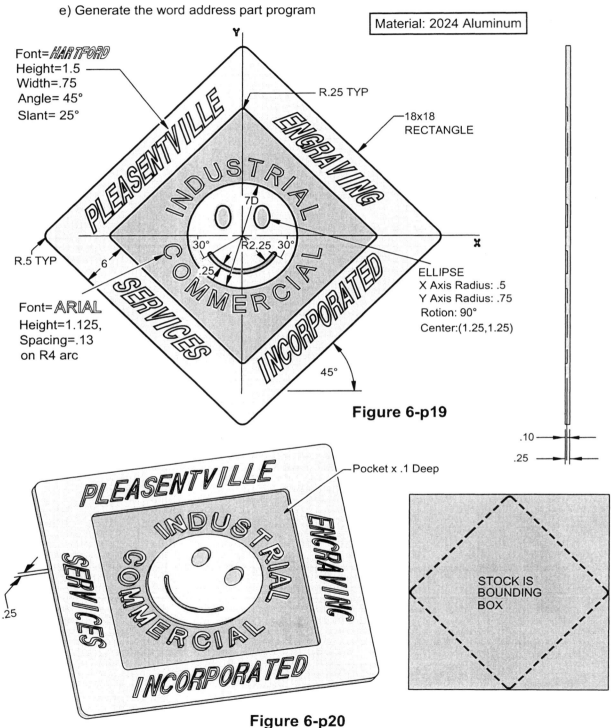

Material: 2024 Aluminum

Font=*HARTFORD*
Height=1.5
Width=.75
Angle= 45°
Slant= 25°

R.25 TYP

18x18
RECTANGLE

7D

30°   R2.25   30°
.25

R.5 TYP   6

ELLIPSE
X Axis Radius: .5
Y Axis Radius: .75
Rotion: 90°
Center:(1.25,1.25)

Font=ARIAL
Height=1.125,
Spacing=.13
on R4 arc

45°

Figure 6-p19

.10
.25

Pocket x .1 Deep

.25

STOCK IS
BOUNDING
BOX

Figure 6-p20

PROCESS PLAN 6P-10

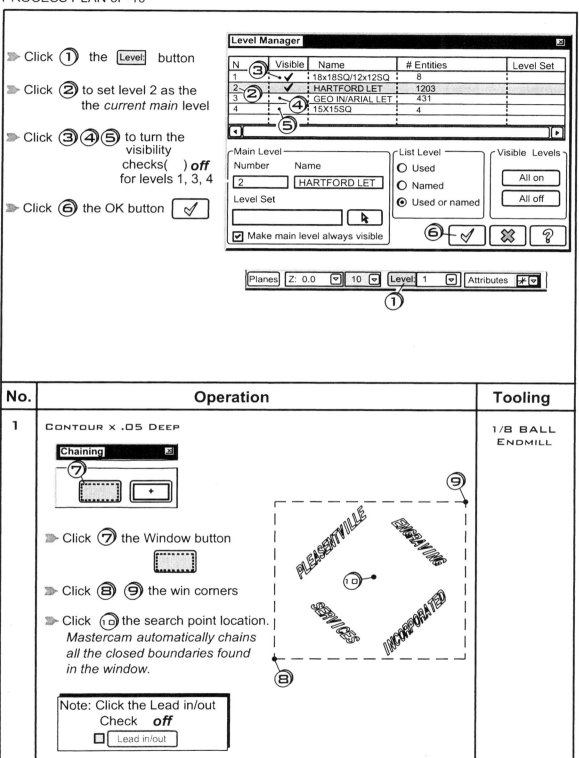

➤ Click ① the [Level:] button

➤ Click ② to set level 2 as the the *current main* level

➤ Click ③④⑤ to turn the visibility checks( ) *off* for levels 1, 3, 4

➤ Click ⑥ the OK button

No.	Operation	Tooling
1	CONTOUR X .05 DEEP	1/8 BALL ENDMILL

➤ Click ⑦ the Window button

➤ Click ⑧ ⑨ the win corners

➤ Click ⑩ the search point location. *Mastercam automatically chains all the closed boundaries found in the window.*

Note: Click the Lead in/out Check *off*
[ ] Lead in/out

PROCESS PLAN 6P-10(*continued*)

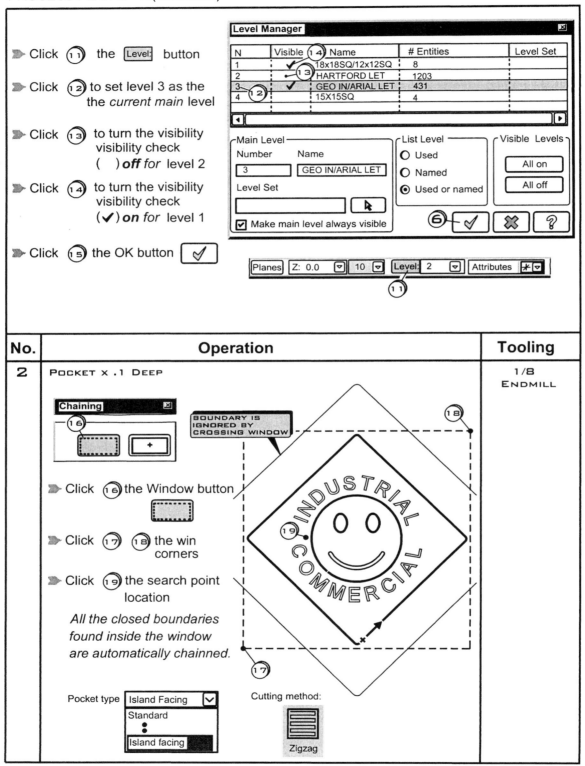

Click ⑪ the [Level:] button

Click ⑫ to set level 3 as the the *current main* level

Click ⑬ to turn the visibility visibility check ( ) *off for* level 2

Click ⑭ to turn the visibility visibility check (✔) *on for* level 1

Click ⑮ the OK button ✓

Level Manager

N	Visible	Name	# Entities	Level Set
1	✔	18x18SQ/12x12SQ	8	
2		HARTFORD LET	1203	
3	✔	GEO IN/ARIAL LET	431	
4		15X15SQ	4	

Main Level
Number   Name
[3]   [GEO IN/ARIAL LET]
Level Set
[   ] ▶
☑ Make main level always visible

List Level
○ Used
○ Named
◉ Used or named

Visible Levels
[All on]
[All off]

[Planes] [Z: 0.0 ▽] [10 ▽] [Level: 2 ▽] [Attributes ✳▽]

No.	Operation	Tooling
2	POCKET X .1 DEEP	1/8 ENDMILL

Chaining

BOUNDARY IS IGNORED BY CROSSING WINDOW

Click ⑯ the Window button

Click ⑰ ⑱ the win corners

Click ⑲ the search point location

All the closed boundaries found inside the window are automatically chainned.

Pocket type [Island Facing ▽]
Standard
⋮
[Island facing]

Cutting method:

Zigzag

PROCESS PLAN 6P-10(*continued*)

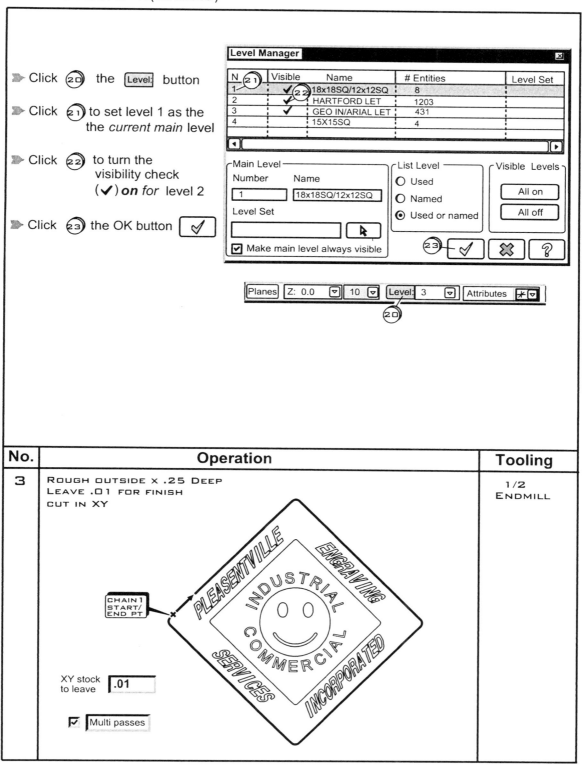

Click ② the Level button

Click ② to set level 1 as the the *current main* level

Click ② to turn the visibility check (✔) *on for* level 2

Click ② the OK button ✓

Level Manager

N	Visible	Name	# Entities	Level Set
1	✔	18x18SQ/12x12SQ	8	
2	✔	HARTFORD LET	1203	
3	✔	GEO IN/ARIAL LET	431	
4		15X15SQ	4	

Main Level

Number Name
1   18x18SQ/12x12SQ

Level Set

☑ Make main level always visible

List Level
○ Used
○ Named
◉ Used or named

Visible Levels
All on
All off

Planes Z: 0.0 ▽ 10 ▽ Level: 3 ▽ Attributes ⊡▽

No.	Operation	Tooling
3	ROUGH OUTSIDE X .25 DEEP LEAVE .01 FOR FINISH CUT IN XY	1/2 ENDMILL

CHAIN 1
START/
END PT

XY stock
to leave   .01

☑ Multi passes

EDITING MACHINING OPERATIONS VIA THE OPERATIONS MANAGER

7-1 Chapter Objectives

After completing this chapter you will be able to:

1. Explain what the operations manager is.

2. State the four parts contained in the Properties folder

3. State the four parts contained in the Operation folder

4. Know what effect associativity has on edited machining operations

5. Specify how to use the operations manager to perform the following functions on operations: create new, move, copy and delete.

5. Understand how to use the operations manager to edit existing toolpaths

6. Explain how to display and print the Operations Manager listing as a Doc file.

7-2 The Operations Manager

The Operations Manager was introduced in Chapters 5 and 6 as a means for selecting toolpaths to be back plotted or verified. It was also used to edit drill tool paths and to direct *Mastercam* to generate a word address part program for a set of tool paths. In this Chapter we will take a closer look at the *Toolpath Manager* contained within the Operations Manager and consider more of its powerful editing features.

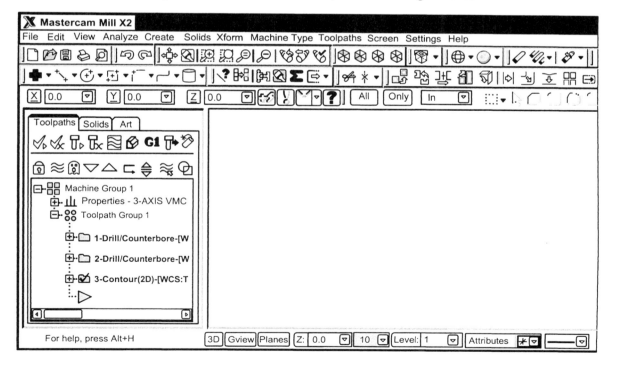

The root folder in the Toolpath Manager is Machine Group. This is created when the operator *selects a machine type from the Machine Type* drop down menu. The machine type defines what *type of machine* is to be used for the job and the file used for *postprocessing.* After Machine Group is created, *Mastercam automatically* generates all the sub-folders named: Machining Properties, Toolpath Group, Operation, Operation Insert Arrow.

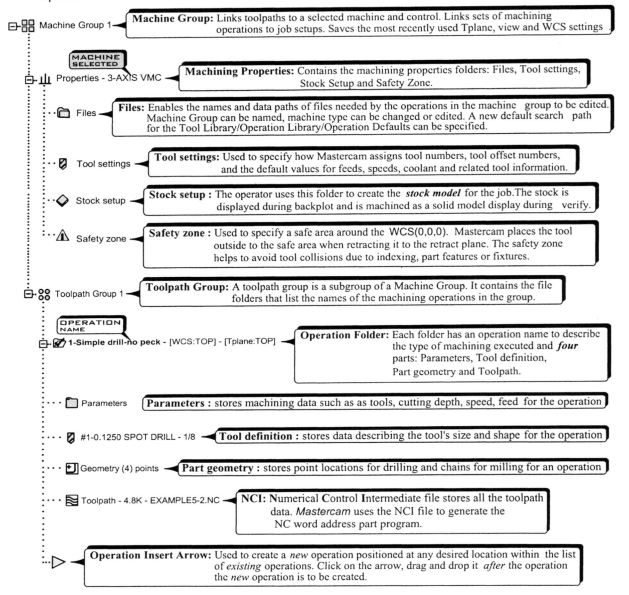

Machine Group 1 — **Machine Group:** Links toolpaths to a selected machine and control. Links sets of machining operations to job setups. Saves the most recently used Tplane, view and WCS settings.

MACHINE SELECTED

Properties - 3-AXIS VMC — **Machining Properties:** Contains the machining properties folders: Files, Tool settings, Stock Setup and Safety Zone.

Files — **Files:** Enables the names and data paths of files needed by the operations in the machine group to be edited. Machine Group can be named, machine type can be changed or edited. A new default search path for the Tool Library/Operation Library/Operation Defaults can be specified.

Tool settings — **Tool settings:** Used to specify how Mastercam assigns tool numbers, tool offset numbers, and the default values for feeds, speeds, coolant and related tool information.

Stock setup — **Stock setup :** The operator uses this folder to create the *stock model* for the job. The stock is displayed during backplot and is machined as a solid model display during verify.

Safety zone — **Safety zone :** Used to specify a safe area around the WCS(0,0,0). Mastercam places the tool outside to the safe area when retracting it to the retract plane. The safety zone helps to avoid tool collisions due to indexing, part features or fixtures.

Toolpath Group 1 — **Toolpath Group:** A toolpath group is a subgroup of a Machine Group. It contains the file folders that list the names of the machining operations in the group.

OPERATION NAME

1-Simple drill-no peck - [WCS:TOP] - [Tplane:TOP] — **Operation Folder:** Each folder has an operation name to describe the type of machining executed and *four* parts: Parameters, Tool definition, Part geometry and Toolpath.

Parameters — **Parameters :** stores machining data such as as tools, cutting depth, speed, feed for the operation

#1-0.1250 SPOT DRILL - 1/8 — **Tool definition :** stores data describing the tool's size and shape for the operation

Geometry (4) points — **Part geometry :** stores point locations for drilling and chains for milling for an operation

Toolpath - 4.8K - EXAMPLE5-2.NC — **NCI:** Numerical Control Intermediate file stores all the toolpath data. *Mastercam* uses the NCI file to generate the NC word address part program.

Operation Insert Arrow: Used to create a *new* operation positioned at any desired location within the list of *existing* operations. Click on the arrow, drag and drop it *after* the operation the *new* operation is to be created.

The Operations Manager is the *control center* for *adding new machining operations* or *changing existing operations for* the *current job*. An operation contains all the information *Mastercam* needs to machine a toolpath.

Associativity

The parts Parameters, Tool definition, Part geometry and Toolpath are all associated or linked together. When the operator enters a different tool and cutting depth in Parameters and adds a drill point in Part geometry *Mastercam* flags these changes. **Associativity** enables the operator to direct *Mastercam* to **automatically regenerate** *all the other parts of the operation as it incorporates the changes*.

States of Associativity

Mastercam will display one of two terms to describe the current state of associativity for an operation as follows:

Clean - A clean operation has *all* its specified parameters *matching* the associated geometry. All *new* operations and those successfully *regenerated* are *clean*.

Dirty - An operation whose parameters have been changed so that they *no longer match* the associated geometry are *dirty*. *Mastercam* will flag each *dirty* operation by placing a **red X** on the operation's *Toolpath icon*.✖ The operation can be changed to **Clean** by clicking on the Regenerate All Dirty Operations icon 🔁 .

7-3 Creating New Operations

New machining operations can be created within the Operations Manager

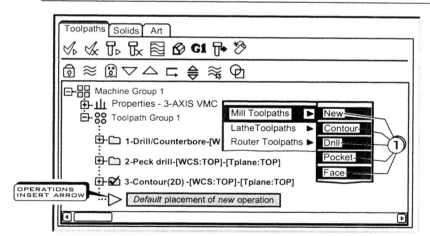

CREATING A NEW OPERATION PLACED AT THE END OF THE EXISTING LIST(DEFAULT)

➤ **Right** Click and move the cursor down to **Mill Toolpaths** over and down to the new machining operation to be created; Click ①

Mastercam will display the **Chaining** dialog box if a Contour, Pocket or Face operation were selected. *Mastercam* displays the **Drill Point Selection** dialog box when the Drill operation is chosen. Refer to Chapters 5 and 6 for further details on creating drill, contour and pocket operations.

CREATING A NEW OPERATION PLACED WITHIN THE EXISTING LIST

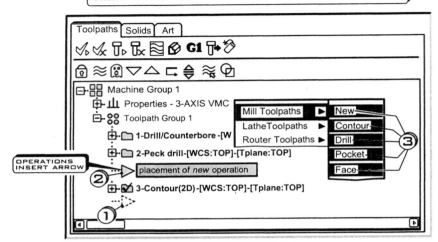

➤ **Left** Click ① on the Operations Insert Arrow and keeping the *left mouse button depressed* move the arrow to ② *after* the operation the new operation is to be created and *release*.

➤ **Right** Click and move the cursor down to **Mill Toolpaths** over and down to the new machining operation to be created; Click ③

Note: the Move Insert Arrow buttons: *Down one item* ▽
Up one item △
can also be used to position the Insert Arrow

7-4 Moving Existing Operations

The move function *changes the existing machining order.* This is useful when roughing finishing and chamfering operations need to be rearranged to minimize tool changes

METHOD A: PICK / PLACE

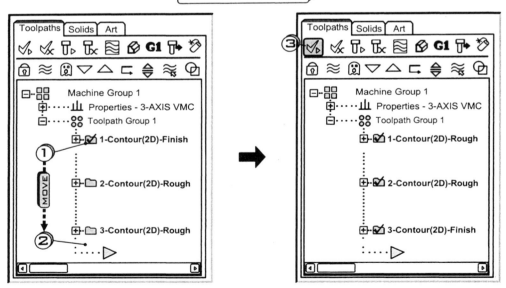

➤ Click ① on the file folder of the operation to be moved and *keeping the Left mouse button depressed* move the folder to the new location ② and release.

➤ Click ③ the Select All button ☑ to *renumber* the operations

METHOD B: OPTION MENU

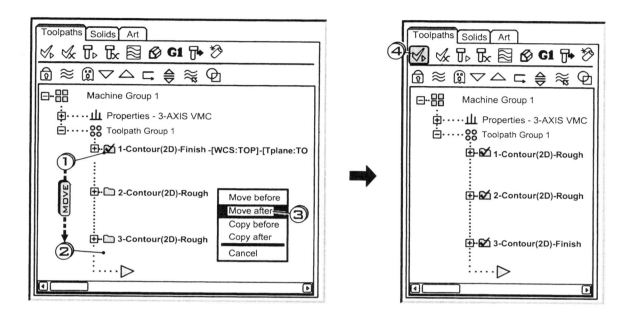

▶ **Right** Click ① on the file folder of the operation to be moved and **keeping the Right mouse button depressed** *drag the folder to the new location* ② *and release*.

▶ Click ③ the desired move selection from the menu

▶ Click ④ the Select All button 🗹 to **renumber** the operations

7-5 Copying Existing Operations

Copying is an *efficiency* tool which is especially useful when *different* machining operations need to be performed on the *same* tool path. A drilling example would be a set of holes that first need to be center drilled then drilled and finally tapped. A milling application would involve roughing a contour with endmill-A using speed-A and feed-A and finishing the contour with endmill-B using speed-B and feed-B.

OPTION MENU

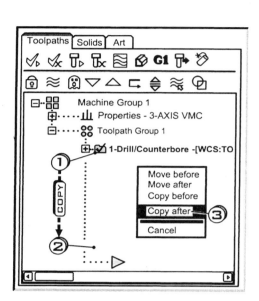

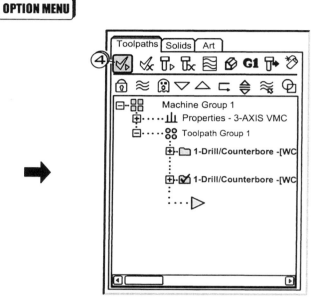

➤ **Right** Click ① on the file folder of the operation to be copied and ***keeping the Right mouse button depressed*** drag the folder to the new location ② and release.

➤ Click ③ the desired Copy selection from the menu

7-6 Adding Comments to Operation File Names

The operator can add comments after an operation file name. This additional display note helps to describe the machining that is performed.

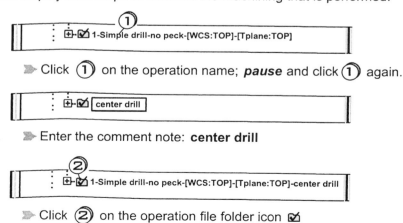

➤ Click ① on the operation name; ***pause*** and click ① again.

➤ Enter the comment note: **center drill**

➤ Click ② on the operation file folder icon ☑

EXAMPLE 7-1

Assume a CAD model the top view shown in Figure 7-1 exists in *Mastercam* and the operation to *center drill* the five holes has *already* been created in the Operations Manager. Use the Copy function as an aid in quickly generating the machining operations needed to produce the drill and tap holes.

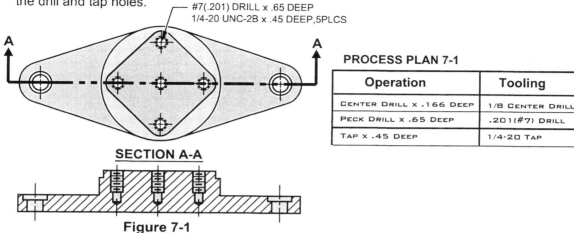

#7(.201) DRILL x .65 DEEP
1/4-20 UNC-2B x .45 DEEP,5PLCS

PROCESS PLAN 7-1

Operation	Tooling
CENTER DRILL x .166 DEEP	1/8 CENTER DRILL
PECK DRILL x .65 DEEP	.201(#7) DRILL
TAP x .45 DEEP	1/4-20 TAP

SECTION A-A

Figure 7-1

A) PECK DRILL THE HOLES .201DIA X .65 DEEP

♦ ENTER THE OPERATIONS MANAGER AND
MAKE A COPY OF THE CENTER DRILL OPERATION

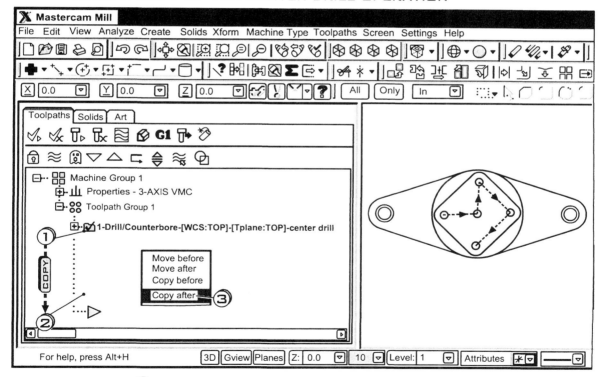

>> *Right* Click ① on the file folder of the operation to be copied and ***keeping the Right mouse button depressed*** *drag the folder to the new location* ② *and release.*

>> Click ③   Copy after   selection from the menu

◆ ENTER THE REQUIRED PECK DRILL PARAMETERS INTO THE COPY

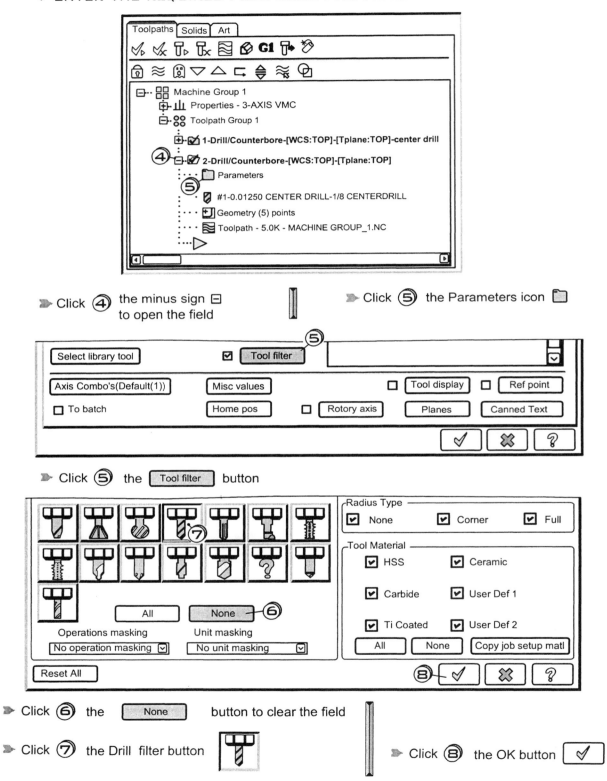

➤ Click ④ the minus sign ⊟ to open the field   ➤ Click ⑤ the Parameters icon 🗀

➤ Click ⑤ the [Tool filter] button

➤ Click ⑥ the [None] button to clear the field

➤ Click ⑦ the Drill filter button   ➤ Click ⑧ the OK button ✓

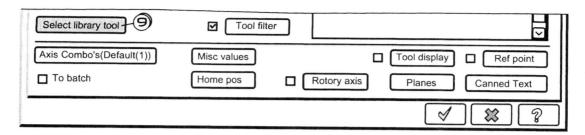

≫ Click ⑨ the [Select library tool] button

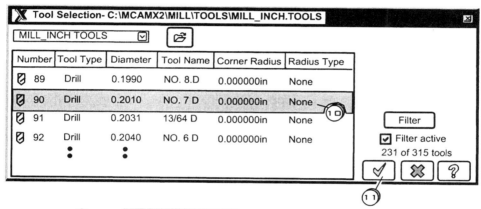

≫ Click ⑩ the .201 Drill Tool

≫ Click ⑪ the OK button ☑ to bring the tool from the library into the part file for the operation.

◆ AUTO ASSIGN SPEEDS FEEDS FOR THE TOOL

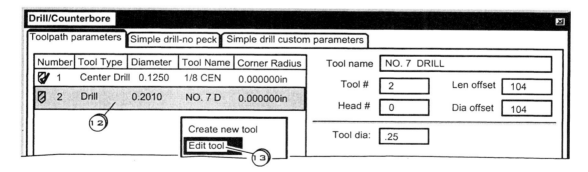

≫ Click ⑫ on the .2010 Drill tool

≫ **Right** Click move the cursor down and Click ⑬ **Edit tool**

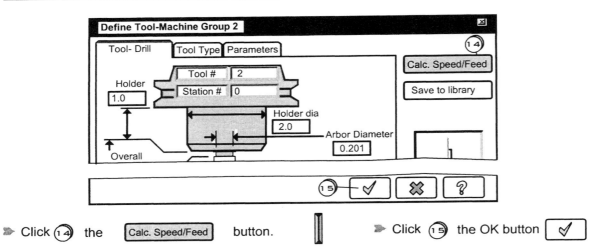

➤ Click ⑭ the [Calc. Speed/Feed] button.   ➤ Click ⑮ the OK button [✓]

◆ ENTER THE REQUIRED .201 DIA HOLE MACHINING PARAMETERS

➤ Click ⑯ [Peck drill-full retract] tab

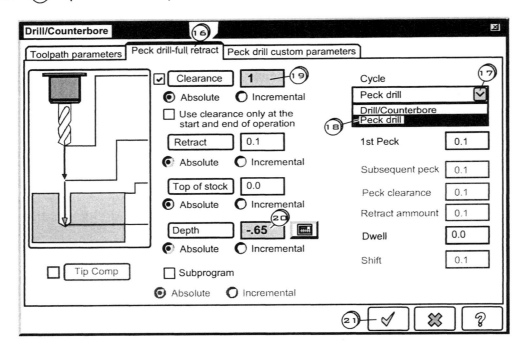

➤ Click ⑰ ☑ the toggle down button

➤ Click ⑱ [Peck drill]

➤ Click ⑲ in the Clearance box;
         enter [1]

➤ Click ⑳ in the Depth box;
         enter [-.65]

➤ Click ㉑ the OK button [✓]

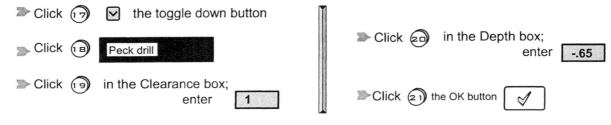

◆ ADD A DISPLAY NOTE TO HELP DESCRIBE THE OPERATION'S MACHINING. REGENERATE THE COPIED OPERATION TO INCORPORATE ANY CHANGES INTO ALL ITS OTHER PARTS VIA ASSOCIATIVITY.

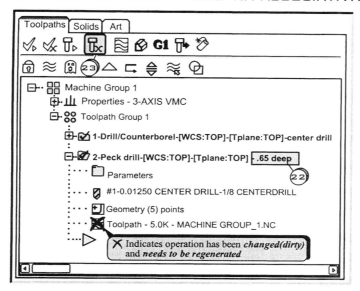

▶ Click ㉒ on the operation name, pause Click ㉒ and enter the note: **- .65 deep**

▶ Click ㉓ the Regen All Dirty Operations button ⌷

C) TAP THE HOLES 1/4-20 X .45 DEEP

◆ MAKE A COPY OF THE PECK DRILL OPERATION

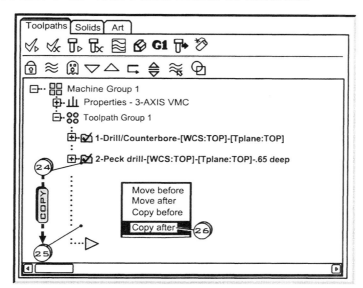

▶ *Right* Click ㉔ on the file folder of the operation to be copied and *keeping the Right mouse button depressed* drag the folder to the new location ㉕ and release.

▶ Click ㉖ **Copy after** selection from the menu

◆ ENTER THE REQUIRED TAPPING PARAMETERS INTO THE COPY

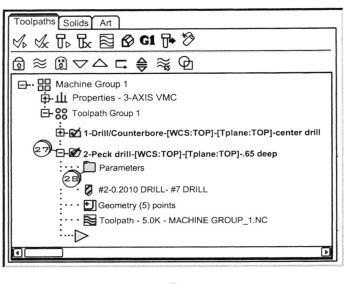

➤ Click ㉗ the minus sign ⊟ to open the field

➤ Click ㉘ the Parameters icon 🗀

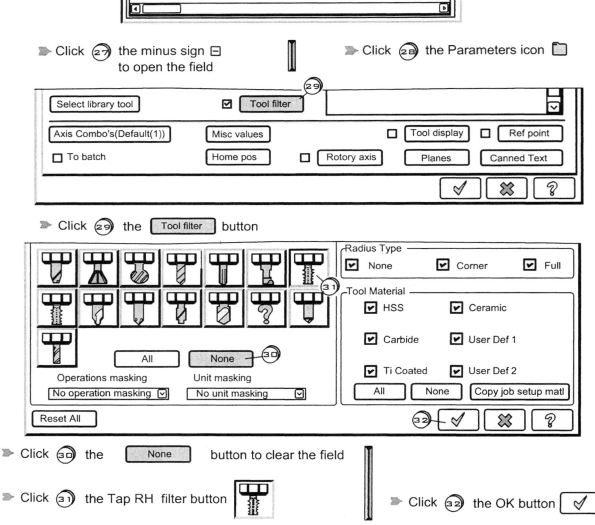

➤ Click ㉙ the Tool filter button

➤ Click ㉚ the None button to clear the field

➤ Click ㉛ the Tap RH filter button

➤ Click ㉜ the OK button ✓

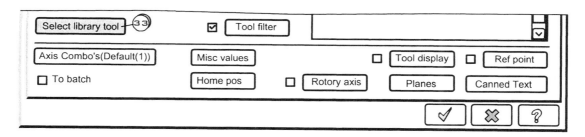

➤ Click ③③ the [Select library tool] button

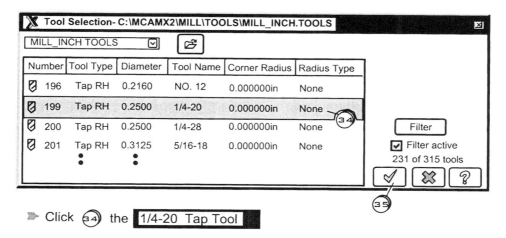

➤ Click ③④ the [1/4-20 Tap Tool]

➤ Click ③⑤ the OK button [✓] to bring the tool from the library into the part file for the operation.

◆ AUTO ASSIGN SPEEDS FEEDS FOR THE TOOL

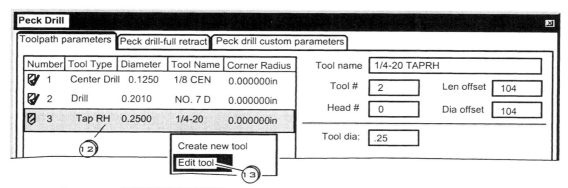

➤ Click ③⑥ on the 1/4-20 Tap tool

➤ *Right* Click move the cursor down and Click ③⑦ [Edit tool]

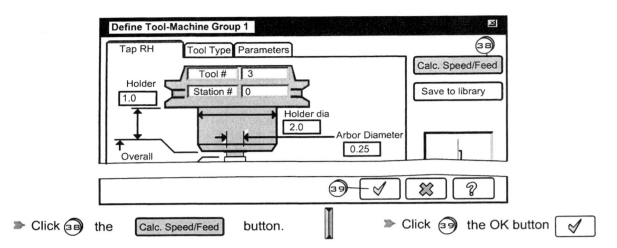

➤ Click ㊳ the [Calc. Speed/Feed] button.   ➤ Click ㊴ the OK button [✓]

◆ ENTER THE REQUIRED PARAMETERS FOR TAPPING THE 1/4-20 THREADS

➤ Click ㊵ [Tapping-feed in, reverse spindle-feed out] tab

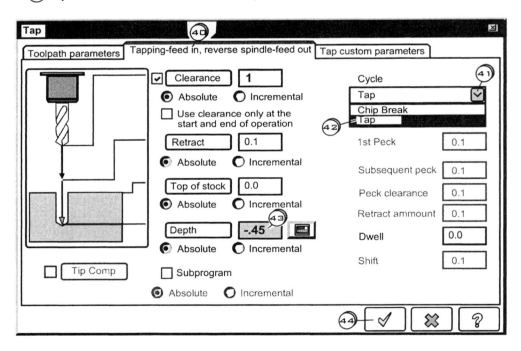

➤ Click ㊶ [▼] the toggle down button   ➤ Click ㊸ in the Depth box;
                                                    enter [-.45]

➤ Click ㊷ [Tap]

➤Click ㊹ the OK button [✓] to specify
1/4-20 tapping parameters for
the operation.

◆ ADD A DISPLAY NOTE TO HELP DESCRIBE THE OPERATION'S MACHINING. REGENERATE THE COPIED OPERATION TO INCORPORATE ANY CHANGES INTO ALL ITS OTHER PARTS VIA ASSOCIATIVITY.

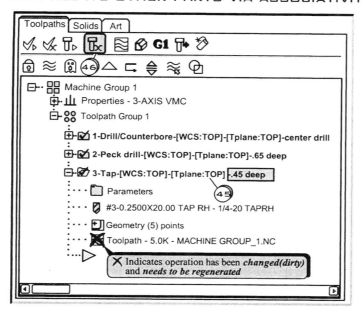

▶ Click ㊺ on the operation name, pause Click ㊺ and enter the note: **- .45 deep**

▶ Click ㊻ the Regen All Dirty Operations button [icon]

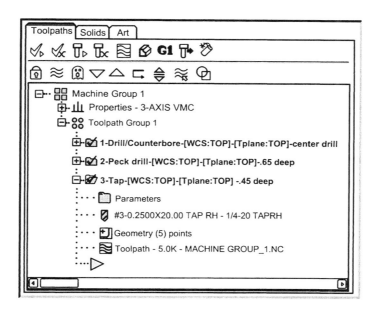

All the machining operations needed to produce the drill and tap holes have now been created in the Operations Manager with the aid of the *Copy* feature.

7-7 Copying Parts of Existing Operations

Any of the *individual parts* of one operation: Parameters, Tools or Part geometry can also be *copied* to another *similar* operation.

Restriction with Parameters: copying can *only* be done between *similar* operation types(contour to contour or pocket to pocket, etc)

Restriction with Part geometry: chains can only be copied between contouring, pocketing or drilling operations.

EXAMPLE 7-2

Copy the Parameters settings from **1-Contour-(2D)** to **2-Contour-(2D)**

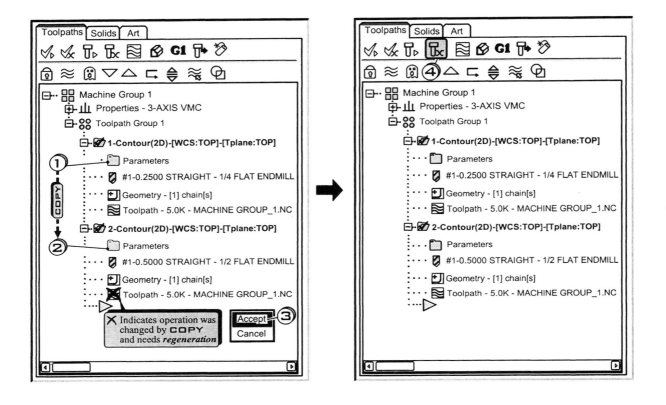

➤ Depress the **Right** mouse button ① on the Parameter icon ☐ to be copied; ***keeping the right mouse button depressed***, move the cursor over the Parameter icon ☐ receiving the copy ② and ***release.***

➤ Click ③ [Accept]

➤ Click ④ the Regenerate All Dirty Operations button 🔲

7-8 Deleting Existing Operations

The Operations Manager is used to delete a *selected* operation a specific *set* of operations or *all* the operations for a job.

DELETING A SELECTED OPERATION

EXAMPLE 7-3

Delete the existing **Contour(2D)** operation shown in Figure 7-2(left).

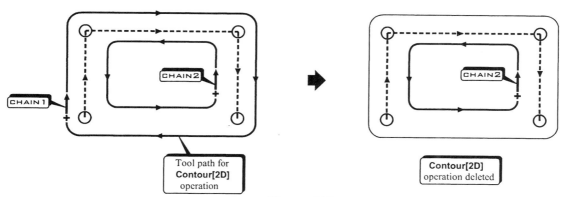

Figure 7-2

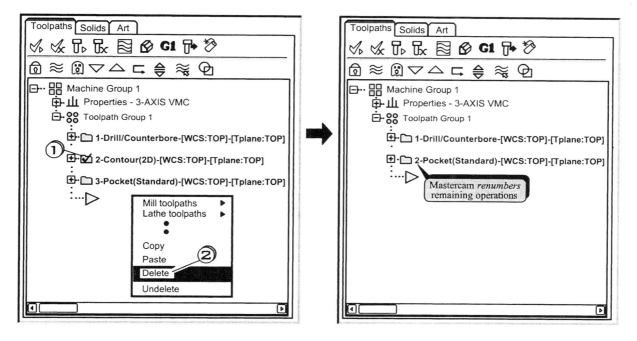

▶ Click ① on the file folder of the operation to be deleted ☑ 2-Contour(2D)-[WCS:TOP]-[Tplane:TOP]

▶ *Right* click , move the cursor down to ▮Delete▮ and Click ②

| DELETING A SET OF OPERATIONS |

EXAMPLE 7-4

Delete the existing **Contour[2D]** and **Pocket[Standard]** operations shown in Figure 7-3(left).

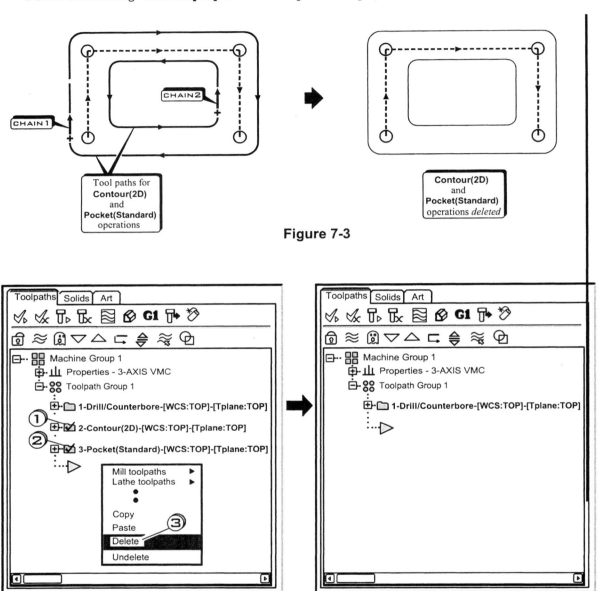

Figure 7-3

➤ Click ① on the file folder of the operation to be deleted ☑ 2-Contour(2D)-[WCS:TOP]-[Tplane:TOP]

➤ Press the [Ctrl] key and *keeping it depressed,* Click ② the next operation file folder ☑ 3-Pocket

➤ *Right* click , move the cursor down to [Delete] and Click ③

| DELETING ALL OPERATIONS |

EXAMPLE 7-5

Delete all the existing tool paths shown below

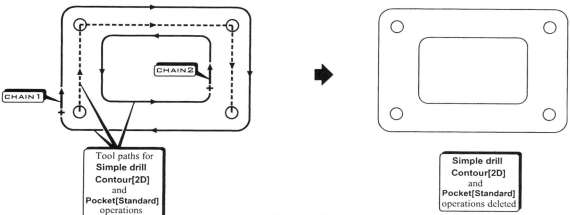

Figure 7-4

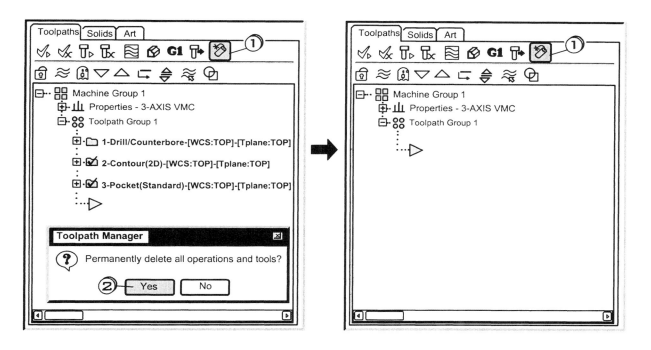

➤ Click ① on the Delete All Operations Groups and Tools icon

➤ Click ② the Yes button

7-9 Editing Existing Toolpaths

Mastercam provides the operator with a full array of features for *editing existing* tool paths created by *chaining*. This allows the operator to quickly respond to design changes that often occur as a part evolves.

All the tool path editing functions based on *chaining* are available from the ***Chain Manager*** dialog box. To enter the Chain Manager follow the steps listed below.

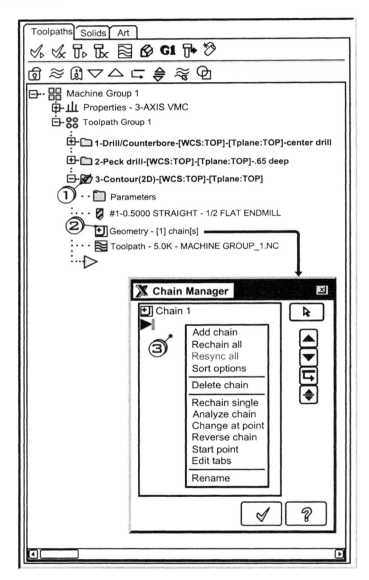

▶ Click ① on the folder of the operation to be edited such that a *check* appears

▶ Click ② on the Part Geometry icon for the operation to be edited   ⊞

▶ ***Right*** Click ③ to display the Chain Manager's editing functions

The applications of the editing functions in the pull down menu will now be considered.

Add chain

Enables the operator to *add more chains to the existing operation.* Additional data *need not* be entered into the *tools* and *parameters* dialog boxes. The Regenerate All Dirty Operations function ⬛, however *must be executed* in the *Operations Manager to update* the changes to all of the other parts of the operation.

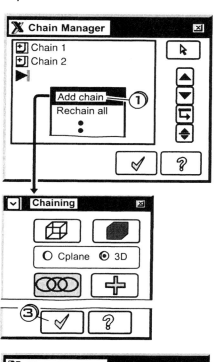

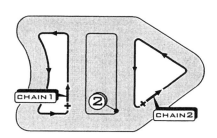

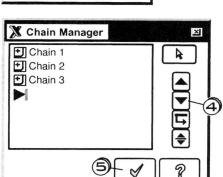

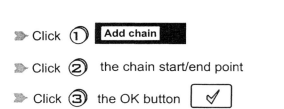

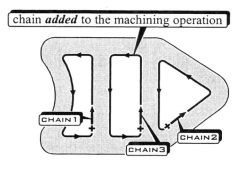

➤ Click ① **Add chain**

➤ Click ② the chain start/end point

➤ Click ③ the OK button ✓

➤ Click ④ the Move Insert Arrow Down One Item button ▼

➤ Click ⑤ the OK button ✓

Rechain all

Deletes all the chains listed in the Chain Manager for an operation and allows the operator to *redefine a new set of chains.*
Follow the operation with Regenerate All Dirty Operations Function

Rechain Single

Deletes an existing chain selected in the Chain Manager for an operation and allows the operator to *redefine it.*
Follow the operation with Regenerate All Dirty Operations Function

Start Point

Enables the operator to return to the graphics window and *reselect a new start point* for a *chain* selected in the Chain Manager for an operation.
Follow the operation with Regenerate All Dirty Operations Function

Delete chain

Deletes a chain selected.
Follow the operation with Regenerate All Dirty Operations Function

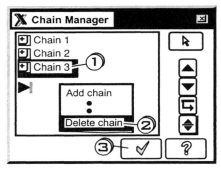

➤ **Right** Click ① on Chain 3 , move the cursor

➤ Click ② **Delete chain**

➤ Click ③ the OK button ✓

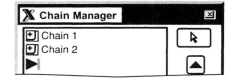

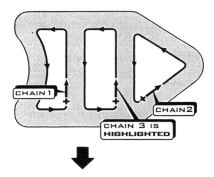

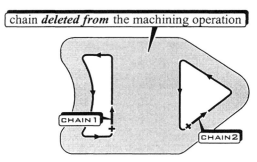

Reverse chain

Changes the direction of chaining for a *chain selected* in the Chain Manager for an operation. Reversing the chaining direction *changes the compensarion* from left to right and vica-versa.

Follow the operation with Regenerate All Dirty Operations Function

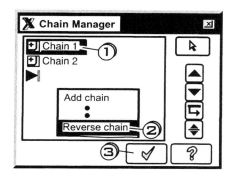

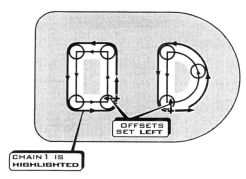

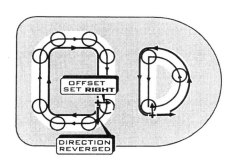

➤ **Right** Click ① on Chain 3 , move the cursor

➤ Click ② Reverse chain

➤ Click ③ the OK button ✓

Change at point

Enables the operator to change the machining parameters at a selected point on the tool path. *This function can only be applied to contour* operations. Follow the operation with Regenerate All Dirty Operations Function

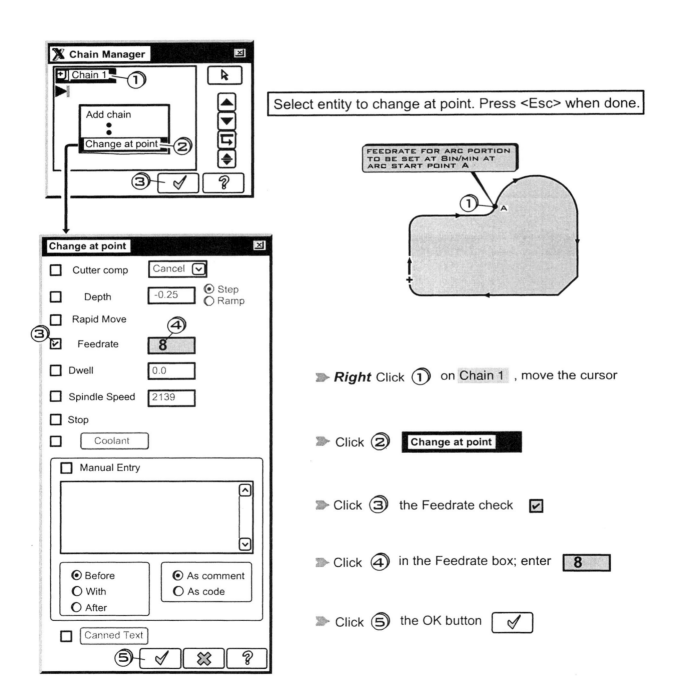

Select entity to change at point. Press <Esc> when done.

FEEDRATE FOR ARC PORTION TO BE SET AT 8IN/MIN AT ARC START POINT A

▶ ***Right*** Click ① on Chain 1 , move the cursor

▶ Click ② **Change at point**

▶ Click ③ the Feedrate check ☑

▶ Click ④ in the Feedrate box; enter 8

▶ Click ⑤ the OK button ✓

Analyze all   **Analyze Single**

Enables the operator to direct *Mastercam* to examine *all chains* or a *selected chain* in an operation for *two* types of *problems*: *errors* due to entities that *overlap* and *errors* caused by *reversal* in *chaining direction*. The *maximum length* that *defines a short entity* is also set in the Analyze Chain dialog box Follow the operation with Regenerate All Dirty Operations Function

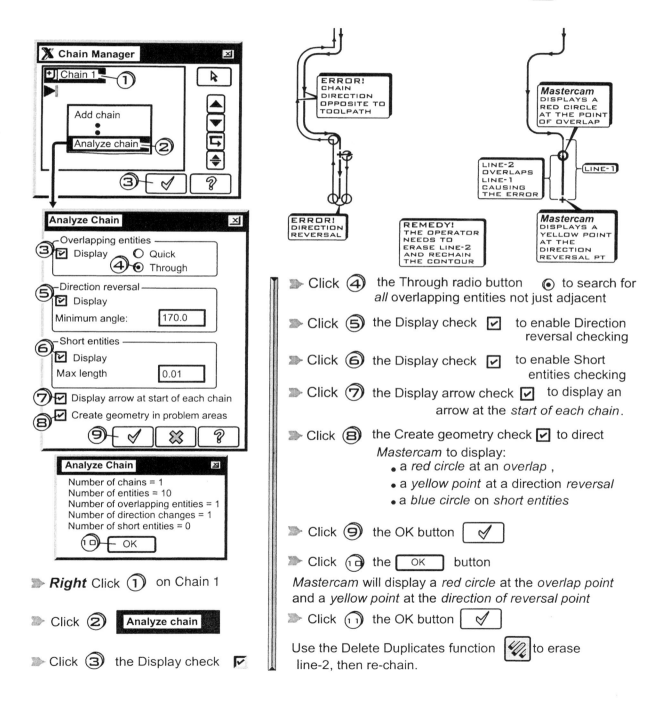

➤ Click ④ the Through radio button ◉ to search for *all* overlapping entities not just adjacent

➤ Click ⑤ the Display check ☑ to enable Direction reversal checking

➤ Click ⑥ the Display check ☑ to enable Short entities checking

➤ Click ⑦ the Display arrow check ☑ to display an arrow at the *start of each chain*.

➤ Click ⑧ the Create geometry check ☑ to direct *Mastercam* to display:
   • a *red circle* at an *overlap* ,
   • a *yellow point* at a direction *reversal*
   • a *blue circle* on *short entities*

➤ Click ⑨ the OK button ✓

➤ Click ⑩ the OK button

Mastercam will display a *red circle* at the *overlap point* and a *yellow point* at the *direction of reversal point*

➤ Click ⑪ the OK button ✓

Use the Delete Duplicates function to erase line-2, then re-chain.

➤ **Right** Click ① on Chain 1

➤ Click ② **Analyze chain**

➤ Click ③ the Display check ☑

7-10 Changing the Chaining Order

The *order* in which the chains are listed in the Chain Manager determines the *machining order* within an operation. The operator can edit the machining order by changing the order in which the chains are listed.

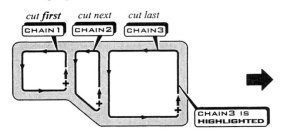
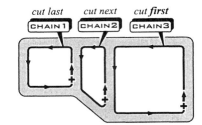

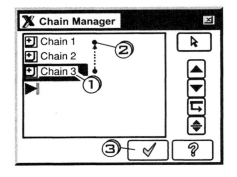
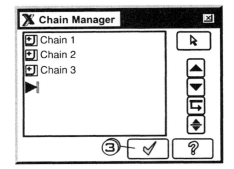

➤ Click ① the chain to be moved; **keeping** the **left** mouse button **depressed** move the chain to its **new** **location** ② and **release**

➤ Click ③ the OK button

➤ Click the Regen All Dirty Operations  button in the Operations Manager.

Rename

Allows the operator to rename an existing chain selected.

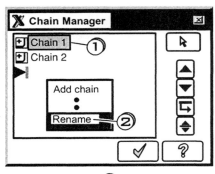

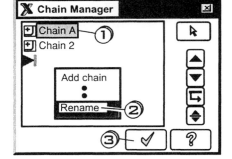

➤ **Right** Click ① on Chain 1

➤ Click ② **Rename**

➤ Enter the new chain name: Chain A

➤ Click ①

➤ Click ③ the OK button

7-11 Expanding and Collapsing the Operations Display Listings

The operations listing for a group can be *fully expanded* or *collapsed* as needed.

DISPLAY EXPANDED　　　　　　**DISPLAY COLLAPSED**

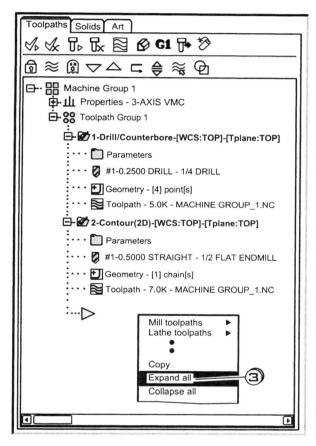
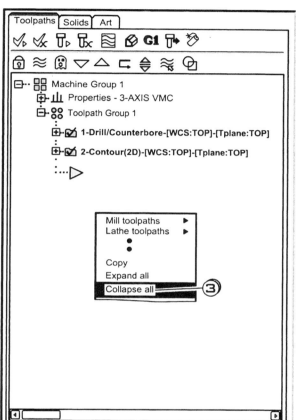

➤ **Right** Click and move the cursor down to 　　　➤ **Right** Click and move the cursor down to

　Expand all ; Click ① 　　　　　　　　　Collapse all ; Click ①

7-12 Displaying the Operations Manager Listings as a Doc File

The Doc file function is used to direct *Mastercam* to produce a *printable document* ASCII file named OPSLIST.TXT from the *display* shown in the Operations Manager..

VIEWING AND PRINTING THE DOC FILE

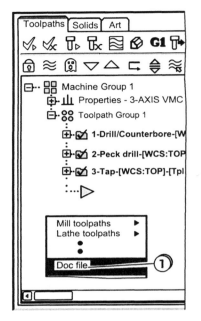

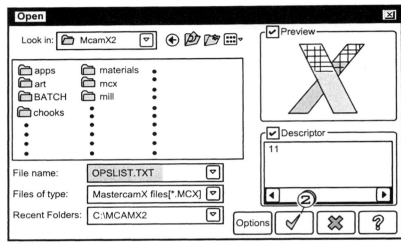

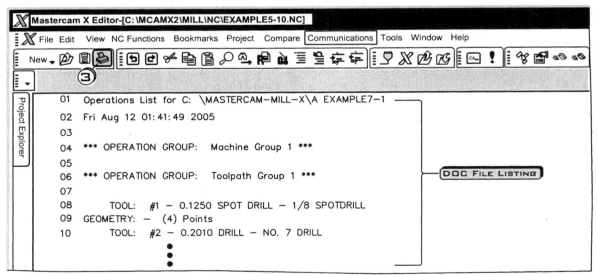

DOC FILE LISTING

▶ *Right* click , move the cursor down
     to Doc file and Click ①

▶ Click ② the OK button   ▶ Click ③ the Print button

EXERCISES

TUTORIAL PRACTICE EXERCISE

 7-1) The operations for producing the part shown in Figure 7p-1(left). currently exist in the Operations Manager. Edit the existing operations to produce the re-designed part as shown in Figure 7p-1(right)

All pockets .188 deep

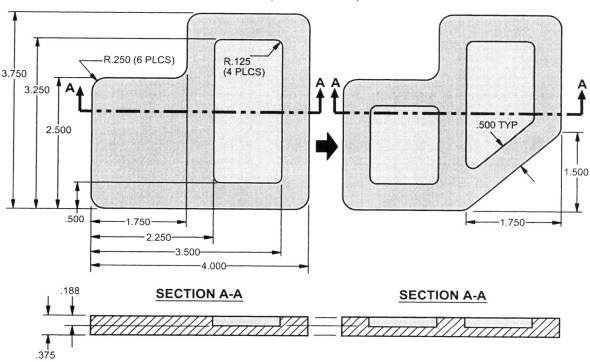

SECTION A-A   **SECTION A-A**

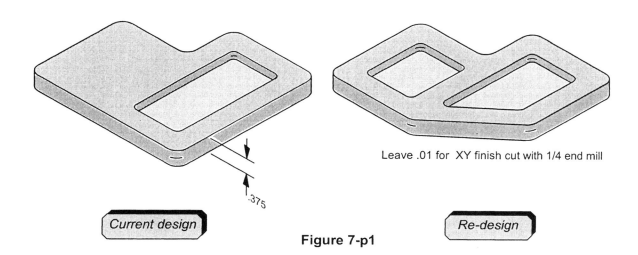

Leave .01 for XY finish cut with 1/4 end mill

Current design   Re-design

Figure 7-p1

A) COPY THE FILE EX7-1 IN THE FOLDER 📁 CHAPTER7 ON CD AT THE
BACK OF THIS TEXT TO YOUR DIRECTORY C:\MCAMX\MCX\JVAL-MILL

your initials

OPEN THE FILE FROM YOUR DIRECTORY.

Refer to Chapter 1, p1-14 for a discussion on the creation of your own directory.

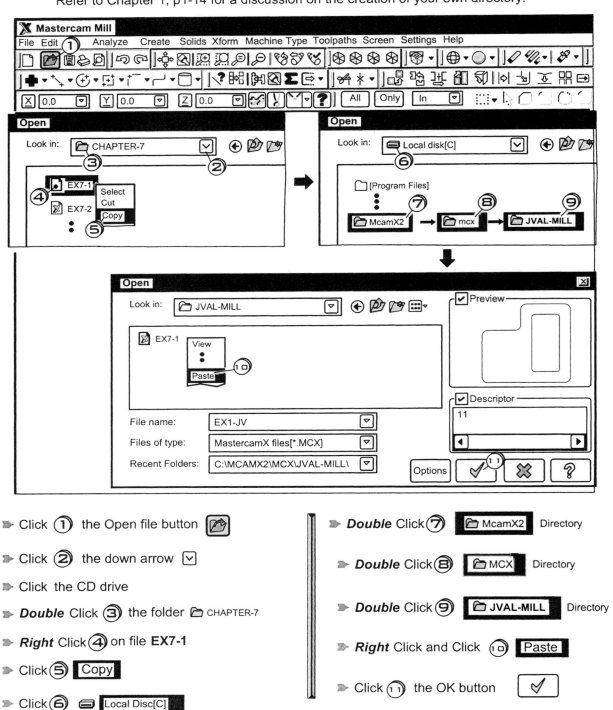

➤ Click ① the Open file button 📂

➤ Click ② the down arrow ☑

➤ Click the CD drive

➤ **Double** Click ③ the folder 📁 CHAPTER-7

➤ **Right** Click ④ on file **EX7-1**

➤ Click ⑤ [Copy]

➤ Click ⑥ 💾 [Local Disc[C]]

➤ **Double** Click ⑦ [📁 McamX2] Directory

➤ **Double** Click ⑧ [📁 MCX] Directory

➤ **Double** Click ⑨ [📁 JVAL-MILL] Directory

➤ **Right** Click and Click ⑩ [Paste]

➤ Click ⑪ the OK button ✓

B) EDIT THE PART GEOMETRY

♦ CREATE THE 1.75 X 1.5 CHAMFER

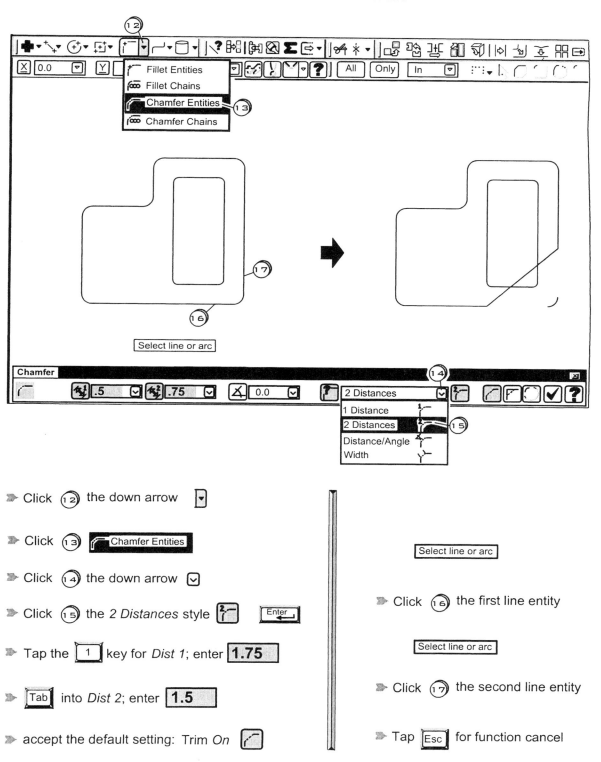

Select line or arc

➤ Click ⑫ the down arrow ▾

➤ Click ⑬ [Chamfer Entities]

➤ Click ⑭ the down arrow ▾

➤ Click ⑮ the *2 Distances* style [chamfer icon] [Enter]

➤ Tap the [1] key for *Dist 1*; enter **1.75**

➤ [Tab] into *Dist 2*; enter **1.5**

➤ accept the default setting: Trim *On* [icon]

Select line or arc

➤ Click ⑯ the first line entity

Select line or arc

➤ Click ⑰ the second line entity

➤ Tap [Esc] for function cancel

♦ CREATE .5 OFFSET

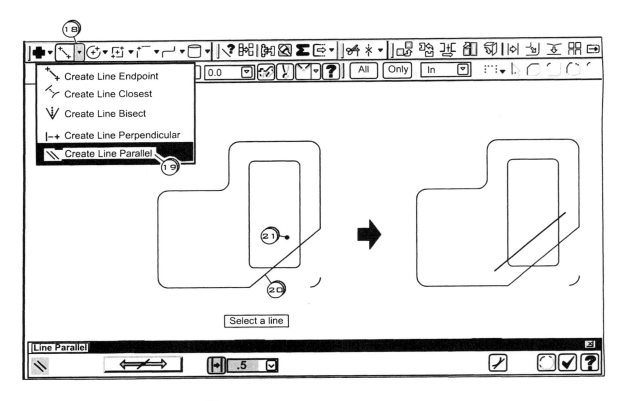

Select a line

> Click ⑱ the down arrow

> Click ⑲ ◼ Create Line Parallel

> Tap the D key for *offset distance*

> Enter .5

  Select a line

> Click ⑳ near the line entity

  Indicate the offset direction

> Click ㉑   the side on which the parallel line is to be created

> Press Esc to cancel the function

♦ CREATE .125R AND .25R FILLETS

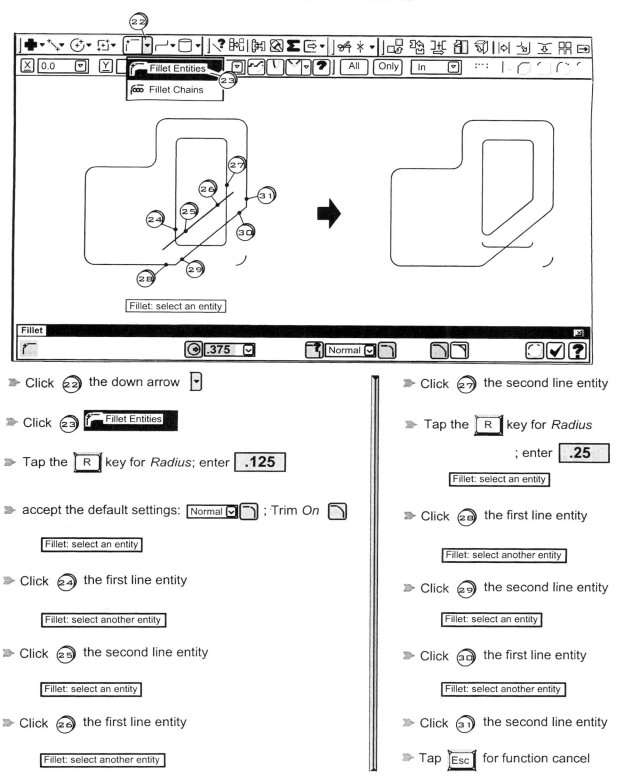

> Click (22) the down arrow ▼

> Click (23) [🔲 Fillet Entities]

> Tap the [R] key for *Radius*; enter [.125]

> accept the default settings: [Normal ▼][🔲] ; Trim *On* [🔲]

[Fillet: select an entity]

> Click (24) the first line entity

[Fillet: select another entity]

> Click (25) the second line entity

[Fillet: select an entity]

> Click (26) the first line entity

[Fillet: select another entity]

> Click (27) the second line entity

> Tap the [R] key for *Radius* ; enter [.25]

[Fillet: select an entity]

> Click (28) the first line entity

[Fillet: select another entity]

> Click (29) the second line entity

[Fillet: select an entity]

> Click (30) the first line entity

[Fillet: select another entity]

> Click (31) the second line entity

> Tap [Esc] for function cancel

♦ CREATE THE 1.25 x 1.5 RECTANGLE

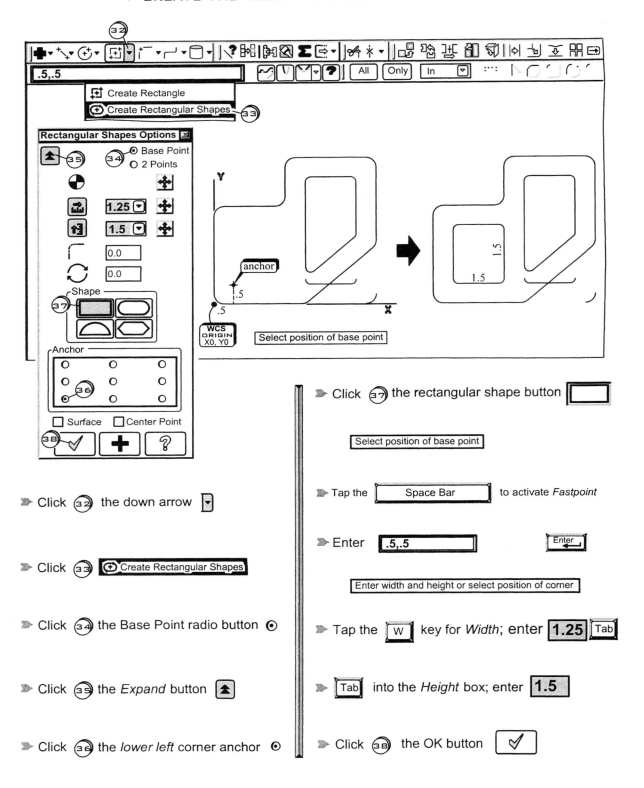

➣ Click ③② the down arrow ▾

➣ Click ③③ ⊕ Create Rectangular Shapes

➣ Click ③④ the Base Point radio button ⊙

➣ Click ③⑤ the *Expand* button ⬆

➣ Click ③⑥ the *lower left* corner anchor ⊙

➣ Click ③⑦ the rectangular shape button ▭

Select position of base point

➣ Tap the [Space Bar] to activate *Fastpoint*

➣ Enter [.5,.5]   [Enter ⏎]

Enter width and height or select position of corner

➣ Tap the [W] key for *Width*; enter [1.25] [Tab]

➣ [Tab] into the *Height* box; enter [1.5]

➣ Click ③⑧ the OK button [✓]

♦ DELETE ALL UNNECESSARY GEOMETRY FROM THE CAD MODEL

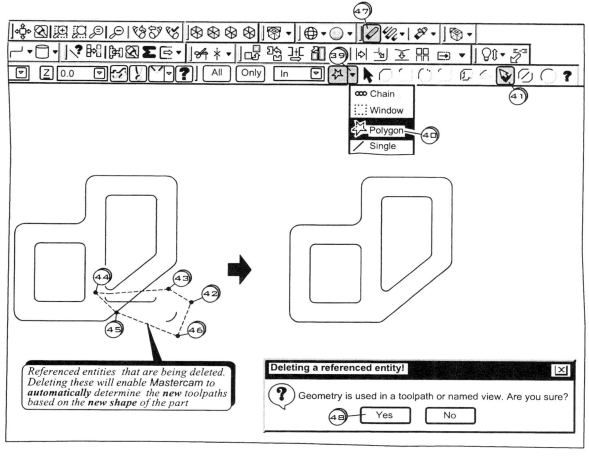

*Referenced entities that are being deleted. Deleting these will enable Mastercam to **automatically** determine the **new toolpaths** based on the **new shape** of the part*

Deleting a referenced entity!

Geometry is used in a toolpath or named view. Are you sure?

Yes   No

➤ Click (39) the down arrow

➤ Click (40) [Polygon] Polygon window

➤ Click (41) the toggle verify selection button

➤ Click (42) (43) (44) (45) (46) the window corners [Enter]

➤ Click (47) the Delete key or tap the [Del] key

➤ Click (48) the [Yes] key

C) ADD THE ADDITIONAL 1.25 x 1.5 POCKET CHAIN

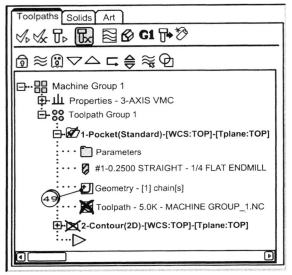

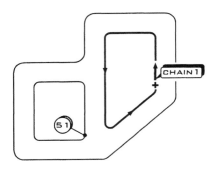

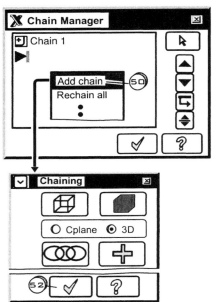

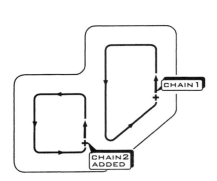

▶ Click ④⑨ on the Part Geometry icon
for the operation to be edited 🔲

▶ **Right** Click move the cursor;
Click ⑤⓪ **Add chain**

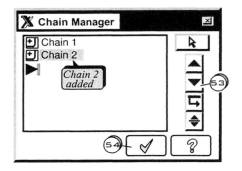

▶ Click ⑤① the chain start/end point

▶ Click ⑤② the OK button ✓

▶ Click ⑤③ the Move Insert Arrow
Down One Item button ▼

▶ Click ⑤④ the OK button ✓

D) Edit Contour[2D], Leave .01 for XY Finish Cut

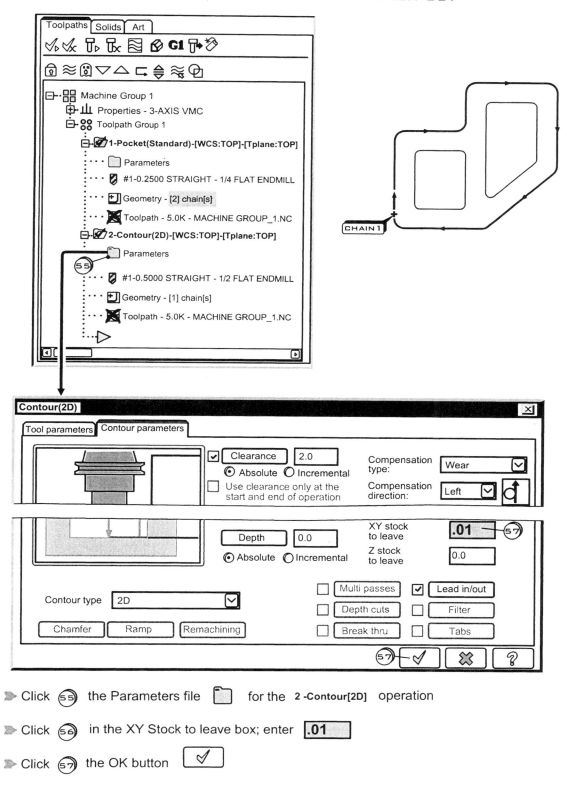

➤ Click ⑤⑤ the Parameters file 📁 for the **2 -Contour[2D]** operation

➤ Click ⑤⑥ in the XY Stock to leave box; enter **.01**

➤ Click ⑤⑦ the OK button ✓

E) FINISH CUT THE OUTSIDE WITH A 1/4 END MILL TOOL
- ENTER THE OPERATIONS MANAGER
- MAKE A COPY OF THE CONTOUR[2D] OPERATION
- EDIT THE PARAMETERS OF THE COPY[REPLACE 1/2ENDMILL TOOL WITH A 1/4 ENDMILL TOOL]

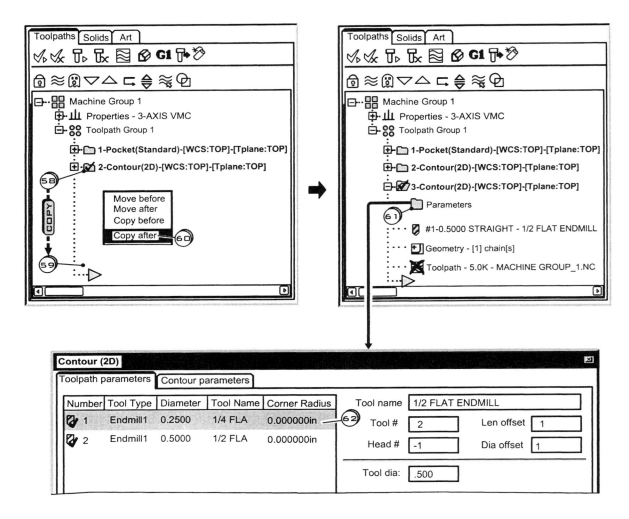

▶ **Right** Click ⑤⑧ on the file folder **2 Contour[2D]** *keeping the Right mouse button depressed* move the folder to the new location ⑤⑨ and release.

▶ Click ⑥⓪   **Copy after**

▶ Click ⑥① the Parameters icon 📁 for the **3 -Contour[2D]** operation

▶ Click ⑥② on the existing  tool

◆ EDIT THE CONTOUR PARAMETERS[SET THE FINISH CUT PARAMETERS]

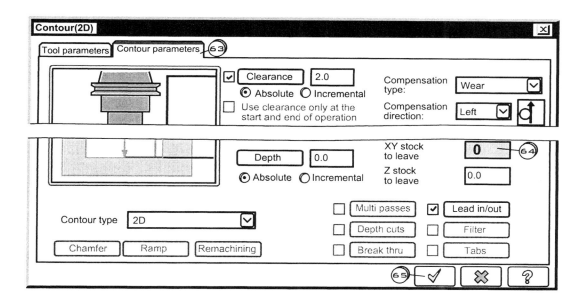

⟫ Click (63) the [Contour parameters] tab

⟫ Click (64) in the XY Stock to leave box; enter [0]

⟫ Click (65) the OK button [✓]

F) EXECUTE THE REGEN ALL DIRTY OPERATIONS FUNCTION SO THAT *Mastercam* **CAN AUTOMATICALLY UPDATE ALL THE PARTS OF EACH EDITED OPERATION VIA ASSOCATIVITY.**

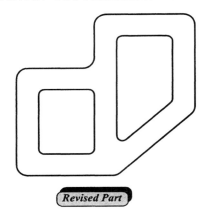

Revised Part

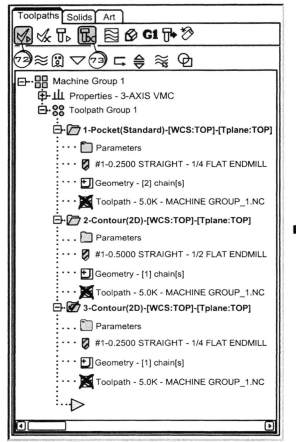 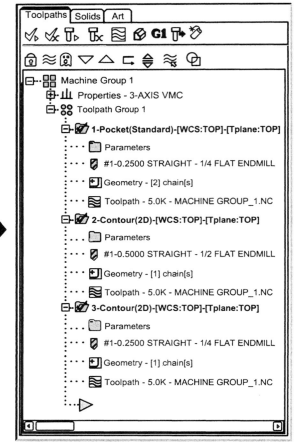

All Dirty Operations Regenerated

➤ Click ⑦² the Select All Operations button

➤ Click ⑦³ the Regenerate All Dirty Operations button

   The machining operations required for the re-desined part have now been generated.

7-2) The operations for producing the part shown in Figure 7p-2(left). currently exist in the Operations Manager. Copy the file **EX7-2** in the folder ◻**CHAPTER7** from the CD into the **JVAL-MILL** subdirectory on C drive. Open the file and edit the existing operations to produce the re-designed part as shown in Figure 7p-2(right)

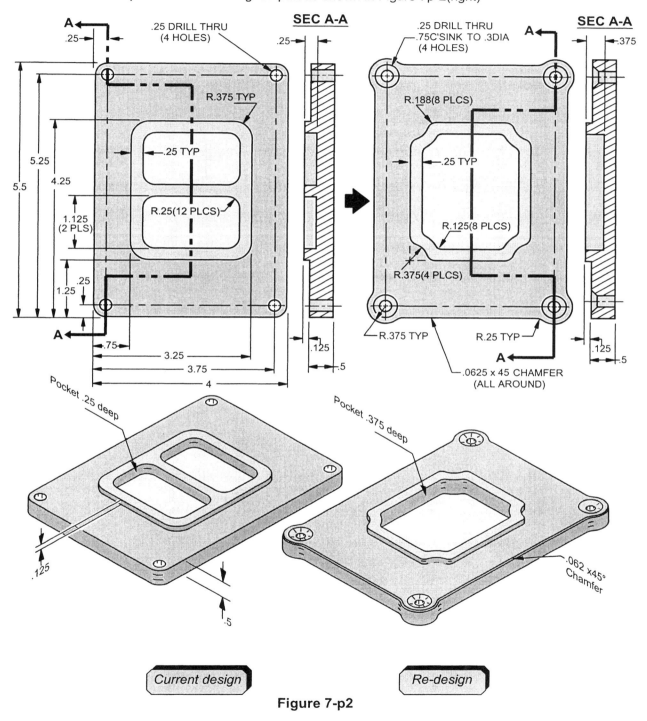

Figure 7-p2

7-3) The operations for producing the part shown in Figure 7p-3(left). currently exist in the Operations Manager. Copy the file **EX7-3** in the folder **⊐CHAPTER7** from the CD into the **JVAL-MILL** subdirectory on C drive. Open the file and edit the existing operations to produce the redesigned part as shown in Figure 7p-3(right)

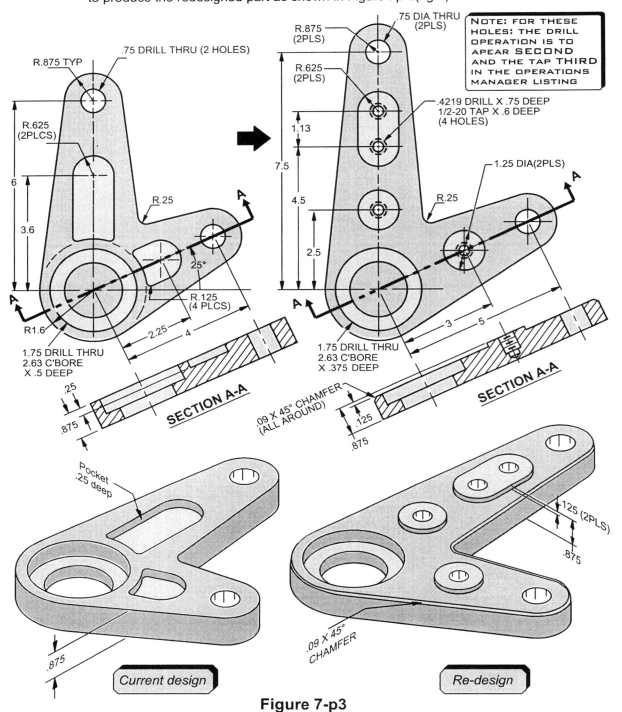

Figure 7-p3

7-4) The operations for producing the part shown in Figure 7p-4(left). currently exist in the Operations Manager. Copy the file **EX7-4** in the folder **⌂CHAPTER7** from the CD into the **JVAL-MILL** subdirectory on C drive. Open the file and edit the existing operations to produce the redesigned part as shown in Figure 7p-4(right)

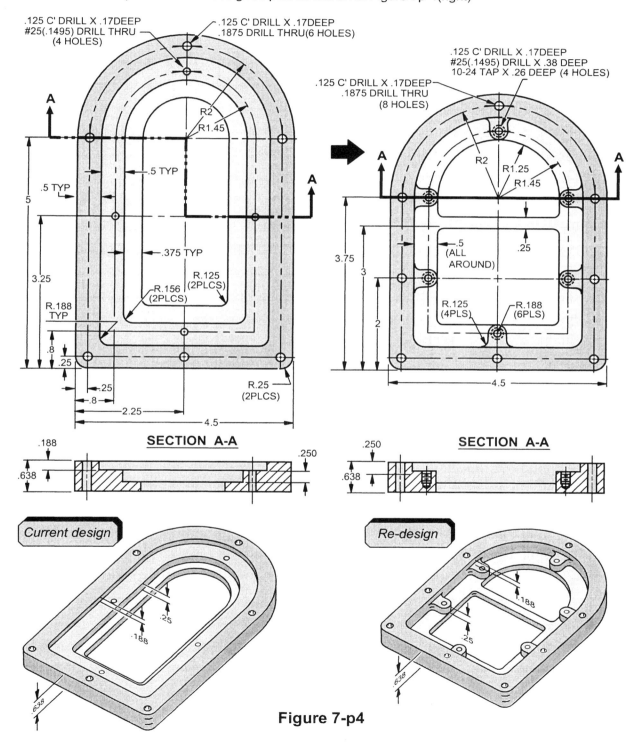

Figure 7-p4

a) Delete the *inner slot* geometry

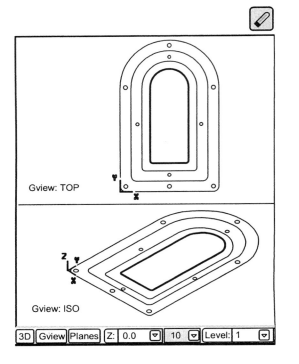

b) Use the *Xform, Translate, Rectang* commands to move the windowed geometry down by 1.25

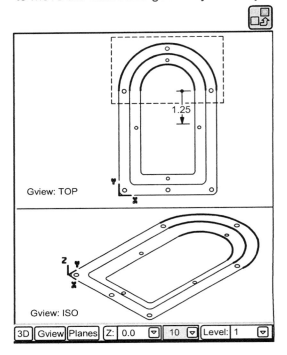

d) Use the *Xform, Offset Contour, Copy* command to create an copy offset .5 from the *inner contour*

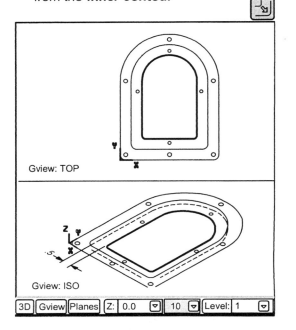

c) Use the *Modify, Trim, 1 entity* commands to trim the *extended line portions*

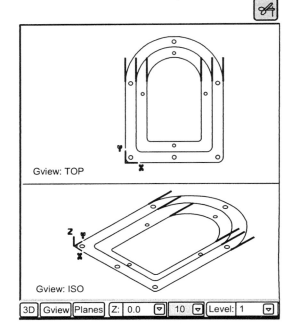

e) Create the .188R corner fillets at a Z depth of -.188

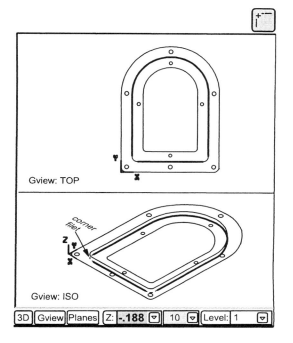

f) Use the *Create, Arc, Polar, Center* commands to create the .188R arc at a depth of -.188

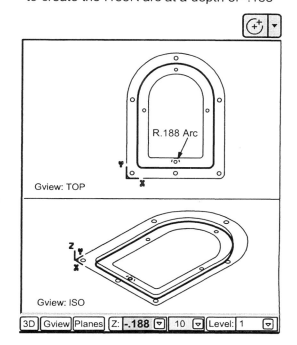

h) Use the *Xform, Translate, Between pts* commands to translate a copy of the tab between the hole centers

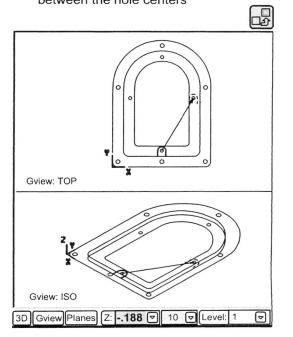

g) Use the *Create, Line, Polar,* commands to create the .312 long tangent lines

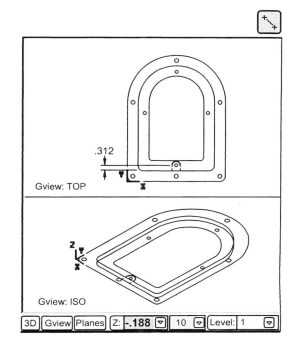

i) Use the **Xform, Rotate,** commands to rotate the tab 90°

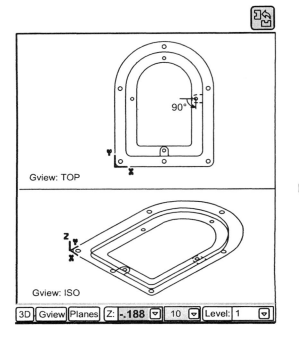

j) Use the **Xform, Mirror,** commands commands to create mirror a copy of the tab geometry

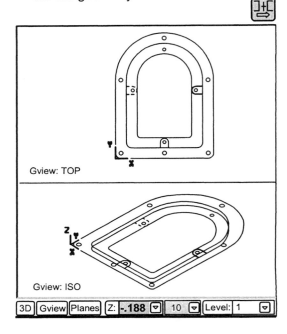

l) Use the **Xform, Translate, Rectang,** commands to move a *copy* of the windowed geometry down by 1.75

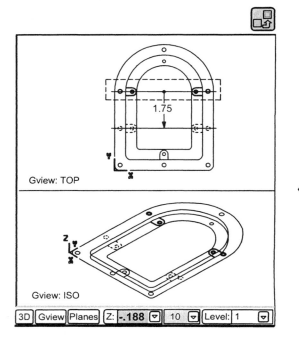

k) Use the **Xform, Translate, Rectang,** commands to move the windowed geometry up by .5

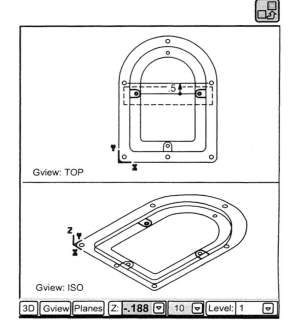

m) Use the **Xform, Rotate** commands to rotate by 90° a *copy* of the windowed geometry.

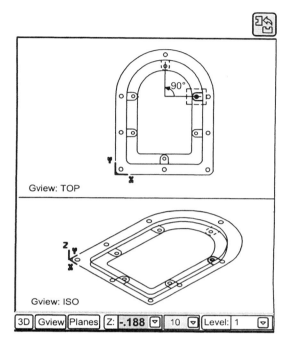

p) Use **Xform, Offset** commands to create the top line of the first slot boundary. Use the Isometric view to click the line at -.188 depth

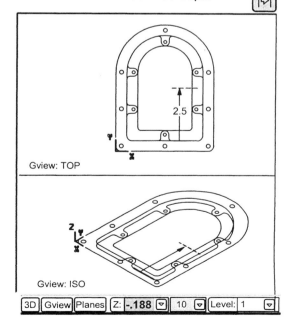

n) Use the **Modify, Trim, Divide** commands to remove the **geometry between the the lines of each tab**. Be sure to work in the *isometric view* to click on the geometry at the depth of -.188.

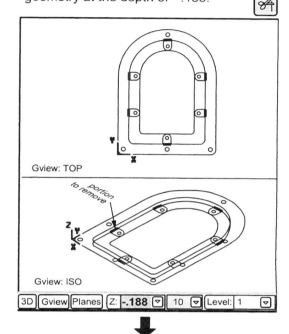

o) Use the **Create, Fillet** commands to create the .125R tab fillets at a depth of -.188. Work in the isometric view to click on the depth geometry.

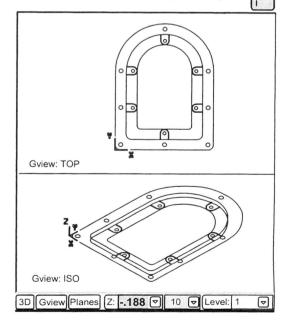

q) Use **Modify, Trim** commands
   to extend the top boundary line

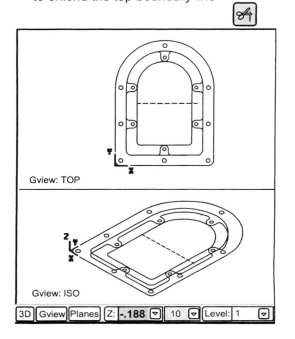

r) Use **Xform, Offset** commands to create
   the bottom line for the second slot boundary

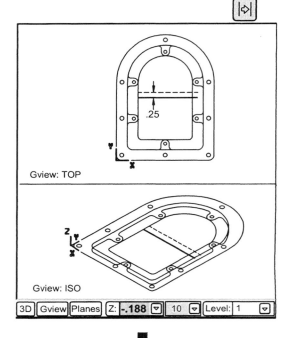

t) Use **Create, Fillet,** commands to
   insert the .125R fillets for the slots
   and complete the new edited shape

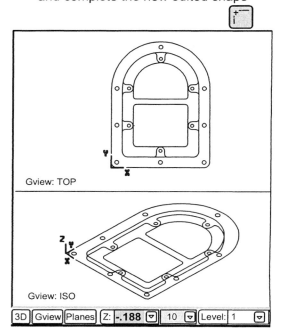

s) Use **Modify, Trim, Divide** commands to
   remove the excess geometry between slots

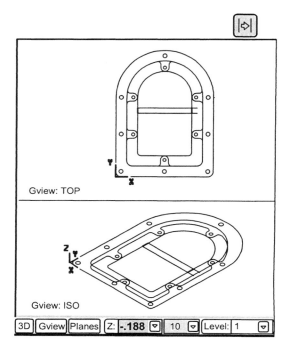

CHAPTER - 8

USING TRANSFORM TO TRANSLATE, ROTATE OR MIRROR EXISTING TOOLPATHS

8-1 Chapter Objectives

After completing this chapter you will be able to:

1. Understand the transform toolpaths function.

2. Use the transform toolpaths function to translate existing toolpaths.

3. Know how to use the transform toolpaths function to rotate toolpaths.

4. Use the transform toolpaths function to mirror existing toolpaths.

5. Explain how to convert a transform toolpath into new geometry and operations.

8-2 The Transform Toolpaths Function

Existing toolpaths can be *translated, rotated or mirrored* via the Transform Toolpaths function. When a toolpath is edited by using Transform all the other information in the operation associated with the toolpath is also updated when Regenerate Selected Operations 🔲 is executed. Thus *associativity is maintained*.

It should be noted that a transform toolpath is **not an independent toolpath** but is **linked to the original** from which it was copied. If the *original is edited* in any way via the chain manager **all the transform copies will automatically be edited as well** .
Many parts contain repetitious features that appear in grid or mirror patterns. Situations also arise where a set of identical parts or left and right hand versions of a part are to be machined in one setup. In these and many other cases Transform toolpath can be used to cut down on CAD geometry creation and dramatically simplify the job of chaining toolpaths.

The Transform Operations Parameters dialog box can be entered as shown below:

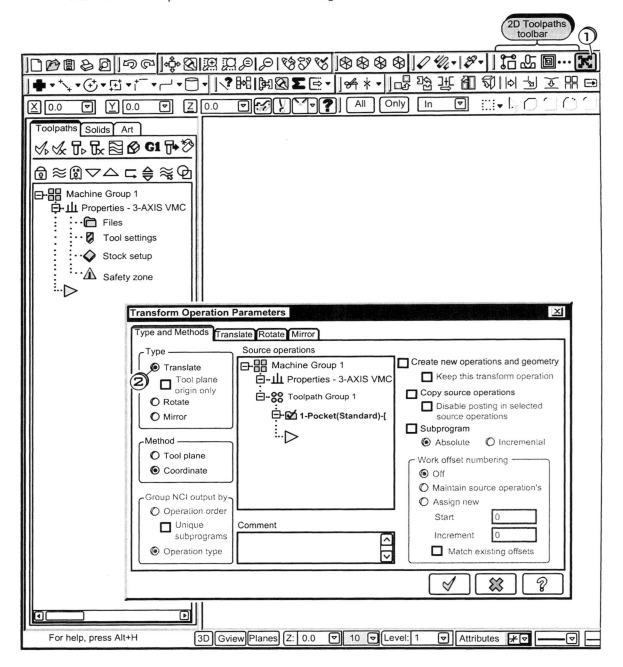

➤ Click ① the Transform Toolpath button [⟨icon⟩] contained in the 2D toolpaths toolbar.

8-3 Translating Existing Toolpaths

The operator clicks the [Translate] tab in the Transform Operations Parameters dialog box to execute move or copy operations on existing toolpaths.

BETWEEN POINTS

The Operator enters the **XY location of the existing toolpath** [From pt...] and **the XY location where the copy is to be placed** [To pt...]. Mastercam determines the required distance. The integer value entered in Step specifies the number of copies to make. Each copy is separated by the distance computed..

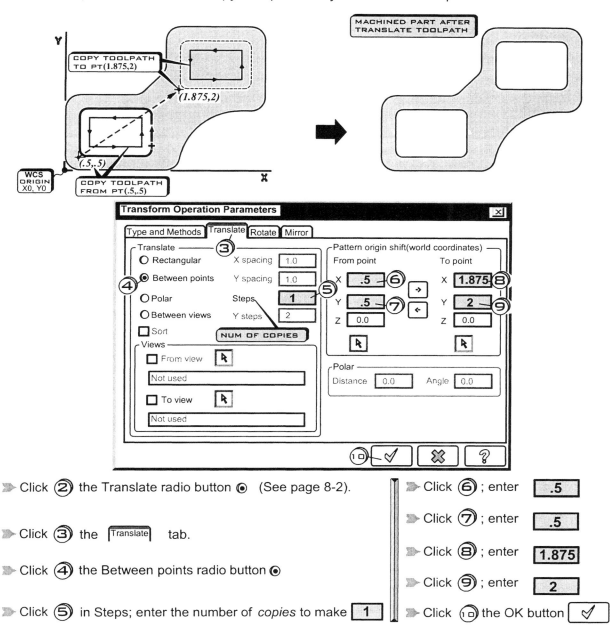

▶ Click ② the Translate radio button ◉ (See page 8-2).

▶ Click ③ the [Translate] tab.

▶ Click ④ the Between points radio button ◉

▶ Click ⑤ in Steps; enter the number of *copies* to make ☐ 1 ☐

▶ Click ⑥ ; enter ☐ .5 ☐

▶ Click ⑦ ; enter ☐ .5 ☐

▶ Click ⑧ ; enter ☐ 1.875 ☐

▶ Click ⑨ ; enter ☐ 2 ☐

▶ Click ⑩ the OK button ☐ ✓ ☐

POLAR

The Operator enters the ***distance between each copy and the required angle.***
The integer value entered in Step specifies the number of copies to make.

EXAMPLE 8-2

The Pocket toolpath and machining operation currently exists as shown in the display at left.
Direct *Mastercam* to create copies of the machining operation at the locations displayed right.

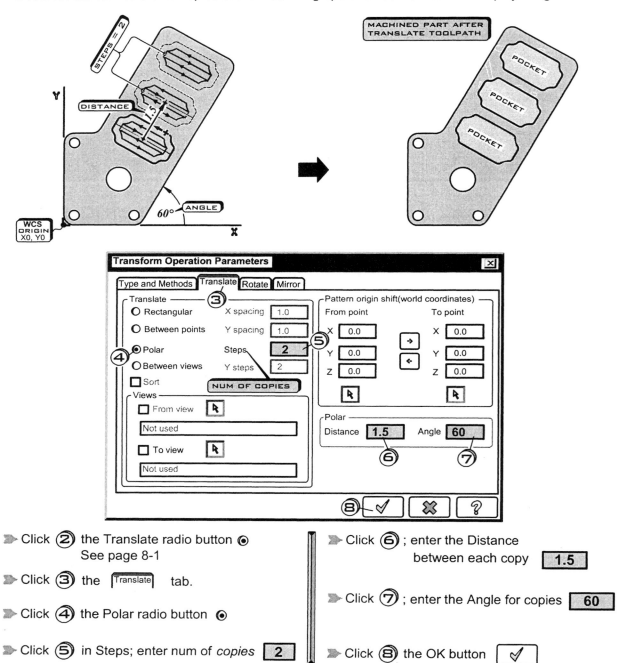

▶ Click ② the Translate radio button ◉
      See page 8-1

▶ Click ③ the |Translate| tab.

▶ Click ④ the Polar radio button ◉

▶ Click ⑤ in Steps; enter num of *copies* [2]

▶ Click ⑥ ; enter the Distance
      between each copy [1.5]

▶ Click ⑦ ; enter the Angle for copies [60]

▶ Click ⑧ the OK button [✓]

RECTANGULAR

The Operator enters the *distance between each copy in the X-direction, distance between each copy in the Y-direction, number of copies to make in the X-direction and number of copies make in the Y-direction.*

EXAMPLE 8-3

Given the slot toolpath and machining operation displayed left. Direct *Mastercam* to create arrayed copies of the slot machining operation the locations displayed right.

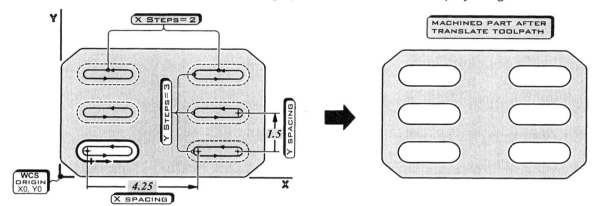

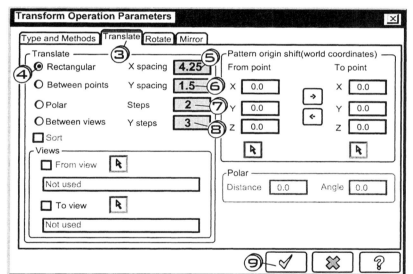

➤ Click ② the Translate radio button ⊙
    See page 8-2

➤ Click ③ the [Translate] tab.

➤ Click ④ the Rectangular radio button ⊙

➤ Click ⑤ in X spacing; enter the distance between each copy in the X-direction **4.25**

➤ Click ⑥ in Y spacing; enter the distance between each copy in the Y-direction **1.5**

➤ Click ⑦ the number of X steps **2**

➤ Click ⑧ the number of Y steps **3**

➤ Click ⑨ the OK button ✓

EXAMPLE 8-4

Given the slot toolpath and machining operation displayed left. Direct *Mastercam* to create arrayed copies of the slot machining operation the locations displayed right.

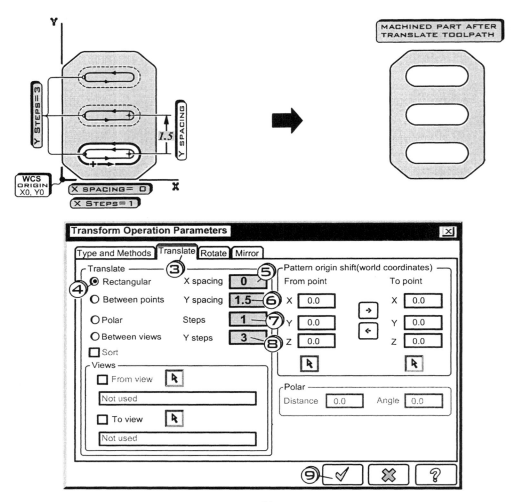

> Click ② the Translate radio button ⦿
> See page 8-2

> Click ③ the Translate tab.

> Click ④ the Rectangular radio button ⦿

> Click ⑤ in X spacing; enter the distance
> between each copy in the
> X -direction   0

> Click ⑥ in Y spacing; enter the distance
> between each copy in the
> Y -direction   1.5

> Click ⑦ the number of X steps   2

> Click ⑧ the number of Y steps   3

> Click ⑨ the OK button   ✓

8-4 Rotating Existing Toolpaths

The rotation function is especially useful for producing parts that have *repeat patterns arranged symmetrically about a point* . In these cases the operator can identify the base pattern create its corresponding toolpath then rotate the toolpath about the required point to produce the complete symmetrical shape.

ROTATE POINT

The Operator enters the ***location of the point about which rotation is to occur,*** the ***Number of steps(copies) to make,*** the ***Start angle for the first copy*** and the ***Rotation angle between each copy.***

EXAMPLE 8-5

The Operator has created the slot toolpath and machining operation as displayed left.
Direct *Mastercam* to create rotated copies of the machining operation as displayed right.

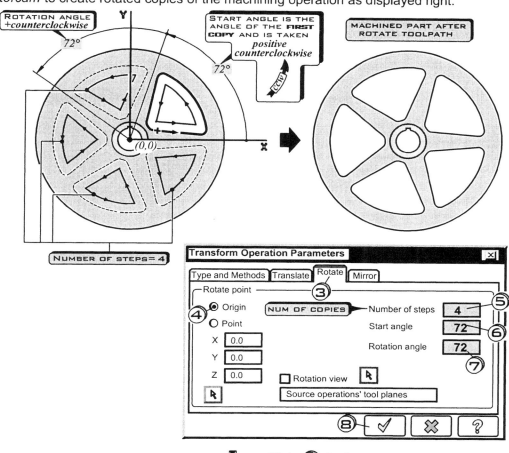

➤ Click ② the Rotate radio button ◉

➤ Click ③ the [Rotate] tab.

➤ Click ④ the Origin radio button ◉
to rotate about X0Y0

➤ Click ⑤ in Steps; enter the number
of *copies* to make ⬛ 4

➤ Click ⑥ in Start angle; enter ⬛ 72

➤ Click ⑦ in Rotation angle; enter ⬛ 72

➤ Click ⑧ the OK button ✓

8-5 Mirroring Existing Toolpaths

The mirror function is used in to mirror existing toolpaths about a line or axis. It simplifies the job of generating toolpaths for parts that have *repeat patterns arranged symmetrically about an axis or line* . It is also used to quickly generate *toolpaths for left and right hand versions* of the same part in one setup.

The Operator *clicks the entity about which mirroring is to occur*

EXAMPLE 8-6

The Pocket toolpath currently exists as shown in the left display. Direct *Mastercam* to create a mirror copy of the pocketing operation as displayed right.

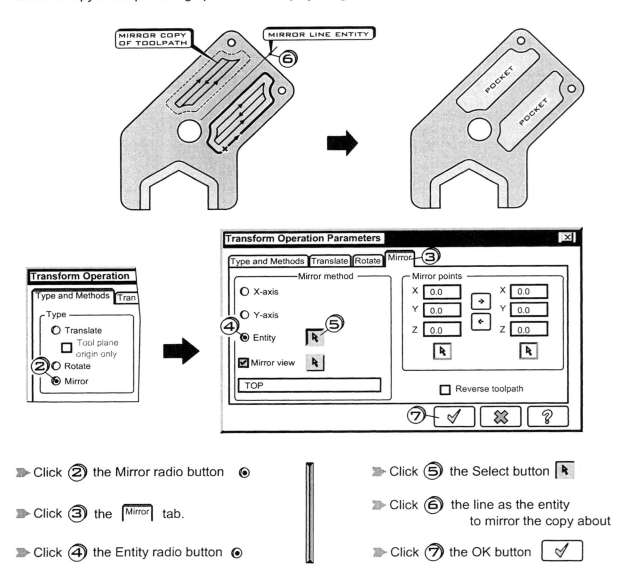

➤ Click ② the Mirror radio button ◉

➤ Click ③ the Mirror tab.

➤ Click ④ the Entity radio button ◉

➤ Click ⑤ the Select button

➤ Click ⑥ the line as the entity to mirror the copy about

➤ Click ⑦ the OK button

Note: the mirror function will **mirror the direction of the toolpath** and thus should be used with caution.

The existing slot toolpath shown in Figure 8-1 is directed to the **left** of upward tool motion(climb milling) but the mirrored toolpath is directed to the **right** of upward tool motion(conventional milling).

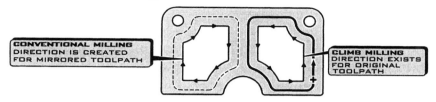

Figure 8-1

In cases like these where *climb milling* is desired for *both slots* it is recommended *not to use the mirror function*. Instead the approach should be to to *mirror the slot geometry* using the **Xform** , **Mirror,** functions. The operator can then chain the second slot in the proper direction for climb milling.

8-6 Converting a Transform into New Geometry and Operations

Transform toolpaths created by the Translate, Rotate or Mirror functions *cannot be edited individually but are tied to the original existing toolpath(s)* from which they were copied. Editing can only be accomplished by *changing the original* existing toolpath(s). When this is done, the changes made to the original(s) will automatically be passed on to their transform copies. This section will consider how to convert a transform toolpath such that it will be listed as a *new and independent operation* in the Operations Manager. The operator can then *edit the operation individually* as needed.

EXAMPLE 8-7

The Original and transform toolpaths have been created in *Mastercam* to produce the part shown in Figure8-2(left) . Edit the existing operations to produce the re-designed part shown in Figure 8-2(right).

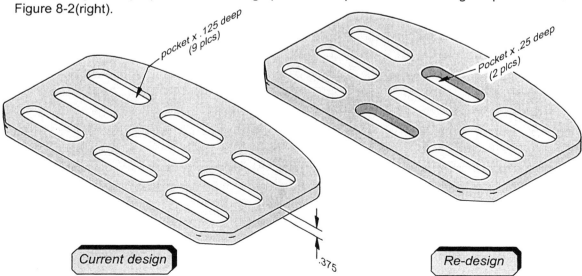

Figure 8-2

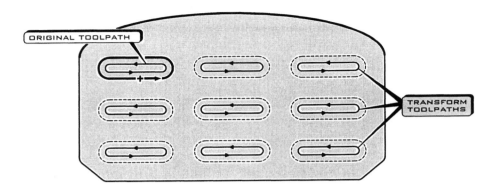

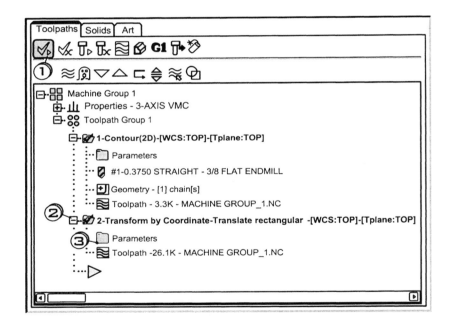

⯈ Click ①   the select all operation icon   🔲

⯈ Click ②   to change plus sign ⊞ to minus sign ⊟ and open folder

⯈ Click ③   on the Parameters icon for the transform toolpaths 🗀 .
   This will open the **Transform Operation Parameters** dialog box
   as shown on page **8-11** .

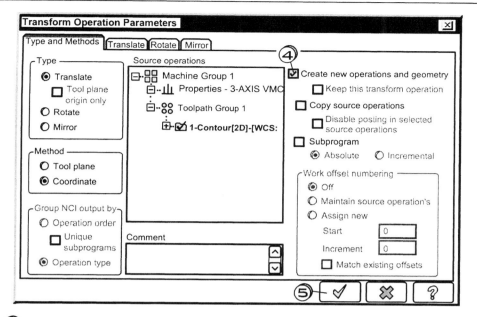

➤ Click ④ the check ☑ to enable **Create new operations and geometry**

➤ Click ⑤ the OK button ✓

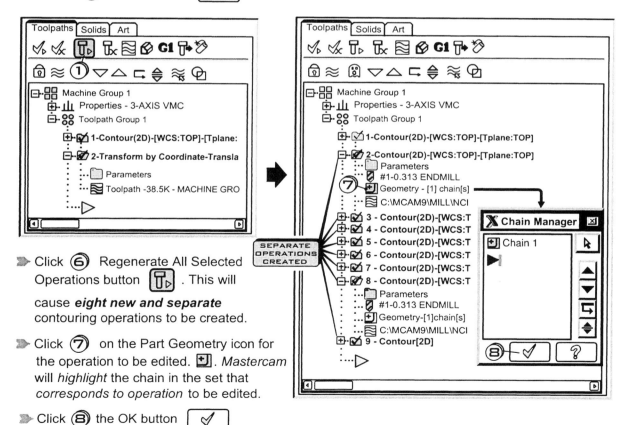

➤ Click ⑥ Regenerate All Selected Operations button 🔧. This will cause *eight new and separate* contouring operations to be created.

➤ Click ⑦ on the Part Geometry icon for the operation to be edited. 🔲. *Mastercam* will *highlight* the chain in the set that *corresponds to operation* to be edited.

➤ Click ⑧ the OK button ✓

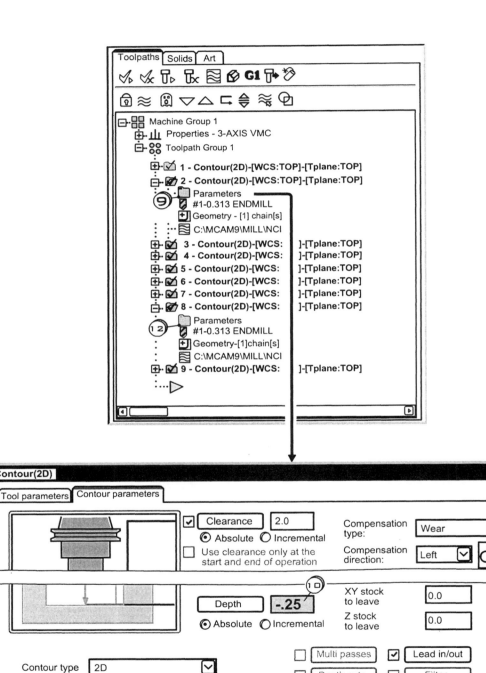

➤ Click ⑨ on the Parameters icon for the operation **2- Contour[2D]** 🗁

➤ Click ⑩ in the Depth box and enter the new depth for the operation ⸢ **-.25** ⸥

➤ Click ⑪ the OK button ⸢ ✓ ⸥

➤ Click ⑫ on the Parameters icon for the operation **8- Contour[2D]** 🗁

➤ Click ⑩ in the Depth box and enter the new depth for the operation ⸢ **-.25** ⸥

➤ Click ⑪ the OK button ⸢ ✓ ⸥

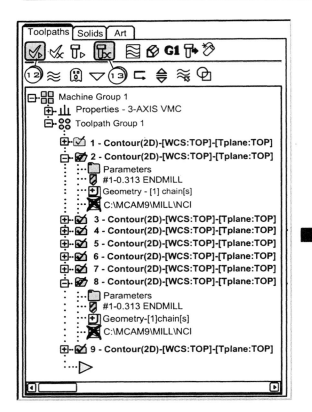
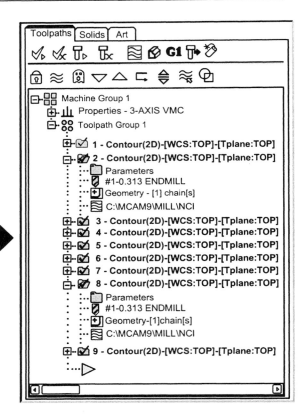

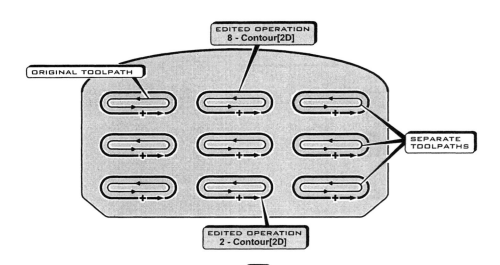

➤ Click ⑫ Select All Operations

➤ Click ⑬ Regen All Dirty Operations

EXERCISES

TUTORIAL PRACTICE EXERCISE

 8-1) The current machining operations in the Operations Manager produce the part features as shown in Figure 8p-1(left). Copy the file **EX8-1** in the folder ⌐CHAPTER8 from the CD into the **JVAL-MILL** subdirectory on C drive. Open the file and use the transform functions to produce the complete part as shown in Figure 8p-1(right)

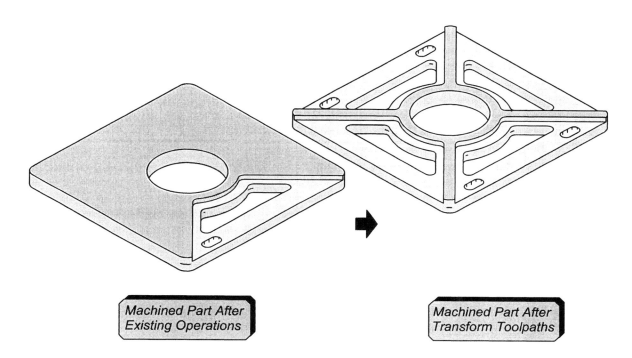

Machined Part After
Existing Operations

Machined Part After
Transform Toolpaths

Figure 8-p1

A) COPY THE FILE EX8-1 FROM THE CD AT THE BACK OF THIS
   TEXT TO YOUR DIRECTORY   C:\MCAMX2\MCX\JVAL-MILL

your initials

OPEN THE FILE FROM YOUR DIRECTORY.
   Refer to Chapter 1 p1-24 for a discussion on the creation of your own directory.

B) OPEN THE TRANSFORM OPERATIONS PARAMETERS DIALOG BOX

    ◆ IDENTIFY THE OPERATIONS TO BE ROTATED; SELECT ROTATE

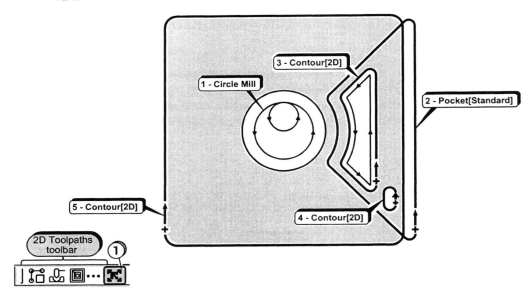

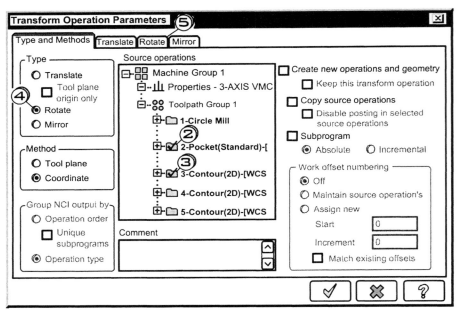

➤ Click ① the Transform Toolpath button 🔀

➤ Click ② check on ☑ for **2-Pocket(Standard)**

➤ Depress the [Ctrl] key and *keeping*
                     it depressed

➤ Click ③ check on ☑ for **3-Contour(2D)**

➤ Click ④ the Rotate radio button ◉

➤ Click ⑤ the [Rotate] tab.

◆ ENTER THE ROTATE TOOLPATH PARAMETERS

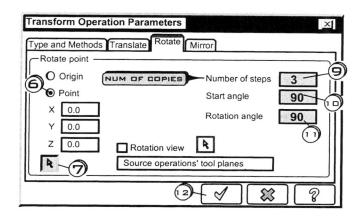

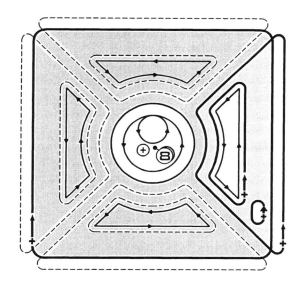

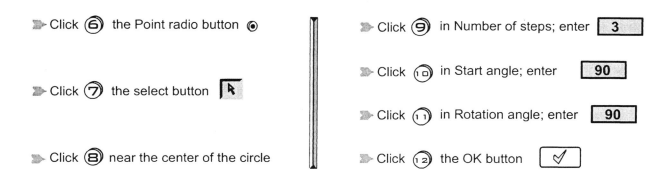

▶ Click ⑥ the Point radio button ◉

▶ Click ⑦ the select button

▶ Click ⑧ near the center of the circle

▶ Click ⑨ in Number of steps; enter 3

▶ Click ⑩ in Start angle; enter 90

▶ Click ⑪ in Rotation angle; enter 90

▶ Click ⑫ the OK button

◆ IDENTIFY THE OPERATIONS TO BE TRANSLATED; SELECT TRANSLATE

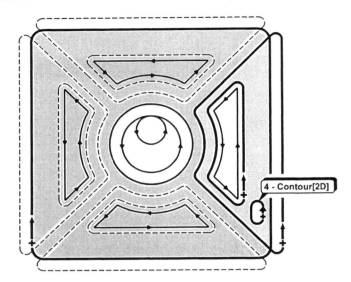

4 - Contour[2D]

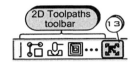

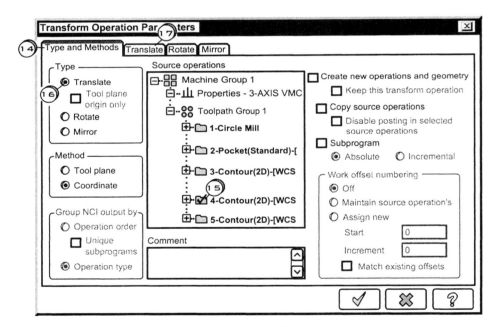

➤ Click ⑬ the Transform Toolpath button

➤ Click ⑭ the [Type and Methods] tab.

➤ Click ⑮ check on ☑ for **4-Contour(2D)**

➤ Click ⑯ the Translate radio button ◉

➤ Click ⑰ the [Translate] tab.

◆ ENTER THE TRANSLATE TOOLPATH PARAMETERS

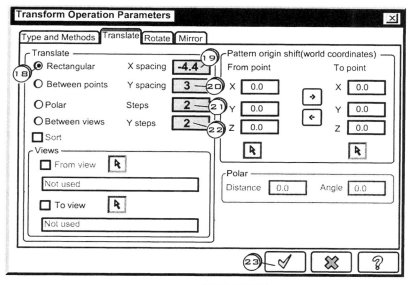

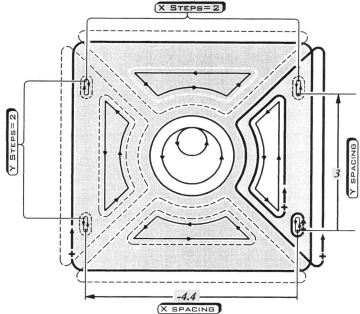

➤ Click ⒅ the Rectangular radio button ⦿

➤ Click ⒆ in X spacing; enter the distance between each copy in the X -direction **-4.4**

➤ Click ⒇ in Y spacing; enter the distance between each copy in the Y -direction **3**

➤ Click ㉑ the number of X steps **2**

➤ Click ㉒ the number of Y steps **2**

➤ Click ㉓ the OK button ✓

8-2) Copy CAD model file **EX8-2** in the folder ⌷**CHAPTER8** from the CD into the **JVAL-MILL** subdirectory on C drive. Open the file and generate a part program for executing the machining listed in PROCESS PLAN 8P-1.

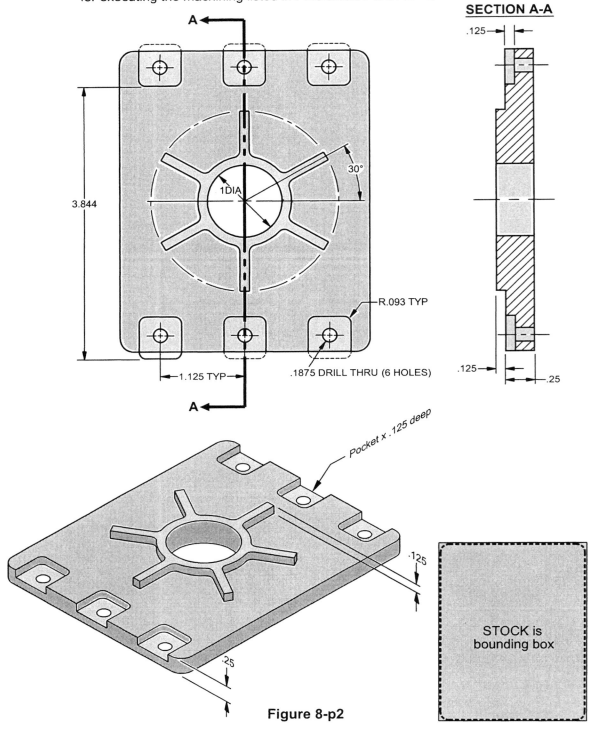

SECTION A-A

.125

3.844

1DIA

30°

R.093 TYP

1.125 TYP

.1875 DRILL THRU (6 HOLES)

.125

.25

Pocket x .125 deep

.125

.25

STOCK is bounding box

Figure 8-p2

PROCESS PLAN 8P-1

No.	Operation	Tooling
1	CENTER DRILL x .16 DEEP (6 HOLES)	1/8 CENTER DRILL
2	PECK DRILL THRU(6 HOLES) (6 HOLES)	3/16 DRILL
3	CIRCLE MILL 1 DIA THRU	1/2 END MILL
4	POCKET x .125 DEEP (BEGIN CLEANING THE TOP SURFACE) Pocket type: Facing ⌄ Standard Facing ⋮ Cutting method: Constant Overlap Spiral	1/2 END MILL

PROCESS PLAN 8P-1(*continued*)

No.	Operation	Tooling
5	POCKET X .125 DEEP Pocket type [Standard ▼] Cutting method: Constant Overlap Spiral	3/16 END MILL
6	TRANSFORM BY COORDINATE- ROTATE ABOUT POINT (FINISH CLEANING THE TOP SURFACE) 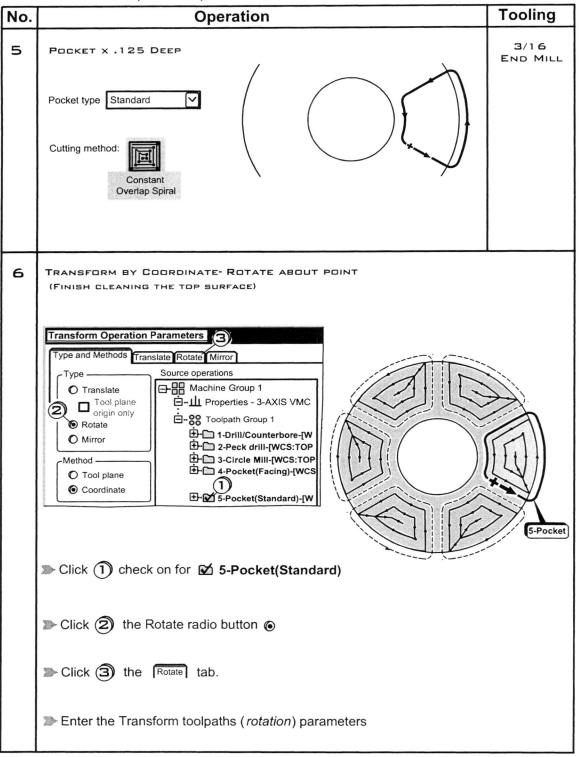 ➤ Click ① check on for ☑ **5-Pocket(Standard)** ➤ Click ② the Rotate radio button ◉ ➤ Click ③ the [Rotate] tab. ➤ Enter the Transform toolpaths (*rotation*) parameters	

PROCESS PLAN 8P-1(*continued*)

No.	Operation	Tooling
7	POCKET X .125 DEEP Pocket type Standard Cutting method: Constant Overlap Spiral 	3/16 END MILL
8	TRANSFORM BY COORDINATE- TRANSLATE	

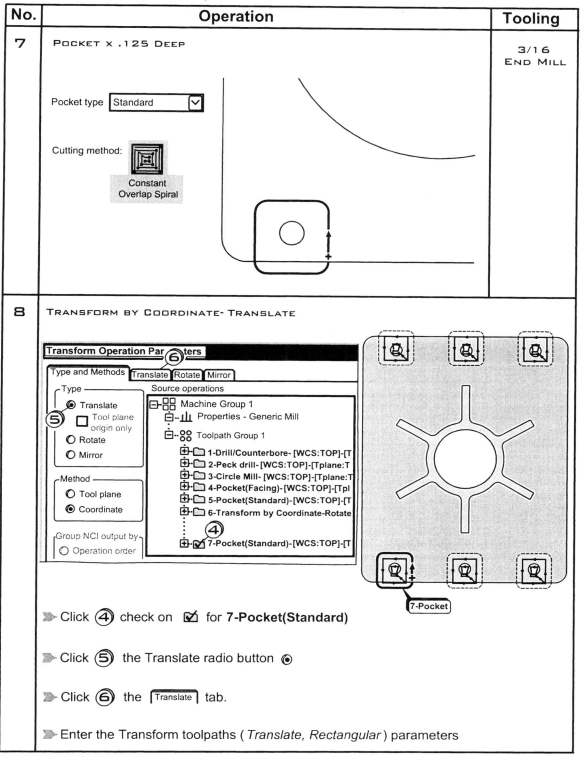

⫸ Click ④ check on ☑ for **7-Pocket(Standard)**

⫸ Click ⑤ the Translate radio button ◉

⫸ Click ⑥ the [Translate] tab.

⫸ Enter the Transform toolpaths (*Translate, Rectangular*) parameters

8-3) Copy CAD model file **EX8-3** in the folder ☐**CHAPTER8** from the CD into the **JVAL-MILL** subdirectory on C drive. Open the file and generate a part program for executing the machining listed in PROCESS PLAN 8P-2.

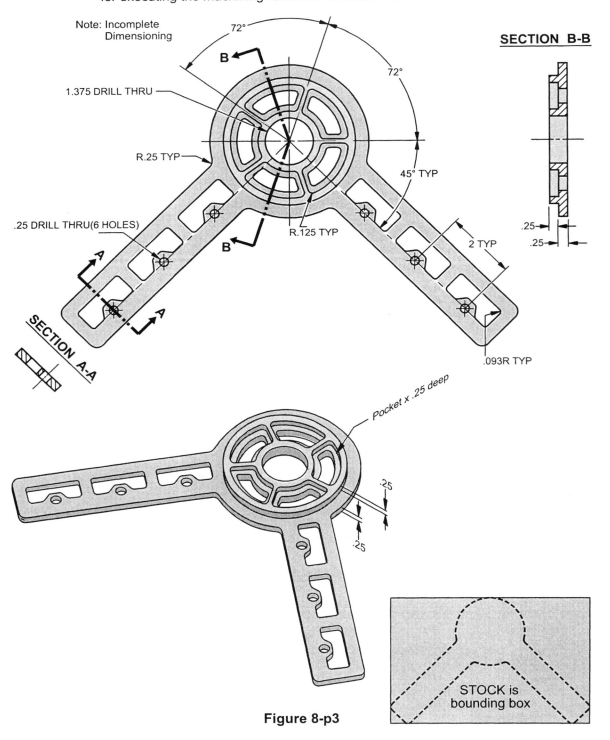

SECTION B-B

Note: Incomplete Dimensioning

72°

72°

B

1.375 DRILL THRU

R.25 TYP

45° TYP

.25 DRILL THRU(6 HOLES)

R.125 TYP

B

2 TYP

.25

.25

SECTION A-A

.093R TYP

Pocket x .25 deep

.25

.25

STOCK is bounding box

Figure 8-p3

PROCESS PLAN 8P-2

No.	Operation	Tooling
1	CENTER DRILL X .16 DEEP (6 HOLES) 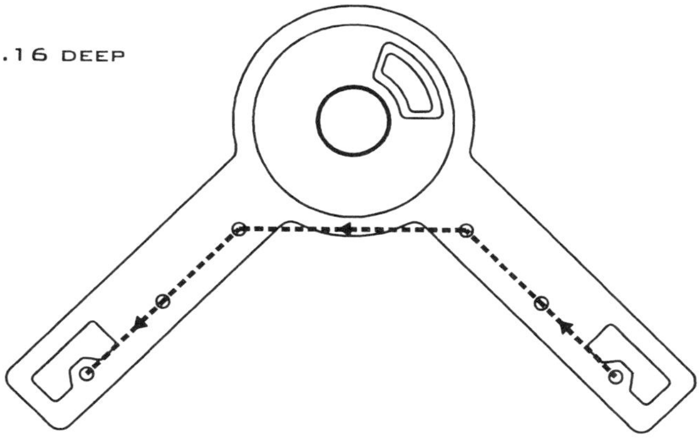	1/8 CENTER DRILL
2	PECK DRILL THRU(6 HOLES) 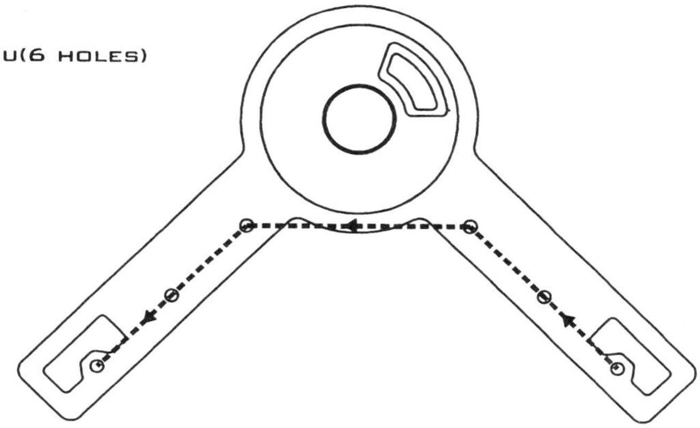	1/4 DRILL
3	POCKET X .25 DEEP (CLEAN THE TOP SURFACE) 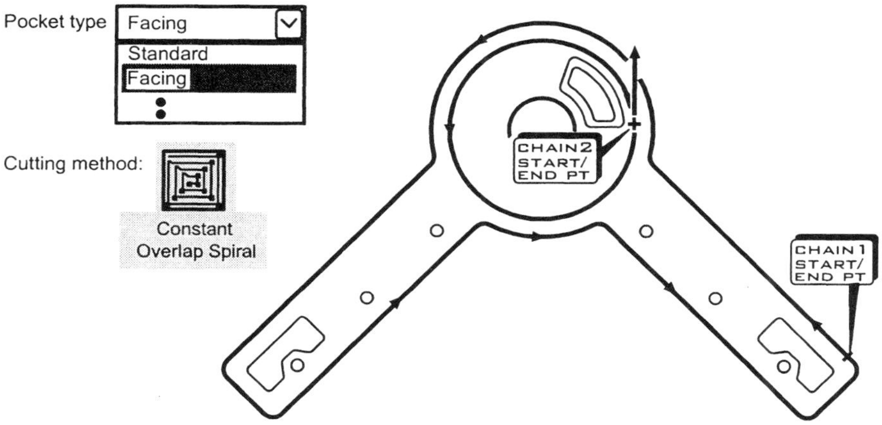	1/2 END MILL

Pocket type: Facing

Standard
Facing
⋮

Cutting method:

Constant
Overlap Spiral

CHAIN2 START/END PT

CHAIN1 START/END PT

PROCESS PLAN 8P-2(*continued*)

No.	Operation	Tooling
4	CIRCLE MILL 1.375 DIA THRU	1/2 END MILL
5	POCKET X .25 DEEP Pocket type [Standard ▽] Cutting method: Constant Overlap Spiral	1/4 END MILL
6	ROUGH AND FINISH SLOT X .25 DEEP	3/16 END MILL
7	TRANSFORM BY COORDINATE- ROTATE ABOUT ORIGIN 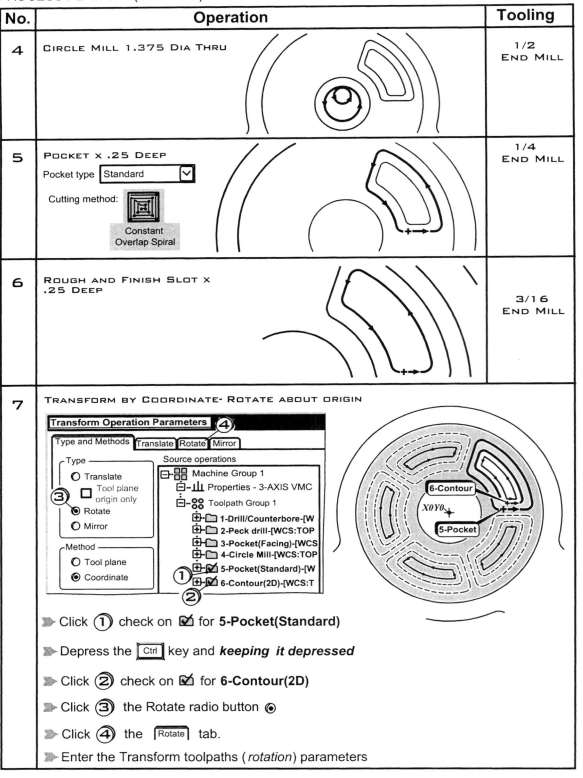 ⇒ Click ① check on ☑ for **5-Pocket(Standard)** ⇒ Depress the [Ctrl] key and *keeping it depressed* ⇒ Click ② check on ☑ for **6-Contour(2D)** ⇒ Click ③ the Rotate radio button ◉ ⇒ Click ④ the [Rotate] tab. ⇒ Enter the Transform toolpaths (*rotation*) parameters	

PROCESS PLAN 8P-2(*continued*)

No.	Operation	Tooling
8	ROUGH AND FINISH SLOT[RIGHT] × .25 DEEP 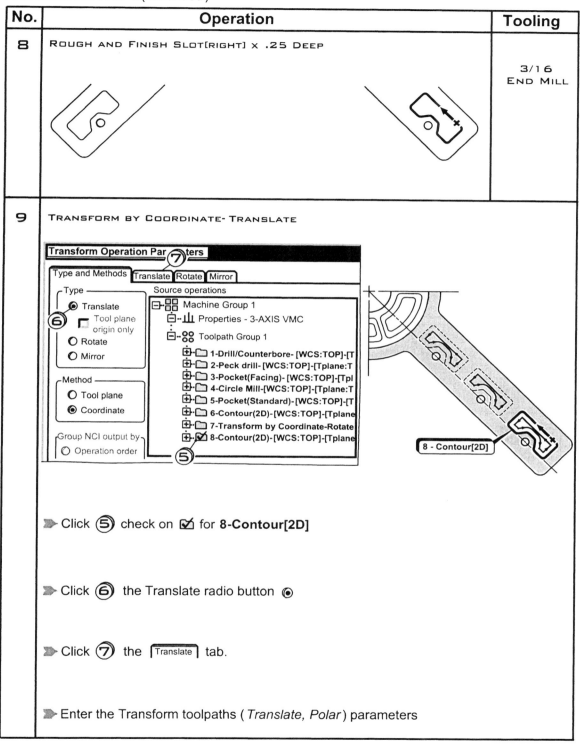	3/16 END MILL
9	TRANSFORM BY COORDINATE- TRANSLATE	

Click ⑤ check on ☑ for **8-Contour[2D]**

Click ⑥ the Translate radio button ⊙

Click ⑦ the ⌐Translate⌐ tab.

Enter the Transform toolpaths (*Translate, Polar*) parameters

PROCESS PLAN 8P-2(*continued*)

No.	Operation	Tooling
1 0	ROUGH AND FINISH SLOT[LEFT] × .25 DEEP	3/16 END MILL
1 1	TRANSFORM BY COORDINATE- TRANSLATE	

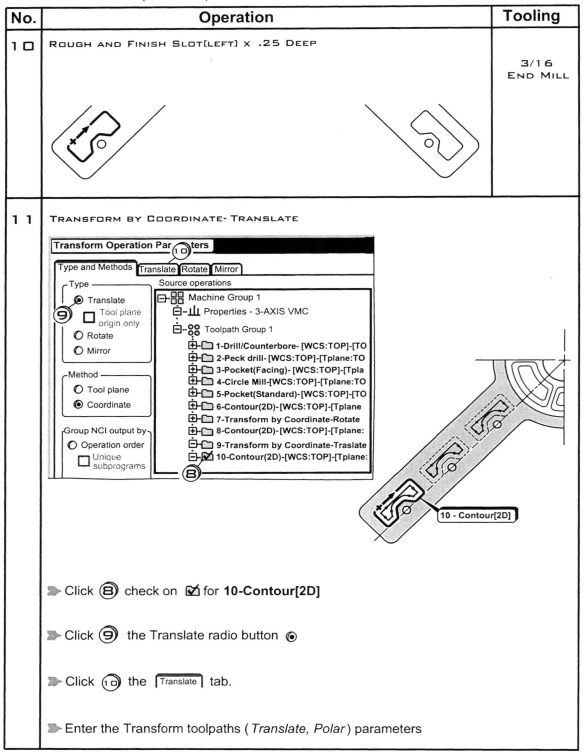

➤ Click ⑧ check on ☑ for **10-Contour[2D]**

➤ Click ⑨ the Translate radio button ◉

➤ Click ⑩ the ⌐Translate⌐ tab.

➤ Enter the Transform toolpaths (*Translate, Polar*) parameters

8-4) Open the CAD model file **EX3-7JV** created in exercise 3-7 in Chapter 3
Generate a part program for machining four copies of the same part in one
table setup at the CNC machine as shown in Figure 8p-4.
The required operations that must be created in the Operations Manager are listed in
PROCESS PLAN 8P-3.

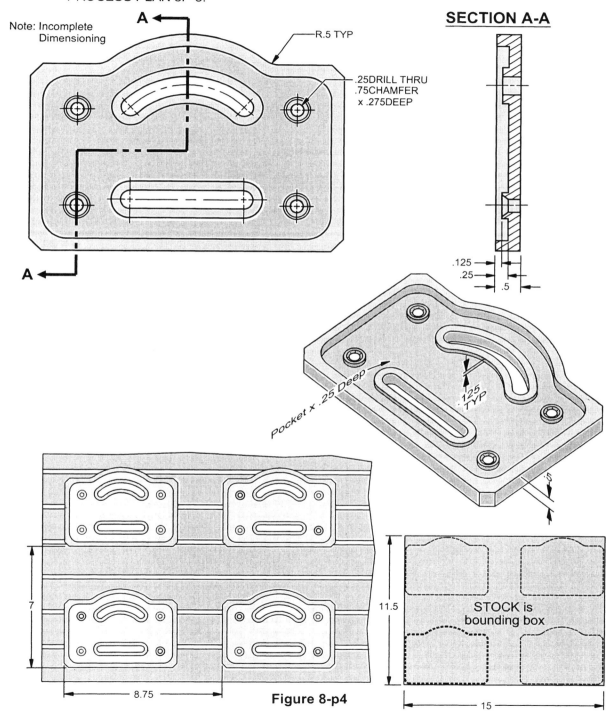

Figure 8-p4

PROCESS PLAN 8P-3

No.	Operation	Tooling
1	POCKET[ISLAND FACING] LEAVE .01 FOR FINISH CUT IN XY AND Z ➤ Click ① the Area button ➤ Click ② the location of the point *Mastercam* will **automatically chain** all the closed boundaries in the **area surrounding the point.** ➤ Click ③ the OK button Pocket type [Island facing ▼] [Island facing] Cutting method: Parallel Spiral Clean Corners	3/8 END MILL
2	CENTER DRILL X .285 DEEP (4 HOLES)	1/8 CENTER DRILL
3	PECK DRILL THRU(4 HOLES)	1/4 DRILL
4	CHAMFER X .275 DEEP(4 HOLES)	1 CHAMFER MILL

PROCESS PLAN 8P-3(*continued*)

No.	Operation	Tooling
5	ROUGH AND FINISH SLOTS x .5 DEEP 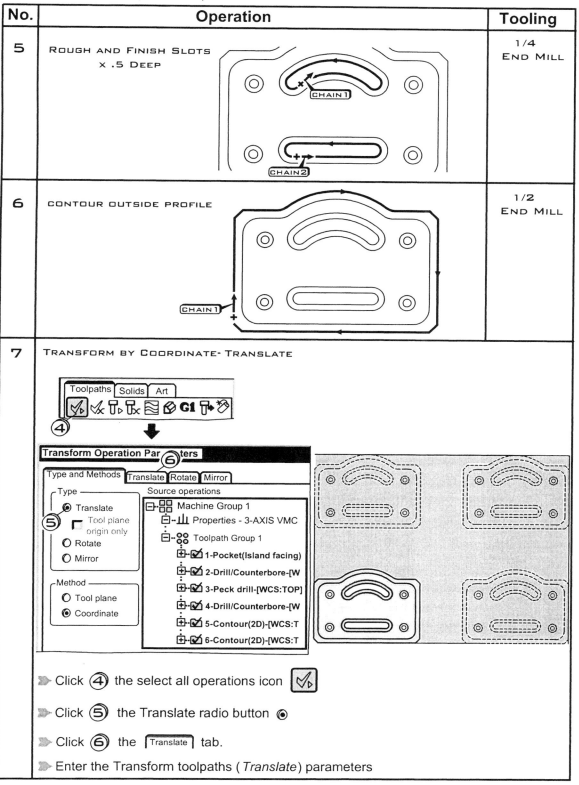	1/4 END MILL
6	CONTOUR OUTSIDE PROFILE	1/2 END MILL
7	TRANSFORM BY COORDINATE- TRANSLATE	

➤ Click ④ the select all operations icon

➤ Click ⑤ the Translate radio button ◉

➤ Click ⑥ the ⌐Translate⌐ tab.

➤ Enter the Transform toolpaths (*Translate*) parameters

CHAPTER - 9

USING A LIBRARY TO SAVE AND IMPORT MACHINING OPERATIONS

9-1 Chapter Objectives

After completing this chapter you will be able to:

1. Understand the advantages of using *Mastercam*'s operations library

2. Know how to save operations to the operations library

3. Explain how to import operations from the operations library.

4. State how to edit data in the operations library.

9-2 Advantages of Using Mastercam's Operations Library

Mastercam's Operations Library is an important productivity booster. It enables the operator to catalog and store an operation or sequence of operations under a single group name. For example, the group name stored in the library **TAP_1/4-20** could contain the center drilling, drilling and tapping operations required for producing the hole. Later, the operator can import the group **TAP_1/4-20** from the library and apply it to any other subsequent part as needed. Producing holes that require center drilling, drilling, tapping, reaming, chamfering and circle milling, etc account for a majority of the machining in many shops. Creating a library of these repetitive machining operations will cut down on needless duplication of effort.

9-3 Saving Operations to the Operations Library

EXAMPLE 9-1

Create a convenient naming code for each group. Create the operations associated with that group. See Table 9-1. Save the groups to the operations library for use in other parts.

Table 9-1

Group Name	Operation(s)	Tooling
CBORE-D5/16-E3/8x1/4 *c'bore is produced* / *first use 5/16 drill thru* / *then use a 3/8 Endmill to machine c'bore to 1/4 deep*	CENTER DRILL X .2 DEEP PECK DRILL THRU CIRCLE MILL X .25 DEEP	1/8 CTR DRILL 5/16 DRILL 3/8 ENDMILL
POC-E3/16x1/8 *pocket is produced* / *use a 3/16 Endmill machine pocket to 1/8 deep*	ROUGH AND FINISH POCKET X .125 DEEP. LEAVE .01 FOR FINISH CUT IN XY AND Z	3/16 ENDMILL

Note: **D** signifies *drill;* **E** *signifies end mill*

A) CREATE THE MINIMUM GEOMETRY NEEDED TO ESTABLISH
   AND VERIFY THE TOOLPATHS FOR EACH GROUP
   ◆ CREATE A .25DIA CIRCLE, AND A 1 × 1 RECTANGLE CENTERED
     WITHIN A 2 × 2 RECTANGLE

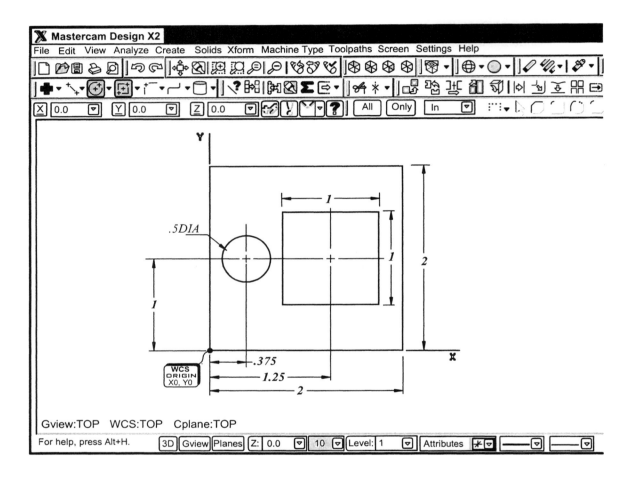

B) GENERATE OPERATIONS FOR GROUP CBORE-D5/16-E3/8x1/4
◆ CREATE THE 1/8 CENTER DRILL TOOLPATH

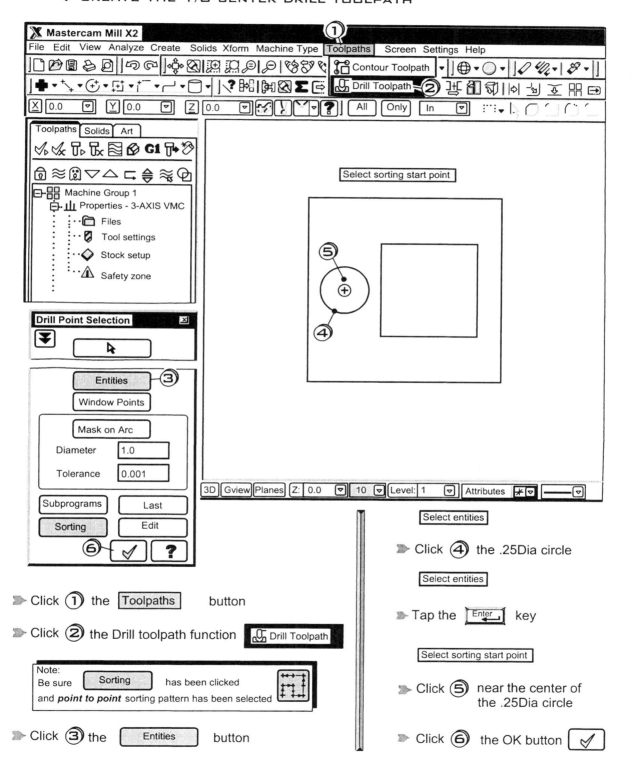

➤ Click ① the **Toolpaths** button

➤ Click ② the Drill toolpath function **Drill Toolpath**

> Note:
> Be sure **Sorting** has been clicked
> and *point to point* sorting pattern has been selected

➤ Click ③ the **Entities** button

➤ Click ④ the .25Dia circle

➤ Tap the **Enter** key

➤ Click ⑤ near the center of the .25Dia circle

➤ Click ⑥ the OK button ✓

◆ OBTAIN THE NEEDED .125 DIA CENTER DRILL TOOL

Mastercam will activate and display the Tool Parameters Tab.

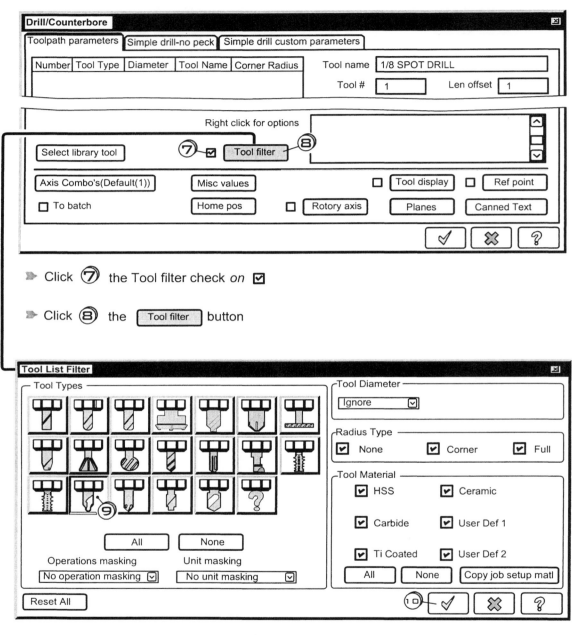

➤ Click ⑦ the Tool filter check *on* ☑

➤ Click ⑧ the [Tool filter] button

➤ Click ⑨ the Center Drill filter button

➤ Click ⑩ the OK button ✓

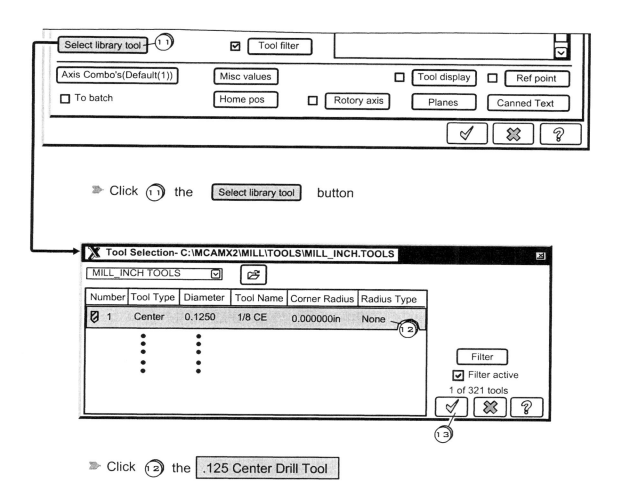

➤ Click ⑪ the Select library tool button

➤ Click ⑫ the .125 Center Drill Tool

➤ Click ⑬ the OK button ✓ to bring the tool from the library into your part file

♦   ENTER THE REQUIRED .125 DIA CENTER DRILL MACHINING PARAMETERS

➤ Click ⑭ | Simple drill-no peck | tab

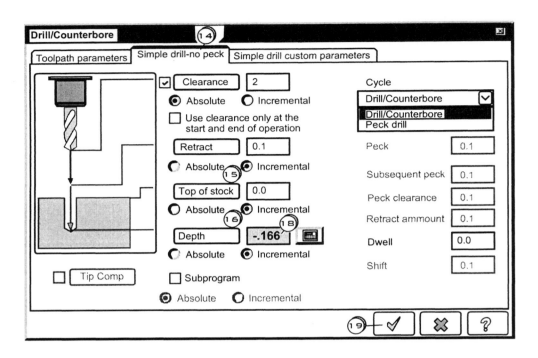

➤ Click ⑭   in the check on ☑ in the Clearance box

➤ Click ⑮   the Incremental radio button on ⊙ for | Retract |

➤ Click ⑯   the Incremental radio button on ⊙ for | Top of stock |

➤ Click ⑰   the Incremental radio button on ⊙ for | Depth |

Note: selecting Incremental for | Retract | | Top of stock | | Depth |
will insure that Mastercam will **begin drilling the hole at the Z-depth of the arc clicked.**
This will hold true if the operation is applied to any other parts with arcs at
different Z-depths.

➤ Click ⑱   in the Depth box; enter **-.166**

➤ Click ⑲   the OK button ✓ to *create* the operation 📁 **1 - Drill/Counterbore**

◆ CREATE THE .313 PECK DRILL THRU TOOLPATH

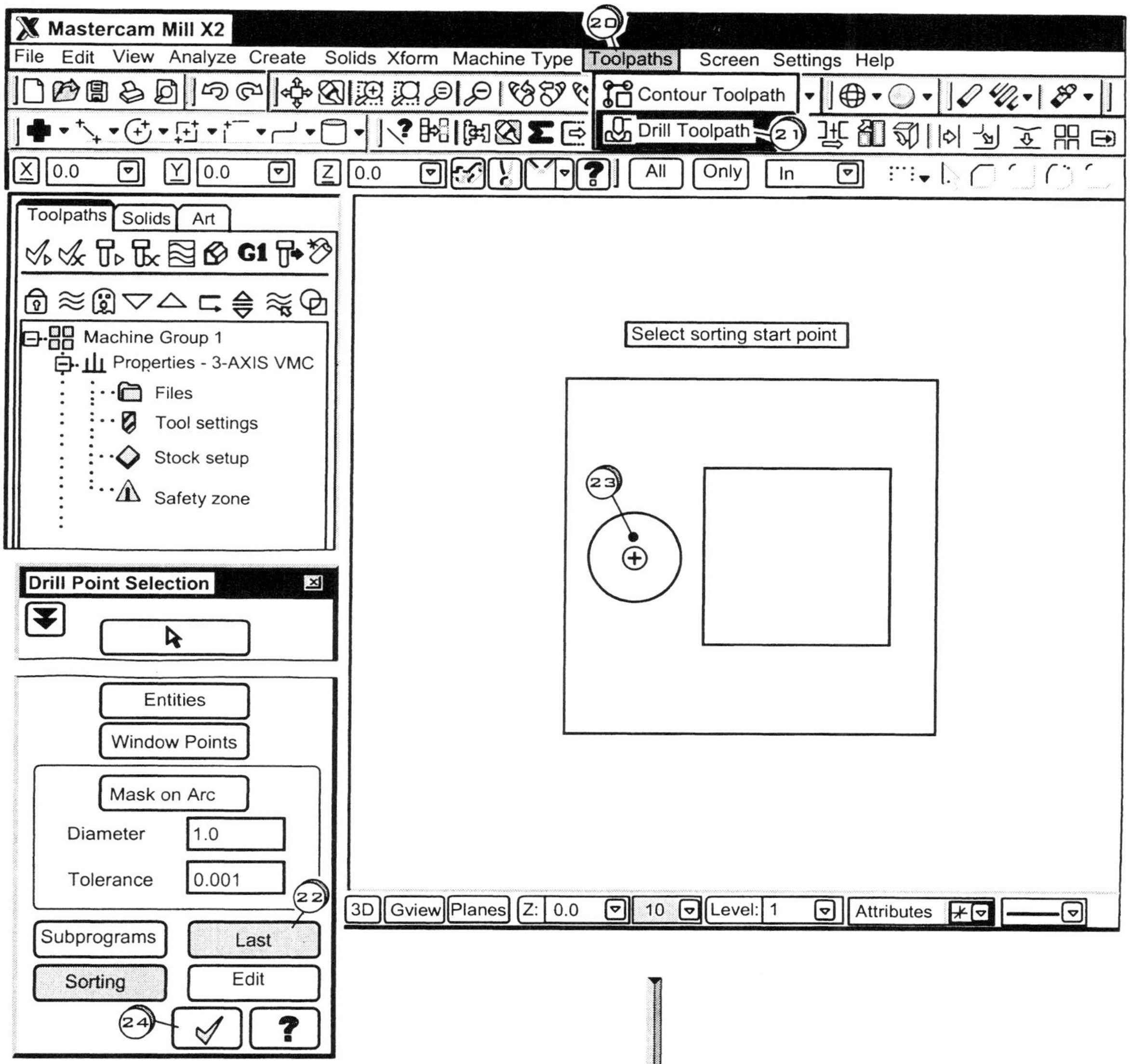

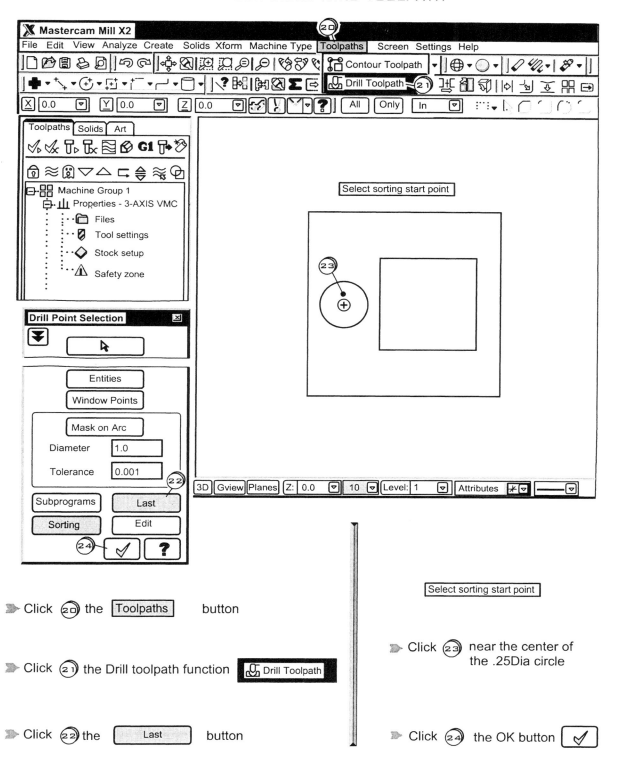

- Click (20) the [Toolpaths] button
- Click (21) the Drill toolpath function [Drill Toolpath]
- Click (22) the [Last] button

[Select sorting start point]

- Click (23) near the center of the .25Dia circle
- Click (24) the OK button [✓]

◆ OBTAIN THE NEEDED .313 DIA DRILL TOOL

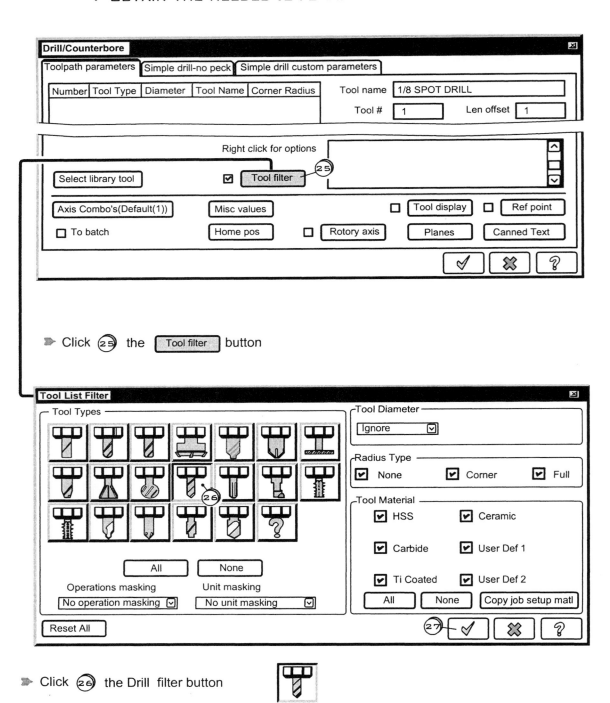

➥ Click ㉕ the [Tool filter] button

➥ Click ㉖ the Drill filter button

➥ Click ㉗ the OK button

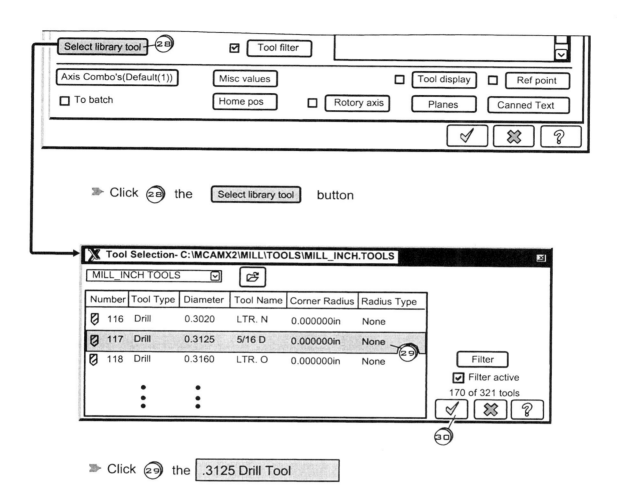

➤ Click ㉘ the [Select library tool] button

➤ Click ㉙ the [.3125 Drill Tool]

➤ Click ㉚ the OK button [✓] to bring the tool from the library into your part file

◆   ENTER THE REQUIRED .313 DIA PECK DRILL MACHINING PARAMETERS

▶ Click ㉛ | Simple drill-no peck | tab

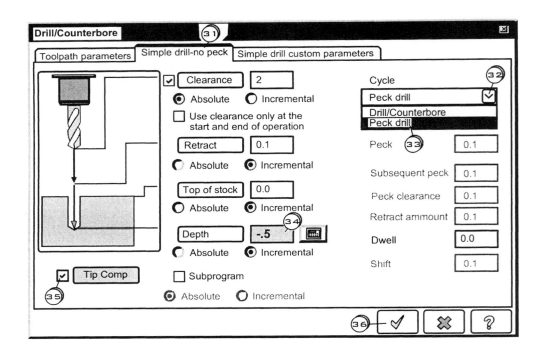

▶ Click ㉜ ☑ the toggle down button

▶ Click ㉝ **Peck drill**

▶ Click ㉞ in the Depth box; enter **-.5**

▶ Click ㉟ the tip comp check *on* ☑ when drilling *thru* holes

▶ Click ㊱ the OK button ☑ to *create* the operation 📂 **2 - Peck drill**

◆ CREATE THE CIRCLE MILL TOOLPATH

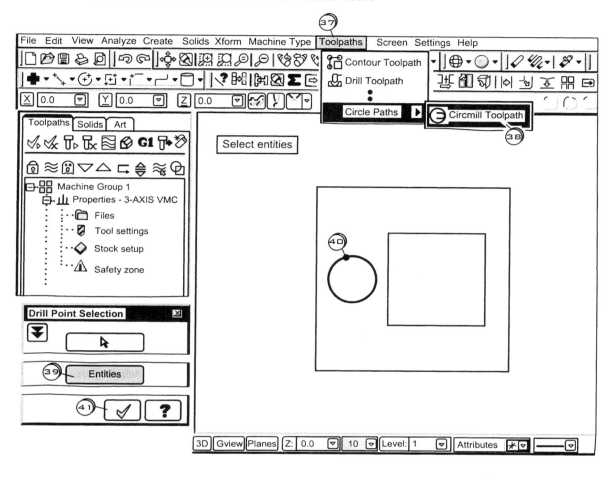

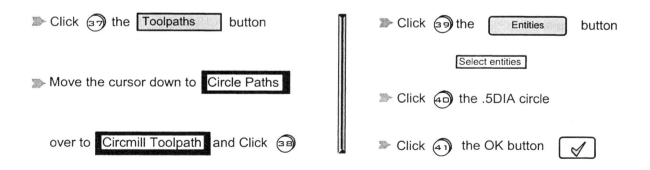

➤ Click ㊲ the [Toolpaths] button

➤ Move the cursor down to [Circle Paths]

over to [Circmill Toolpath] and Click ㊳

➤ Click ㊴ the [Entities] button

[Select entities]

➤ Click ㊵ the .5DIA circle

➤ Click ㊶ the OK button ☑

◆ OBTAIN A 3/8 DIA FLAT END MILL TOOL FROM THE TOOL LIBRARY

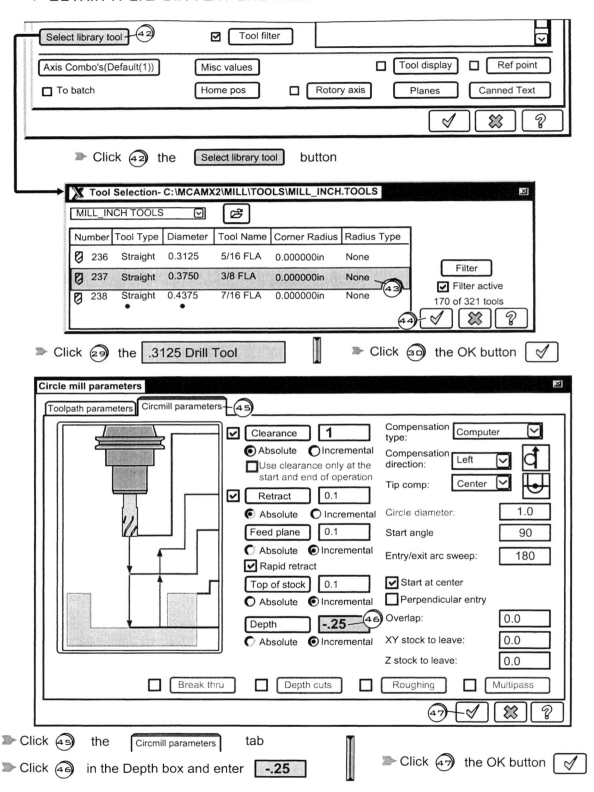

➤ Click ④② the **Select library tool** button

➤ Click ② the **.3125 Drill Tool**   ➤ Click ③ the OK button ✓

➤ Click ④⑤ the **Circmill parameters** tab

➤ Click ④⑥ in the Depth box and enter **-.25**   ➤ Click ④⑦ the OK button ✓

c) GENERATE OPERATIONS FOR GROUP POC-E3/16x1/8

◆ CREATE THE POCKET TOOLPATH

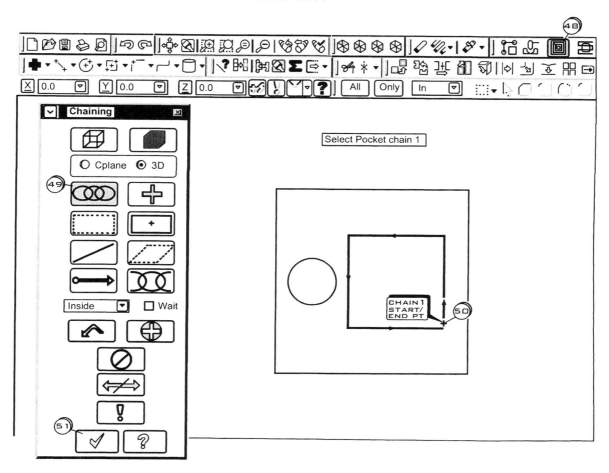

➤ Click ④⑧ the Pocket Toolpaths button

➤ Click ④⑨ on the chain button

➤ Click ⑤⓪ on the entity to specify the chain **start/end** point

➤ Click ⑤① the OK button

◆ OBTAIN A 3/16 DIA FLAT END MILL TOOL FROM THE TOOL LIBRARY

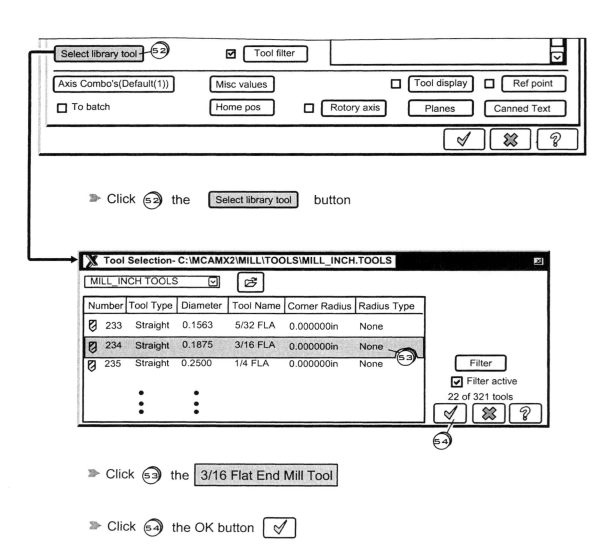

➤ Click ⑤② the [Select library tool] button

➤ Click ⑤③ the [3/16 Flat End Mill Tool]

➤ Click ⑤④ the OK button [✓]

▶ Click ⑤⑤ [Pocketing parameters] tab

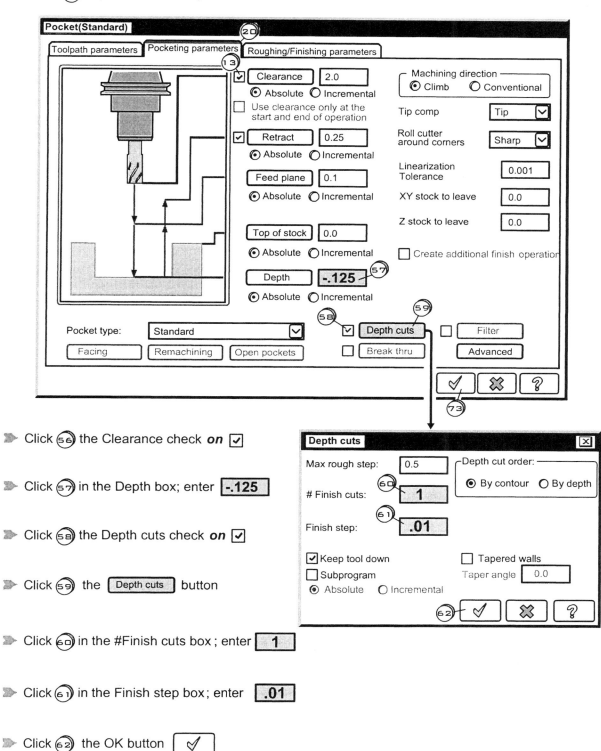

▶ Click ⑤⑥ the Clearance check **on** ☑

▶ Click ⑤⑦ in the Depth box; enter **-.125**

▶ Click ⑤⑧ the Depth cuts check **on** ☑

▶ Click ⑤⑨ the [Depth cuts] button

▶ Click ⑥⓪ in the #Finish cuts box ; enter **1**

▶ Click ⑥① in the Finish step box; enter **.01**

▶ Click ⑥② the OK button ☑

➤ Click ⑥③ [Roughing/Finishing parameters] tab.

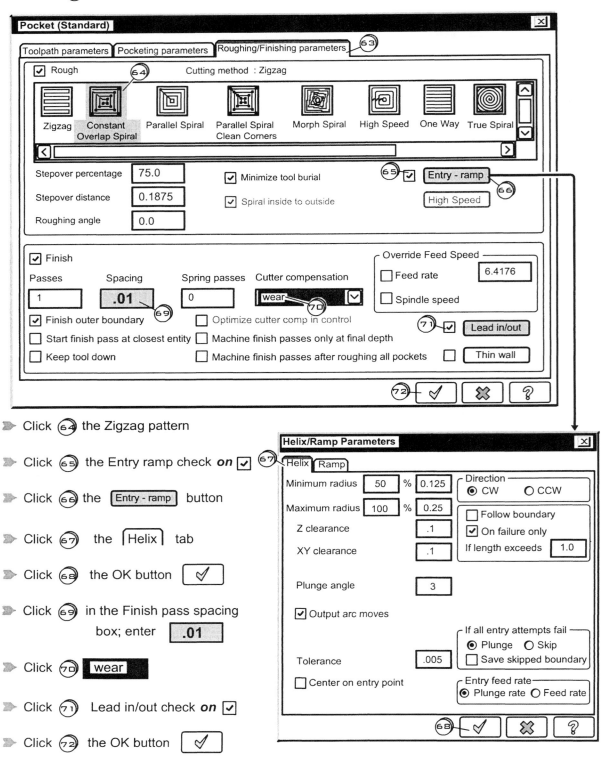

➤ Click ⑥④ the Zigzag pattern

➤ Click ⑥⑤ the Entry ramp check **on** ☑

➤ Click ⑥⑥ the [Entry - ramp] button

➤ Click ⑥⑦ the [Helix] tab

➤ Click ⑥⑧ the OK button ☑

➤ Click ⑥⑨ in the Finish pass spacing box; enter .01

➤ Click ⑦⓪ wear

➤ Click ⑦① Lead in/out check **on** ☑

➤ Click ⑦② the OK button ☑

D) VERIFY THE TOOLPATHS FOR ALL THE MACHINING OPERATIONS

◆ SELECT SIZE OF THE STOCK

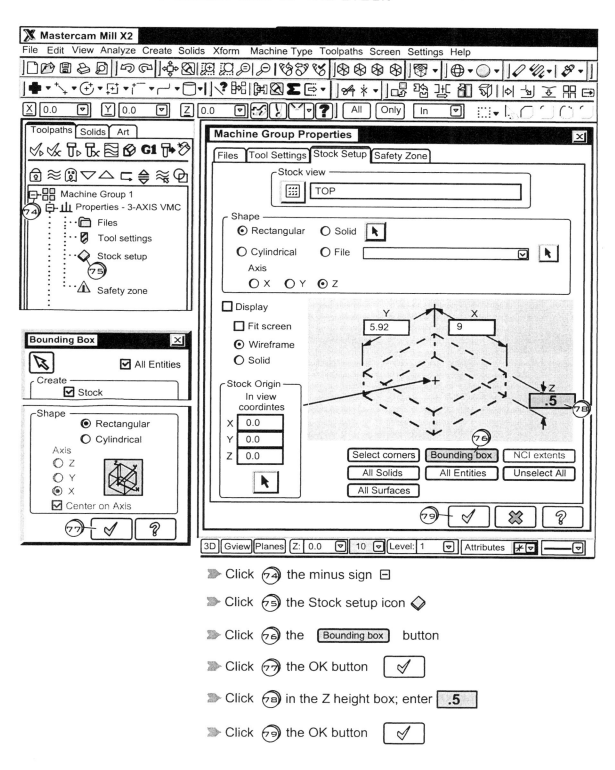

⮞ Click ⑦④ the minus sign ⊟

⮞ Click ⑦⑤ the Stock setup icon ◇

⮞ Click ⑦⑥ the [Bounding box] button

⮞ Click ⑦⑦ the OK button ✓

⮞ Click ⑦⑧ in the Z height box; enter **.5**

⮞ Click ⑦⑨ the OK button ✓

◆ DIRECT *Mastercam* TO ANIMATE THE MACHINING OF THE GROUPS
CBORE-D5/16-E3/8X1/4 AND POC-E3/16X1/8

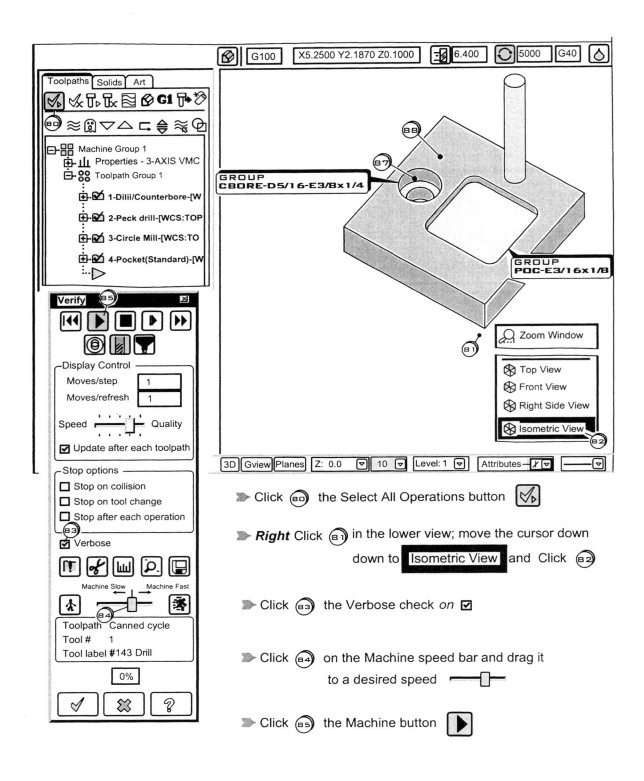

➣➣ Click ⑧⓪ the Select All Operations button

➣➣ **Right** Click ⑧① in the lower view; move the cursor down
down to **Isometric View** and Click ⑧②

➣➣ Click ⑧③ the Verbose check *on* ☑

➣➣ Click ⑧④ on the Machine speed bar and drag it
to a desired speed

➣➣ Click ⑧⑤ the Machine button ▶

♦ DIRECT *Mastercam* TO SECTION THE PART TO VISUALLY INSPECT THE CORRECTNESS OF THE MACHINING FOR EACH GROUP

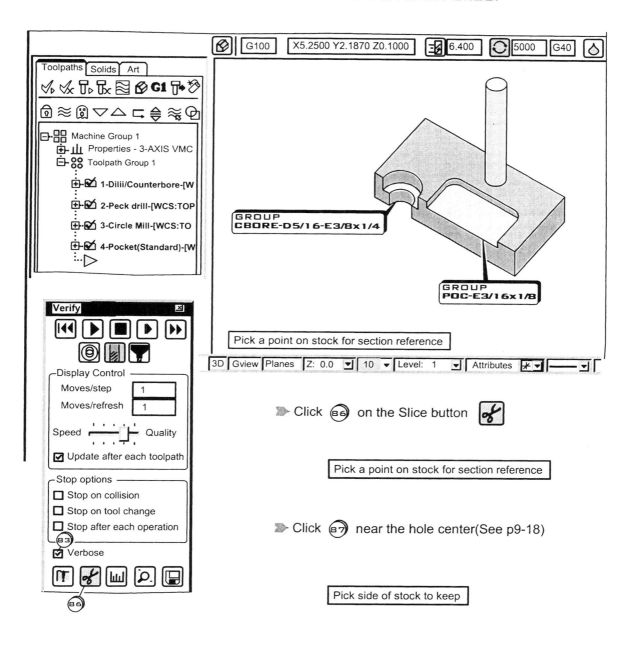

▶ Click ⑧⑥ on the Slice button

Pick a point on stock for section reference

▶ Click ⑧⑦ near the hole center(See p9-18)

Pick side of stock to keep

▶ Click ⑧⑧ the portion of the part to *remain*

(See p9-18)

E) Save the operations generated in step B to the operations Library under the group name CBORE-D5/16-E3/8x1/4

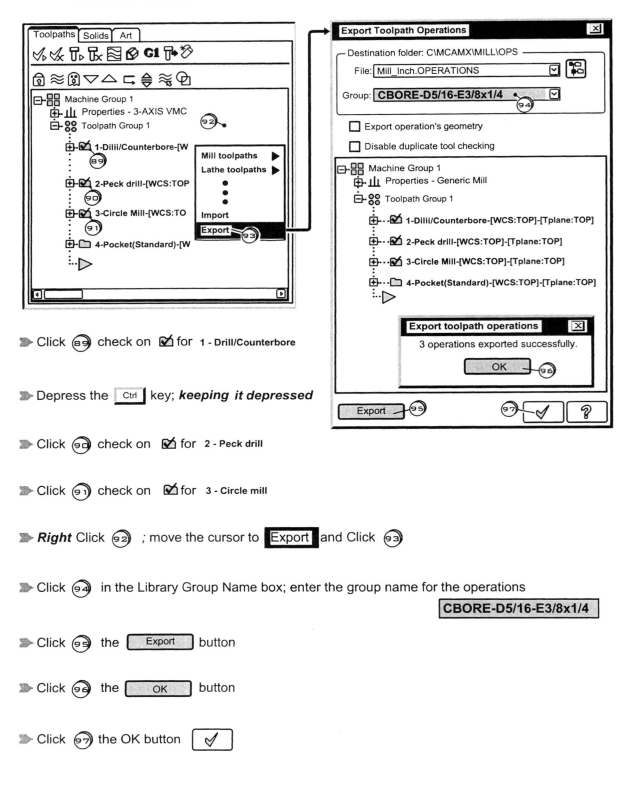

▶ Click ⑧⑨ check on ☑ for 1 - Drill/Counterbore

▶ Depress the ⎡Ctrl⎤ key; *keeping it depressed*

▶ Click ⑨⓪ check on ☑ for 2 - Peck drill

▶ Click ⑨① check on ☑ for 3 - Circle mill

▶ *Right* Click ⑨② ; move the cursor to ⎡Export⎤ and Click ⑨③

▶ Click ⑨④ in the Library Group Name box; enter the group name for the operations

                                                          CBORE-D5/16-E3/8x1/4

▶ Click ⑨⑤ the ⎡ Export ⎤ button

▶ Click ⑨⑥ the ⎡ OK ⎤ button

▶ Click ⑨⑦ the OK button ⎡ ✓ ⎤

F) SAVE THE OPERATIONS GENERATED IN STEP C TO THE OPERATIONS LIBRARY UNDER THE GROUP NAME POC-E3/16x1/8

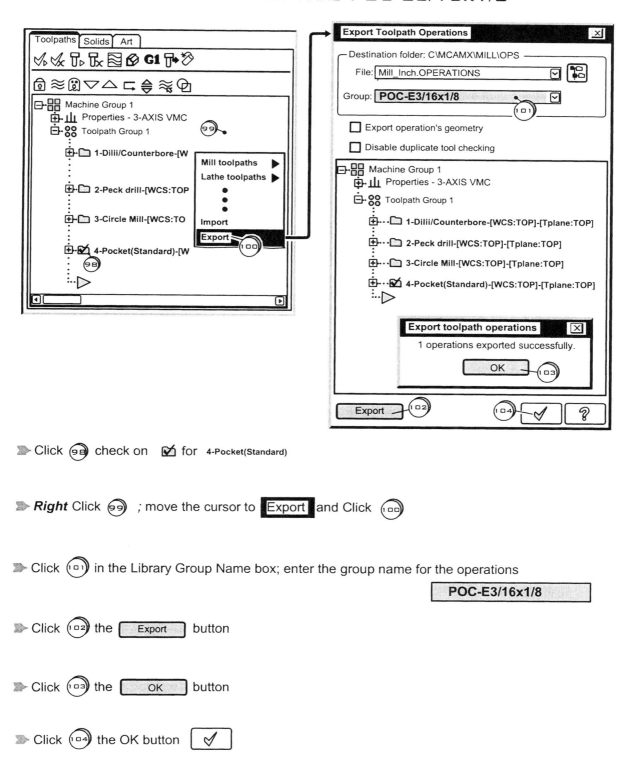

➤ Click ⟨98⟩ check on ☑ for 4-Pocket(Standard)

➤ *Right* Click ⟨99⟩ ; move the cursor to Export and Click ⟨100⟩

➤ Click ⟨101⟩ in the Library Group Name box; enter the group name for the operations

POC-E3/16x1/8

➤ Click ⟨102⟩ the Export button

➤ Click ⟨103⟩ the OK button

➤ Click ⟨104⟩ the OK button ☑

G) Verify Operations Library has been Updated to Include the New Groups CBORE-D5/15-E3/8x1/4 and POC-E3/16x1/8

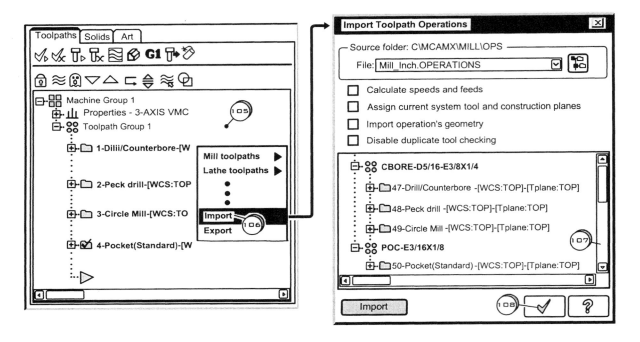

➤ **Right** Click (105) ; move the cursor to **Import** and Click (106)

➤ Click (107) on the scroll button and drag it down to the bottom of the listing

> *Mastercam* will display the *latest* library
> listing which should include the new
> groups **CBORE-D5/16-E3/8x1/4** and **POC -E3/16x1/8**

➤ Click (108) the OK button ✓

9-4 Importing Operations from the Operations Library

Mastercam provides complete *flexibility* in importing the data stored in the operations library as follows: import of a *single* operation

- import of *any number* of operations clicked
- import of *all* the operations associated with a *group* clicked
- import of *all* the operations associated with *any number* of *groups* clicked.

It was shown in section 9-2 that only a minimum of dummy geometry was needed to create the required operations. In this section it will be demonstrated that the groups and associated operations in the operations library can be quickly applied to a wide array of parts having different physical shapes, thicknesses and materials.

EXAMPLE 9-2

For the part shown in Figure9-1. Use the groups created and saved in the operations library in Example 9-1 as an aid in quickly generating the the required machining operations as outlined in PROCESS PLAN 9-1. Assume the CAD model already exists and is stored as file **EXAMPLE9-2** in the **JVAL-MILL** subdirectory.

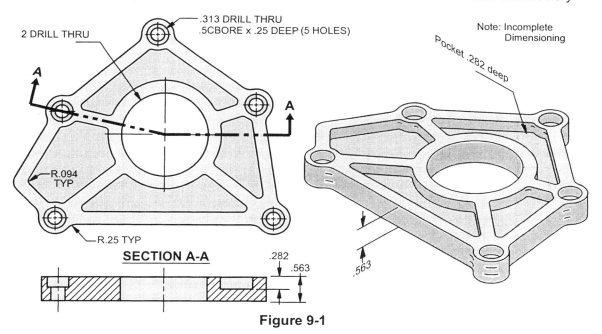

Figure 9-1

PROCESS PLAN 9-1

No.	Operation	Tooling
1	CENTER DRILL x .2 DEEP(ALL HOLES)	1/8 CTR DRILL
2	PECK DRILL THRU(ALL HOLES)	5/16 DRILL
3	CIRCLE MILL x .25 DEEP(ALL HOLES)	3/8 ENDMILL
4	ROUGH AND FINISH POCKET x .282 DEEP LEAVE .01 FOR FINISH CUT IN XY AND Z.	3/16 ENDMILL

CBORE-D5/16-E3/8x1/4

POC-E3/16x1/8

A) OPEN THE CAD FILE EXAMPLE9-2 CONTAINING THE PART
   GEOMETRY

B) IMPORT THE GROUP CBORE-D5/16-E3/8X1/4 FROM THE
   OPERATIONS LIBRARY

   ◆ WITHIN THE OPERATIONS MANAGER SELECT IMPORT

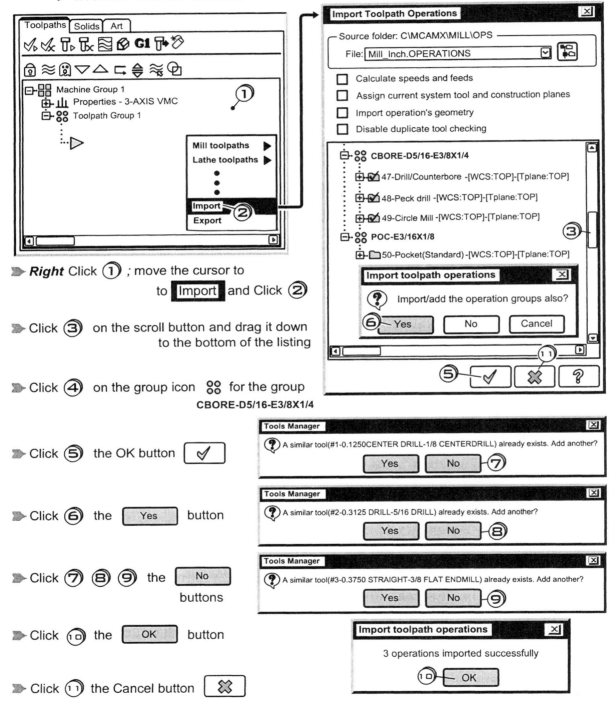

➤ **Right** Click ① ; move the cursor to
   to Import and Click ②

➤ Click ③ on the scroll button and drag it down
   to the bottom of the listing

➤ Click ④ on the group icon 88 for the group
   CBORE-D5/16-E3/8X1/4

➤ Click ⑤ the OK button ✓

➤ Click ⑥ the Yes button

➤ Click ⑦ ⑧ ⑨ the No
   buttons

➤ Click ⑩ the OK button

➤ Click ⑪ the Cancel button ✖

> *Note: By default, Mastercam will assign **no drill points** and **no chains** to **any** operations **imported** from the Operations library*

C) ADD DRILL POINTS TO THE DRILLING OPERATIONS

♦ ADD DRILL POINTS TO OPERATION: 1 - Drill/Counterbore

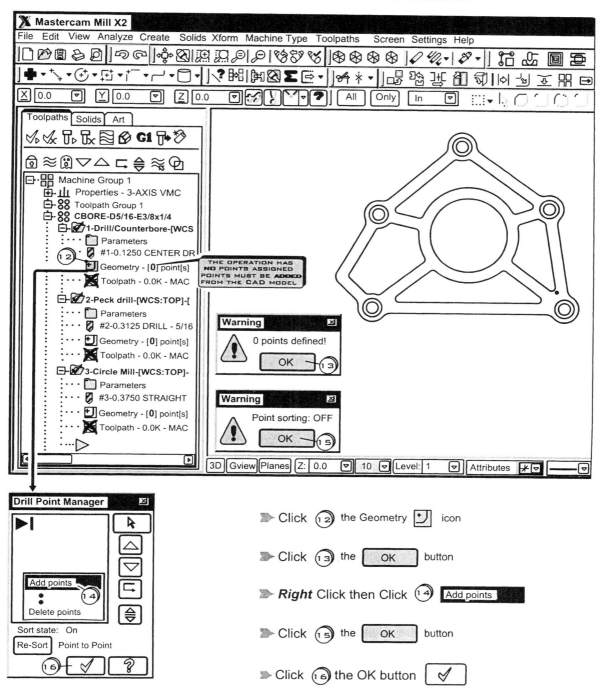

➤ Click ⑫ the Geometry ⊡ icon

➤ Click ⑬ the OK button

➤ ***Right*** Click then Click ⑭ Add points

➤ Click ⑮ the OK button

➤ Click ⑯ the OK button ✓

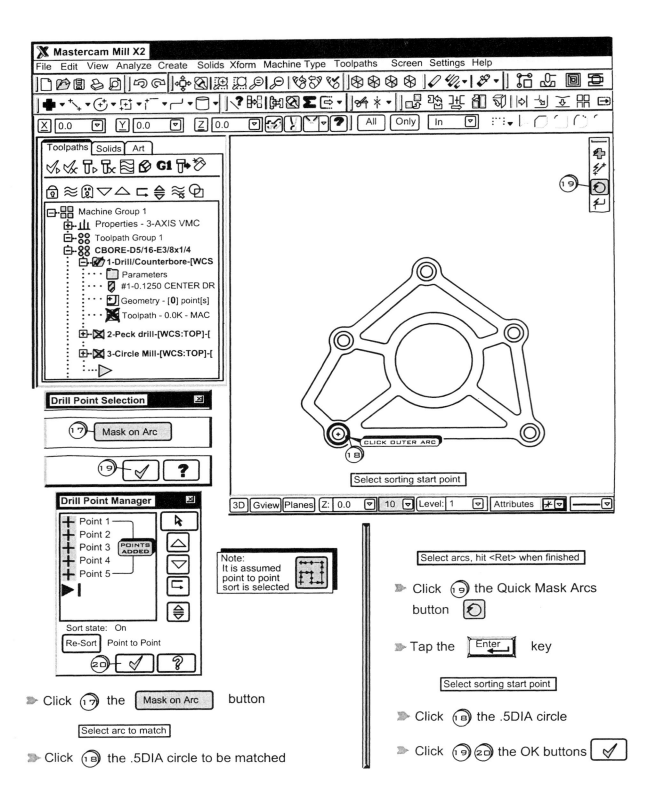

Select arcs, hit <Ret> when finished

➤ Click (19) the Quick Mask Arcs
button

➤ Tap the [Enter] key

Select sorting start point

➤ Click (18) the .5DIA circle

➤ Click (19)(20) the OK buttons ✓

➤ Click (17) the [Mask on Arc] button

Select arc to match

➤ Click (18) the .5DIA circle to be matched

◆ ADD DRILL POINTS TO THE OPERATIONS: 2 - Peck drill- full retract AND 3- Circle Mill

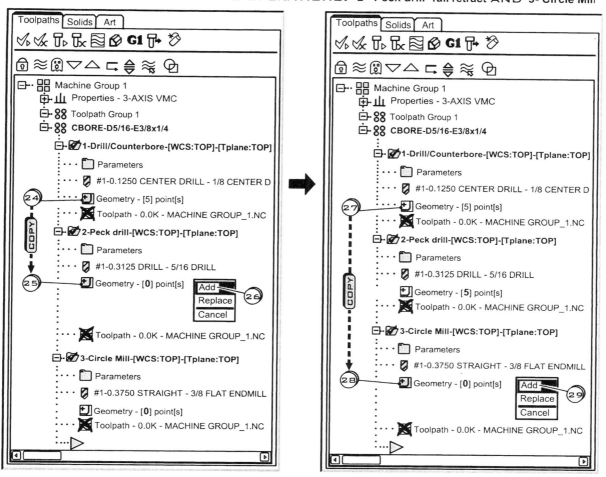

➤ Click ㉔ on the Part geometry icon 🔲 for the operation **1 - Drill/Counterbore**
keeping the left mouse button depressed move the icon over the
geometry icon for the operation ㉕ **2 - Peck drill** and *release*

➤ Click ㉖ **Add** *Mastercam* will *add a copy* of the five drill points from operation **1**
to operation **2**

➤ Click ㉗ on the Part geometry icon 🔲 for the operation **1 - Drill/Counterbore**
keeping the left mouse button depressed move the icon over the
geometry icon for the operation ㉘ **3 - Circle Mill** and *release*

➤ Click ㉙ **Add** *Mastercam* will *also add a copy* of the five drill points from operation **1**
to operation **3**

C) ADJUST THE DRILL DEPTH TO THE MATERIAL THICKNESS OF .563IN

◆ EDIT THE **2-Peck drill** OPERATION PARAMETERS

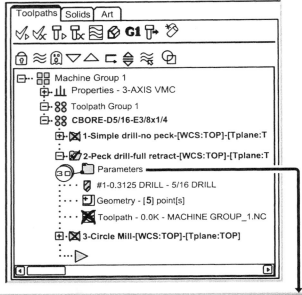

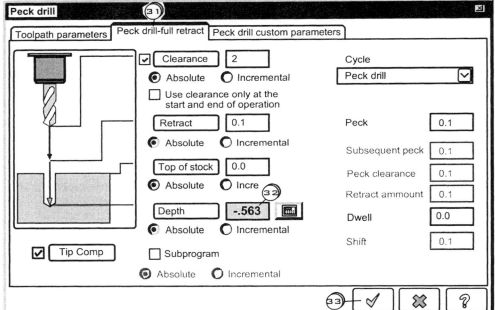

Click ③⓪ the Parameters icon 📁 for the **2 - Peck drill** operation
to enter the .313Dia peck drill machining parameters

Click ③① the [Peck drill-full retract] tab.

Click ③② in the Depth box; enter [**-.563**]

Click ③③ the OK button [✓]

D) IMPORT THE GROUP POC-E3/16x1/8 FROM THE OPERATIONS LIBRARY

 ◆ FROM THE OPERATIONS MANAGER SELECT IMPORT

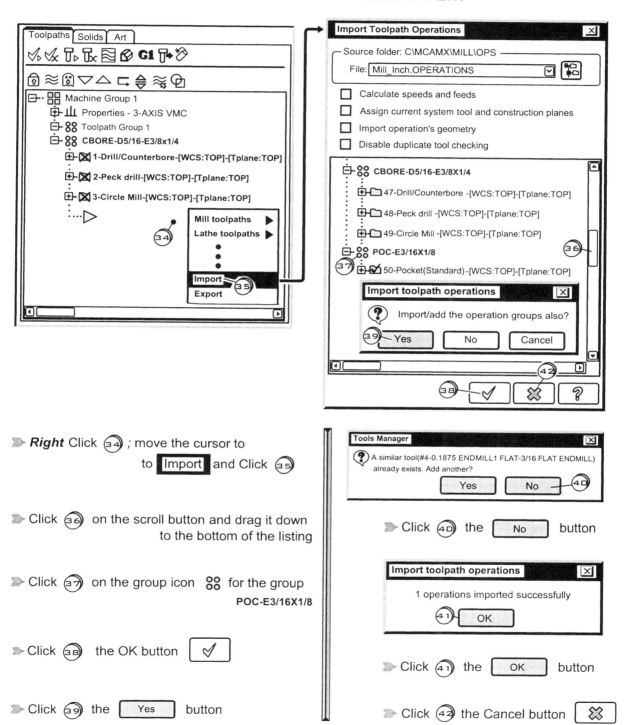

▶ **Right** Click ③④ ; move the cursor to to Import and Click ③⑤

▶ Click ③⑥ on the scroll button and drag it down to the bottom of the listing

▶ Click ③⑦ on the group icon 88 for the group
POC-E3/16X1/8

▶ Click ③⑧ the OK button ✓

▶ Click ③⑨ the Yes button

▶ Click ④⓪ the No button

▶ Click ④① the OK button

▶ Click ④② the Cancel button ✗

E) ADD CHAINS TO OPERATION 4- Pocket(Standard)

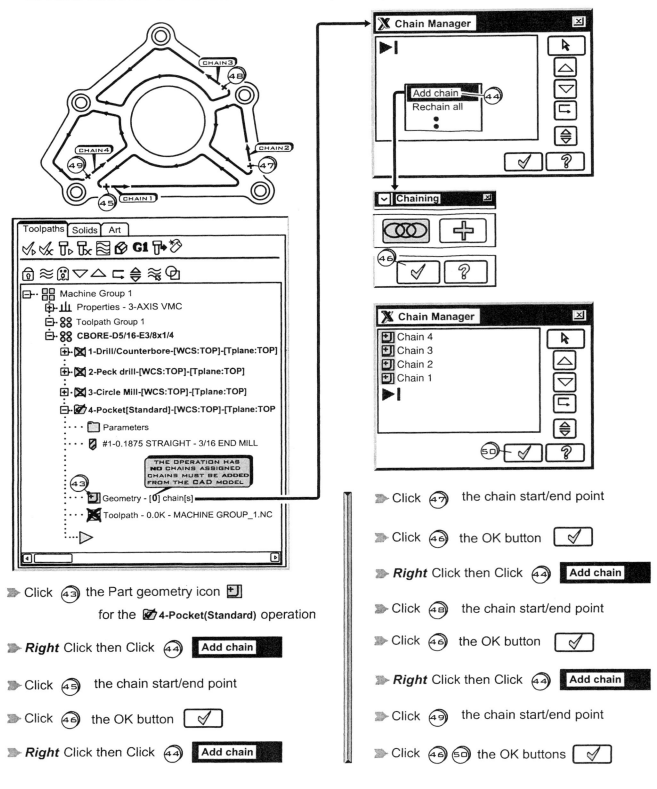

➤ Click ㊽ the Part geometry icon 🔳

        for the 📝 **4-Pocket(Standard)** operation

➤ *Right* Click then Click ㊽   **Add chain**

➤ Click ㊺   the chain start/end point

➤ Click ㊻   the OK button ✓

➤ *Right* Click then Click ㊽   **Add chain**

➤ Click ㊼   the chain start/end point

➤ Click ㊻   the OK button ✓

➤ *Right* Click then Click ㊽   **Add chain**

➤ Click ㊽   the chain start/end point

➤ Click ㊻   the OK button ✓

➤ *Right* Click then Click ㊽   **Add chain**

➤ Click ㊾   the chain start/end point

➤ Click ㊻ ㊿   the OK buttons ✓

F) ADJUST THE POCKETING DEPTH TO .282

◆ EDIT THE **4-Pocket(Standard)** OPERATION PARAMETERS

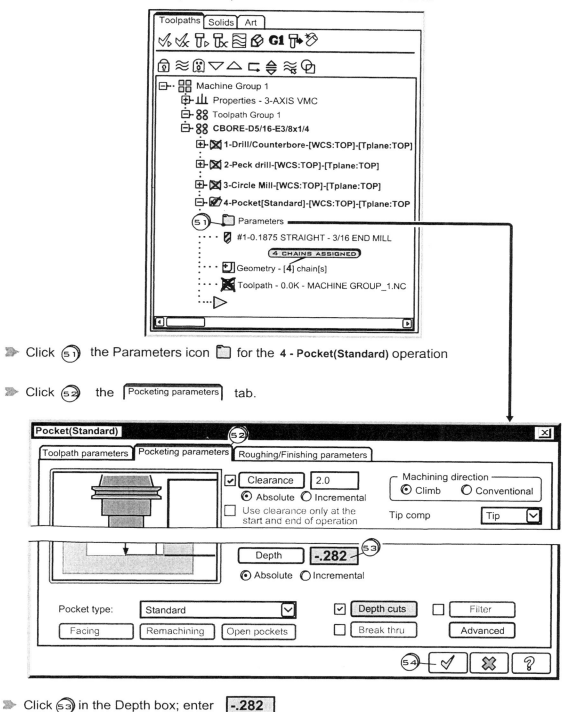

➤ Click ⑤① the Parameters icon 📁 for the **4 - Pocket(Standard)** operation

➤ Click ⑤② the Pocketing parameters tab.

➤ Click ⑤③ in the Depth box; enter **-.282**

➤ Click ⑤④ the OK button ✓

G) REGENERATE THE TOOLPATHS TO REFLECT THE LATEST CHANGES

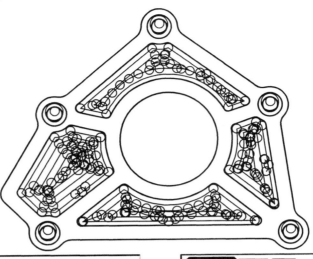

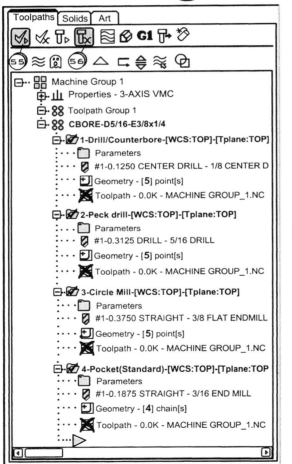

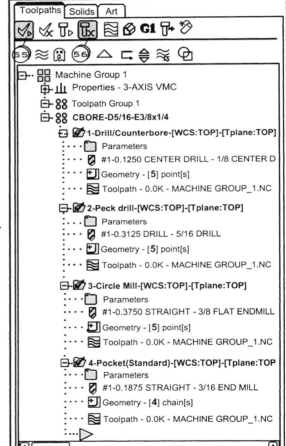

➤ Click ⑤⑤ the Select all operations button

➤ Click ⑤⑥ the *Regenerate all dirty*
               operations button

All the machining operations as outlined in
PROCESS PLAN 9-1 have now been
created in *Mastercam's* Operations Manager.

9-5 Editing Data in the Operations Library

This section will present techniques for editing the data contained in *Mastercam*'s Operations Library. This includes *deleting* unwanted groups and operations as well as *creating new* data by *copying* and editing existing groups and operations.

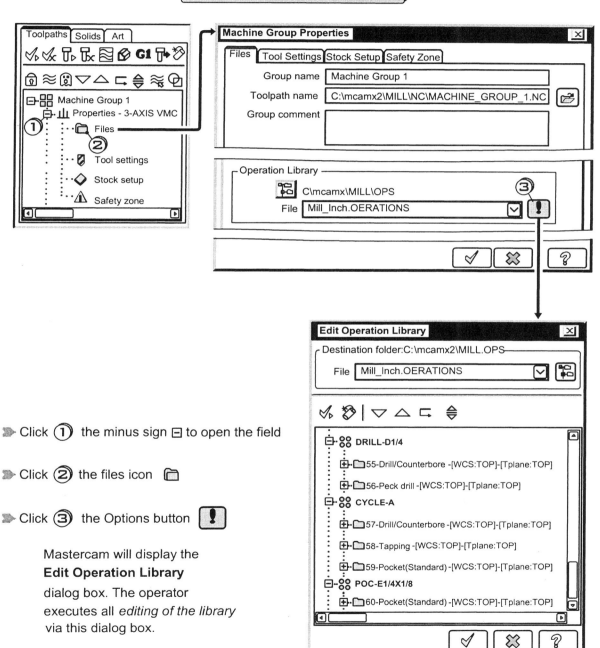

Click ① the minus sign ⊟ to open the field

Click ② the files icon 🗀

Click ③ the Options button [❗]

Mastercam will display the **Edit Operation Library** dialog box. The operator executes all *editing of the library* via this dialog box.

DELETING GROUPS AND OPERATIONS

EXAMPLE 9-3

Use the Library Operations Editor to delete the group **CYCLE-A** and its associated operatons from *Mastercam*'s Operations Library.

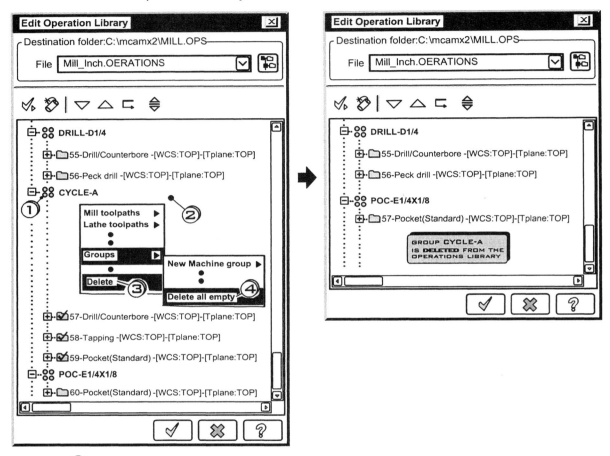

➤ Click ① on the group icon 🔳 for the group **CYCLE-A**

➤ *Right* Click ② *;* move the cursor down to Delete and Click ③

All the operations in **CYCLE-A** will be *deleted* and the group will be *empty*

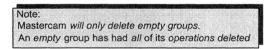

> Note:
> Mastercam *will only delete empty groups.*
> An *empty* group has had *all* of its *operations deleted*

➤ *Right* Click ② *;* move the cursor down to Groups over and down to
Delete all empty *;* Click ④

the group **CYCLE-A** will be *deleted* from the operation library

COPYING/EDITING EXISTING GROUPS AND OPERATIONS

EXAMPLE 9-4

Use the Library Operations Editor to create a new group in the library **POC-E3/8x1/8** by copying and editing the existing group **POC-E1/4x1/8** .

A) GENERATE A COPY OF THE GROUP POC-E1/4x1/8

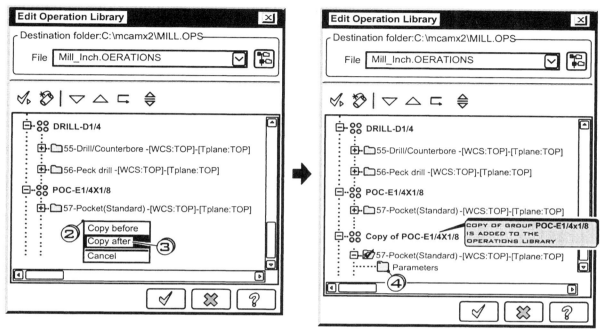

➤ Press the **Right** mouse button over the **POC-E1/4x1/8** group icon ①

       keeping it depressed *drag* the cursor down to ② and release

➤ Click ③ **Copy after** *keeping it depressed*

B) EDIT THE COPY(REPLACE THE 1/4 ENDMILL WITH A 3/8 ENDMILL TOOL)

➤ Click ④ the Parameters icon 🗂 for the 58-Pocket(Standard) operation

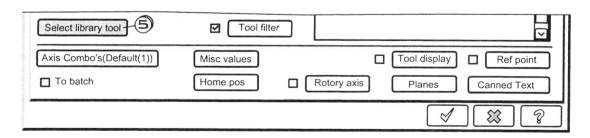

➤ Click ⑤ the Select library tool button

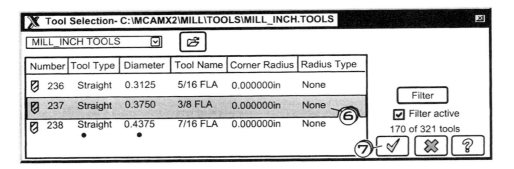

➣ Click ⑥ the 3/8 FLA Tool

➣ Click ⑦ the OK button ☑

c) RENAME THE EDITED GROUP TO POC-E3/8x1/4

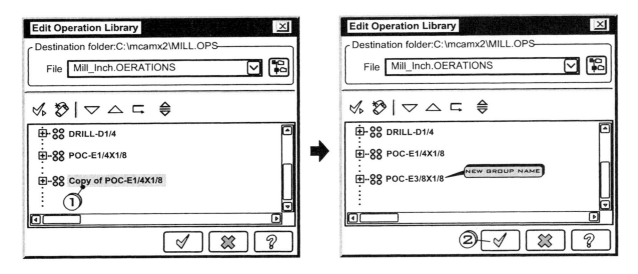

➣ Click ① Copy of POC-E1/4 x 1/8 on the group name, pause Click ①

➣ Enter the new name POC-E3/8X1/8 [Enter↵]

f) SAVE THE CHANGES TO THE OPERATIONS LIBRARY

➣ Click ② the OK button ☑

All the changes will then be *saved* to the Operations Library

EXERCISES
 9-1)
PART A:

Create a dummy .25Dia circle and use it to generate a new group **TAP1/4-20BLIND** containing the set of operations listed in Table 9p-1.

Save the new group **TAP1/4-20BLIND** to Mastercam's Operations Library.

Table 9P-1

Group Name	Operation(s)	Tooling
TAP1/4-20BLIND .401 .14 .653	CENTER DRILL x .2 DEEP PECK DRILL x .601DEEP TAP 1/4-20UNC x .401DEEP C'SINK 90° TO .28 DIA	1/8 CTR DRILL #7(.201) DRILL 1/4-20UNC TAP 1/2 CHAMFER MILL

PART B:

The CAD model file **EX9-1** is provided on the CD in the folder ▢**CHAPTER9** on the CD. Copy it into the **JVAL-MILL** subdirectory on C drive. Open it and generate a part program for executing the machining outlined in PROCESS PLAN 9P-1. The part shown in Figure 9p-1 is to be produced.

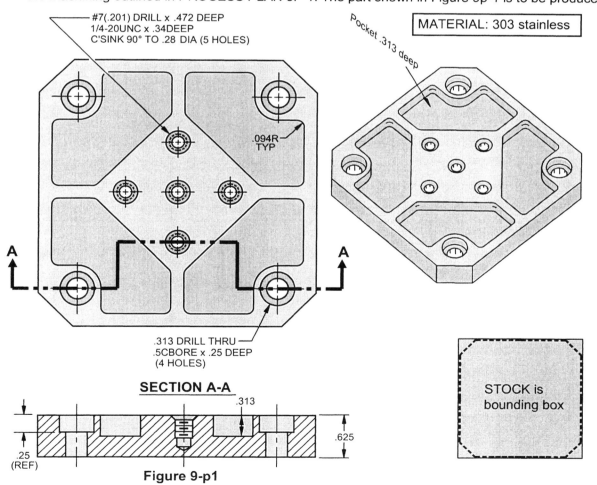

#7(.201) DRILL x .472 DEEP
1/4-20UNC x .34DEEP
C'SINK 90° TO .28 DIA (5 HOLES)

Pocket .313 deep

MATERIAL: 303 stainless

.094R
TYP

.313 DRILL THRU
.5CBORE x .25 DEEP
(4 HOLES)

SECTION A-A

.313

.625

.25
(REF)

Figure 9-p1

STOCK is
bounding box

PROCESS PLAN 9P-1

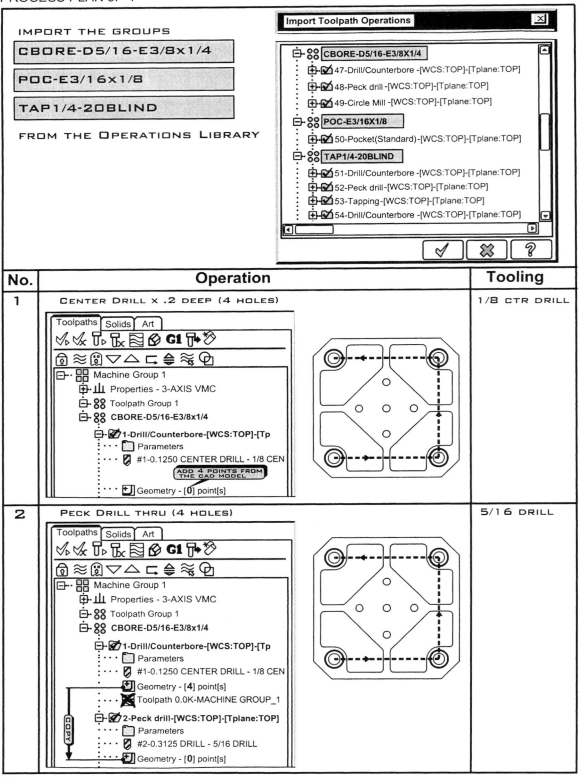

No.	Operation	Tooling
1	CENTER DRILL X .2 DEEP (4 HOLES)	1/8 CTR DRILL
2	PECK DRILL THRU (4 HOLES)	5/16 DRILL

PROCESS PLAN 9P-1(*continued*)

No.	Operation	Tooling
3	CIRCLE MILL X .25 DEEP (4 HOLES)	3/8 ENDMILL
4	ROUGH AND FINISH POCKET X .313 DEEP LEAVE .01 FOR FINISH CUT IN X,Y AND Z	3/16 ENDMILL
5	CENTER DRILL X .2 DEEP (5 HOLES)	1/8 CTR DRILL

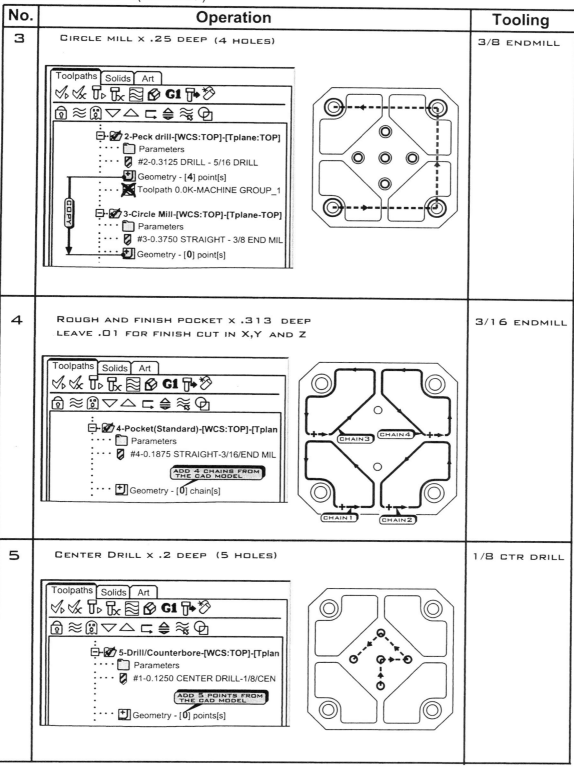

PROCESS PLAN 9P-1(*continued*)

No.	Operation	Tooling
6	PECK DRILL X .472 DEEP (5 HOLES)	#7(.201)DRILL
7	TAP X .34 DEEP (5 HOLES)	1/4-20 TAPRH
8	C'SINK 90° TO .28 DIA (5 HOLES)	1/2 CHAMFER MILL

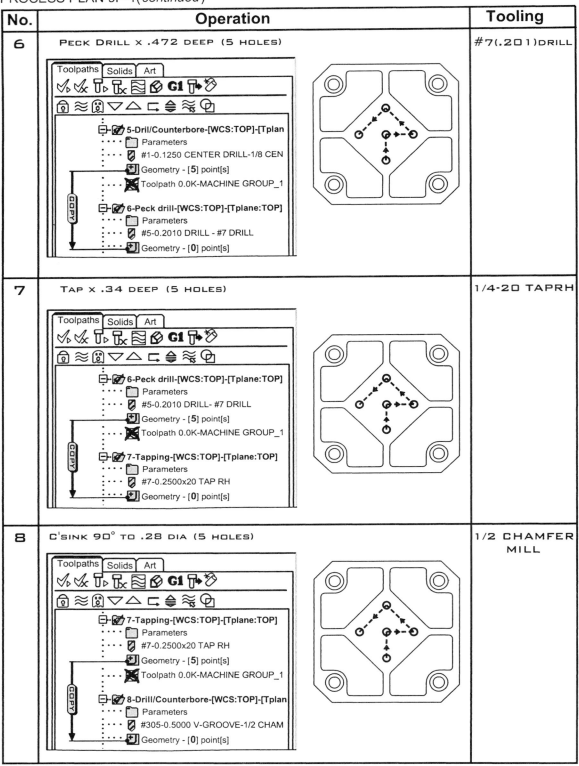

9-2) Enter the Operations Library Editor. Generate a new group **TAP#8-32BLIND** by copying and editing the existing group **TAP1/4-20BLIND**. The new group **TAP#8-32BLIND** is to contain the set of operations listed in Table 9p-2. Save all the groups to Mastercam's Operations Library.

PART A:

Table 9P-2

Group Name	Operation(s)	Tooling
TAP#8-32BLIND .08 .289 .466	CENTER DRILL x .2 DEEP PECK DRILL x .414 DEEP TAP #8-32UNF x .289 DEEP C'SINK 90° TO .16 DIA	1/8 CTR DRILL #29(.136) DRILL #8-32UNF TAP 1/2 CHAMFER MILL

PART B:

The CAD model file **EX9-2** is in the folder ☐ **CHAPTER9** on the CD . Copy it into the subdirectory **JVAL-MILL** on C drive. Open it and follow the PROCESS PLAN 9P2 as a guide in generating a part program to produce the part shown in Figure 9p-2

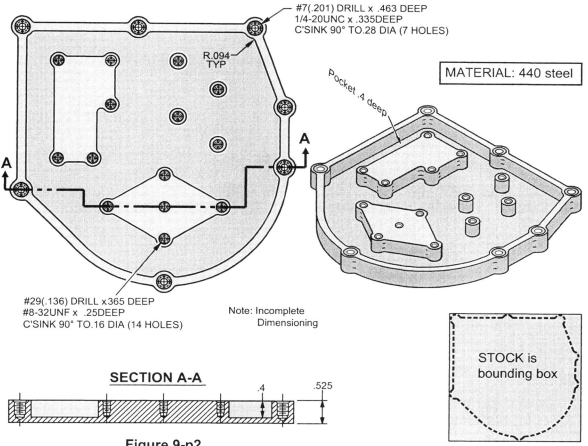

#7(.201) DRILL x .463 DEEP
1/4-20UNC x .335 DEEP
C'SINK 90° TO .28 DIA (7 HOLES)

R.094 TYP

MATERIAL: 440 steel

Pocket .4 deep

A

A

#29(.136) DRILL x 365 DEEP
#8-32UNF x .25 DEEP
C'SINK 90° TO .16 DIA (14 HOLES)

Note: Incomplete Dimensioning

STOCK is bounding box

SECTION A-A

.4   .525

Figure 9-p2

PROCESS PLAN 9P-2

IMPORT THE GROUPS

POC-E3/16x1/8

TAP1/4-20BLIND

TAP#8-32BLIND

FROM THE OPERATIONS LIBRARY

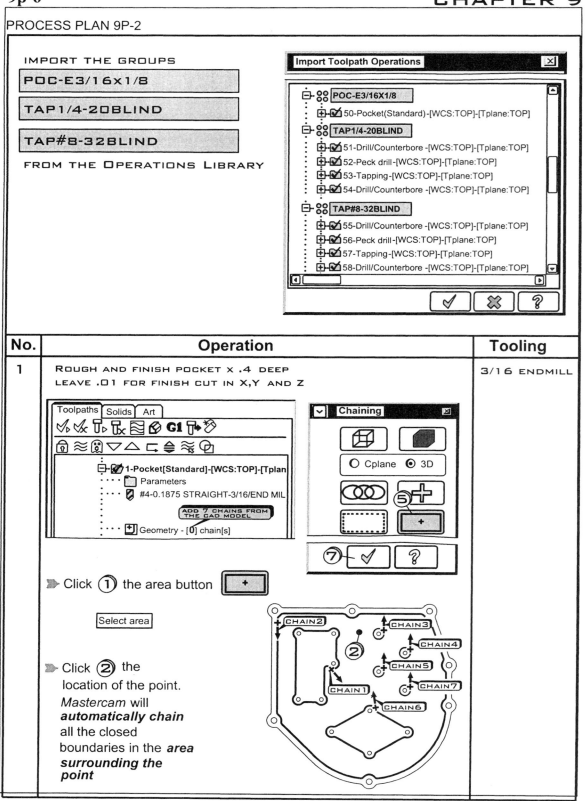

Import Toolpath Operations

POC-E3/16X1/8
  50-Pocket(Standard)-[WCS:TOP]-[Tplane:TOP]
TAP1/4-20BLIND
  51-Drill/Counterbore -[WCS:TOP]-[Tplane:TOP]
  52-Peck drill-[WCS:TOP]-[Tplane:TOP]
  53-Tapping-[WCS:TOP]-[Tplane:TOP]
  54-Drill/Counterbore -[WCS:TOP]-[Tplane:TOP]
TAP#8-32BLIND
  55-Drill/Counterbore -[WCS:TOP]-[Tplane:TOP]
  56-Peck drill-[WCS:TOP]-[Tplane:TOP]
  57-Tapping-[WCS:TOP]-[Tplane:TOP]
  58-Drill/Counterbore -[WCS:TOP]-[Tplane:TOP]

No.	Operation	Tooling
1	ROUGH AND FINISH POCKET X .4 DEEP LEAVE .01 FOR FINISH CUT IN X,Y AND Z	3/16 ENDMILL

Toolpaths Solids Art

1-Pocket[Standard]-[WCS:TOP]-[Tplan
  Parameters
  #4-0.1875 STRAIGHT-3/16/END MIL
  ADD 7 CHAINS FROM THE CAD MODEL
  Geometry - [0] chain[s]

Chaining

○ Cplane ● 3D

5

7 ✓ ?

➤ Click ① the area button [+]

Select area

➤ Click ② the
location of the point.
Mastercam will
automatically chain
all the closed
boundaries in the *area
surrounding the
point*

CHAIN2 CHAIN3
CHAIN4
② CHAIN5 CHAIN7
CHAIN1
CHAIN6

PROCESS PLAN 9P-2(*continued*)

No.	Operation	Tooling
2	CENTER DRILL X .2 DEEP (7 HOLES) 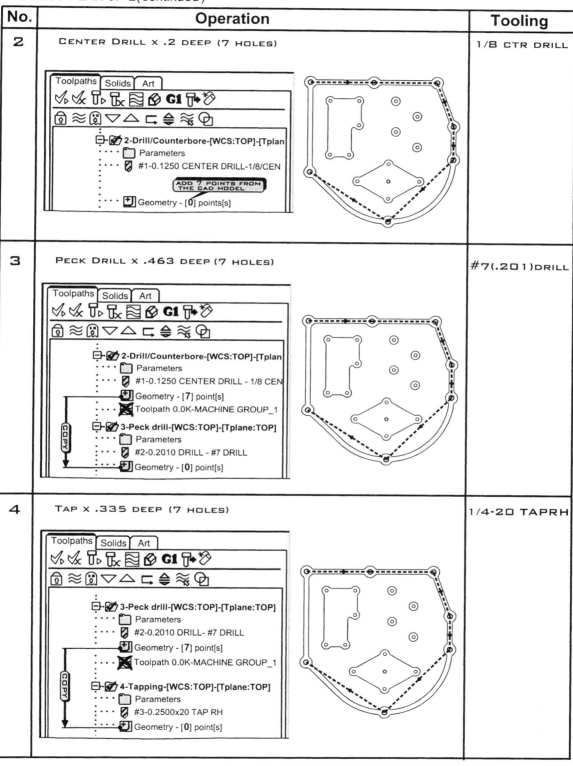	1/8 CTR DRILL
3	PECK DRILL X .463 DEEP (7 HOLES)	#7(.201)DRILL
4	TAP X .335 DEEP (7 HOLES)	1/4-20 TAPRH

PROCESS PLAN 9P-2(*continued*)

No.	Operation	Tooling
5	C'SINK 90° TO .28 DIA(7 HOLES)	1/2 CHAMFER MILL
6	CENTER DRILL X .2 DEEP (14 HOLES)	1/8 CTR DRILL
7	PECK DRILL X .365 DEEP (14 HOLES)	#29(.136) DRILL

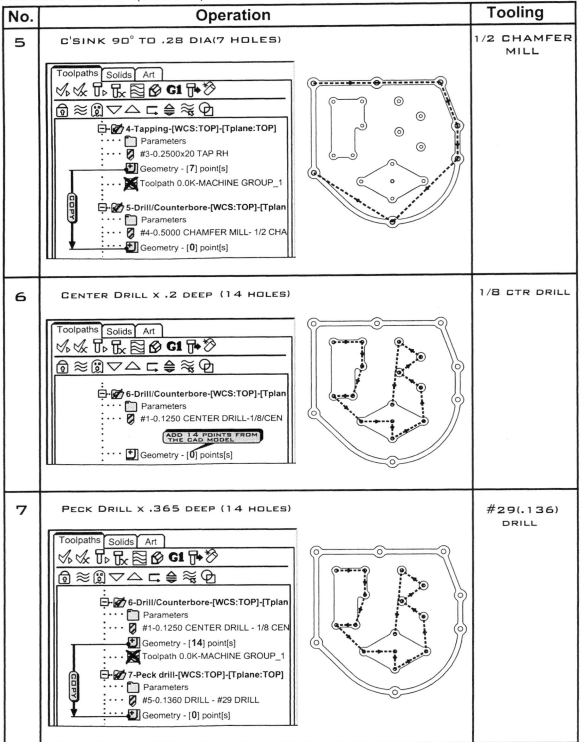

PROCESS PLAN 9P-2(*continued*)

No.	Operation	Tooling
8	TAP X .25 DEEP (14 HOLES)	#8-32 TAPRH
9	C'SINK 90° TO .16 DIA (14 HOLES)	1/2 CHAMFER MILL

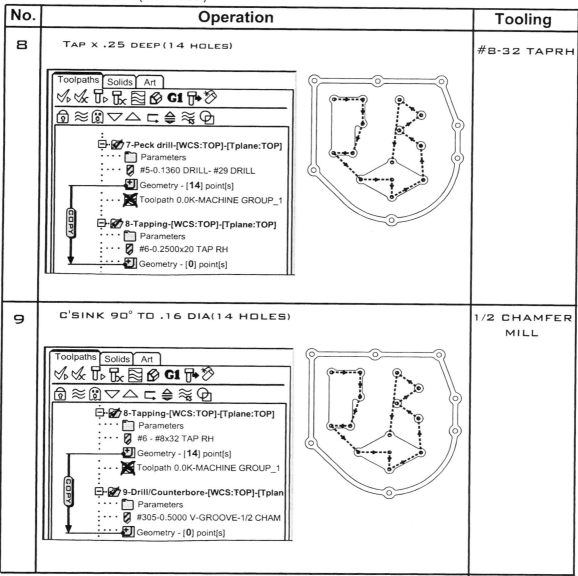

9-3)   Enter the Operations Library Editor. Generate a new group **DRILL-D1/5-CHAMIL1X.719** by copying and editing the existing group **TAP#8-32BLIND**. The new group **DRILL-D1/2-CHAMIL1X.72** is to contain the set of operations listed in Table 9p-3.
Save all the groups to *Mastercam's* Operations Library.

PART A:

Table 9P-3   Note: use the *Mastercam* tool library | BIG_Inch.Tools |

Group Name	Operation(s)	Tooling
DRILL-D1/2-CSINK1X.72	CENTER DRILL X .25 DEEP	1/4 CTR DRILL
	PECK DRILL THRU	1/2 DRILL
	DRILL X .72 DEEP	1 CSINK

PART B:

The CAD model file **EX9-3** is in the folder 📁**CHAPTER9** on the CD . Copy it into the **JVAL-MILL** subdirectory on C drive. Open it and follow the PROCESS PLAN 9P-3 as a guide in generating a part program to produce the part shown in Figure 9p-3

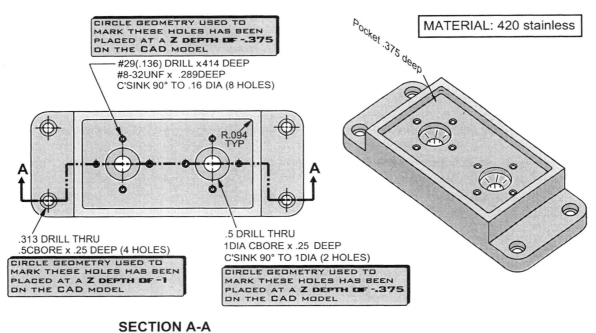

MATERIAL: 420 stainless

CIRCLE GEOMETRY USED TO MARK THESE HOLES HAS BEEN PLACED AT A **Z DEPTH OF -.375** ON THE **CAD** MODEL

#29(.136) DRILL x414 DEEP
#8-32UNF x .289DEEP
C'SINK 90° TO .16 DIA (8 HOLES)

R.094 TYP

.313 DRILL THRU
.5CBORE x .25 DEEP (4 HOLES)

CIRCLE GEOMETRY USED TO MARK THESE HOLES HAS BEEN PLACED AT A **Z DEPTH OF -1** ON THE **CAD** MODEL

.5 DRILL THRU
1DIA CBORE x .25 DEEP
C'SINK 90° TO 1DIA (2 HOLES)

CIRCLE GEOMETRY USED TO MARK THESE HOLES HAS BEEN PLACED AT A **Z DEPTH OF -.375** ON THE **CAD** MODEL

SECTION A-A

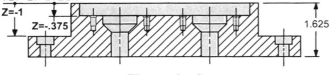

TOP OF STOCK
Z=-1
Z=-.375
1.625

STOCK is bounding box

Figure 9-p3

PROCESS PLAN 9P-3

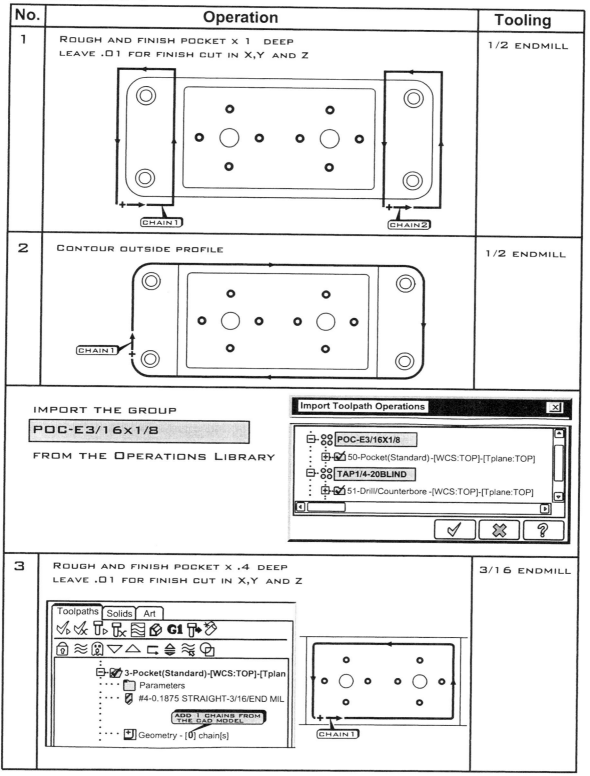

No.	Operation	Tooling
1	ROUGH AND FINISH POCKET X 1 DEEP LEAVE .01 FOR FINISH CUT IN X,Y AND Z	1/2 ENDMILL
2	CONTOUR OUTSIDE PROFILE	1/2 ENDMILL
	IMPORT THE GROUP POC-E3/16X1/8 FROM THE OPERATIONS LIBRARY	
3	ROUGH AND FINISH POCKET X .4 DEEP LEAVE .01 FOR FINISH CUT IN X,Y AND Z	3/16 ENDMILL

PROCESS PLAN 9P-3(*continued*)

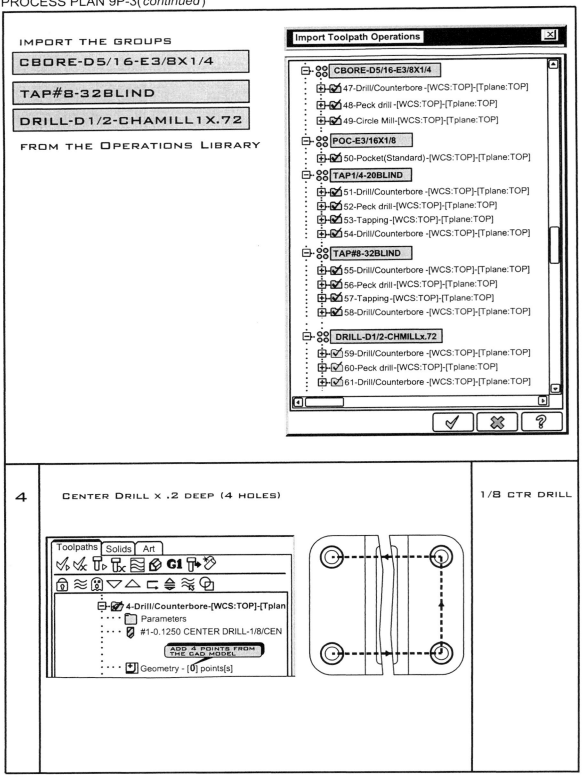

IMPORT THE GROUPS

CBORE-D5/16-E3/8X1/4

TAP#8-32BLIND

DRILL-D1/2-CHAMILL1X.72

FROM THE OPERATIONS LIBRARY

Import Toolpath Operations

- CBORE-D5/16-E3/8X1/4
  - 47-Drill/Counterbore -[WCS:TOP]-[Tplane:TOP]
  - 48-Peck drill -[WCS:TOP]-[Tplane:TOP]
  - 49-Circle Mill -[WCS:TOP]-[Tplane:TOP]
- POC-E3/16X1/8
  - 50-Pocket(Standard)-[WCS:TOP]-[Tplane:TOP]
- TAP1/4-20BLIND
  - 51-Drill/Counterbore -[WCS:TOP]-[Tplane:TOP]
  - 52-Peck drill-[WCS:TOP]-[Tplane:TOP]
  - 53-Tapping-[WCS:TOP]-[Tplane:TOP]
  - 54-Drill/Counterbore -[WCS:TOP]-[Tplane:TOP]
- TAP#8-32BLIND
  - 55-Drill/Counterbore -[WCS:TOP]-[Tplane:TOP]
  - 56-Peck drill-[WCS:TOP]-[Tplane:TOP]
  - 57-Tapping-[WCS:TOP]-[Tplane:TOP]
  - 58-Drill/Counterbore -[WCS:TOP]-[Tplane:TOP]
- DRILL-D1/2-CHMILLx.72
  - 59-Drill/Counterbore -[WCS:TOP]-[Tplane:TOP]
  - 60-Peck drill-[WCS:TOP]-[Tplane:TOP]
  - 61-Drill/Counterbore -[WCS:TOP]-[Tplane:TOP]

| 4 | CENTER DRILL X .2 DEEP (4 HOLES) | 1/8 CTR DRILL |

Toolpaths | Solids | Art

- 4-Drill/Counterbore-[WCS:TOP]-[Tplan
  - Parameters
  - #1-0.1250 CENTER DRILL-1/8/CEN
  - ADD 4 POINTS FROM THE CAD MODEL
  - Geometry - [0] points[s]

PROCESS PLAN 9P-3(*continued*)

No.	Operation	Tooling
5	PECK DRILL THRU (4 HOLES) 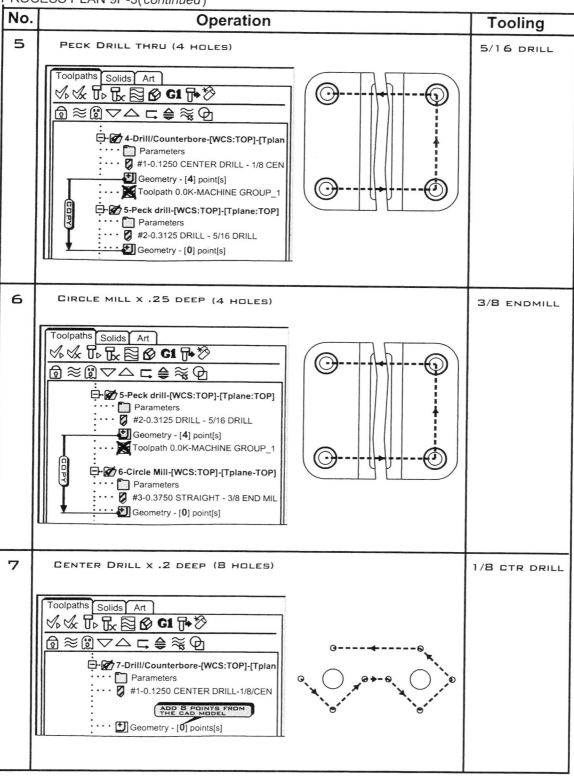	5/16 DRILL
6	CIRCLE MILL X .25 DEEP (4 HOLES)	3/8 ENDMILL
7	CENTER DRILL X .2 DEEP (8 HOLES)	1/8 CTR DRILL

PROCESS PLAN 9P-3(*continued*)

No.	Operation	Tooling
8	Peck Drill x .365 deep (8 holes)	#29(.136) DRILL
9	Tap x .25 deep (8 holes)	#8-32 TAPRH
10	C'sink 90° to .16 dia (8 holes)	1/2 CHAMFER MILL

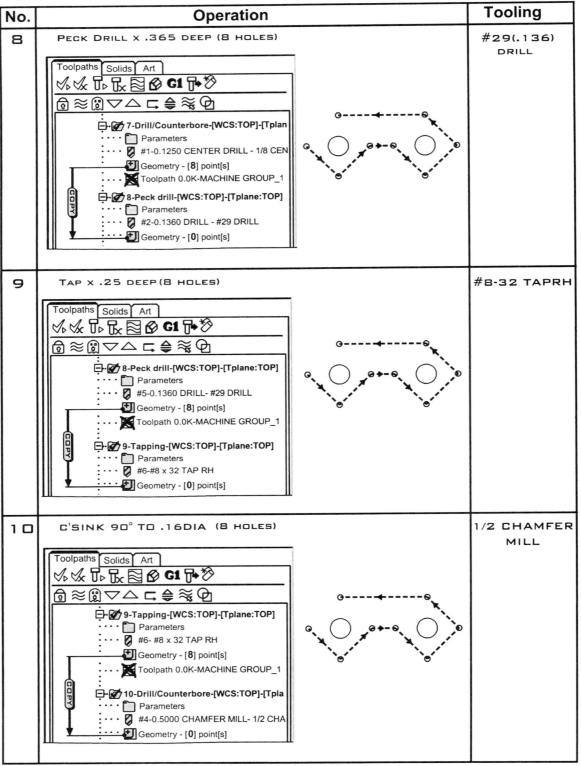

PROCESS PLAN 9P-3(*continued*)

No.	Operation	Tooling
11	CENTER DRILL X .25 DEEP (2 HOLES)	1/4 CTR DRILL
12	PECK DRILL THRU (2 HOLES)	1/2 DRILL
13	DRILL X .72 DEEP (2 HOLES)	MILL 1 CHAMFER

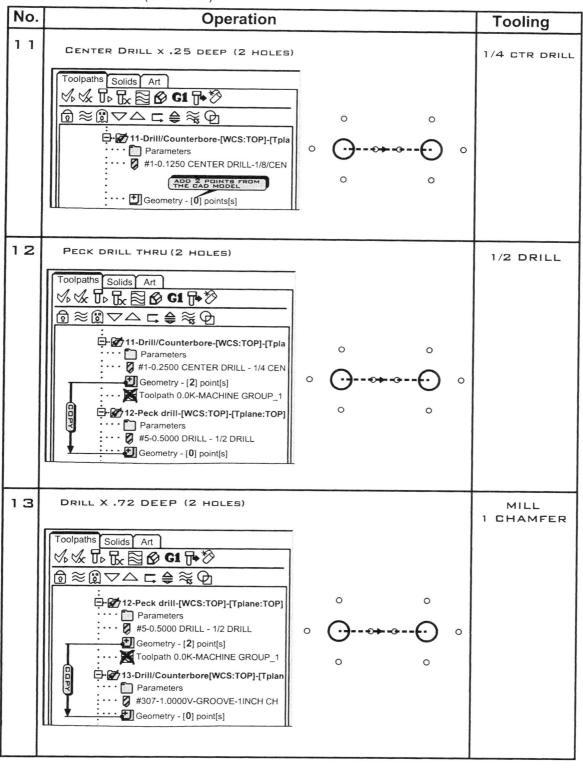

CHAPTER - 10

USING TABS AND WORK OFFSETS

10-1 Chapter Objectives

After completing this chapter you will be able to:

1. Understand the uses of *Mastercam X2*'s tab function.

2. Know how to specify tab parameters to create various types of tabs.

3. Explain the methods of placing tabs on CAD models.

4. Know how to edit tabs.

5. Explain how to use the View Manager to create and set WCSs, Tplanes and their origins for work offsets.

6. Know how to assign generic work offset numbers in the view manager.

7. Understand how to select previously created work offsets using the View Manager.

10-2 Applications of *Mastercam X2*'s Tab Function

Mastercam X2's Tab function is used to *create undercut areas* of a toolpath. These areas can be used to help hold down a nested pattern of parts that are to be cut from a single sheet of metal. Necessary *jumps around clamps and other holding devices* can also be created via the tab function.

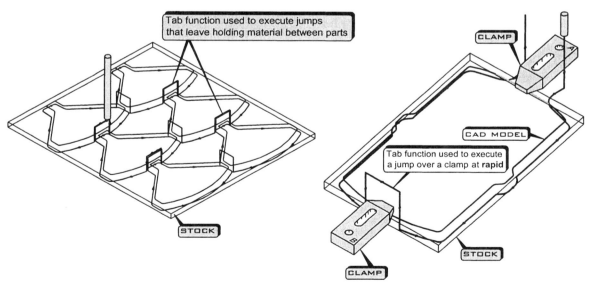

Machining multiple or nested parts   Clamp avoidance

Figure 10-1

10-3 Specifying Tab Parameters

The Tabs dialog box contains the parameters for specifting various types of tabs. It is entered by clicking the Tabs button in the Contour 2D dialog box.

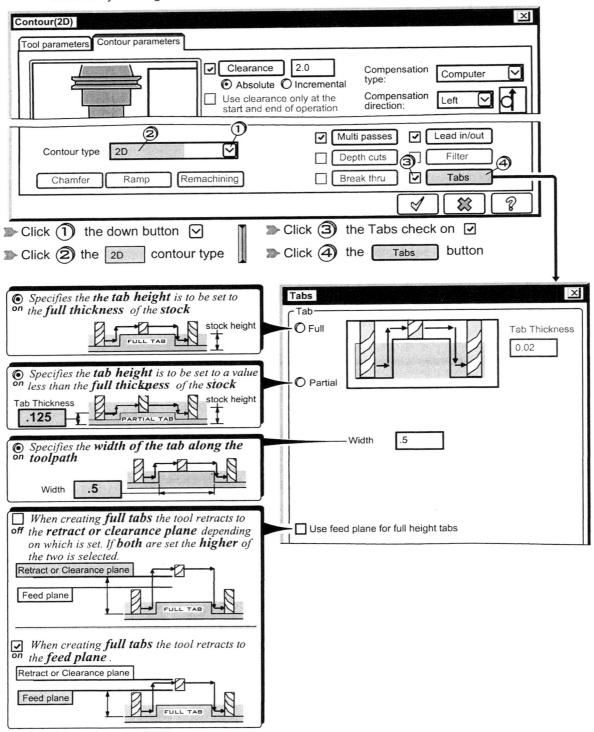

▶ Click ① the down button ☑

▶ Click ② the 2D contour type

▶ Click ③ the Tabs check on ☑

▶ Click ④ the Tabs button

---

on ⦿ *Specifies the **the tab height** is to be set to the **full thickness** of the **stock***

stock height

FULL TAB

---

on ⦿ *Specifies the **tab height** is to be set to a value less than the **full thickness** of the **stock***

Tab Thickness
.125

stock height

PARTIAL TAB

---

on ⦿ *Specifies the **width of the tab along the toolpath***

Width **.5**

---

off ☐ *When creating **full tabs** the tool retracts to the **retract or clearance plane** depending on which is set. If **both** are set the **higher** of the two is selected.*

Retract or Clearance plane

Feed plane

FULL TAB

---

on ☑ *When creating **full tabs** the tool retracts to the **feed plane**.*

Retract or Clearance plane

Feed plane

FULL TAB

Tabs

Tab

○ Full

○ Partial

Tab Thickness
0.02

Width .5

☐ Use feed plane for full height tabs

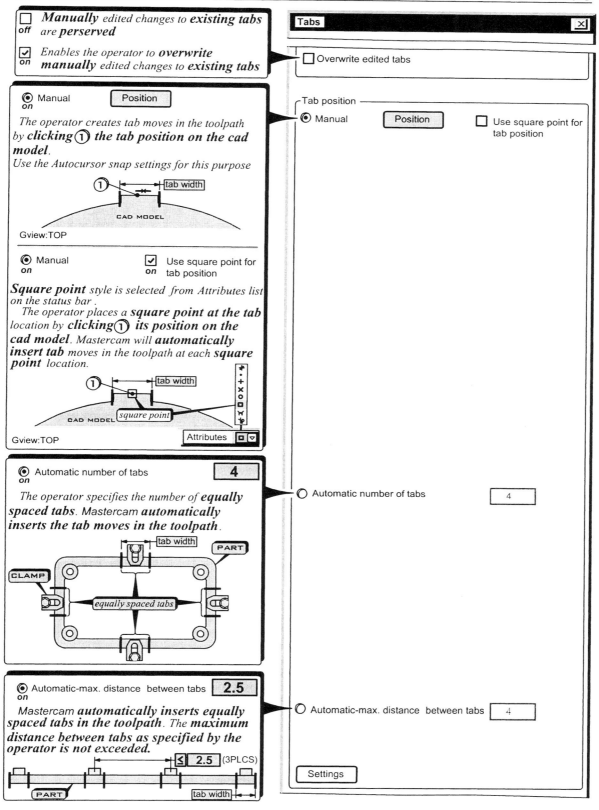

□ **Manually** *edited changes to* **existing tabs**
off *are* **perserved**

☑ *Enables the operator to* **overwrite**
on **manually** *edited changes to* **existing tabs**

◉ Manual [Position]
on

The operator creates tab moves in the toolpath
by **clicking** ① **the tab position on the cad**
model.
Use the Autocursor snap settings for this purpose

① ├──┤ tab width
CAD MODEL
Gview:TOP

◉ Manual ☑ Use square point for
on on tab position

Square point *style is selected from Attributes list*
on the status bar .
  The operator places a **square point at the tab**
location by **clicking** ① **its position on the**
cad model. *Mastercam will* **automatically**
insert tab *moves in the toolpath at each* **square**
point *location.*

① ├──┤ tab width
square point
CAD MODEL
Attributes
Gview:TOP

◉ Automatic number of tabs [4]
on

The operator specifies the number of **equally**
spaced tabs. *Mastercam* **automatically**
inserts the tab moves in the toolpath.

tab width
PART
CLAMP
equally spaced tabs

◉ Automatic-max. distance between tabs [2.5]
on

Mastercam **automatically inserts equally**
spaced tabs *in the toolpath*. *The* **maximum**
distance between tabs as specified by the
operator is not exceeded.

≤ [2.5] (3PLCS)
PART tab width

Tabs ⊠

□ Overwrite edited tabs

┌ Tab position ─────────
◉ Manual [Position] □ Use square point for
  tab position

○ Automatic number of tabs [4]

○ Automatic-max. distance between tabs [4]

[Settings]

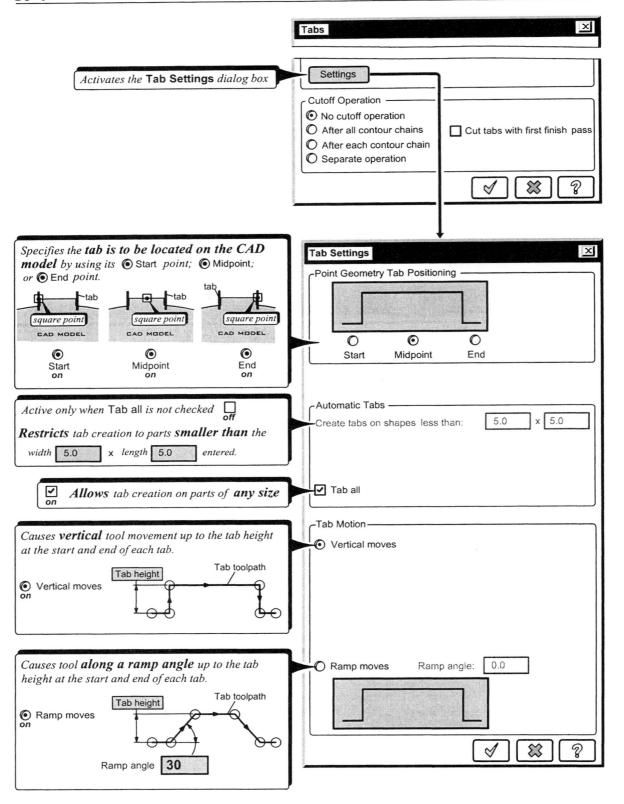

Tabs

Activates the **Tab Settings** *dialog box* → Settings

Cutoff Operation
- ● No cutoff operation
- ○ After all contour chains
- ○ After each contour chain
- ○ Separate operation

☐ Cut tabs with first finish pass

Specifies the **tab is to be located on the CAD model** *by using its* ● Start *point;* ● Midpoint; *or* ● End *point.*

Start **on** Midpoint **on** End **on**

Tab Settings

Point Geometry Tab Positioning

○ Start   ● Midpoint   ○ End

Active only when Tab all *is not checked* ☐ **off**

Restricts *tab creation to parts* **smaller than** *the* width **5.0** x *length* **5.0** *entered.*

Automatic Tabs
Create tabs on shapes less than: **5.0** x **5.0**

☑ **on** **Allows** *tab creation on parts of* **any size**

☑ Tab all

Causes **vertical** *tool movement up to the tab height at the start and end of each tab.*

● Vertical moves **on**

Tab height   Tab toolpath

Tab Motion
● Vertical moves

Causes tool **along a ramp angle** *up to the tab height at the start and end of each tab.*

● Ramp moves **on**

Tab height   Tab toolpath

Ramp angle **30**

○ Ramp moves   Ramp angle: 0.0

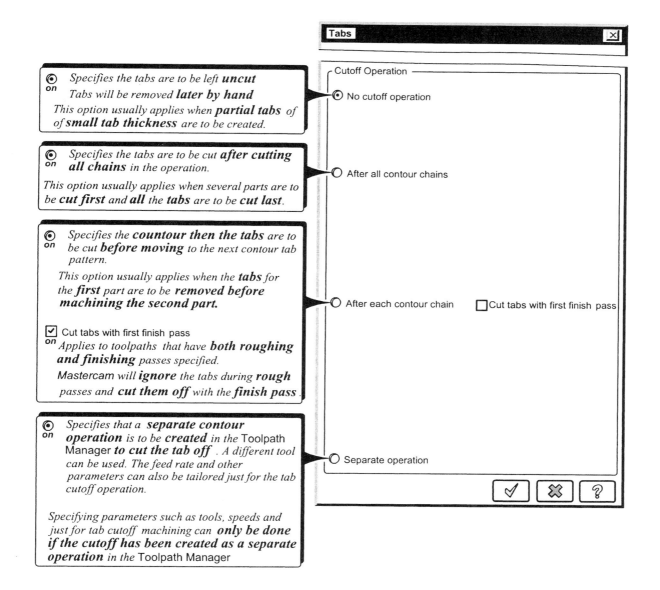

Specifies the tabs are to be left **uncut**
on Tabs will be removed **later by hand**
This option usually applies when **partial tabs** *of*
of **small tab thickness** *are to be created.*

Specifies the tabs are to be cut **after cutting**
on **all chains** *in the operation.*
This option usually applies when several parts are to
be **cut first** *and* **all** *the* **tabs** *are to be* **cut last**.

Specifies the **countour then the tabs** *are to*
on *be cut* **before moving** *to the next contour tab*
pattern.
This option usually applies when the **tabs** *for*
the **first** *part are to be* **removed before**
machining the second part.

☑ Cut tabs with first finish pass
on *Applies to toolpaths that have* **both roughing**
and finishing *passes specified.*
Mastercam will **ignore** *the tabs during* **rough**
passes and **cut them off** *with the* **finish pass** .

Specifies that a **separate contour**
on **operation** *is to be* **created** *in the* Toolpath
Manager **to cut the tab off** . *A different tool*
can be used. The feed rate and other
parameters can also be tailored just for the tab
cutoff operation.

Specifying parameters such as tools, speeds and
just for tab cutoff machining can **only be done**
if the cutoff has been created as a separate
operation *in the* Toolpath Manager

Tabs [×]

┌ Cutoff Operation ──────────

⦿ No cutoff operation

○ After all contour chains

○ After each contour chain   ☐ Cut tabs with first finish pass

○ Separate operation

EXAMPLE 10-1

The pattern shown in **Figure10-2** is to be cut from a 4.5 x 4 x .125 piece of brass.
Assume the CAD model for the pattern has already been created. Use *Mastercam's*
tab function as an aid in accomplishing the required machining.

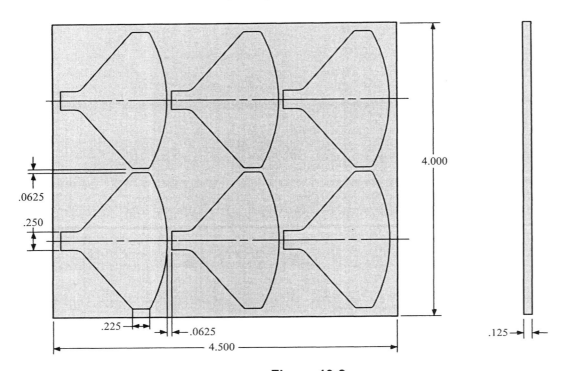

Figure 10-2

A) SELECT SQUARE POINT FROM THE ATTRIBUTE LIST ON THE STATUS BAR

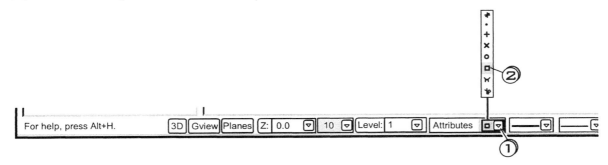

➤ Click ① the Attributes down arrow ▽

➤ Click ② the square point style ☐

B) INSERT SQUARE POINTS AT THE TAB LOCATIONS

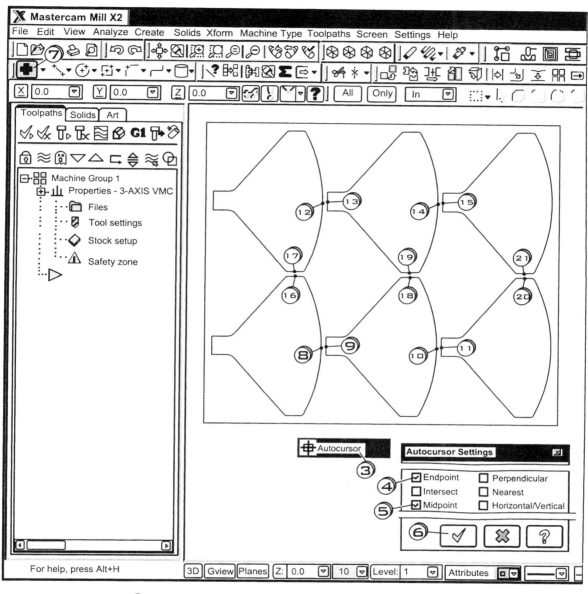

▷ *Right* Click; Click ③ **Autocursor**

▷ Click ④ ⑤ the checks on ☑ for Endpoint, Midpoint

▷ Click ⑥ the OK ✓ button

▷ Click ⑦ the Create Point Position button ➕

▷ Click ⑧ ⑨ ⑩ ⑪ ⑫ ⑬ ⑭ ⑮ ⑯ ⑰ ⑱ ⑲ ⑳ ㉑ Tap Esc key

C) CREATE MACHINING CONTOURS BY CHAINING

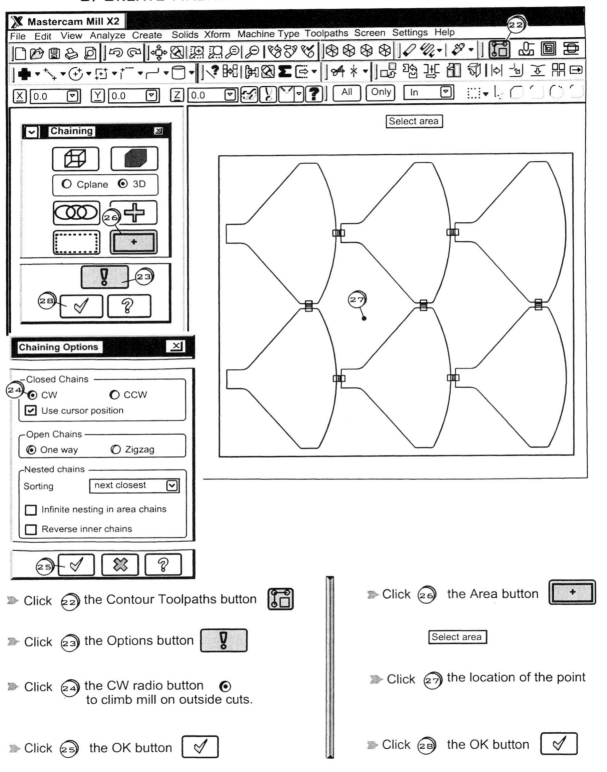

➤ Click ㉒ the Contour Toolpaths button

➤ Click ㉓ the Options button

➤ Click ㉔ the CW radio button ⊙
to climb mill on outside cuts.

➤ Click ㉕ the OK button

➤ Click ㉖ the Area button

Select area

➤ Click ㉗ the location of the point

➤ Click ㉘ the OK button

D) Obtain the 1/8 End Mill tool from the tools library

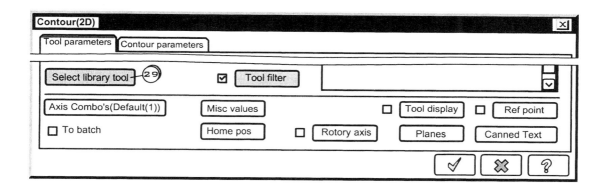

➤ Click ㉙ the [Select library tool] button

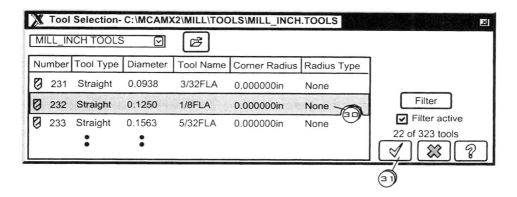

Click ㉚ on the 1/8 FLA end mill tool

Click ㉛ the OK button ✓

E) Enter the Contour(2D) and Tab Parameters

➤ Click ③② the [Contour parameters] tab.

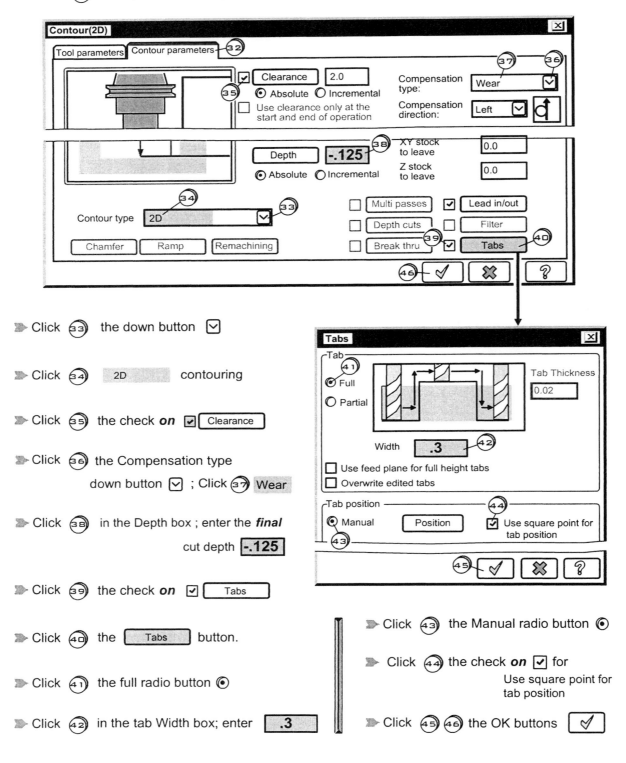

➤ Click ③③ the down button ☑

➤ Click ③④   2D   contouring

➤ Click ③⑤ the check **on** ☑ Clearance

➤ Click ③⑥ the Compensation type
          down button ☑ ; Click ③⑦ Wear

➤ Click ③⑧ in the Depth box ; enter the **final**
          cut depth -.125

➤ Click ③⑨ the check **on** ☑ Tabs

➤ Click ④⓪ the Tabs button.

➤ Click ④① the full radio button ⊙

➤ Click ④② in the tab Width box; enter .3

➤ Click ④③ the Manual radio button ⊙

➤ Click ④④ the check **on** ☑ for
          Use square point for
          tab position

➤ Click ④⑤ ④⑥ the OK buttons ☑

F) DELETE Chain 1 WHICH IS NOT TO BE CUT

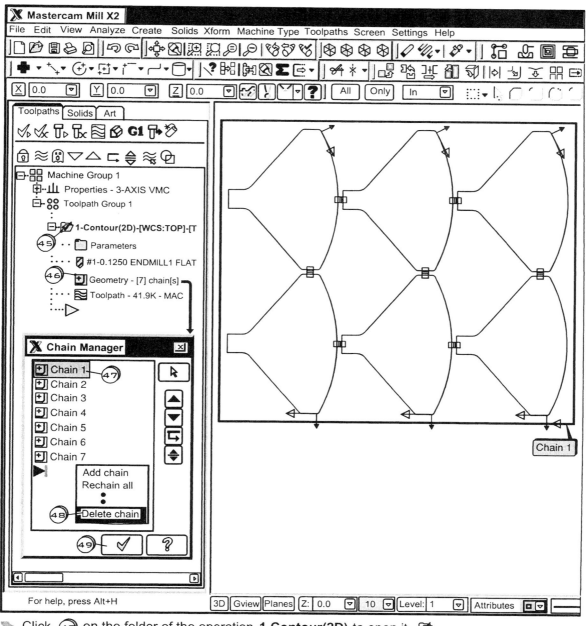

➤ Click (47) on the folder of the operation **1 Contour(2D)** to open it 📝

➤ Click (48) on the part geometry icon for the operation ⊞

➤ **Right** Click (49) on Chain 1 , the chain to be **deleted**

, move the cursor down and Click (50) Delete chain

➤ Click (51) the OK button ✓

G) REGENERATE ALL DIRTY OPERATIONS

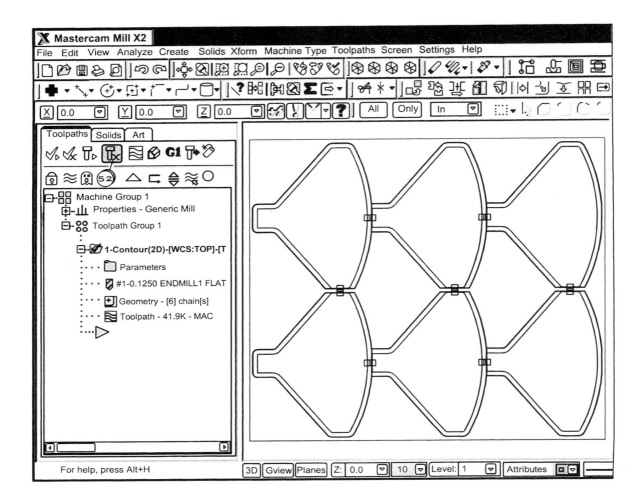

➤ Click ⑤₂ the Regenerate all dirty operations button

H) Backplot the Contour(2D) operation

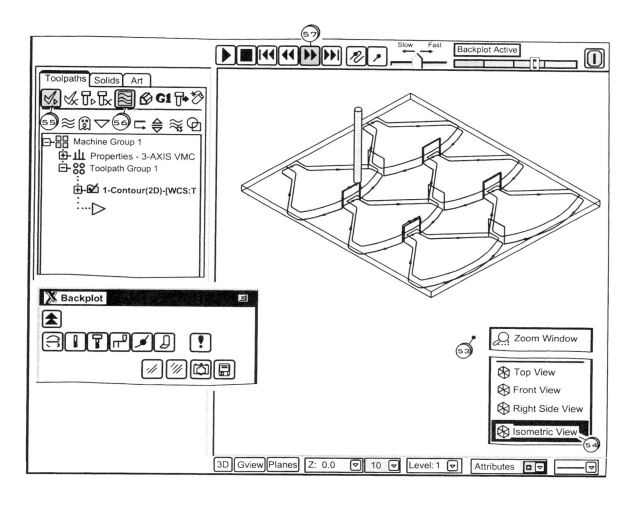

➤ **Right** Click ⑤③ in the lower view; move the cursor down to Isometric View and Click ⑤④

➤ Click ⑤⑤ the Select all operations button

➤ Click ⑤⑥ the Backplot button

➤ *Continuously* Click ⑤⑦ the Step forward button to see a *step by step* moverment of the tool along the specified contour toolpath

10-4 Editing Tabs

Existing Tabs on a contour toolpath can be edited in one of *two* ways.

- Editing *all* the tabs in a toolpath
- Editing *individual* tabs in a toolpath

EDITING *ALL* THE TABS IN THE TOOLPATH

EXAMPLE 10-2

Change the tab width of *all* the tabs machined in EXAMPLE 10-1 to .35..

➤ Click ① the Parameters icon 🗀

➤ Click ② the [Tabs] button.

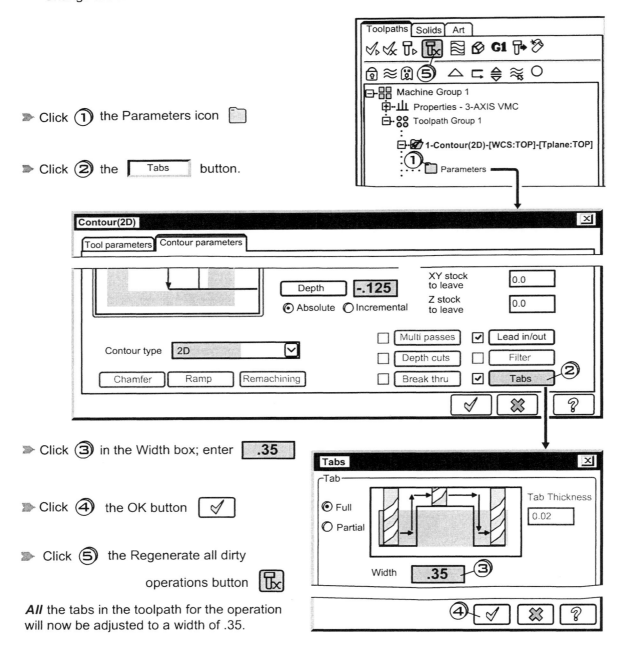

➤ Click ③ in the Width box; enter [.35]

➤ Click ④ the OK button [✓]

➤ Click ⑤ the Regenerate all dirty

operations button 🗔

All the tabs in the toolpath for the operation
will now be adjusted to a width of .35.

EDITING *INDIVIDUAL* TABS IN THE TOOLPATH

EXAMPLE 10-3

Edit the **tab in Chain 2**, for EXAMPLE 10-2. The new tab width is to be .35.

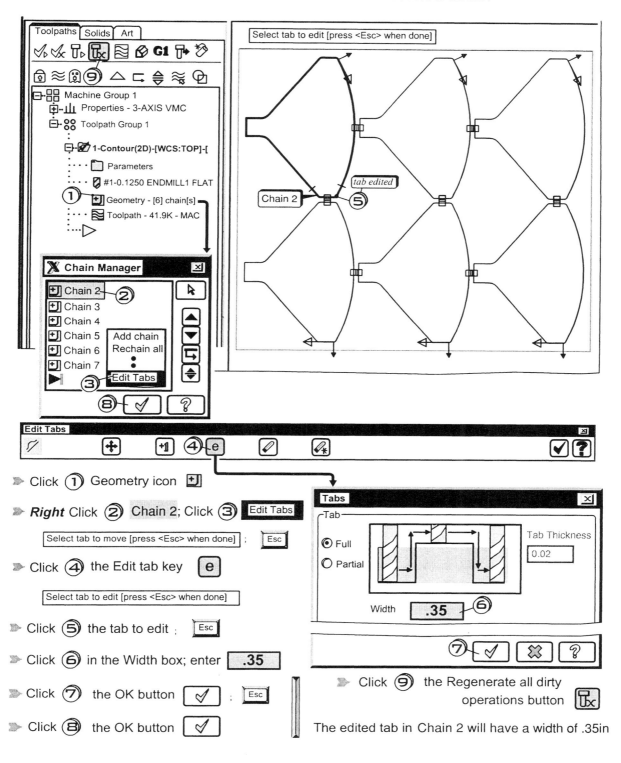

≫ Click ① Geometry icon

≫ ***Right*** Click ② Chain 2; Click ③ Edit Tabs

Select tab to move [press <Esc> when done] ; Esc

≫ Click ④ the Edit tab key [e]

Select tab to edit [press <Esc> when done]

≫ Click ⑤ the tab to edit ; Esc

≫ Click ⑥ in the Width box; enter .35

≫ Click ⑦ the OK button ✓ ; Esc

≫ Click ⑧ the OK button ✓

≫ Click ⑨ the Regenerate all dirty operations button

The edited tab in Chain 2 will have a width of .35in

10-5 Work Offsets

The work offset feature in most controllers enables the operator to set up parts in multiple fixtures on the machine table. Each part has its own offset number. Fanuc controllers use G54, G55, G56, G57, G58, G59 codes. Fadal recognizes E1,E2, etc. Normally up to six parts can be set up but optional packages can be purchased that allow more setups.

The operator must input the program zero into the machine control unit for **each work offset** by measuring the distance *from machine zero to the part's program zero.*

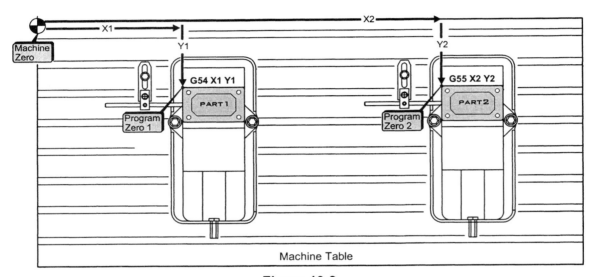

Figure 10-3

10-6 Applying *Mastercam X2* to Work Offsets

Mastercam X2 recommends using the **View Manager to create work offsets**. The dialog box enables the operator to create **new views representing work offsets**. Recall from the definition of a view given in Chapter 1, p 1-20, views are used as devices to set the **orientations and origins** of WCS's, Cplanes and Tplanes. The approach will be as follows:

◆ Create and name a *new view for each fixture* holding a part (view name:G54,G55,etc)

◆ Specify the *origin for each new* view by clicking on a point on each fixture

◆ Use the **Work Offset#** field in the View Manager dialog box to indicate *which work offset*(G54,G55,etc) is to be assigned to the currently *selected* view.

◆ To cut part1 in fixture1, *first select the appropriate new fixture view* created (G54). The program origin will automatically be shifted to that defined in the work offset for part1 All the *Mastercam X2* functions can then be used to execute the machining operation. *Mastercam* will generate the correct word address program during postprocessing.

EXAMPLE 10-4

A run of 400 pieces is to be made for the part shown in Figure 10-4.

The production plan calls for executing the machining operations 1-4 given in the SETUP/PROCESS PLAN 10-1 on the 6" x 4" x 1" stock held in vise1. *Without stopping*, the spindle rapids to the partially machined part in vise2 and executes operations 5-6 . The program is stopped the *finished part is removed from vise2.* The patrially machined part in vise 1 is flipped and clamped in vise2. A new blank is clamped in vise1 and the next run begins.

Establish a work offset at each of the vises. Use the offset origins when executing the operations 1-4 and 5-6.

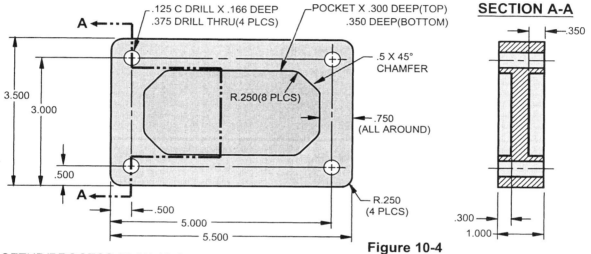

Figure 10-4

SETUP/PROCESS PLAN 10-1

Pgm Origin	Setup Plan	No.	Operation	Tooling
G54 ◉ LOWER LEFT CORNER OF PART IN VISE1	STOCK(6" x 4" x 1") HELD IN VISE1	1	CENTER DRILL x 166 DEEP (4PLCS)	1/8 C' DRILL
		2	PECK DRILL THRU(4PLCS)	3/8 DRILL
		3	ROUGH AND FINISH POCKET x .30 DEEP LEAVE .01 FOR FINISH CUT IN XY AND Z	1/2 FLAT END MILL
		4	CONTOUR OUTSIDE x .65DEEP LEAVE .01 FOR FINISH CUT IN XY AND Z	1/2 FLAT END MILL
G55 ◉ LOWER LEFT CORNER OF PART IN VISE2	FLIPPED OVER PARTIALLY MACHINED PART HELD IN VISE2	5	ROUGH AND FINISH POCKET x .35 DEEP LEAVE .01 FOR FINISH CUT IN XY AND Z	14 FLAT END MILL
		6	CONTOUR OUTSIDE x .65DEEP LEAVE .01 FOR FINISH CUT IN XY AND Z	1/2 FLAT END MILL

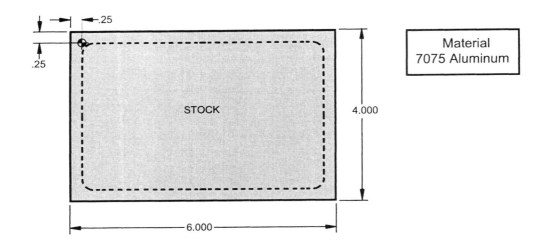

.25

.25

STOCK

4.000

6.000

Material
7075 Aluminum

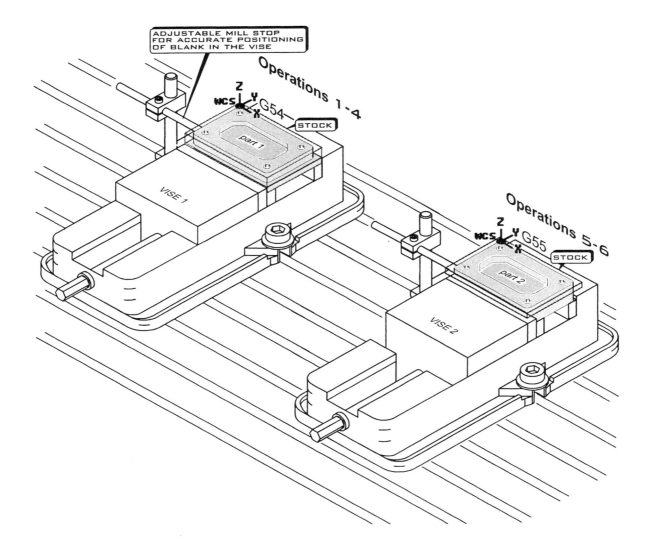

ADJUSTABLE MILL STOP
FOR ACCURATE POSITIONING
OF BLANK IN THE VISE

Operations 1-4

Z
WCS Y G54
X

STOCK

part 1

VISE 1

Operations 5-6

Z
WCS Y G55
X

STOCK

part 2

VISE 2

A) OPEN THE View Manager DIALOG BOX

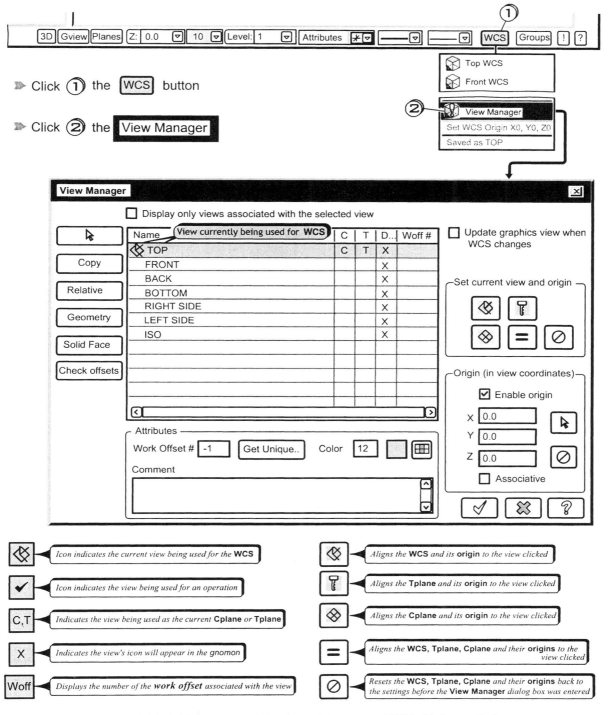

> Click ① the [WCS] button

> Click ② the [View Manager]

Name	C	T	D...	Woff #
TOP	C	T	X	
FRONT			X	
BACK			X	
BOTTOM			X	
RIGHT SIDE			X	
LEFT SIDE			X	
ISO			X	

Icon indicates the current view being used for the **WCS**

Icon indicates the view being used for an operation

Indicates the view being used as the current **Cplane** *or* **Tplane**

Indicates the view's icon will appear in the gnomon

Displays the number of the **work offset** *associated with the view*

Aligns the **WCS** *and its* **origin** *to the view clicked*

Aligns the **Tplane** *and its* **origin** *to the view clicked*

Aligns the **Cplane** *and its* **origin** *to the view clicked*

Aligns the **WCS, Tplane, Cplane** *and their* **origins** *to the view clicked*

Resets the **WCS, Tplane, Cplane** *and their* **origins** *back to the settings before the* **View Manager** *dialog box was entered*

Refer to pages 1-19, 1-20, 1-21 for an explaination of the terms **WCS, Tplane, Cplane.**

B) CREATE G54 AS A NEW TOP VIEW

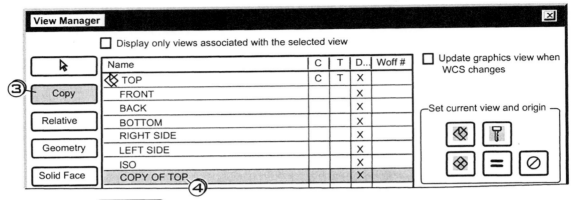

➤ Click ③ the [Copy] button.

The TOP view will be copied. This is correct since the new view being used to create G54 is in the same plane as TOP.

➤ Click ④ on the name of the view COPY OF TOP ; enter the new name G54 [Enter]

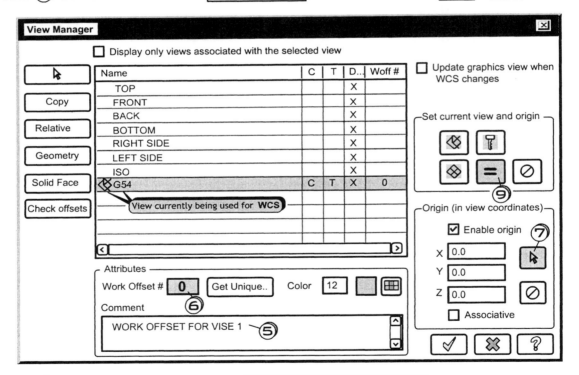

➤ Click ⑤ in the Comment box and enter WORK OFFSET FOR VISE 1

➤ Click ⑥ in the Work Offset # box ; enter [0]

This specifies the *lowest or first* offset code(0=G54) is to be created when the view G54 is selected for an operation.

> Note:
> The default value **-1** appearing in the Work Offset # box will cause most posts to generate *no* offset codes
> When working in *Mastercam*, the operator enters a single generic code in the Work Offset # box to specify
> work offsets. The value **0** specifies the first or lowest. Each higher offset is incremented by 1 in succession.
> The postprocessor for a particular machine tool will translate these values into the appropriate offset code
> for that controller. For Fanuc controls 0=G54, 1=G55, 2=G56, etc. Fadal controls will post the code as follows
> 1=E1, 2=E2, 3=E3, etc.

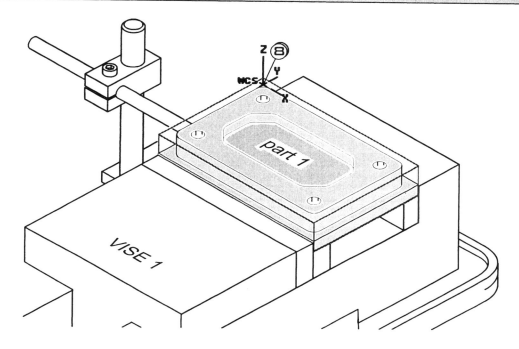

➤ Click ⑦ the Select new origin for the selected view button

Select a point

➤ Click ⑧ on the *existing* point.

This will specify the origin for the new view named G54. The origin for the WCS
associated with that view will be set at that point.

➤ Click ⑨ the Set all button.

This will make sure the **WCS**, Tplane, Cplane have the same *origin and alignment*
with respect to the new view G54.

C) CREATE G55 AS A NEW TOP VIEW

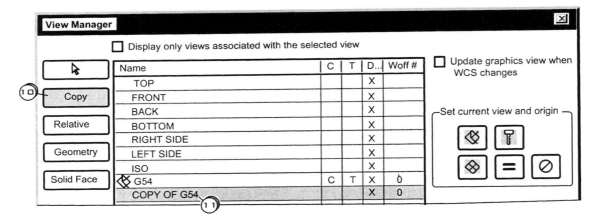

➤ Click ⑩ the [Copy] button.

➤ Click ⑪ on the name of the view `COPY OF G54` ; enter the new name `G55` [Enter]

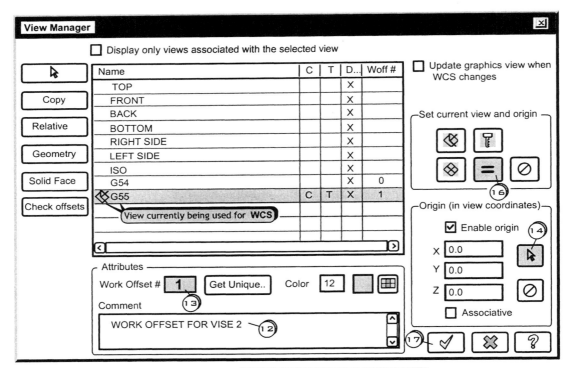

➤ Click ⑫ in the Comment box and enter WORK OFFSET FOR VISE 2

➤ Click ⑬ in the Work Offset # box ; enter [1]

   This specifies the *next generic number in sequence following 0 or the second
   offset code* code(1=G55) is to be created when the view G55 is selected for the operation.

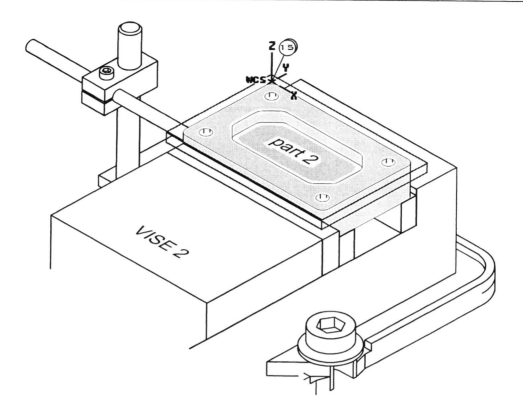

➤ Click ⑭ the Select new origin for the selected view button

Select a point

➤ Click ⑮ on the *existing* point.
    This will specify the origin for the new view named G55. The origin for the WCS
    associated with that view will be set at that point.

➤ Click ⑯ the Set all button.

    This will make sure the WCS, Tplane, Cplane have the same *origin and alignment*
    with respect to the new view G55.

D) ENTER *Mastercam's* MACHINE GROUP PROPERTIES DIALOG BOX

◆ TURN OFF THE VISIBILITY OF ALL LEVELS
EXECPT FOR THE LEVEL NAMED PART-1

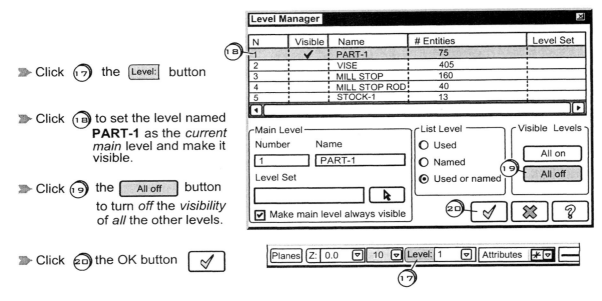

➤ Click ⑰ the Level: button

➤ Click ⑱ to set the level named **PART-1** as the *current main* level and make it visible.

➤ Click ⑲ the All off button to turn *off* the *visibility* of *all* the other levels.

➤ Click ⑳ the OK button ✓

◆ SPECIFY THE SIZE OF THE STOCK FOR PART-1 (Refer to p 5-17)

E) ENTER *Mastercam's* MATERIAL MANAGER DIALOG BOX

◆ SPECIFY THE STOCK MATERIAL (Refer to p 5-18)

F) SET G54 AS THE CURRENT WORKING VIEW. ALIGN THE WCS, Cplanes, Tplanes AND THEIR ORIGINS WITH THIS VIEW.

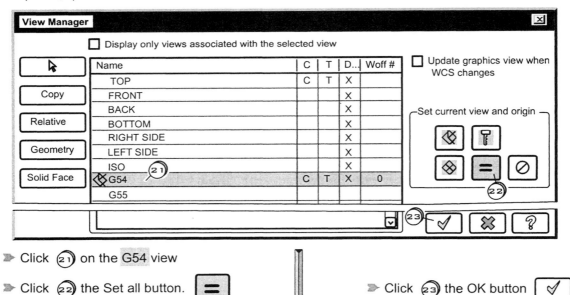

➤ Click ㉑ on the G54 view

➤ Click ㉒ the Set all button. =

➤ Click ㉓ the OK button ✓

G) GENERATE OPERATIONS 1-4 ON PART1

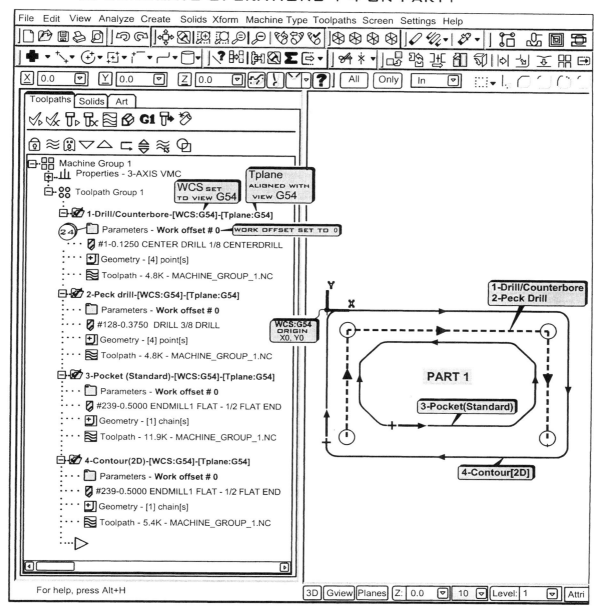

Mastercam posts the **WCS** , **Tplane** and **Work offset** settings for each operation in the **Operations Manager** dialog box.

Enter the **Toolpath Coordinate System** dialog box to see a *detail listing* of the view's **WCS**, **Tplane**, and **Cplane** and **Work offset** settings for *any operation clicked on* in the **Operations Manager** dialog box.

➤ Click ②④ on the Parameters icon 📁 for operation **1-Drill/Counterbore**

➤ Click ㉕ on the [Planes] button to display the **Toolpath Coordinate System** dialog box.

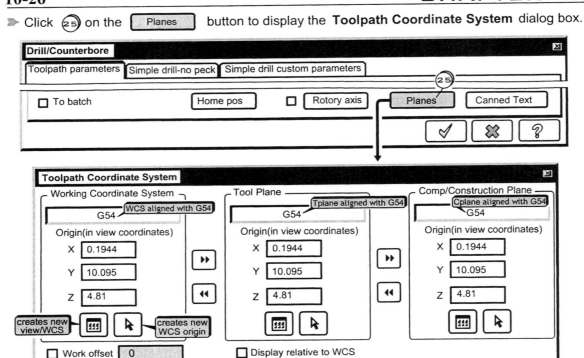

➤ Click ㉖ ㉗ the expand button *twice* [▸▸] and see a display indicating the **WCS, Tplane**

and **Cplane** all have the same settings and the work offset value for the operation is **0**.

➤ Click ㉘ the OK button [✓]

H) SET G55 AS THE CURRENT WORKING VIEW. ALIGN THE WCS, Cplanes, Tplanes **AND THEIR ORIGINS WITH THIS VIEW.**

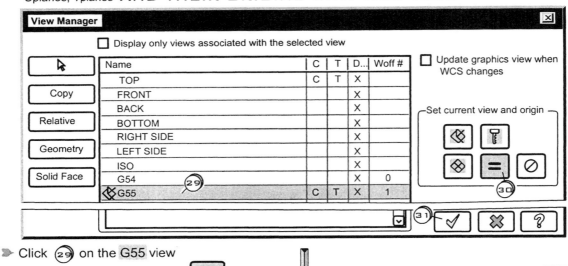

➤ Click ㉙ on the G55 view

➤ Click ㉚ the Set all button. [=]

➤ Click ㉛ the OK button [✓]

I) TURN ON THE VISIBILITY OF THE LEVEL NAMED PART-2

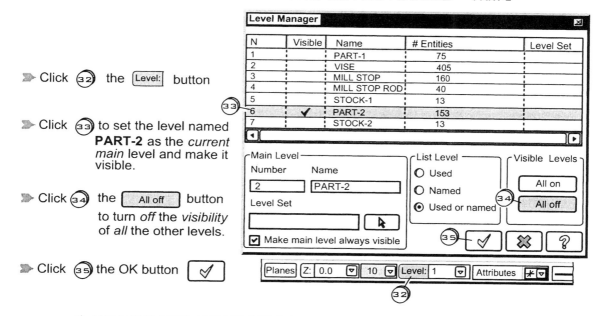

▶ Click ③② the [Level:] button

▶ Click ③③ to set the level named **PART-2** as the *current main* level and make it visible.

▶ Click ③④ the [All off] button to turn *off* the *visibility* of *all* the other levels.

▶ Click ③⑤ the OK button

J) GENERATE OPERATIONS 5-6 ON PART2

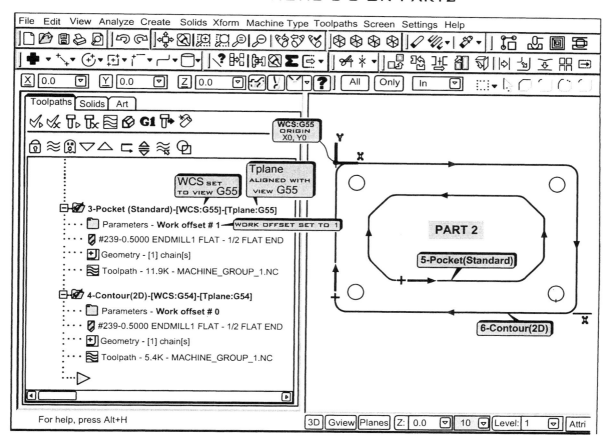

K) TURN ON THE VISIBILITY OF ALL THE LEVELS

➤ Click ③⑤ the Level: button

➤ Click ③⑥ the All on button.

➤ Click ③⑦ the OK button ✓

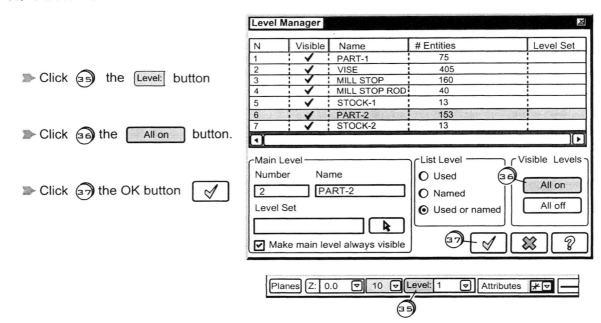

L) BACKPLOT ALL THE OPERATIONS

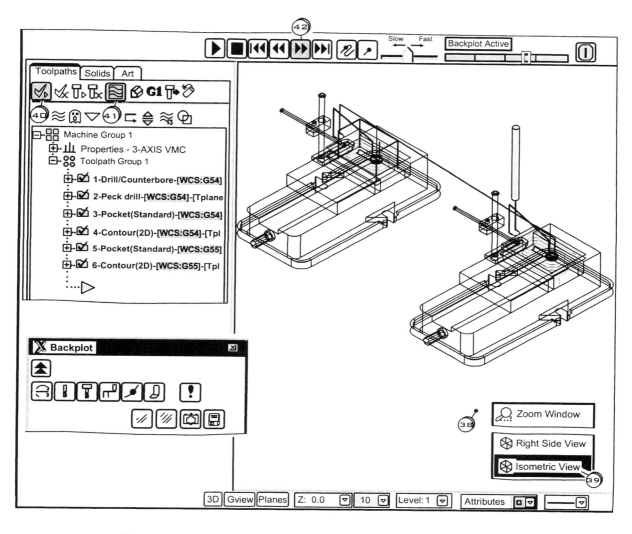

➤ *Right* Click ㊳ in the lower view; move the cursor down to Isometric View and Click ㊴

➤ Click ㊵ the Select all operations button

➤ Click ㊶ the Backplot button

➤ *Continuously* Click ㊷ the Step forward button to see a *step by step* moverment of the tool along the specified contour toolpath

EXERCISES

10-1) Six parts are to be cut from a single sheet of 303 Steel 8.250" x 6.750" x .125". Create a CAD model file **EX10-1JV** by using the dimensions given in Figure 10p-1.

or

Copy the file **EX10-1JV** from the CD provided at the back of this text located in the folder ⬜ CHAPTER 10 . PROCESS PLAN 10P-1 outlines the steps.

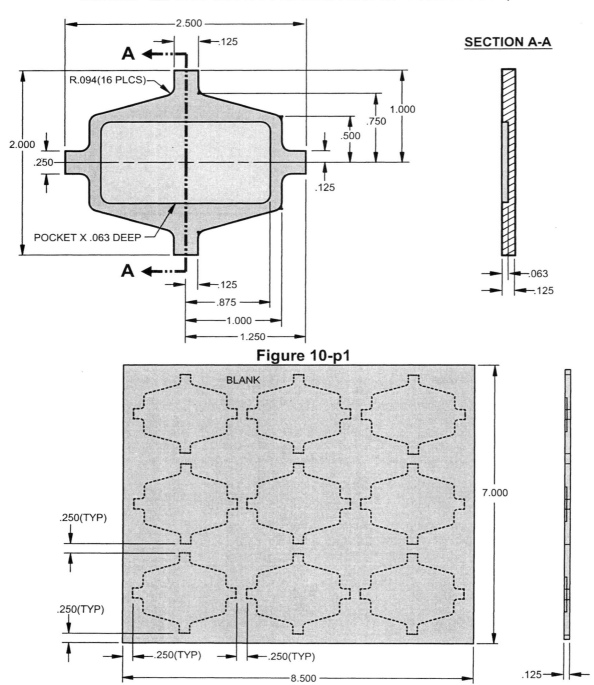

Figure 10-p1

PROCESS PLAN 10P-1

No.	Operation	Tooling
1	ROUGH AND FINISH POCKET x .25 DEEP LEAVE .01 FOR FINISH CUT IN XY AND Z	3/16 FLAT END MILL

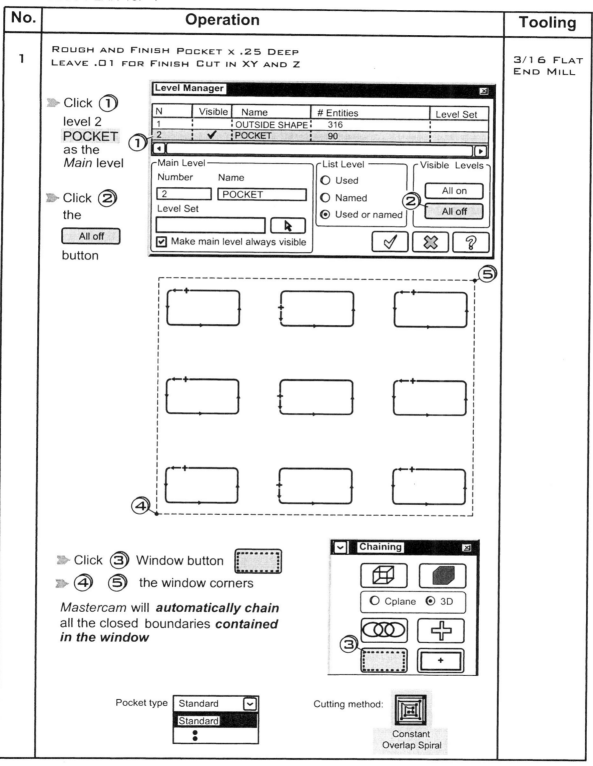

➤ Click ① level 2 POCKET as the *Main* level

➤ Click ② the **All off** button

Level Manager

N	Visible	Name	# Entities	Level Set
1		OUTSIDE SHAPE	316	
2	✓	POCKET	90	

Main Level
Number: 2 Name: POCKET
Level Set:
☑ Make main level always visible

List Level
○ Used
○ Named
◉ Used or named

Visible Levels
All on
All off

➤ Click ③ Window button
➤ ④ ⑤ the window corners

Mastercam will **automatically chain** all the closed boundaries **contained in the window**

Chaining
○ Cplane ◉ 3D

Pocket type Standard
Standard

Cutting method:
Constant Overlap Spiral

PROCESS PLAN 10P-1

No.	Operation	Tooling
2	ROUGH OUTSIDE X .125 DEEP LEAVE .01 FOR FINISH CUT IN XY AND Z LEAVE PARTIAL TABS .02 THICK TO HOLD PARTS AFTER MACHINING	3/16 FLAT END MILL

➤ Click ⑥
level 1
OUTSIDE-
SHAPE
as the
Main level

➤ Click ⑦
the

All off

button

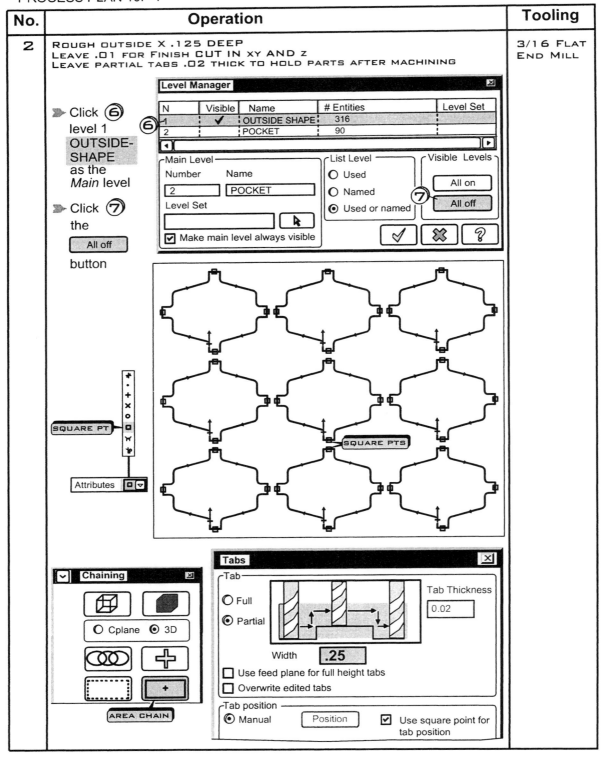

EXERCISES

10-2) The part shown in Figure 10p-2 is to be cut from a blank of 7075 Aluminum
5 1/4 x 3 1/4 x 1/2 .
Create a CAD model file **EX10-2JV** by using the dimensions given in Figure 10-p2
or
Copy the file **EX10-2JV** from the CD provided at the back of this text located in
the folder ▭ CHAPTER 10. Follow PROCESS PLAN 10P-2.

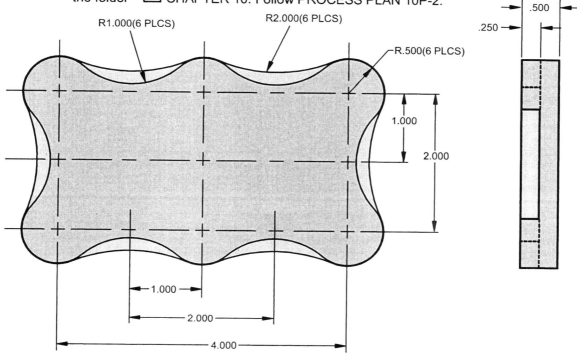

Figure 10-p2

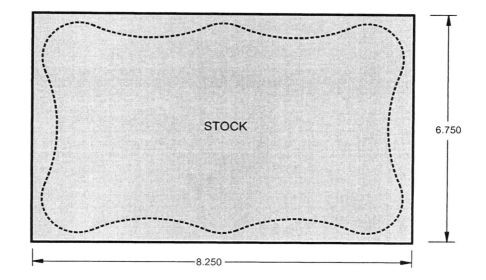

PROCESS PLAN 10P-2

No.	Operation	Tooling
1	ROUGH CONTOUR GEO1A-POS1 X .5 DEEP WITH CLAMPS IN POS1 LEAVE .01 FOR FINISH CUT IN XY AND Z	1/2 FLAT END MILL

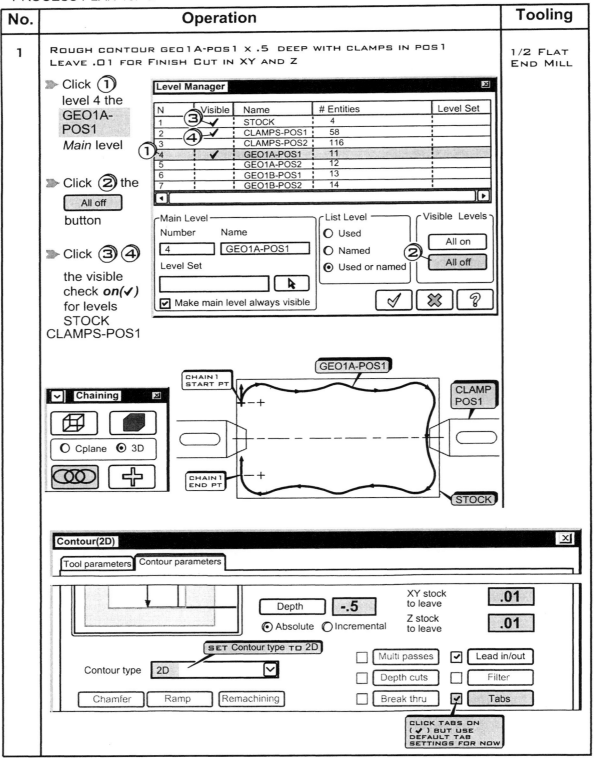

➤ Click ① level 4 the **GEO1A-POS1** *Main* level

➤ Click ② the [All off] button

➤ Click ③ ④ the visible check *on(✓)* for levels STOCK CLAMPS-POS1

Level Manager

N	Visible	Name	# Entities	Level Set
1	✓	STOCK	4	
2	✓	CLAMPS-POS1	58	
3		CLAMPS-POS2	116	
4	✓	GEO1A-POS1	11	
5		GEO1A-POS2	12	
6		GEO1B-POS1	13	
7		GEO1B-POS2	14	

Main Level
Number: 4   Name: GEO1A-POS1
Level Set

☑ Make main level always visible

List Level
○ Used
○ Named
◉ Used or named

Visible Levels
[All on]
[All off]

Chaining
○ Cplane ◉ 3D

CHAIN1 START PT

GEO1A-POS1

CLAMP POS1

CHAIN1 END PT

STOCK

Contour(2D)

Tool parameters | Contour parameters

Depth [-.5]
◉ Absolute ○ Incremental

XY stock to leave [.01]
Z stock to leave [.01]

SET Contour type TO 2D

Contour type [2D ▼]

[Chamfer] [Ramp] [Remachining]

☐ Multi passes   ☑ Lead in/out
☐ Depth cuts   ☐ Filter
☐ Break thru   ☑ Tabs

CLICK TABS ON (✓) BUT USE DEFAULT TAB SETTINGS FOR NOW

PROCESS PLAN 10P-2

No.	Operation	Tooling

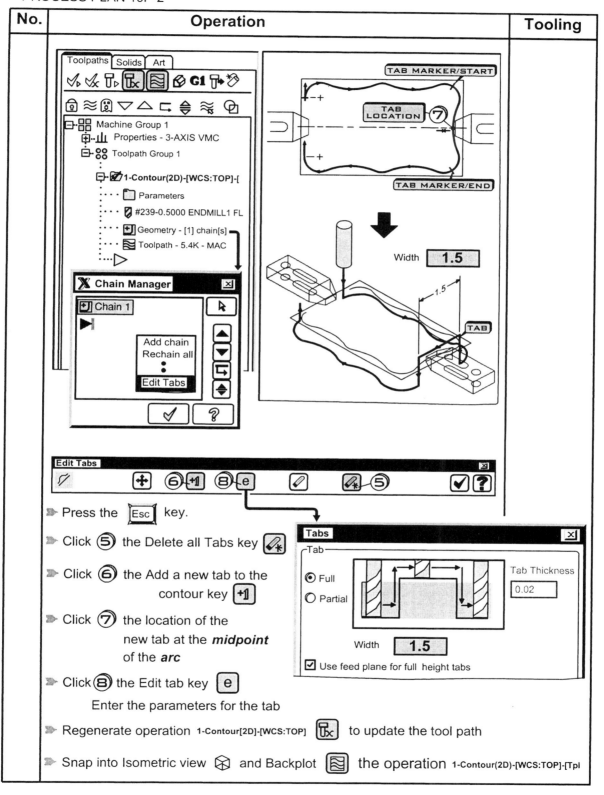

➤ Press the [Esc] key.

➤ Click ⑤ the Delete all Tabs key

➤ Click ⑥ the Add a new tab to the
contour key [+1]

➤ Click ⑦ the location of the
new tab at the *midpoint*
of the *arc*

➤ Click ⑧ the Edit tab key [e]

    Enter the parameters for the tab

➤ Regenerate operation 1-Contour[2D]-[WCS:TOP] to update the tool path

➤ Snap into Isometric view and Backplot the operation 1-Contour(2D)-[WCS:TOP]-[Tpl

PROCESS PLAN 10P-2

No.	Operation	Tooling
2	ROUGH CONTOUR GEO1B-POS1 X .25 DEEP WITH CLAMPS IN POS1 LEAVE .01 FOR FINISH CUT IN XY AND Z	1/2 FLAT END MILL

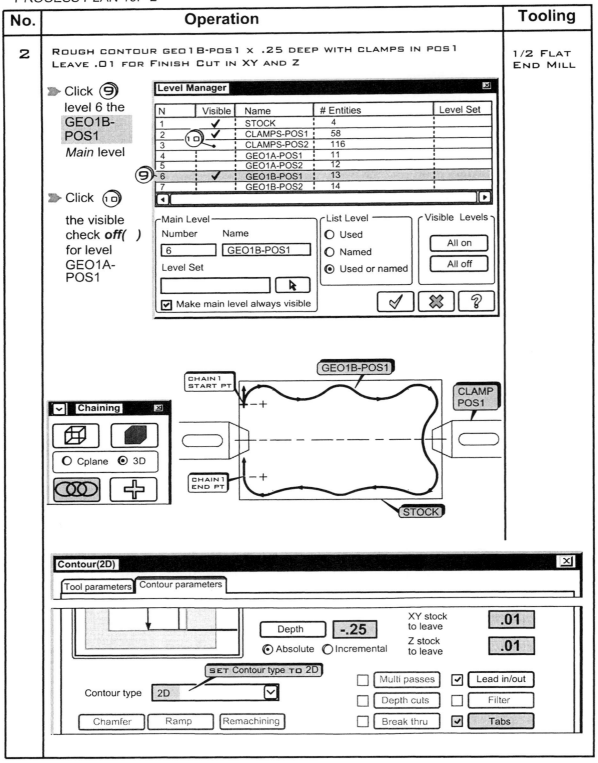

➤ Click ⑨ level 6 the GEO1B-POS1 *Main* level

➤ Click ⑩ the visible check *off( )* for level GEO1A-POS1

Level Manager

N	Visible	Name	# Entities	Level Set
1	✓	STOCK	4	
2	✓	CLAMPS-POS1	58	
3		CLAMPS-POS2	116	
4		GEO1A-POS1	11	
5		GEO1A-POS2	12	
6	✓	GEO1B-POS1	13	
7		GEO1B-POS2	14	

Main Level
Number: 6   Name: GEO1B-POS1
Level Set:

List Level
○ Used
○ Named
◉ Used or named

Visible Levels
All on
All off

☑ Make main level always visible

Chaining
○ Cplane ◉ 3D

CHAIN1 START PT
GEO1B-POS1
CLAMP POS1
CHAIN1 END PT
STOCK

Contour(2D)

Tool parameters | Contour parameters

Depth [-.25]
◉ Absolute ○ Incremental

XY stock to leave [.01]
Z stock to leave [.01]

SET Contour type TO 2D

Contour type [2D]

Chamfer | Ramp | Remachining

☐ Multi passes   ☑ Lead in/out
☐ Depth cuts   ☐ Filter
☐ Break thru   ☑ Tabs

PROCESS PLAN 10P-2

No.	Operation	Tooling

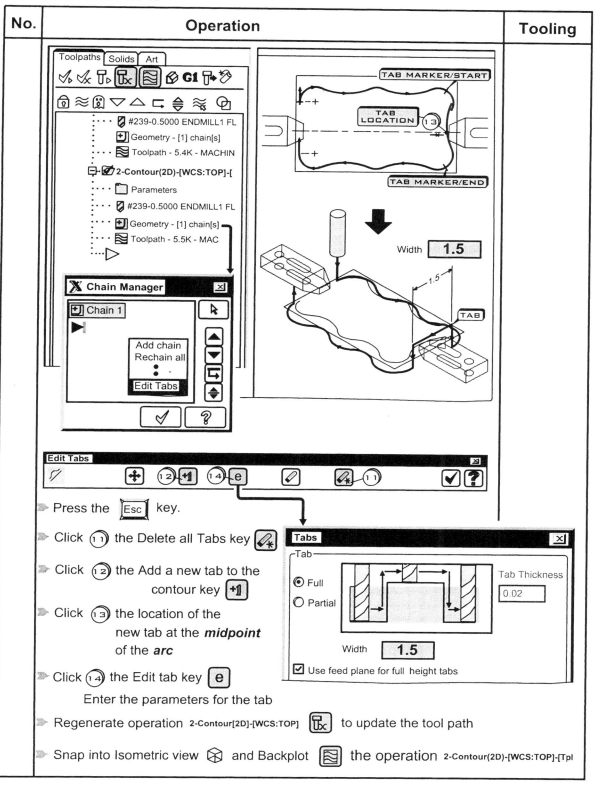

⟫ Press the ⬚Esc⬚ key.

⟫ Click ⑪ the Delete all Tabs key ✐*

⟫ Click ⑫ the Add a new tab to the
contour key ⊞1

⟫ Click ⑬ the location of the
new tab at the **midpoint**
of the **arc**

⟫ Click ⑭ the Edit tab key e
Enter the parameters for the tab

⟫ Regenerate operation 2-Contour[2D]-[WCS:TOP] 🗍x to update the tool path

⟫ Snap into Isometric view ⬡ and Backplot ≋ the operation 2-Contour(2D)-[WCS:TOP]-[Tpl

PROCESS PLAN 10P-2

No.	Operation	Tooling
3	ROUGH CONTOUR GEO1A-POS2 x .5 DEEP WITH CLAMPS IN POS2 LEAVE .01 FOR FINISH CUT IN XY AND Z	1/2 FLAT END MILL

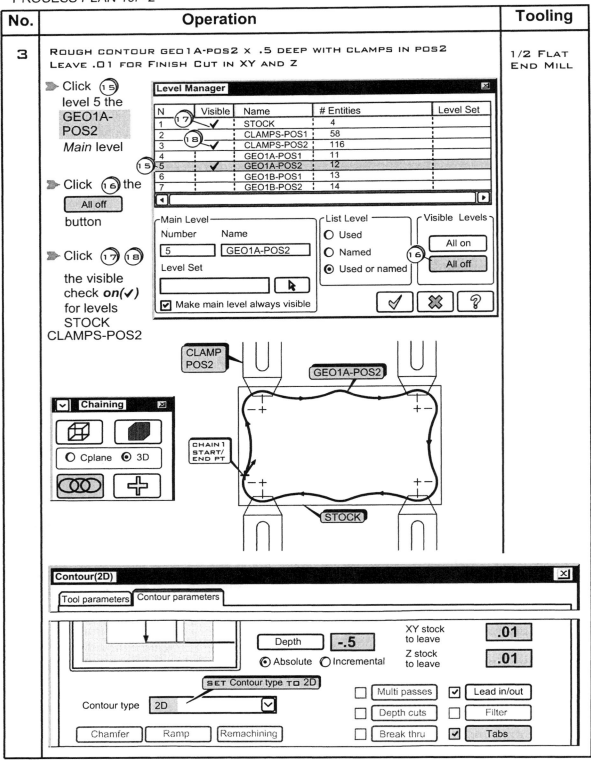

➤ Click ⑮ level 5 the **GEO1A-POS2** *Main* level

➤ Click ⑯ the All off button

➤ Click ⑰ ⑱ the visible check *on(✓)* for levels STOCK CLAMPS-POS2

PROCESS PLAN 10P-2

No.	Operation	Tooling

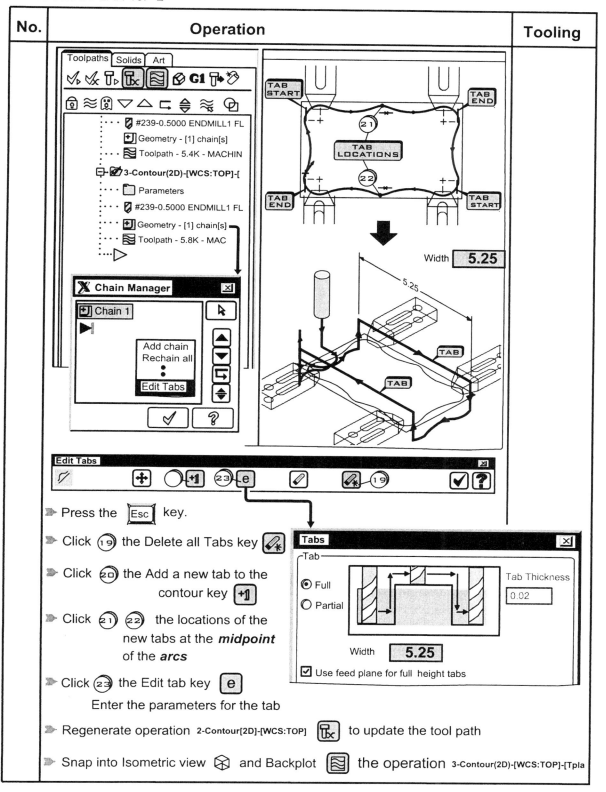

➤ Press the ⌊Esc⌋ key.

➤ Click ⑲ the Delete all Tabs key

➤ Click ⑳ the Add a new tab to the
contour key [+1]

➤ Click ㉑ ㉒ the locations of the
new tabs at the **midpoint**
of the **arcs**

➤ Click ㉓ the Edit tab key [e]
Enter the parameters for the tab

➤ Regenerate operation 2-Contour[2D]-[WCS:TOP] to update the tool path

➤ Snap into Isometric view ⬡ and Backplot ≋ the operation 3-Contour(2D)-[WCS:TOP]-[Tpla

PROCESS PLAN 10P-2

No.	Operation	Tooling
4	ROUGH CONTOUR GEO1B-POS2 x .25 DEEP WITH CLAMPS IN POS2 LEAVE .01 FOR FINISH CUT IN XY AND Z	1/2 FLAT END MILL

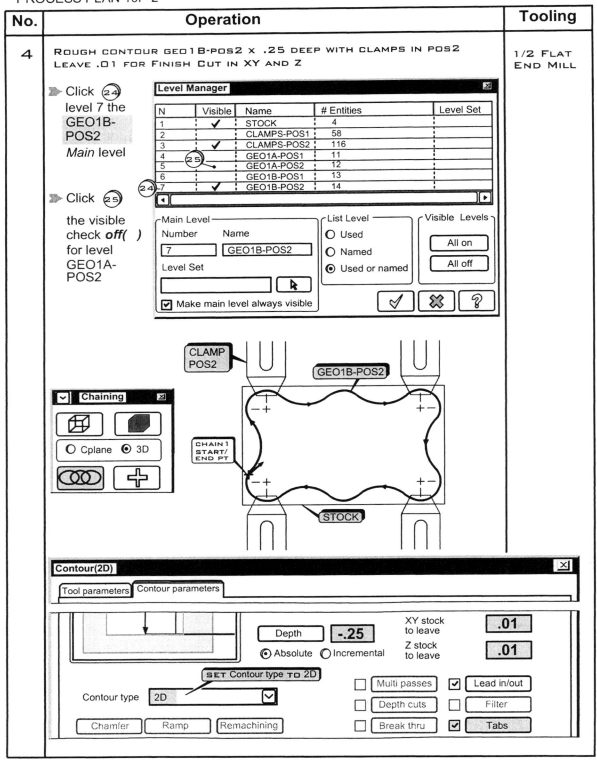

▶ Click ㉔ level 7 the GEO1B-POS2 *Main* level

▶ Click ㉕ the visible check *off( )* for level GEO1A-POS2

Level Manager

N	Visible	Name	# Entities	Level Set
1	✓	STOCK	4	
2		CLAMPS-POS1	58	
3	✓	CLAMPS-POS2	116	
4		GEO1A-POS1	11	
5		GEO1A-POS2	12	
6		GEO1B-POS1	13	
7	✓	GEO1B-POS2	14	

Main Level

Number: 7   Name: GEO1B-POS2

Level Set

☑ Make main level always visible

List Level
○ Used
○ Named
◉ Used or named

Visible Levels
All on
All off

Chaining

○ Cplane ◉ 3D

CLAMP POS2

GEO1B-POS2

CHAIN 1 START/ END PT

STOCK

Contour(2D)

Tool parameters | Contour parameters

Depth -.25
◉ Absolute ○ Incremental

XY stock to leave .01
Z stock to leave .01

SET Contour type TO 2D

Contour type 2D

Chamfer | Ramp | Remachining

☐ Multi passes   ☑ Lead in/out
☐ Depth cuts   ☐ Filter
☐ Break thru   ☑ Tabs

PROCESS PLAN 10P-2

No.	Operation	Tooling

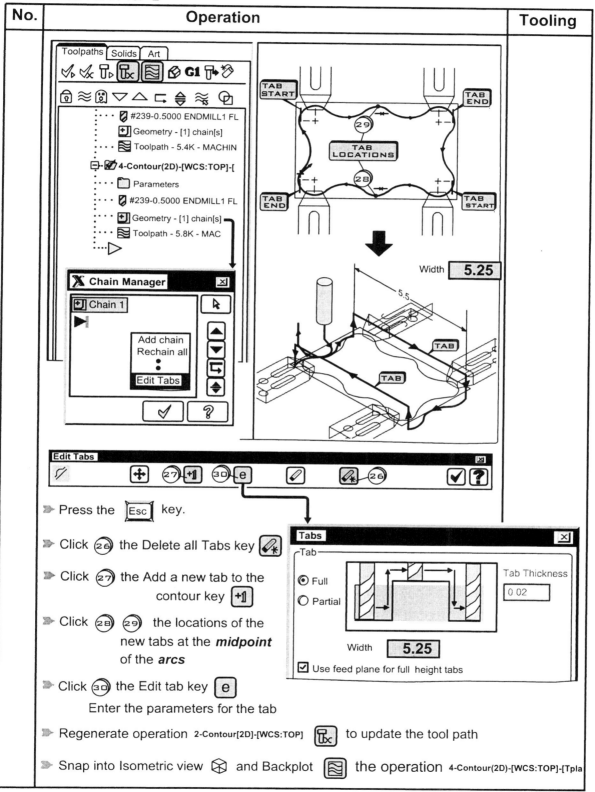

➤ Press the [Esc] key.

➤ Click (26) the Delete all Tabs key

➤ Click (27) the Add a new tab to the
        contour key [+1]

➤ Click (28) (29) the locations of the
        new tabs at the *midpoint*
        of the *arcs*

➤ Click (30) the Edit tab key [e]
        Enter the parameters for the tab

➤ Regenerate operation 2-Contour[2D]-[WCS:TOP] to update the tool path

➤ Snap into Isometric view and Backplot the operation 4-Contour(2D)-[WCS:TOP]-[Tpla

10-3) A run of 500 pieces is to be made for the part shown in Figure 10-p3.
The production plan calls for executing the machining operations 1-5 given in the
SETUP/PROCESS PLAN 10P-3 on a 3.5" x 6" x .75" blank(part1) held in a vise. After
partially machining part1 in the vise, the spindle rapids to the machined blank held on
the fixture plate(part2) and executes the remaining operations 6 and 7. The program
then stops, the *finished part* is removed from the *fixture plate*. The partially machined
blank is removed from the vise and secured in the fixture plate. A new blank is placed
in the vise, the program is started again *and a new run begins.* See page 10p-14.

Copy the file **EX10-3** in the folder ☐**CHAPTER10** from the CD into the **JVAL-MILL**
subdirectory on C drive. Open the file and create the machining operations as listed in
SETUP/PROCESSS PLAN 10P-3. *Backplot* all the operations to verify their correctiness.

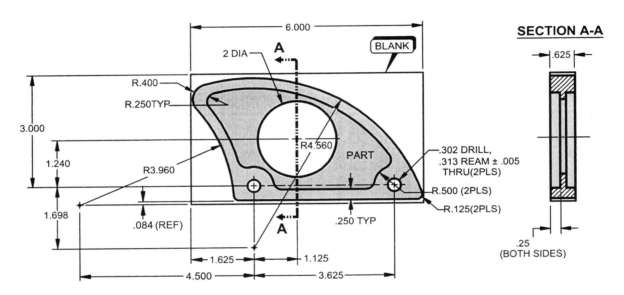

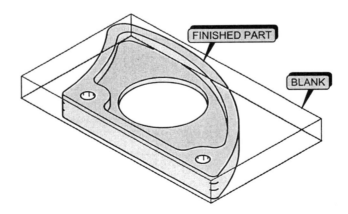

Figure 10-p3

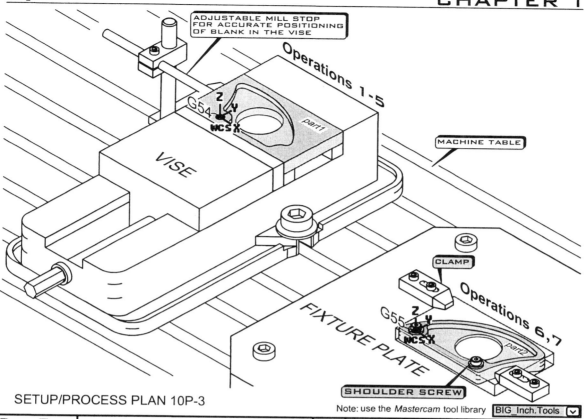

SETUP/PROCESS PLAN 10P-3

Note: use the *Mastercam* tool library | BIG Inch Tools ⌄ |

Pgm Zero	Setup Plan	No.	Operation	Tooling
G54 ◕ LOWER LEFT CORNER OF PART IN VISE	BLANK IS HELD IN A VISE stop PART 1 .500 VISE ← 1.625	1	CENTER DRILL x 166 DEEP (2 PLCS)	1/8 C' DRILL
		2	PECK DRILL THRU (2 PLCS)	.302 LTR N DRILL
		3	REAM THRU (2 PLCS)	5/16 REAMER
		4	ROUGH AND FINISH POCKET x .25 DEEP LEAVE .01 FOR FINISH CUT IN XY AND Z	1/2 FLAT END MILL
		5	CIRCLE MILL THRU	1/2 FLAT END MILL
G55 ◕ LOWER LEFT DRILL HOLE IN PART ON FIXTURE PLATE	BLANK IS FLIPPED OVER, AND HELD ON A FIXTURE PLATE WITH SHOULDER SCREWS AND CLAMPS. FIXTURE PLATE PART 2 .500 ← .250 .625 →	6	ROUGH AND FINISH POCKET x .25 DEEP LEAVE .01 FOR FINISH CUT IN XY AND Z	1/4 FLAT END MILL
		7	ROUGH OUTSIDE x .625 DEEP LEAVE .01 FOR FINISH CUT IN XY	1/4 FLAT END MILL

EXERCISES

10-4)   The part shown in Figure 10-p4 is to be machined in a 600 piece run.
The SETUP/PROCESS PLAN 10P-4 specifies that operations 1-6 are to
executed on a 41/2 x 31/2 x 1 Aluminum blank(part1) held in vise1. After partially
machining part1 in vise1, the spindle rapids to the machined blank held in vise2 and
executes the remaining operations 7-8. The program then stops the *finished part is
removed from vise 2*. The machined blank is removed from vise1 and placed
in vise2. A new blank is placed in vise1 and the program is started again for the *next* run.
Copy the file **EX10-4** in the folder ⊟**CHAPTER10** from the CD into the **JVAL-MILL**
subdirectory on C drive. Open the file and create the machining operations as listed in
SETUP/PROCESSS PLAN 10P-4. *Backplot* all the operations to verify their correctiness.

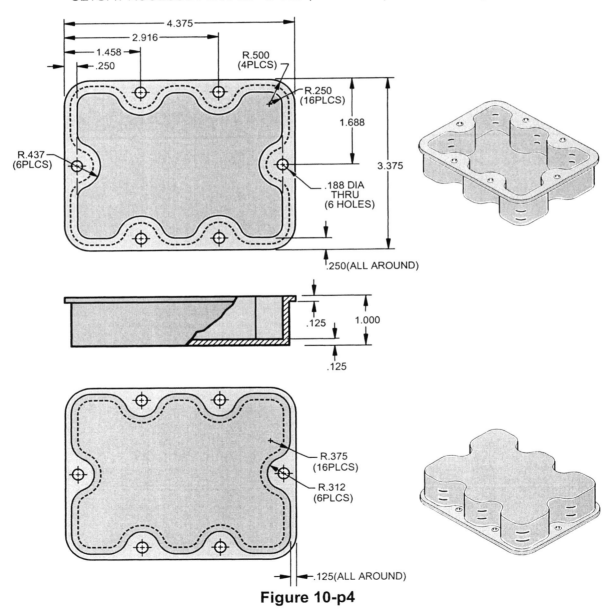

Figure 10-p4

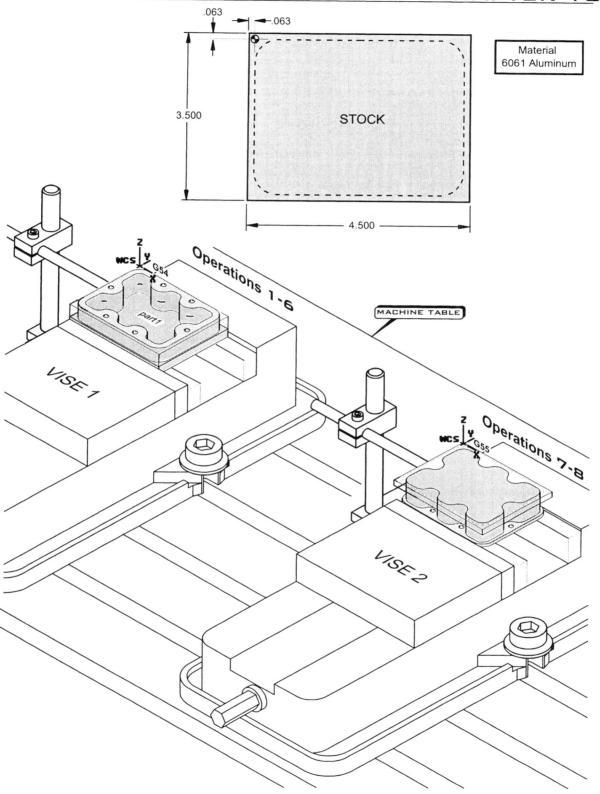

.063

.063

3.500

STOCK

4.500

Material
6061 Aluminum

Operations 1-6

WCS

Z

Y

X

G54

part1

MACHINE TABLE

VISE 1

Operations 7-8

WCS

Z

Y

X

G55

VISE 2

SETUP/PROCESS PLAN 10P-4

Pgm Zero	Setup Plan	No.	Operation	Tooling
G54 ◑ LOWER LEFT CORNER OF PART IN VISE	STOCK (4.5" x 3.5" x 1") IS HELD IN VISE 1 ⟵ .063 ⟵ 1.000 .063 PART 1 stop VISE 1 ⟵ .750	1	ROUGH OUTSIDE PROFILE X .65 DEEP LEAVE .01 FOR FINISH CUT IN XY AND Z	1/2 FLAT END MILL
		2	FINISH OUTSIDE	3/8 FLAT END MILL
		3	ROUGH POCKET X .875 DEEP LEAVE .01 FOR FINISH CUT IN XY AND Z	1/2 FLAT END MILL
		4	FINISH POCKET	3/8 FLAT END MILL
		5	CENTER DRILL X 166 DEEP (6PLCS)	1/8(#4) C' DRILL
		6	PECK DRILL THRU(6PLCS)	3/16 DRILL
G55 ◑ LOWER LEFT CORNER OF PART IN VISE	FLIPPED OVER PARTIALLY MACHINED PART IS HELD IN VISE 2 ⟵ .063 .063 PART 2 stop VISE 2 ⟵ .900	7	ROUGH OUTSIDE PROFILE X .875 DEEP LEAVE .01 FOR FINISH CUT IN XY AND Z	5/8 FLAT END MILL
		8	FINISH OUTSIDE	1/2 FLAT END MILL

EXERCISES

10-5)   Four parts are to be machined from an 8.875" x 5" x 1", 304 a Stainless blank.
The part dimensions are shown in Figure 10-p5
The SETUP/PROCESS PLAN 10P-5 specifies that operations 1, 5, 9, 13, 17 *execute four times for each tool.* Each time an operation is *repeated a different work offset is used.* The program then stops, two(2) screws are added to each part. The program is started again and the operation 21 executes four times *for each tool and work offset* to complete the production of all the parts.
Copy the file **EX10-5** in the folder ⌥**CHAPTER10** from the CD into the **JVAL-MILL** subdirectory on C drive. Open the file and create the machining operations as listed in SETUP/PROCESSS PLAN 10P-5. *Backplot* all the operations to verify their correctness.

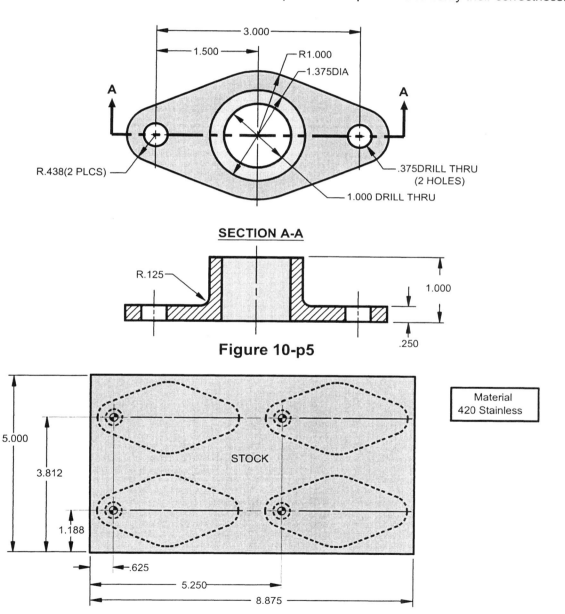

Figure 10-p5

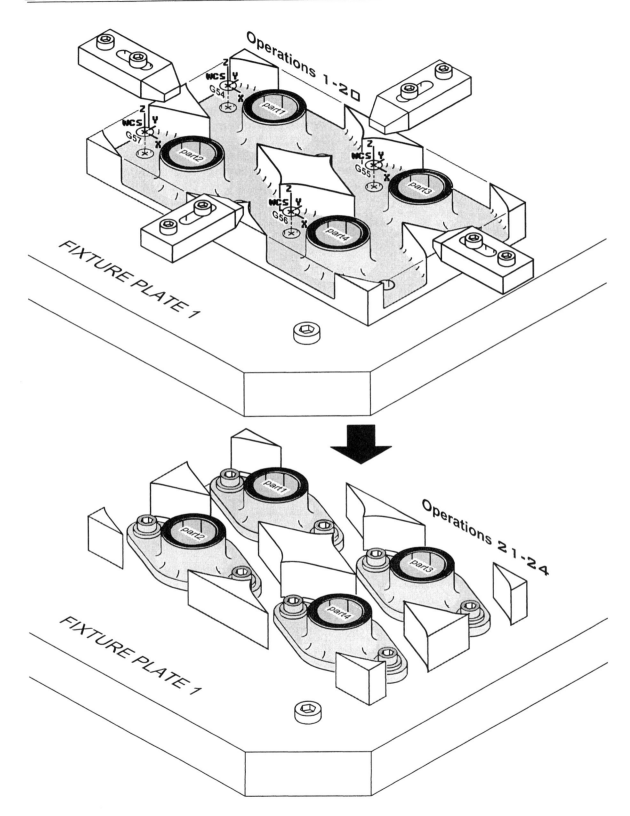

PROCESS PLAN 10P-5

No.	Operation	Tooling
1	CENTER DRILL 3/8 DIA(3) HOLES X .17 DEEP	1/8 (#4) CENTER DRILL
2	•	
3	•	
4	•	

➤ Click ① level 4 the **PART OPS (1-20)** *Main* level

➤ Click ② the [All off] button

➤ Click ③④ the visible check **on(✔)** for levels STOCK CLAMPS

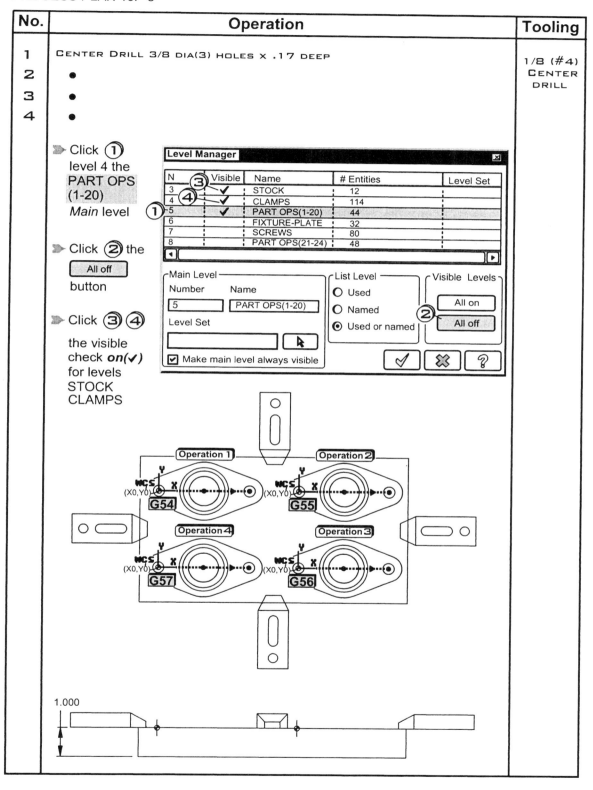

PROCESS PLAN 10P-5

No.	Operation	Tooling
	Use the **View Manager** to create the work offsets **G54, G55, G56, G57** as views 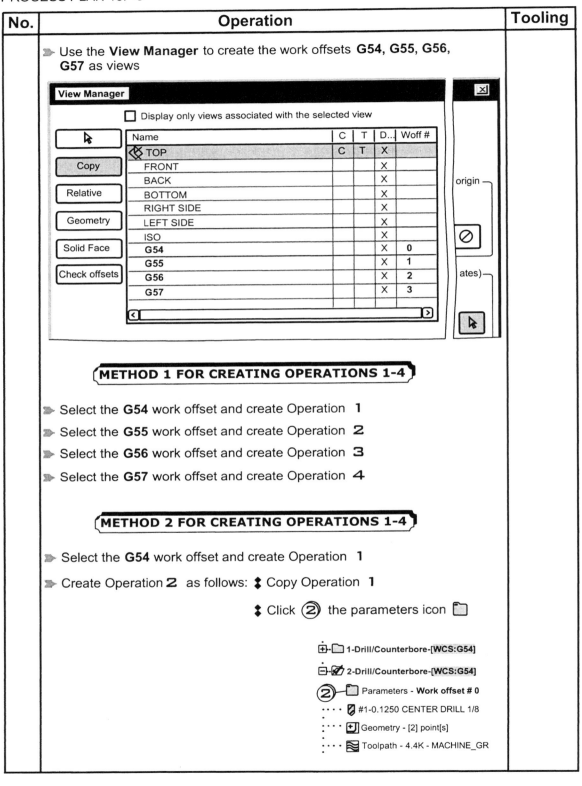	

METHOD 1 FOR CREATING OPERATIONS 1-4

Select the **G54** work offset and create Operation 1

Select the **G55** work offset and create Operation 2

Select the **G56** work offset and create Operation 3

Select the **G57** work offset and create Operation 4

METHOD 2 FOR CREATING OPERATIONS 1-4

Select the **G54** work offset and create Operation 1

Create Operation 2 as follows: Copy Operation 1

Click ② the parameters icon

1-Drill/Counterbore-[WCS:G54]

2-Drill/Counterbore-[WCS:G54]

② Parameters - **Work offset # 0**

#1-0.1250 CENTER DRILL 1/8

Geometry - [2] point[s]

Toolpath - 4.4K - MACHINE_GR

PROCESS PLAN 10P-5

No.	Operation	Tooling
	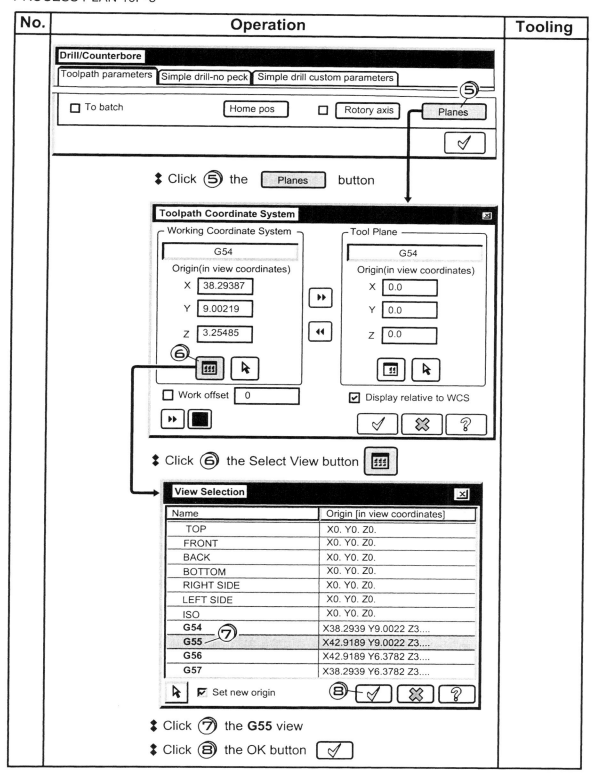	

Drill/Counterbore

Toolpath parameters | Simple drill-no peck | Simple drill custom parameters

☐ To batch   Home pos   ☐ Rotary axis   Planes ⑤

✻ Click ⑤ the Planes button

Toolpath Coordinate System

Working Coordinate System
G54
Origin(in view coordinates)
X 38.29387
Y 9.00219
Z 3.25485
⑥

Tool Plane
G54
Origin(in view coordinates)
X 0.0
Y 0.0
Z 0.0

☐ Work offset 0
☑ Display relative to WCS

✻ Click ⑥ the Select View button

View Selection

Name	Origin [in view coordinates]
TOP	X0. Y0. Z0.
FRONT	X0. Y0. Z0.
BACK	X0. Y0. Z0.
BOTTOM	X0. Y0. Z0.
RIGHT SIDE	X0. Y0. Z0.
LEFT SIDE	X0. Y0. Z0.
ISO	X0. Y0. Z0.
G54 ⑦	X38.2939 Y9.0022 Z3….
G55	X42.9189 Y9.0022 Z3….
G56	X42.9189 Y6.3782 Z3….
G57	X38.2939 Y6.3782 Z3….

☑ Set new origin ⑧

✻ Click ⑦ the **G55** view

✻ Click ⑧ the OK button

PROCESS PLAN 10P-5

No.	Operation	Tooling

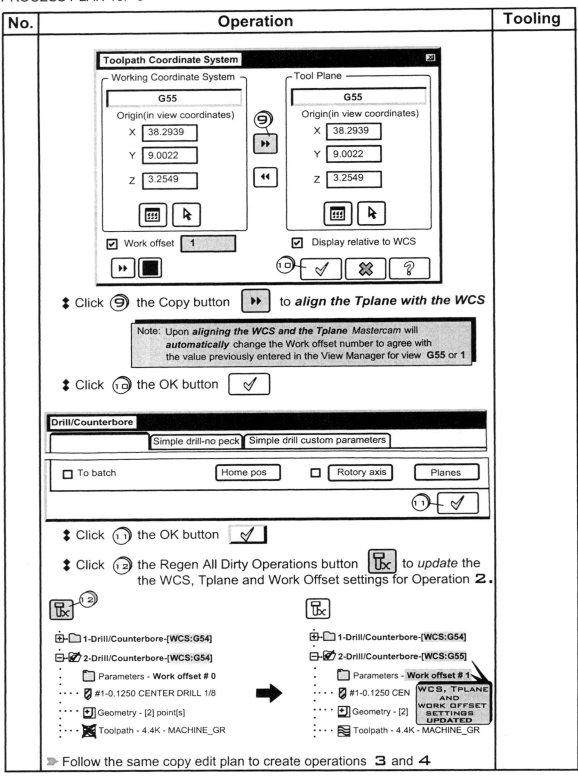

Click ⑨ the Copy button ▶▶ to *align the Tplane with the WCS*

Note: Upon *aligning the WCS and the Tplane* Mastercam will
automatically change the Work offset number to agree with
the value previously entered in the View Manager for view **G55 or 1**

Click ⑩ the OK button ✓

Click ⑪ the OK button ✓

**Click ⑫ the Regen All Dirty Operations button 🔧 to *update* the
the WCS, Tplane and Work Offset settings for Operation 2.**

⊞-📁 1-Drill/Counterbore-[WCS:G54]
⊟-🔩 2-Drill/Counterbore-[WCS:G54]
　　📁 Parameters - **Work offset # 0**
　　🔩 #1-0.1250 CENTER DRILL 1/8
　　⊞ Geometry - [2] point[s]
　　✖ Toolpath - 4.4K - MACHINE_GR

➡

⊞-📁 1-Drill/Counterbore-[WCS:G54]
⊟-🔩 2-Drill/Counterbore-[WCS:G55]
　　📁 Parameters - **Work offset # 1**
　　🔩 #1-0.1250 CEN
　　⊞ Geometry - [2]
　　〰 Toolpath - 4.4K - MACHINE_GR

**WCS, TPLANE
AND
WORK OFFSET
SETTINGS
UPDATED**

➤ Follow the same copy edit plan to create operations 3 and 4

PROCESS PLAN 10P-5

No.	Operation	Tooling

5
6
7
8
	DRILL 3/8 DIA(3) HOLES (THRU)
	  ●
	  ●
	  ●
	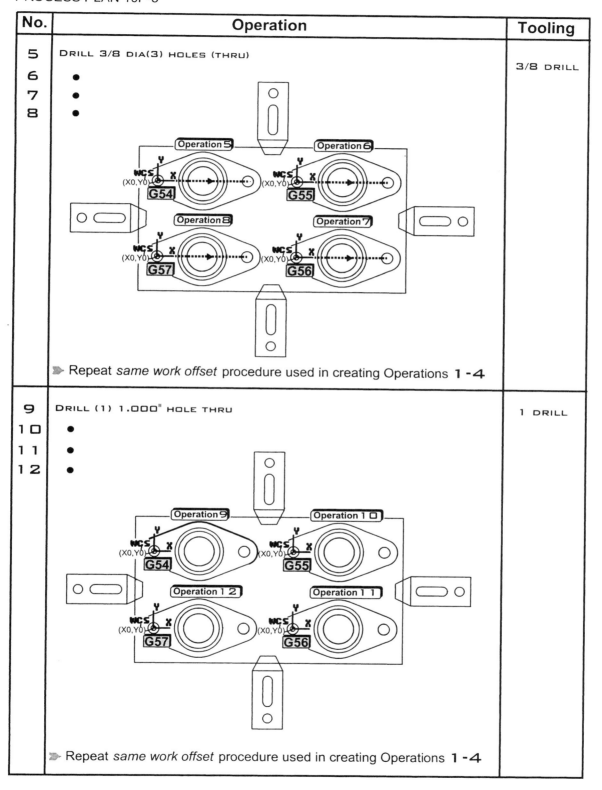
	➤ Repeat *same work offset* procedure used in creating Operations **1-4**
		3/8 DRILL

9
10
11
12
	DRILL (1) 1.000" HOLE THRU
	  ●
	  ●
	  ●
	➤ Repeat *same work offset* procedure used in creating Operations **1-4**
		1 DRILL

PROCESS PLAN 10P-5

No.	Operation	Tooling
1 3 1 4 1 5 1 6	POCKET X .625 DEEP. ● ● ● 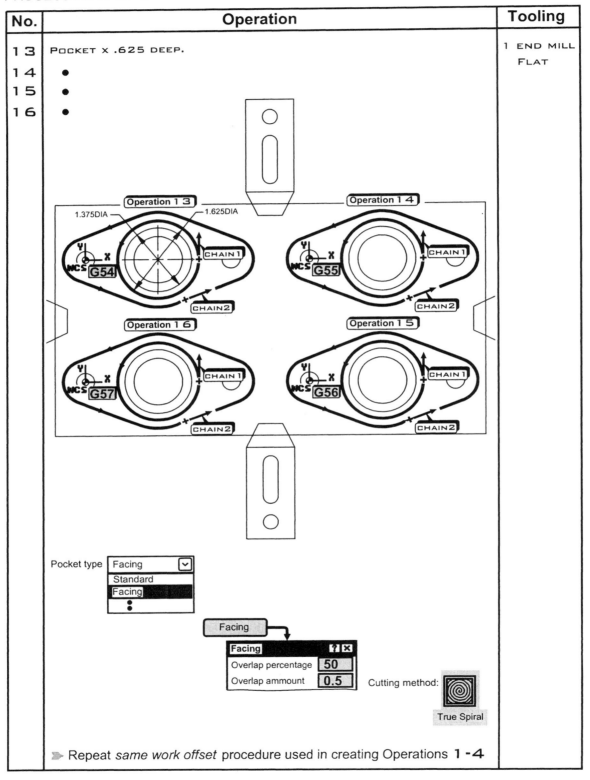	1 END MILL FLAT

➤ Repeat *same work offset* procedure used in creating Operations **1-4**

PROCESS PLAN 10P-5

No.	Operation	Tooling
17	POCKET X .75 DEEP. LEAVE .01 FOR FINISH CUT IN XY.	1 BULL
18	●	END MILL
19	●	.125R
20	●	

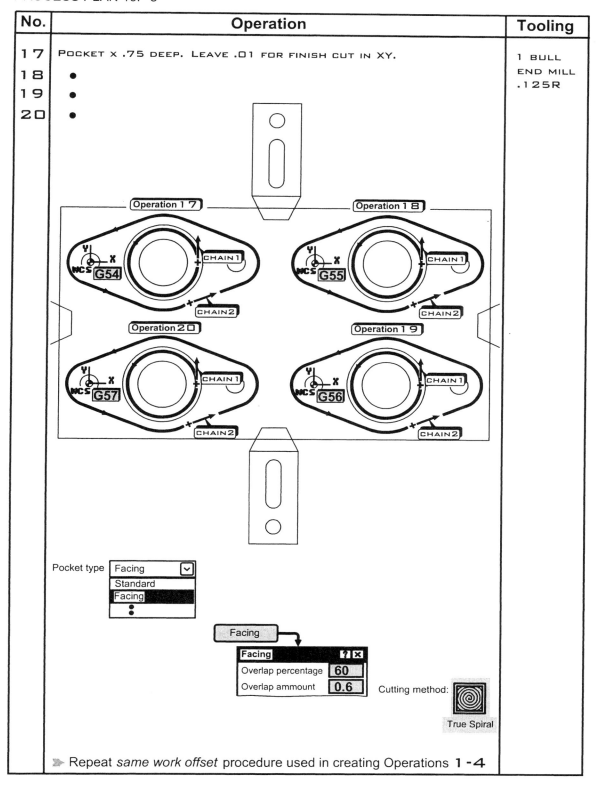

Pocket type [Facing ▾]
  Standard
  Facing
    ●

Facing

Facing	? X
Overlap percentage	60
Overlap ammount	0.6

Cutting method: True Spiral

≫ Repeat *same work offset* procedure used in creating Operations **1-4**

PROCESS PLAN 10P-5

	STOP MACHINE: ADD (2) SCREWS TO EACH PART	
No.	**Operation**	**Tooling**
21	ROUGH OUTSIDE X .25 DEEP. LEAVE .01 FOR FINISH CUT IN XY.	1/2 END MILL FLAT

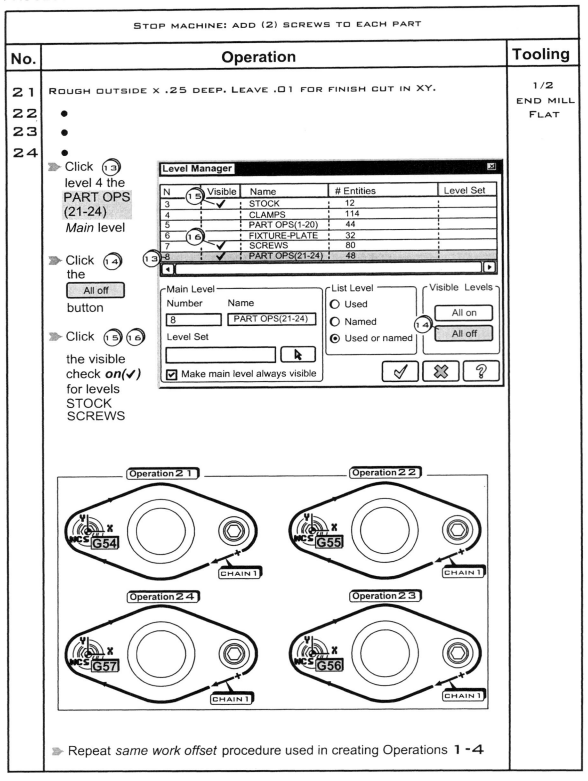

21 ROUGH OUTSIDE X .25 DEEP. LEAVE .01 FOR FINISH CUT IN XY.

22 •

23 •

24 •

➤ Click ⑬ level 4 the PART OPS (21-24) *Main* level

➤ Click ⑭ the [All off] button

➤ Click ⑮ ⑯ the visible check *on(✔)* for levels STOCK SCREWS

Level Manager

N	Visible	Name	# Entities	Level Set
3	✔	STOCK	12	
4		CLAMPS	114	
5		PART OPS(1-20)	44	
6		FIXTURE-PLATE	32	
7	✔	SCREWS	80	
8	✔	PART OPS(21-24)	48	

Main Level — Number: 8 — Name: PART OPS(21-24) — Level Set: ☑ Make main level always visible

List Level — ○ Used — ○ Named — ⊙ Used or named

Visible Levels — [All on] [All off]

Operation 21 — MCS G54 — CHAIN 1
Operation 22 — MCS G55 — CHAIN 1
Operation 24 — MCS G57 — CHAIN 1
Operation 23 — MCS G56 — CHAIN 1

➤ Repeat *same work offset* procedure used in creating Operations **1-4**

CHAPTER - 11

CREATING BASIC SOLID MODELS

11-1 Chapter Objectives

After completing this chapter you will be able to:

1. Understand the uses of the different types of CAD models: wireframe, surface, and solid.
2. Know how to create solid features by extrusion.
3. Understand how to avoid extrusion errors.
4. Know how to create solid features by revolving.
5. Explain the methods of creating fillet and chamfer features on solid models.
6. Describe the methods of editing solid model features.
7. Know how to create solid models via Boolean functions.
8. Understand how to build a history tree for an imported "brick" solid.

11-2 Types of CAD Models

Depending upon its geometric complexity, the type of machining to be executed and other production parameters, the actual part is defined as a simple wire frame model, a surface model or a solid model.

Wireframe Models

- ✧ *Simplest* representation of a part as a collection of geometric elements: lines, arcs, points and splines.

- ✧ *Only* the *boundary or edges of the part can be defined*

- ✧ *Only flat planes* can be defined *between the boundaries* and the *interior is void* of points.

- ✧ *Requires the least* ammount of database.

- ✧ System *does not* create a history tree of all the elements that comprise the model.

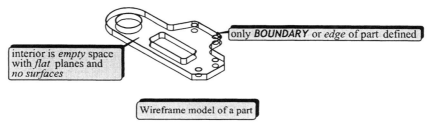

interior is *empty* space with *flat* planes and *no surfaces*

only *BOUNDARY* or *edge* of part defined

Wireframe model of a part

Figure 11-1

Surface Models

- ✧ *More complete* representation of a part as a collection of geometric elements: lines, arcs, points and splines and surfaces of *zero thickness*.

- ✧ *Only* the *boundary* and *surface within boundary can be defined*

- ✧ The *interior is void* of points.

- ✧ System *does not* create a history tree of all the elements that comprise the model.

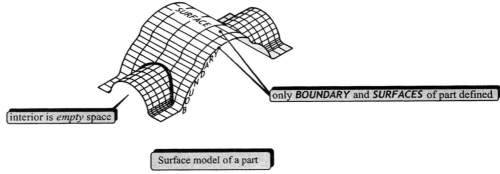

interior is *empty* space

only *BOUNDARY* and *SURFACES* of part defined

Surface model of a part

Figure 11-2

Solid Models

- ✧ *Most Complete* representation of a part as **related set** of solid features

- ✧ The *edges, faces(surfaces) and interior are defined.*

- ✧ Requires the most ammount of database(interior is *entirely filled* with points)

- ✧ System *creates a history tree* of *all the features* that comprise the model and treats it as a *single unified object*.

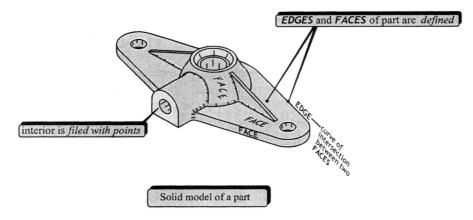

EDGES and *FACES* of part are *defined*

interior is *filed with points*

EDGE - curve of intersection between two FACES

Solid model of a part

Figure 11-3

11-3 Creating the Base Feature of the Solid Model

The *first solid element* used to create the solid model is called the *base feature.* The base operation is listed as the *first operation in the history tree* for the solid in the Solids Manager dialog box. It *cannot be moved or deleted* from the operations list.

11-4 Creating Base Features by Extruding

This process starts with the generation of a *closed 2D wireframe* base curve called a *sketch*. The *orientation and shape* of the curve depends upon the orientation and shape of the *predominant base shape of the actual part* . The sketch is *chained* and extruded along a specified direction by an inputed distance value.

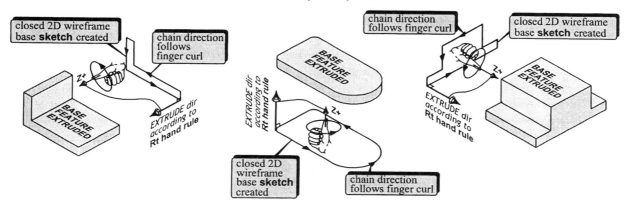

Figure 11-4

Solid Extrude from the Solids Menu

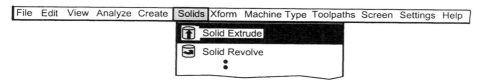

Solid Extrude from the Solids Toolbar

EXAMPLE 11-1

Direct *Mastercam X2* to create the base feature as shown in Fig 11-5.

.25 x 45° CHAMFER

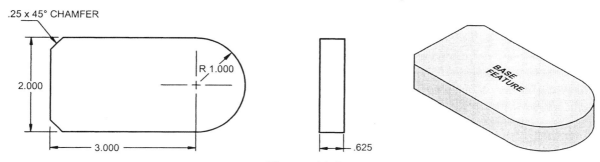

Figure 11-5

a) Create the 3 in x 2 in rectangle.

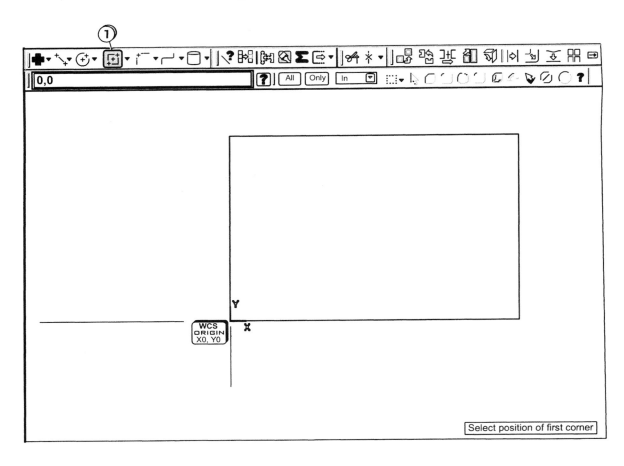

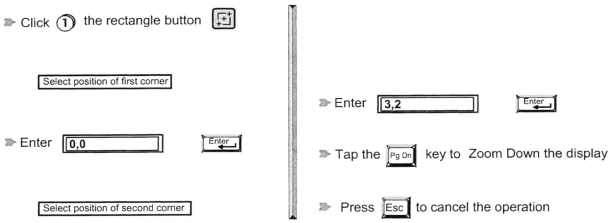

⇒ Click ① the rectangle button

Select position of first corner

⇒ Enter   0,0   Enter

Select position of second corner

⇒ Enter   3,2   Enter

⇒ Tap the Pg Dn key to Zoom Down the display

⇒ Press Esc to cancel the operation

b) Create the R1.000 arc entity.

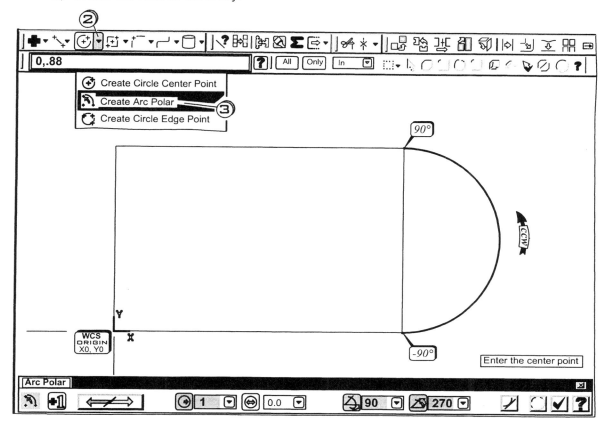

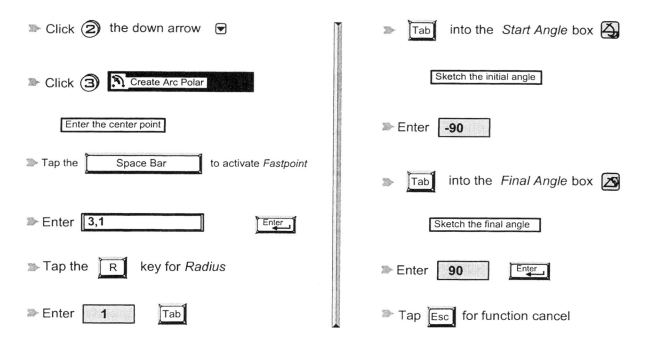

Click ② the down arrow ▼

Click ③ 🔧 Create Arc Polar

Enter the center point

Tap the [Space Bar] to activate *Fastpoint*

Enter | 3,1 | [Enter]

Tap the [R] key for *Radius*

Enter [1] [Tab]

[Tab] into the *Start Angle* box

Sketch the initial angle

Enter [-90]

[Tab] into the *Final Angle* box

Sketch the final angle

Enter [90] [Enter]

Tap [Esc] for function cancel

c) Generate the .25 x45° chamfer and delete the vertical line.

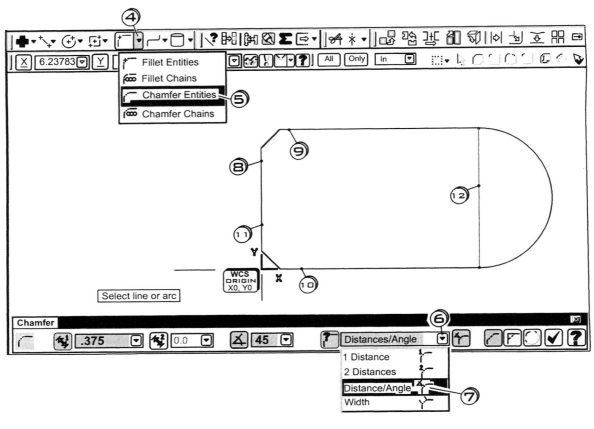

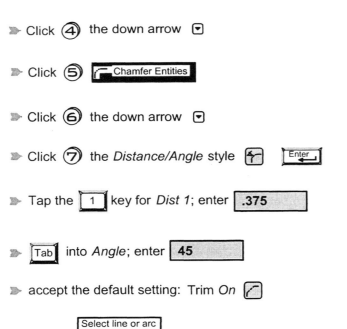

➤ Click ④ the down arrow ▾

➤ Click ⑤ [🔲 Chamfer Entities]

➤ Click ⑥ the down arrow ▾

➤ Click ⑦ the *Distance/Angle* style [↲] [Enter]

➤ Tap the [1] key for *Dist 1*; enter [.375]

➤ [Tab] into *Angle*; enter [45]

➤ accept the default setting: Trim *On* [◠]

[Select line or arc]

➤ Click ⑧ the *first* line entity

[Select line or arc]

➤ Click ⑨ the *second* line entity

[Select line or arc]

➤ Click ⑩ the *first* line entity

[Select line or arc]

➤ Click ⑪ the *second* line entity

➤ Tap [Esc] for function cancel

➤ Click ⑫ on the vertical line

➤ Tap [Del] key to delete it.

d) Extrude the 2D sketch in a direction ***normal to the sketch plane***.

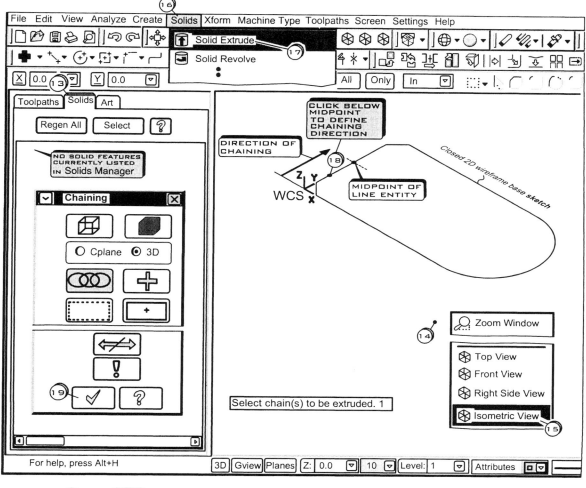

➤ Click ⑬ the ⌐Solids⌐ folder

Note: *No solid features are currently listed in the history tree of Solids Manager dialog box.*

➤ ***Right*** Click ⑭ ; Click ⑮ ⊗ Isometric View

➤ Click ⑯ the ⌐Solids⌐ menu bar.

➤ Click ⑰ 🔼 Solid Extrude

Select chani(s) to be extruded. 1

➤ Click ⑱ *below the midpoint of the line* to specify the chain **start/end** point.

➤ Click ⑲ the OK button ✓

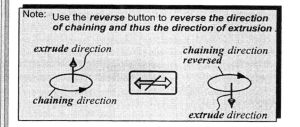

Note: Use the ***reverse*** button to ***reverse the direction of chaining and thus the direction of extrusion***

e) Specify the Type of extrusion(⊙ Create Body) and the extrusion thickness.

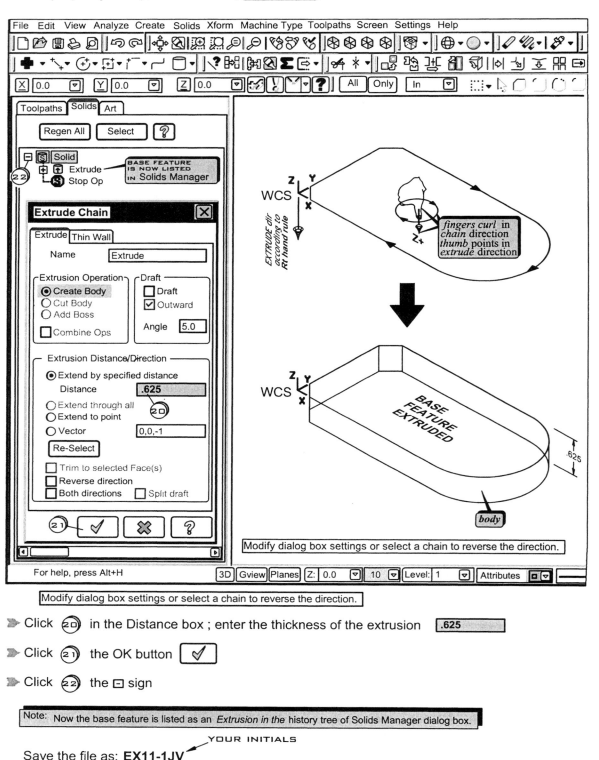

Modify dialog box settings or select a chain to reverse the direction.

➤ Click ㉒ in the Distance box ; enter the thickness of the extrusion   .625

➤ Click ㉑ the OK button   ✓

➤ Click ㉒ the ⊟ sign

Note: Now the base feature is listed as an *Extrusion in the* history tree of Solids Manager dialog box.

YOUR INITIALS

Save the file as: **EX11-1JV**

11-5 Checking when Creating the Solid Model

Recall, a wireframe model is a *separate collection* of lines arcs, splines and points. As such, any one of these entities can be easily accessed and changed. The solid, on the other hand, is treated as a *single object*. Once created, the only way to edit it is to go back up its history tree and edit the elements that comprise it. Depending on model complexity, routine editing operations such as deleting and changing portions of geometry for a wireframe model can become very difficult to do with a solid model.

Thus, it becomes *very important to verify and check the solid model as features are added*. If too many modelling errors accumulate as features are added it may be necessary to scrap the model and begin again.

Check to See If the Solid Model Shades Properly

If the feature has been created properly the solid *should shade properly*. *If Not re-do this feature before proceeding on* .

Check the Dimensions

Spot check the dimensions. If they check out proceed to the next step. *If Not re-do this feature before proceeding on* .

Check the Solid Model by Dynamic Rotation

If the feature has been created properly the solid should look proper in any Gview *If Not re-do this feature before proceeding on* .

d) Extrude the 2D sketch in a direction *normal to the sketch plane*.

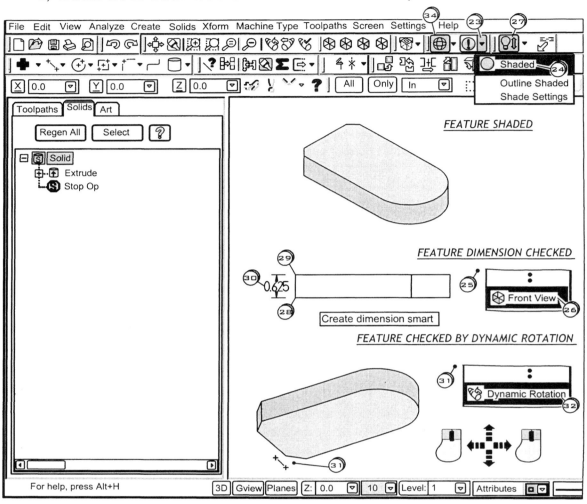

Check to See If the Solid Model Shades Properly

➤ Tap `Alt` **+** `S` keys
             or
➤ Click ㉓ the down arrow ▾

➤ Click ㉔ ⬤ Shaded

Check the Dimensions

➤ *Right* Click ㉕ ;   Click ㉖ ⊕ Front View

➤ Click ㉗ the Smart Dimension button 🔆

➤ Click ㉘ ㉙ ㉚

Check the Model by Dyanmic Rotation

➤ *Right* Click ㉛ ;

     Click ㉜ 🧩 Dynamic Rotation

➤ Click ㉝ the endpoint *to rotate about*

     Move the mouse *left*, *right*,
     up and *down* to effect
     dynamic rotation.

To Return to Wireframe Display

➤ Tap `Alt` **+** `S` keys
             or

➤ Click ㉞ ⊕ the wireframe button

11-6 Creating Base Features by Using Primitives

Primitives are predefined solid shapes such as blocks, cylinders, cones, spheres and torusus. They are a useful approach if the base feature has these shapes.

Solid Primitives from the Create Menu

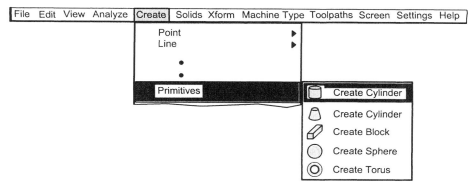

11-7 Creating Base Features by Importing

More parts are coming as solid models from clients. These parts are being created using other CAD packages such as SolidWorks, SolidEdge, Catia, Pro/Engineer, Inventor, Parasolid, SAT, SETP, and others. The process of file conversion was discussed in Chapter 4 , Section 4-5.

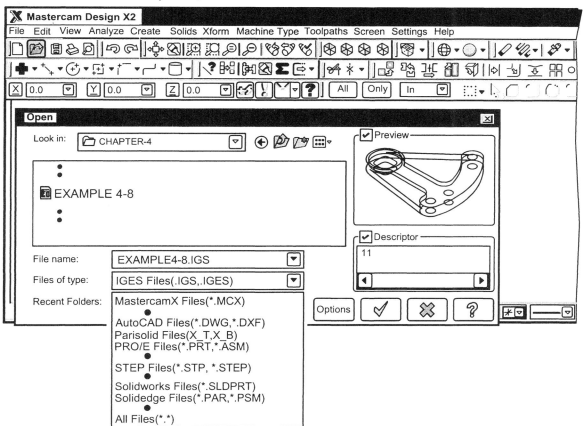

11-8 Creating Additional Features by Extrusion

After establishining the most basic shape of the part via the base feature *additional features are created to successively refine and complete the solid model*. The section of the additional feature is created as a closed 2D wireframe shape and chained. The **Extrude Chain** dialog box is then activated.

It is used to:

◈ extrude the 2D chain along a *specified direction* by *an inputted distance* *to create the additional feature.*

◈ *further refine the base feature by either adding material to it(boss) or removing material from it(cut) using the additional feature.*

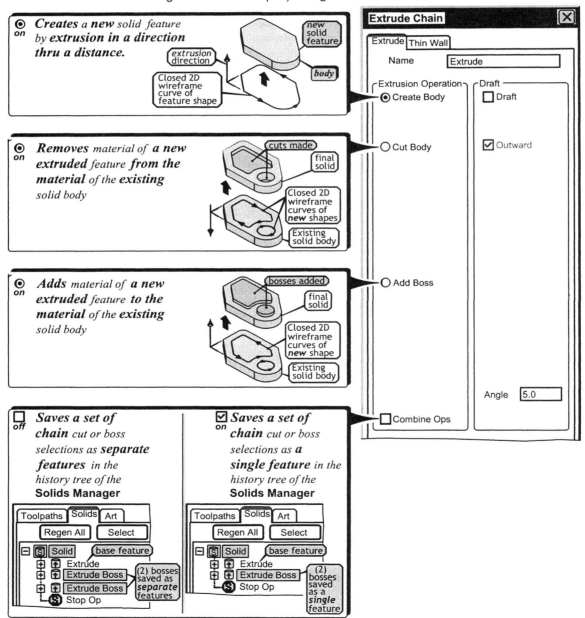

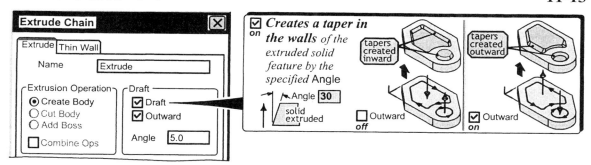

Extrude Chain

Extrude | Thin Wall

Name   Extrude

Extrusion Operation
- ◉ Create Body
- ○ Cut Body
- ○ Add Boss
- ☐ Combine Ops

Draft
- ☑ Draft
- ☑ Outward

Angle 5.0

☑ *Creates a taper in* **the walls** *of the extruded solid feature by the specified* **Angle**

Angle **30**

solid extruded

☐ Outward **off**   ☑ Outward **on**

tapers created inward   tapers created outward

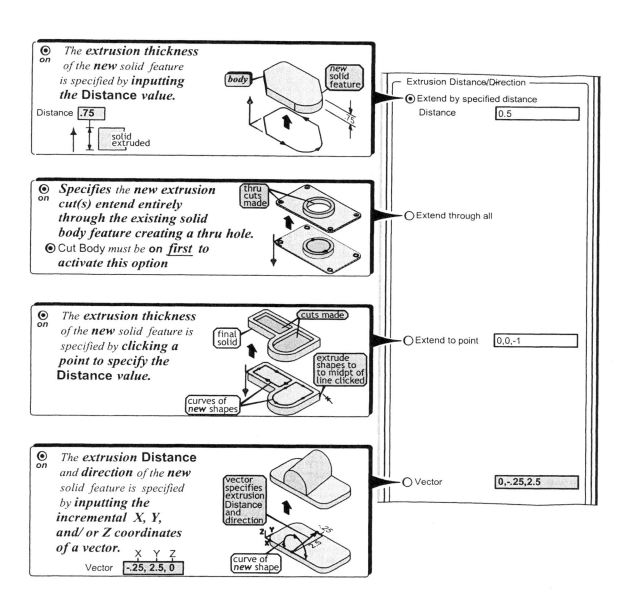

◉ *The* **extrusion thickness** *of the* **new** *solid feature is specified by* **inputting the Distance** *value.*

Distance **.75**

solid extruded

body   new solid feature

Extrusion Distance/Direction

◉ Extend by specified distance

Distance 0.5

◉ *Specifies the* **new extrusion cut(s) entend entirely through the existing solid body feature creating a thru hole.**
◉ **Cut Body** *must be on* **first** *to activate this option*

thru cuts made

○ Extend through all

◉ *The* **extrusion thickness** *of the* **new** *solid feature is specified by* **clicking a point to specify the Distance** *value.*

cuts made

final solid

extrude shapes to to midpt of line clicked

curves of **new** shapes

○ Extend to point   0,0,-1

◉ *The* **extrusion Distance** *and* **direction** *of the* **new** *solid feature is specified by* **inputting the incremental X, Y, and/or Z coordinates** *of a vector.*

     X Y Z
Vector **-.25, 2.5, 0**

vector specifies extrusion Distance and direction

curve of **new** shape

○ Vector   0,-.25,2.5

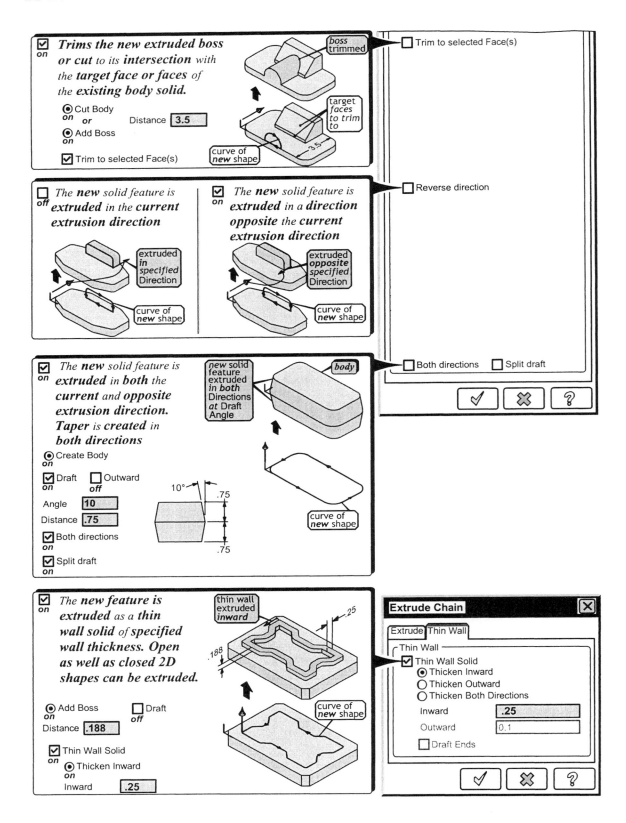

☑ **Trims the new extruded boss or cut** to its **intersection** with the **target face or faces** of the **existing body solid.**
on

⦿ Cut Body
on   or   Distance [3.5]

⦿ Add Boss
on

☑ Trim to selected Face(s)

boss trimmed

target faces to trim to

curve of *new* shape

☐ Trim to selected Face(s)

☐ The **new** *solid feature is* **extruded** in the **current extrusion direction**
off

☑ The **new** *solid feature is* **extruded** in a **direction opposite** the **current extrusion direction**
on

extruded *in* specified Direction

curve of *new* shape

extruded *opposite* specified Direction

curve of *new* shape

☐ Reverse direction

☑ The **new** *solid feature is* **extruded** in **both** the **current** and **opposite extrusion direction. Taper** is **created** in **both** directions
on

⦿ Create Body
on

☑ Draft   ☐ Outward
on   off

Angle [10]

Distance [.75]

☑ Both directions
on

☑ Split draft
on

10°   .75

.75

new solid feature extruded *in both* Directions *at* Draft Angle

body

curve of *new* shape

☐ Both directions   ☐ Split draft

✓   ✗   ?

☑ The **new feature is extruded** as a **thin wall solid** of *specified wall thickness. Open as well as closed 2D shapes can be extruded.*
on

⦿ Add Boss   ☐ Draft
on   off

Distance [.188]

☑ Thin Wall Solid
on   ⦿ Thicken Inward
     on
     Inward [.25]

thin wall extruded *inward*

.25

.188

curve of *new* shape

Extrude Chain   [X]

Extrude | Thin Wall

┌ Thin Wall ─────────────
☑ Thin Wall Solid
  ⦿ Thicken Inward
  ○ Thicken Outward
  ○ Thicken Both Directions

  Inward   [.25]
  Outward   [0.1]

  ☐ Draft Ends

✓   ✗   ?

11-9 Avoiding Extrusion Errors

A solid model must be a **closed** body. When creating an additional feature via *add boss* or *cut body* care must be taken to insure the results will produce a **single, unified, closed** body.

⊙ Add Boss

The 2D closed wireframe contour of the additional feature **must be extruded in the direction of the base feature** and **must intersect the base feature.**

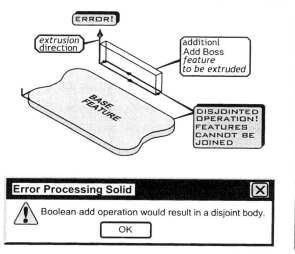
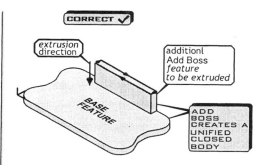

⊙ Cut Body

The 2D closed wireframe contour of the additional feature **must not be extruded such that it cuts the base feature into separate disjointed bodies.**

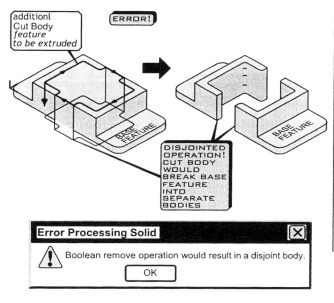
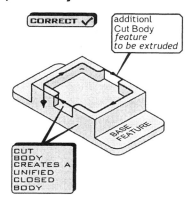

11-10 Creating Additional Features by Revolving

The section of the additional feature is created as a closed 2D wireframe shape and chained. The axis about which the chain is to rotate is specified. The **Extrude Chain** dialog box is then activated.

It is used to:

✧ *create the additional feature by revolving the chain about the specified axis through a given angle.*

✧ *further refine the base feature by either adding material to it(boss) or remove material from it(cut) using the additional feature.*

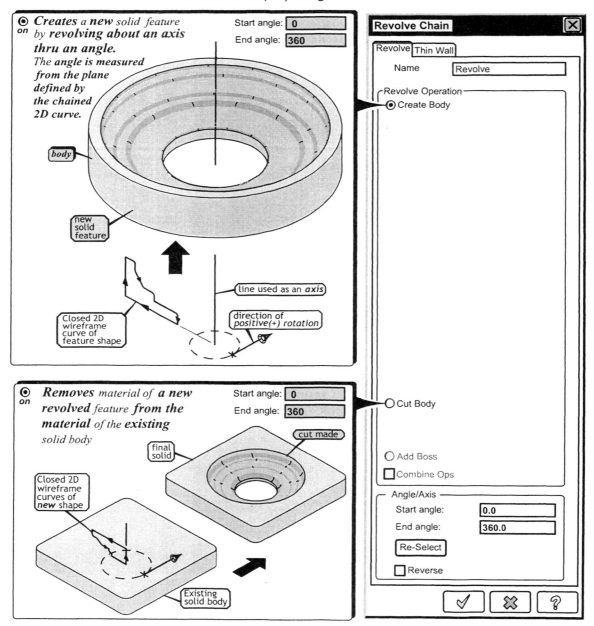

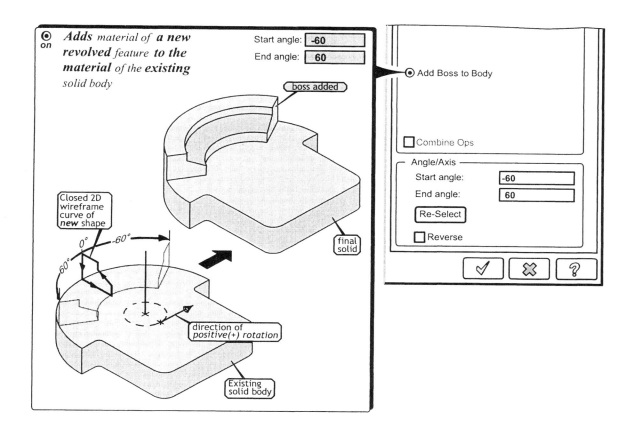

Adds *material of **a new revolved** feature **to the material** of the **existing** solid body*

Start angle: **-60**
End angle: **60**

boss added

Closed 2D wireframe curve of *new* shape

0°   -60°
60°

direction of *positive(+) rotation*

final solid

Existing solid body

● Add Boss to Body

☐ Combine Ops

Angle/Axis
Start angle: -60
End angle: 60
[Re-Select]
☐ Reverse

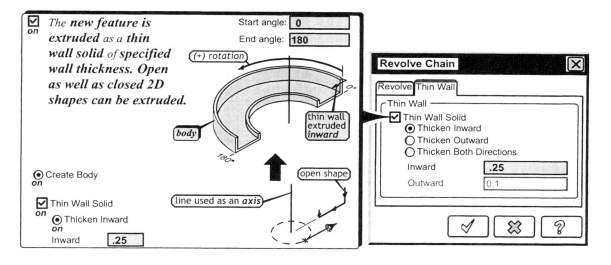

*The **new feature is extruded** as a **thin wall** solid of specified wall thickness. **Open** as well as closed 2D shapes can be extruded.*

Start angle: **0**
End angle: **180**

(+) rotation

thin wall extruded *inward*

body

180°

open shape

line used as an *axis*

● Create Body
on

☑ Thin Wall Solid
on
   ● Thicken Inward
   on
   Inward   .25

Revolve Chain [X]

Revolve | Thin Wall

Thin Wall
☑ Thin Wall Solid
   ● Thicken Inward
   ○ Thicken Outward
   ○ Thicken Both Directions
   Inward   **.25**
   Outward   0.1

11-11 Adding Fillet Features to Solid Models

A *fillet* feature is used to *round* the edges of a solid. A *fillet* is a rounding of the *inside edges* of a solid, and a *round* is a smoothing of the *outside edges* of the solid.

Fillet from the Solids Menu

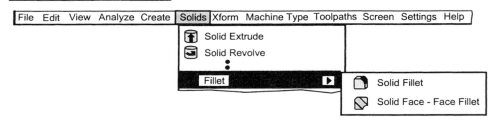

Fillet from the Solids Toolbar

All the **edges clicked** on the solid **are filleted**. Constant Radius specifies that the *same value* inputted for fillet *Radius* is used along *every* selected edge.

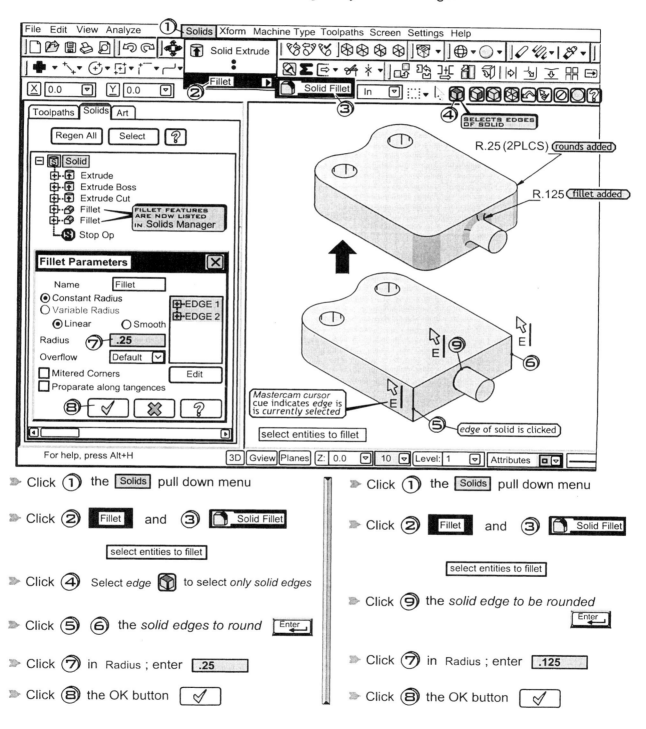

➤ Click ① the Solids pull down menu

➤ Click ② Fillet and ③ Solid Fillet

select entities to fillet

➤ Click ④ Select edge 🔲 to select *only solid edges*

➤ Click ⑤ ⑥ the *solid edges to round* Enter

➤ Click ⑦ in Radius ; enter .25

➤ Click ⑧ the OK button ✓

➤ Click ① the Solids pull down menu

➤ Click ② Fillet and ③ Solid Fillet

select entities to fillet

➤ Click ⑨ the *solid edge to be rounded* Enter

➤ Click ⑦ in Radius ; enter .125

➤ Click ⑧ the OK button ✓

All the ***edges that bound the face clicked*** on the solid ***are filleted***. Constant Radius specifies that the *same value* inputted for fillet *Radius* is used along *every edge that bounds the selected face* .

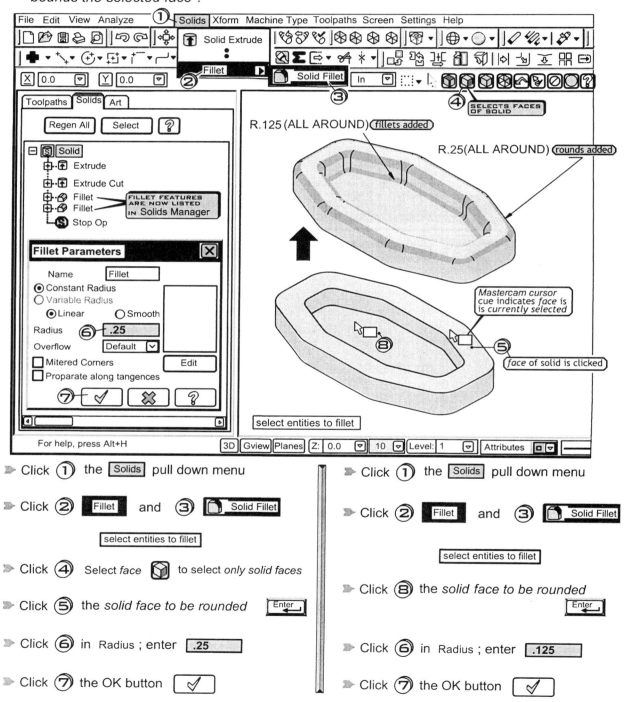

>> Click ① the Solids pull down menu

>> Click ② Fillet and ③ Solid Fillet

select entities to fillet

>> Click ④ Select face to select *only solid faces*

>> Click ⑤ the *solid face to be rounded* Enter

>> Click ⑥ in Radius ; enter .25

>> Click ⑦ the OK button ✓

>> Click ① the Solids pull down menu

>> Click ② Fillet and ③ Solid Fillet

select entities to fillet

>> Click ⑧ the *solid face to be rounded* Enter

>> Click ⑥ in Radius ; enter .125

>> Click ⑦ the OK button ✓

All the **edges that bound the body** of the solid **clicked are filleted**. Constant Radius specifies that the *same value* inputted for fillet *Radius* is used along *every edge that bounds the faces of the body*.

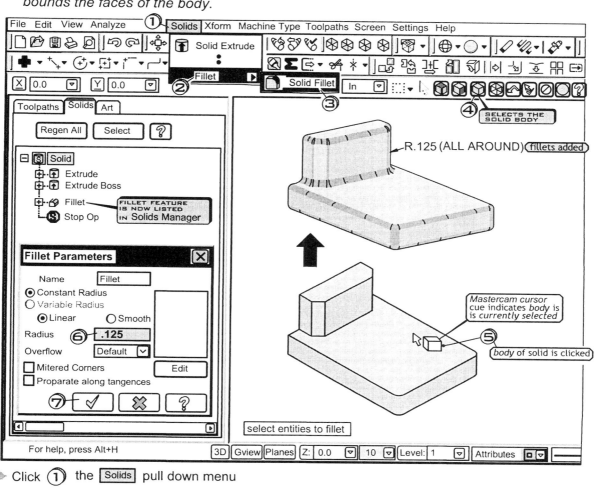

➤ Click ① the **Solids** pull down menu

➤ Click ② **Fillet** and ③ **Solid Fillet**

select entities to fillet

➤ Click ④ Select *body* ⬚ to select *only solid bodies*

➤ Click ⑤ the *solid body to be rounded* **Enter**

➤ Click ⑥ in Radius ; enter **.125**

➤ Click ⑦ the OK button ✓

11-11 Adding Chamfer Features to Solid Models

A *chamfer* feature is used to *bevel* the edges of a solid. The sharp edge is replaced by a *smooth flat inclined crossection*. To break a sharp edge chamfers on the *inside add* material to the solid body while those on the *outside remove* material.

Chamfer from the Solids Menu

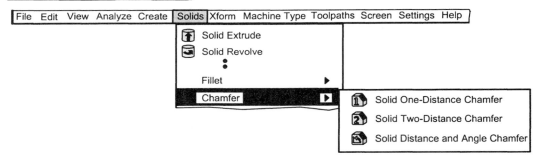

Chamfer from the Solids Toolbar

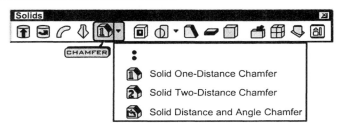

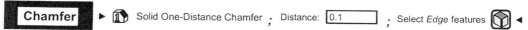

All the **edges clicked** on the solid **are chamfered**. *One-Distance* specifies that the chamfer is offset the **same distance from both faces of the edge clicked.**

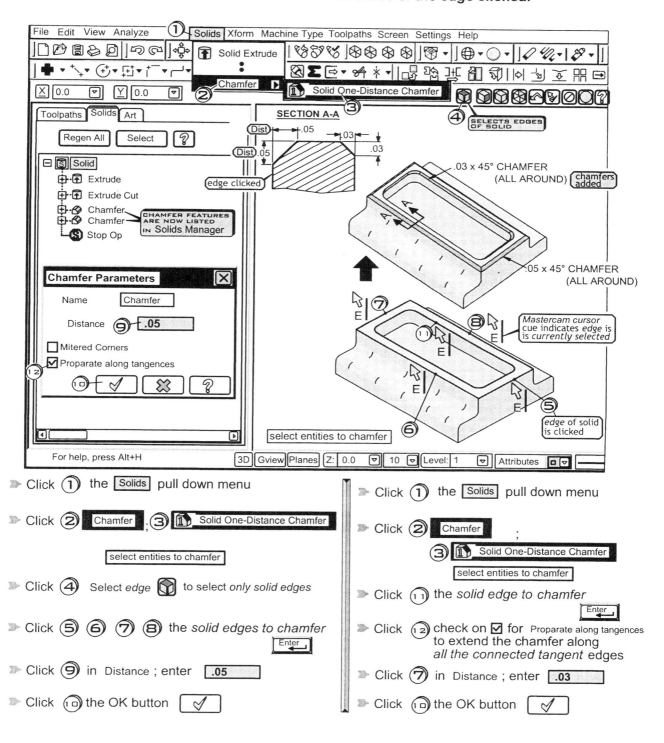

» Click ① the Solids pull down menu

» Click ② Chamfer ; ③ Solid One-Distance Chamfer

select entities to chamfer

» Click ④ Select *edge* to select *only solid edges*

» Click ⑤ ⑥ ⑦ ⑧ the *solid edges to chamfer*

Enter

» Click ⑨ in Distance ; enter .05

» Click ⑩ the OK button

» Click ① the Solids pull down menu

» Click ② Chamfer ;

③ Solid One-Distance Chamfer

select entities to chamfer

» Click ⑪ the *solid edge to chamfer*

Enter

» Click ⑫ check on ☑ for Proparate along tangences to extend the chamfer along *all the connected tangent* edges

» Click ⑦ in Distance ; enter .03

» Click ⑩ the OK button

Chamfer ▶ 🧊 Solid Two-Distance Chamfer ; Distance 1: 0.1 ; Distance 2: 0.1 ; Select *Face* features 🧊 ◀

All the *edges that bound the reference face* of the solid *clicked are chamfered*.
Two-Distances specifies that the chamfer is offset by *Distance 1 from face 1(the reference face clicked)* and *Distance 2 from face 2.*

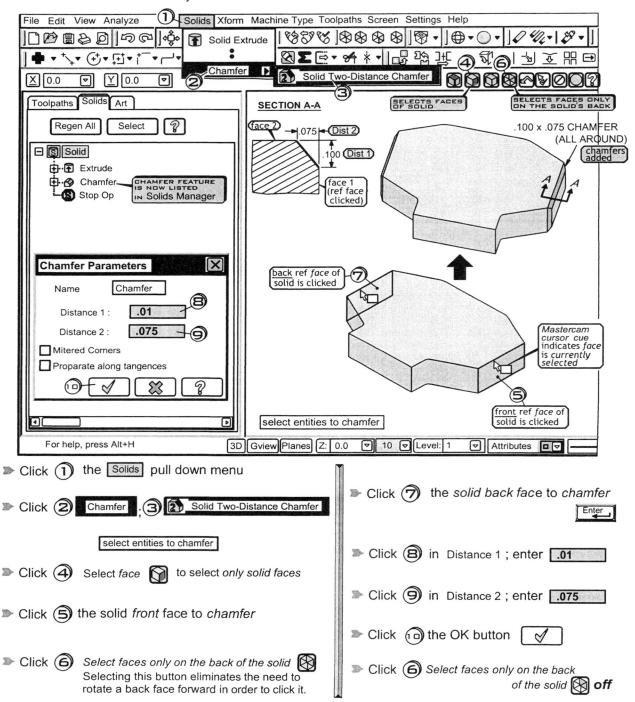

Click ① the Solids pull down menu

Click ② Chamfer , ③ 🧊 Solid Two-Distance Chamfer

select entities to chamfer

Click ④ Select *face* 🧊 to select *only solid faces*

Click ⑤ the solid *front* face to *chamfer*

Click ⑥ *Select faces only on the back of the solid* 🔲
Selecting this button eliminates the need to rotate a back face forward in order to click it.

Click ⑦ the *solid back face* to *chamfer*
Enter ↵

Click ⑧ in Distance 1 ; enter .01

Click ⑨ in Distance 2 ; enter .075

Click ⑩ the OK button ✓

Click ⑥ *Select faces only on the back of the solid* 🔲 off

Chamfer ▶ 🔲 Solid Distance and Angle Chamfer ; Distance : `0.1` ; Angle: `0.1` ; Select *Edge* features 🔲 ◀

All the **edges clicked** on the solid **are chamfered**. The Operator selects the desired reference face bounded by the edge. the chamfer is **offset by a Distance along the reference face and runs at an Angle with respect to the reference face**

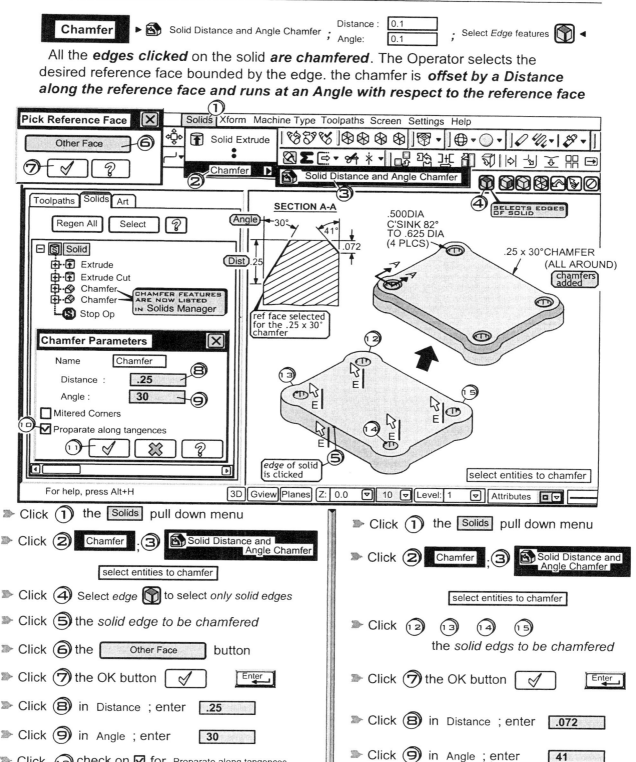

➤ Click ① the Solids pull down menu

➤ Click ② Chamfer , ③ 🔲 Solid Distance and Angle Chamfer

select entities to chamfer

➤ Click ④ Select *edge* 🔲 to select *only solid edges*

➤ Click ⑤ the *solid edge to be chamfered*

➤ Click ⑥ the Other Face button

➤ Click ⑦ the OK button ✓ Enter

➤ Click ⑧ in Distance ; enter `.25`

➤ Click ⑨ in Angle ; enter `30`

➤ Click ⑩ check on ☑ for Proparate along tangences

➤ Click ⑪ the OK button ✓

➤ Click ① the Solids pull down menu

➤ Click ② Chamfer , ③ 🔲 Solid Distance and Angle Chamfer

select entities to chamfer

➤ Click ⑫ ⑬ ⑭ ⑮ the *solid edgs to be chamfered*

➤ Click ⑦ the OK button ✓ Enter

➤ Click ⑧ in Distance ; enter `.072`

➤ Click ⑨ in Angle ; enter `41`

➤ Click ⑪ the OK button ✓

11-12 Editing Solid Model Features

Once a solid model has been created, it can be edited via the Solids Manager.

EXAMPLE 11-2

Given the currently designed solid model shown to the *left* in Figure 11-6 use the Solids Manager to produce the *re-designed version shown to the right.*

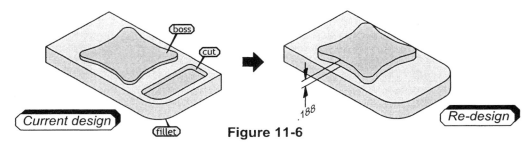

Figure 11-6

Delete the Extrude Cut Feature

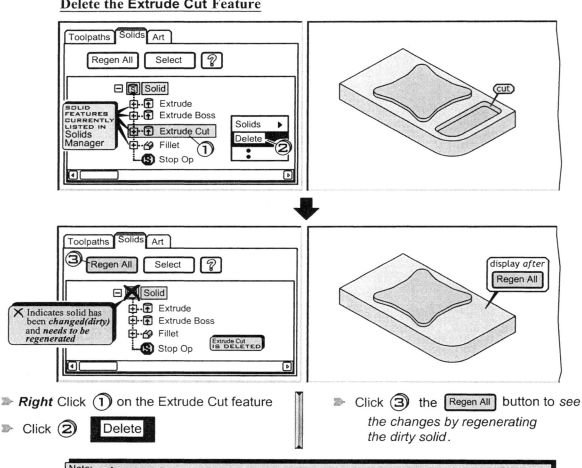

➣ **Right** Click ① on the Extrude Cut feature

➣ Click ② [**Delete**]

➣ Click ③ the [Regen All] button to *see the changes by regenerating the dirty solid.*

Note: ◆ A *Base* Feature *cannot* be deleted

◆ ***Changes made to solids using the Solids Manager cannot be undone*** using the 🔄 button. This includes any deleteions or extrusions.

Add an Additional Fillet

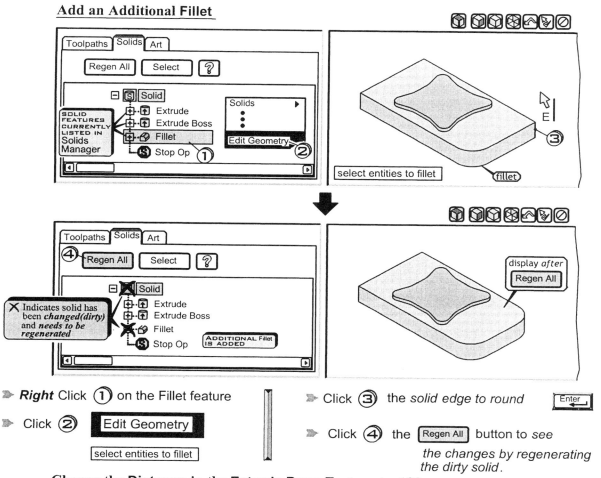

➤ **Right** Click ① on the Fillet feature

➤ Click ② **Edit Geometry**

> select entities to fillet

➤ Click ③ the *solid edge to round* Enter

➤ Click ④ the Regen All button to *see*
the changes by regenerating
the dirty solid.

Change the Distance in the Extrude Boss Feature to .188

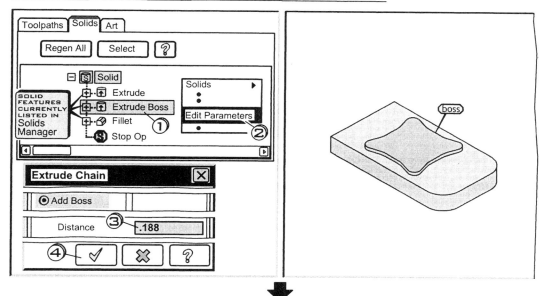

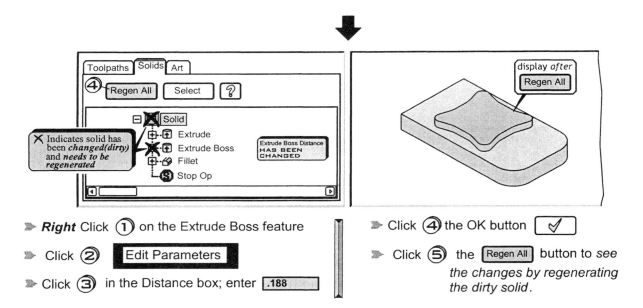

▶ **Right** Click ① on the Extrude Boss feature

▶ Click ② **Edit Parameters**

▶ Click ③ in the Distance box; enter .188

▶ Click ④ the OK button ✓

▶ Click ⑤ the Regen All button to *see the changes by regenerating the dirty solid*.

11-13 Creating Solid Models Via Boolean Functions

A solid model can also be created by using *Boolean functions on combinations of two or more **existing solid bodies** to:*

◇ **add material**
◇ **remove material**
◇ **produce the common material shared by the set of intersecting solids.**

Fillet from the Solids Menu

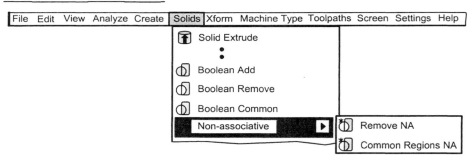

Fillet from the Solids Toolbar

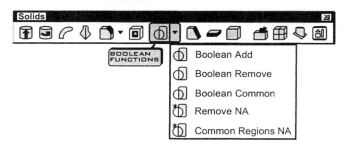

Adds the material of the *sparate solid target* body clicked to the material of the *separate solid tool* bodies clicked. The result produced is a *single unified solid body*.

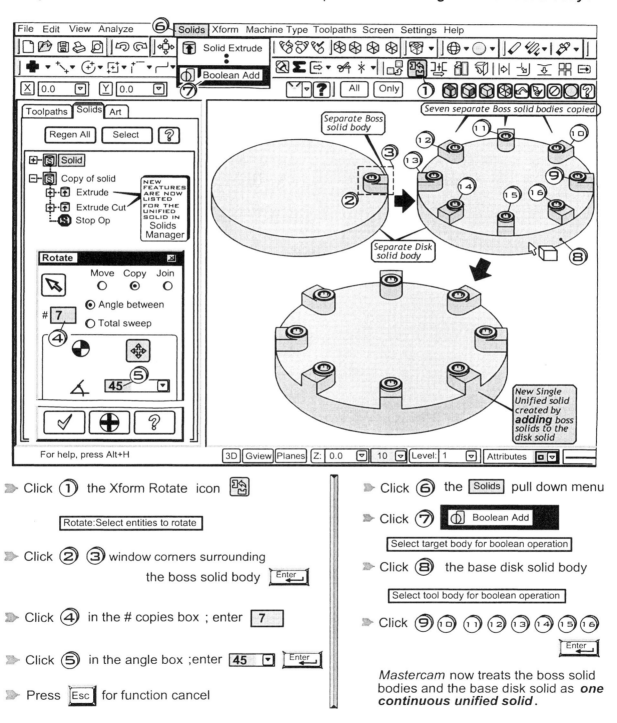

➤ Click ① the Xform Rotate icon

Rotate:Select entities to rotate

➤ Click ② ③ window corners surrounding the boss solid body [Enter]

➤ Click ④ in the # copies box ; enter [7]

➤ Click ⑤ in the angle box ;enter [45 ▼] [Enter]

➤ Press [Esc] for function cancel

➤ Click ⑥ the [Solids] pull down menu

➤ Click ⑦ [Boolean Add]

Select target body for boolean operation

➤ Click ⑧ the base disk solid body

Select tool body for boolean operation

➤ Click ⑨ ⑩ ⑪ ⑫ ⑬ ⑭ ⑮ ⑯ [Enter]

Mastercam now treats the boss solid bodies and the base disk solid as *one continuous unified solid*.

Removes from the separate target body clicked the material of the *sparate solid tool bodies clicked*. The result produced is a *single unified solid body*.

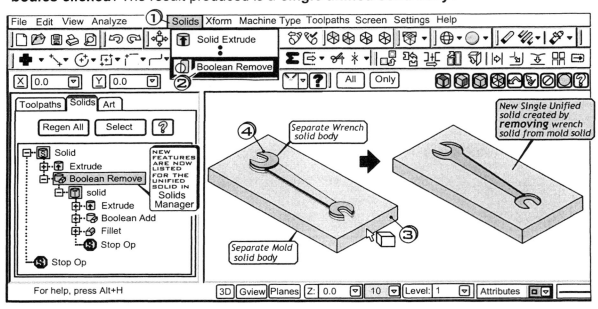

➤ Click ① the [Solids] pull down menu

➤ Click ② [Boolean Remove]

[Select target body for boolean operation]

➤ Click ③ the *mold* solid body

[Select tool body for boolean operation]

➤ Click ④ the *wrench* solid body   [Enter ↵]

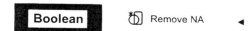

Removes from the separate target body clicked the material of the ***sparate solid tool bodies clicked***. The result produced is a ***single unified solid body***. Target body and/or tool body can be tagged as *non-associative*. ***After the Boolean operation the tagged body is kept as a separate solid and can be operated on.***

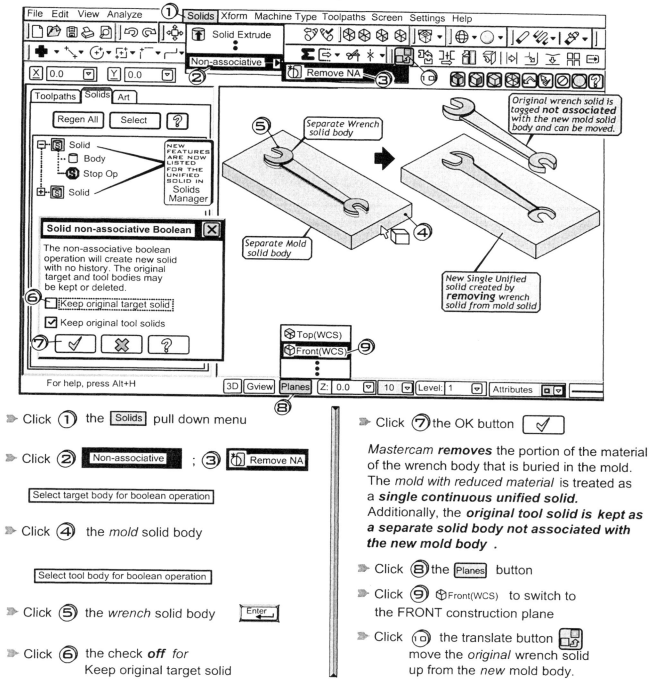

➤ Click ① the `Solids` pull down menu

➤ Click ② `Non-associative` ; ③ `Remove NA`

`Select target body for boolean operation`

➤ Click ④ the *mold* solid body

`Select tool body for boolean operation`

➤ Click ⑤ the *wrench* solid body `Enter`

➤ Click ⑥ the check **off** *for* Keep original target solid

➤ Click ⑦ the OK button `✓`

*Mastercam **removes** the portion of the material of the wrench body that is buried in the mold. The *mold with reduced material* is treated as a **single continuous unified solid.** Additionally, the **original tool solid is kept as a separate solid body not associated with the new mold body** .*

➤ Click ⑧ the `Planes` button

➤ Click ⑨ ⊕Front(WCS) to switch to the FRONT construction plane

➤ Click ⑩ the translate button ⊞ move the *original* wrench solid up from the *new* mold body.

11-14 Building a History Tree for Imported Solids

The operator will find that when a when a solid model created in *Solidworks(.SLDPRT)* is imported to *Mastercam X2*, **its entire history tree is also imported intact.**

A solid model created in other CAD packages such as *AutoCAD(.DWG) , Inventor(.IPT), Parasolid (.X_T), ACIS Kernel(.SAT)* or *IGES(.IGES)*,however, will only import to *Mastercam X2* as a **brick solid**. A **brick solid has only its base feature, "Body", listed in its history tree in the Solids Manager.**

The operator uses the **Find Features** function to *rebuild the history tree of an imported brick solid.*

Searches an imported **brick solid** for specific features such as **fillets and holes** and either **adds them to or removes them from the model's base feature**.

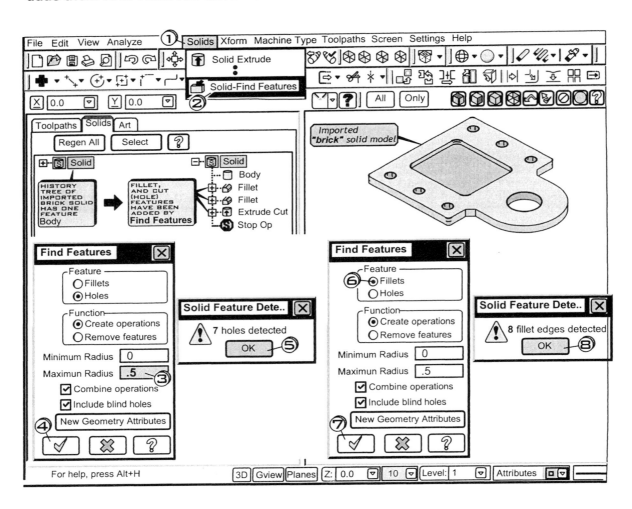

≫ Click ① the [Solids] pull down menu

≫ Click ② [🗂 Solid-Find Features]

≫ Click ③ in Maximum Radius; enter [.5]

≫ Click ④ ⑤ the OK buttons [✓] ; [OK]

Mastercam now searches the imported *brick* solid for all existing *hole features* in the range of 0R to .5R in . Any found are **added to the solid's history tree.**

≫ Click ① the [Solids] pull down menu

≫ Click ② [🗂 Solid-Find Features]

≫ Click ⑥ the ⦿ Fillets radio button **on**

≫ Click ⑦ ⑧ the OK buttons [✓] ; [OK]

Mastercam now searches the imported *brick* solid for all existing *fillet features* in the range of 0R to .5R in . Any found are **added to the solid's history tree.**

EXERCISES

In each exercise create the *solid model* part geometry using *MastercamX2's* Design package.

[**TUTORIAL PRACTICE EXERCISE**]

11-1) Open the file saved in **EXAMPLE 11-1** : [**EX11-1JV**] ←— YOUR INITIALS
        Add the additional features required to complete the part as shown in **Figure 11-p1**

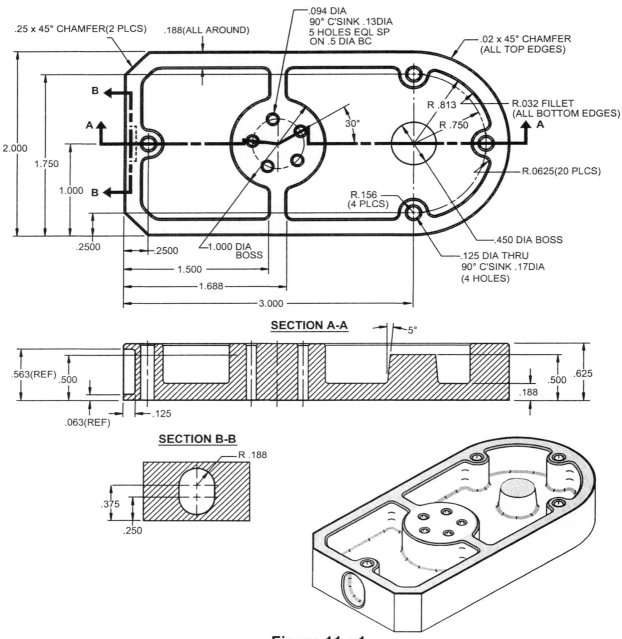

Figure 11-p1

a) Create the .188 in offset.

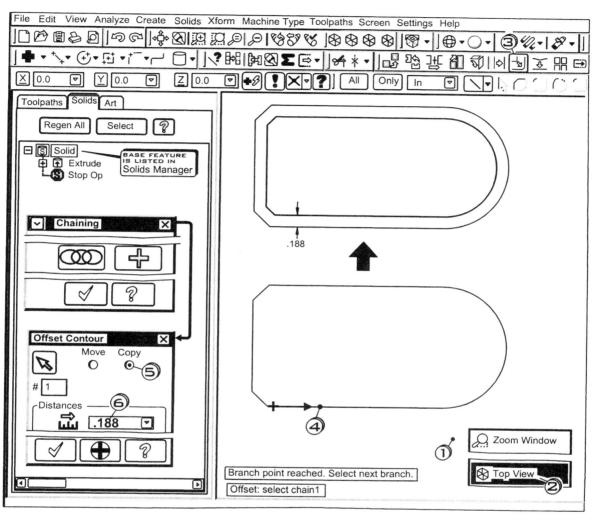

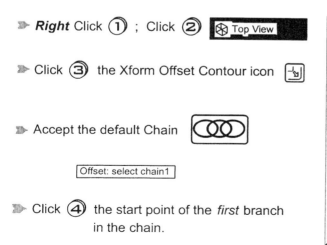

> ❯❯ **Right** Click ① ; Click ② [Top View]
>
> ❯❯ Click ③ the Xform Offset Contour icon
>
> ❯❯ Accept the default Chain
>
> Offset: select chain1
>
> ❯❯ Click ④ the start point of the *first* branch in the chain.

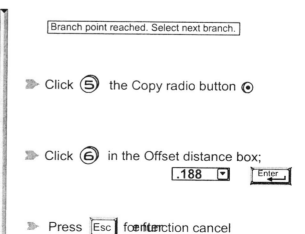

> Branch point reached. Select next branch.
>
> ❯❯ Click ⑤ the Copy radio button ◉
>
> ❯❯ Click ⑥ in the Offset distance box;
>
> .188 ▾   Enter
>
> ❯❯ Press [Esc] for menu/function cancel

b) Construct the 1.5 in and 1.688 in center web lines.

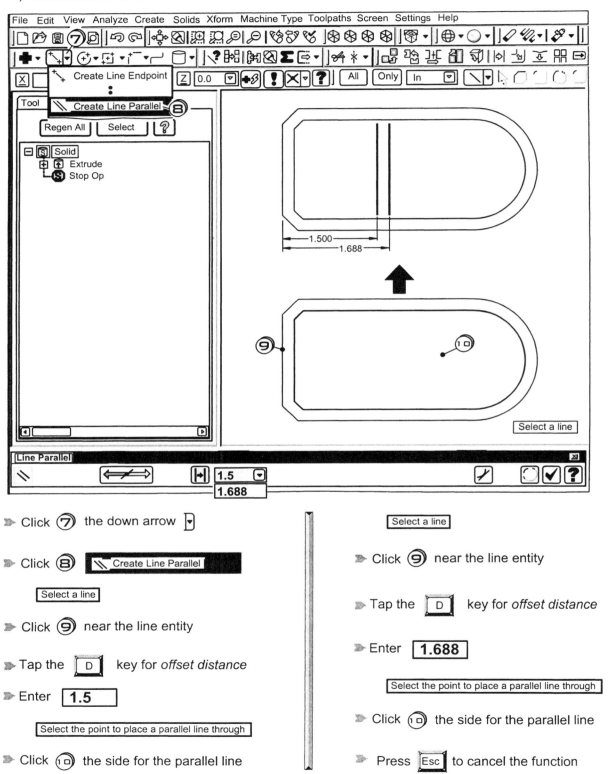

➤ Click ⑦ the down arrow [▾]

➤ Click ⑧ [╲ Create Line Parallel]

[Select a line]

➤ Click ⑨ near the line entity

➤ Tap the [D] key for *offset distance*

➤ Enter [1.5]

[Select the point to place a parallel line through]

➤ Click ⑩ the side for the parallel line

[Select a line]

➤ Click ⑨ near the line entity

➤ Tap the [D] key for *offset distance*

➤ Enter [1.688]

[Select the point to place a parallel line through]

➤ Click ⑩ the side for the parallel line

➤ Press [Esc] to cancel the function

c) Add the 1 in DIA boss circle.

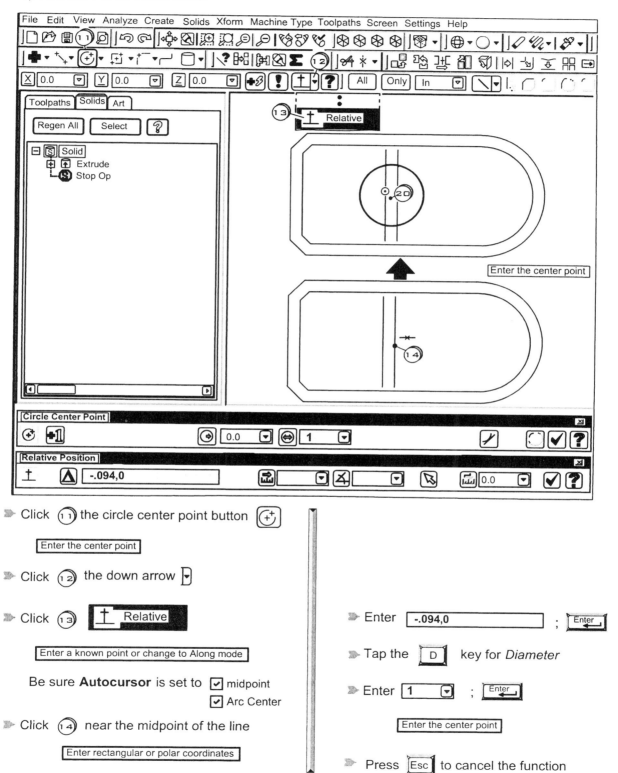

➤ Click ⑪ the circle center point button ⊕⁺

    Enter the center point

➤ Click ⑫ the down arrow ⊡

➤ Click ⑬ | ┼ Relative |

    Enter a known point or change to Along mode

    Be sure **Autocursor** is set to ☑ midpoint
    ☑ Arc Center

➤ Click ⑭ near the midpoint of the line

    Enter rectangular or polar coordinates

➤ Enter | -.094,0 | ; Enter

➤ Tap the D key for *Diameter*

➤ Enter 1 ⊡ ; Enter

    Enter the center point

➤ Press Esc to cancel the function

d) Construct the four .156R tab circles and the corresponding .125 DIA drill thru circles.

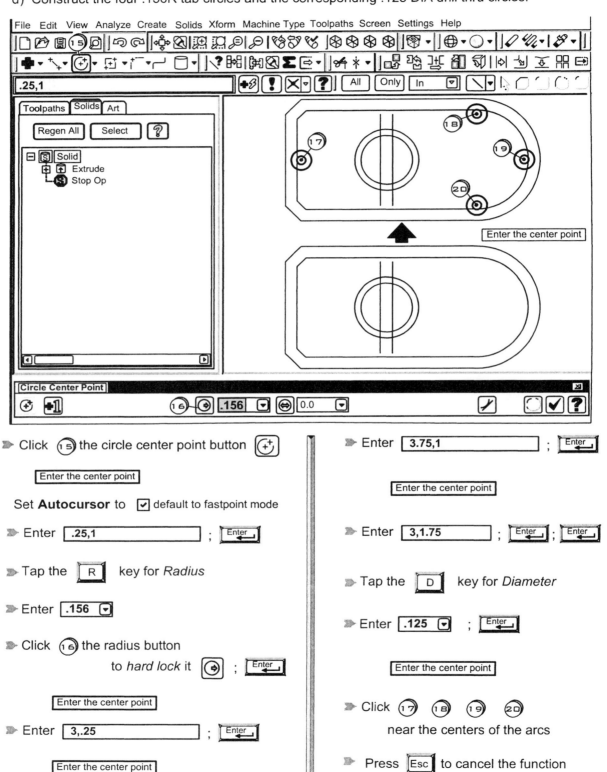

➤ Click ⑮ the circle center point button ⊕

    | Enter the center point |

Set **Autocursor** to ☑ default to fastpoint mode

➤ Enter | .25,1 | ; | Enter |

➤ Tap the | R | key for *Radius*

➤ Enter | .156 ▾ |

➤ Click ⑯ the radius button
               to *hard lock* it ⊙ ; | Enter |

    | Enter the center point |

➤ Enter | 3,.25 | ; | Enter |

    | Enter the center point |

➤ Enter | 3.75,1 | ; | Enter |

    | Enter the center point |

➤ Enter | 3,1.75 | ; | Enter | ; | Enter |

➤ Tap the | D | key for *Diameter*

➤ Enter | .125 ▾ | ; | Enter |

    | Enter the center point |

➤ Click ⑰ ⑱ ⑲ ⑳
    near the centers of the arcs

➤ Press | Esc | to cancel the function

e) Execute the required divide and trimming operations on the boss circle web lines and tab circles.

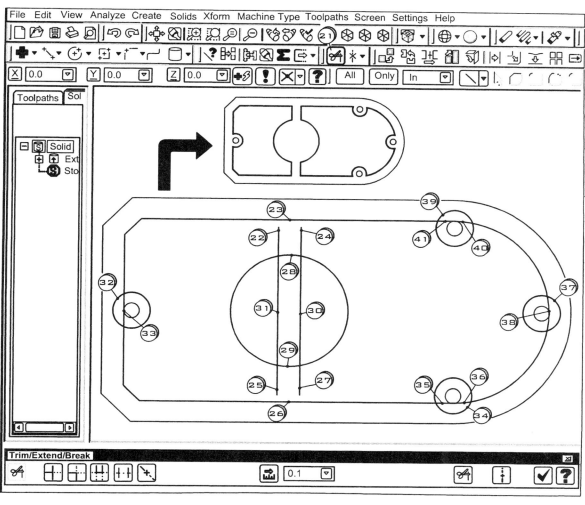

➤ Click ㉑ the Trim icon

> Select the entity to trim/extend

➤ Click ㉒ near the end of the vertical line

> Select the entity to trim/extend to

➤ Click ㉓ near the horizontal line

➤ Click ㉔ ; ㉓

➤ Click ㉕ ; ㉖

➤ Click ㉗ ; ㉖

➤ Tap the ⌑D key for *Divide entities*

> Select the curve to divide

➤ Click ㉘ ㉙ ㉚ ㉛ ㉜ ㉝

㉞ ㉟ ㊱ ㊲ ㊳ ㊴

㊵ ㊶

f) Create the R.063 in fillets.

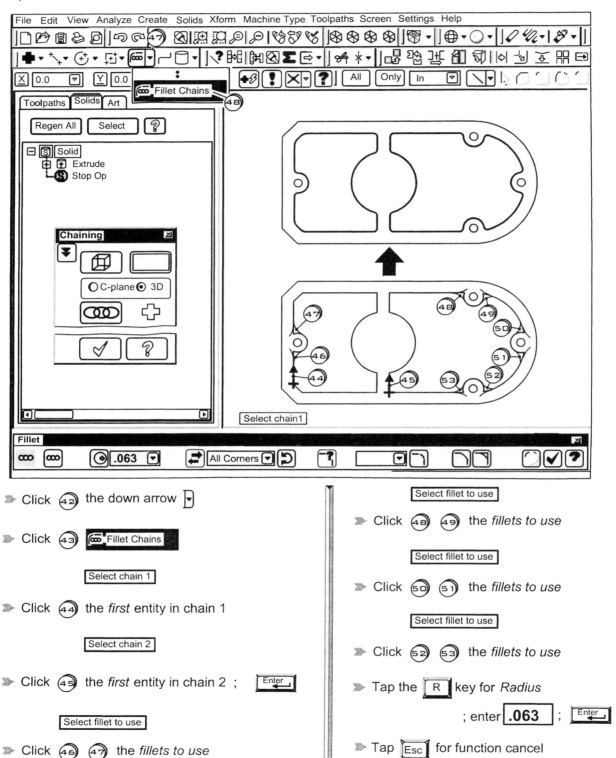

➤ Click ④② the down arrow ⊡

➤ Click ④③ 🔘 Fillet Chains

      Select chain 1

➤ Click ④④ the *first* entity in chain 1

      Select chain 2

➤ Click ④⑤ the *first* entity in chain 2 ; Enter ⏎

      Select fillet to use

➤ Click ④⑥ ④⑦ the *fillets to use*

      Select fillet to use

➤ Click ④⑧ ④⑨ the *fillets to use*

      Select fillet to use

➤ Click ⑤⓪ ⑤① the *fillets to use*

      Select fillet to use

➤ Click ⑤② ⑤③ the *fillets to use*

➤ Tap the [R] key for *Radius*

      ; enter [.063] ; Enter ⏎

➤ Tap [Esc] for function cancel

g) Create the two solid pockets x .437 in deep.

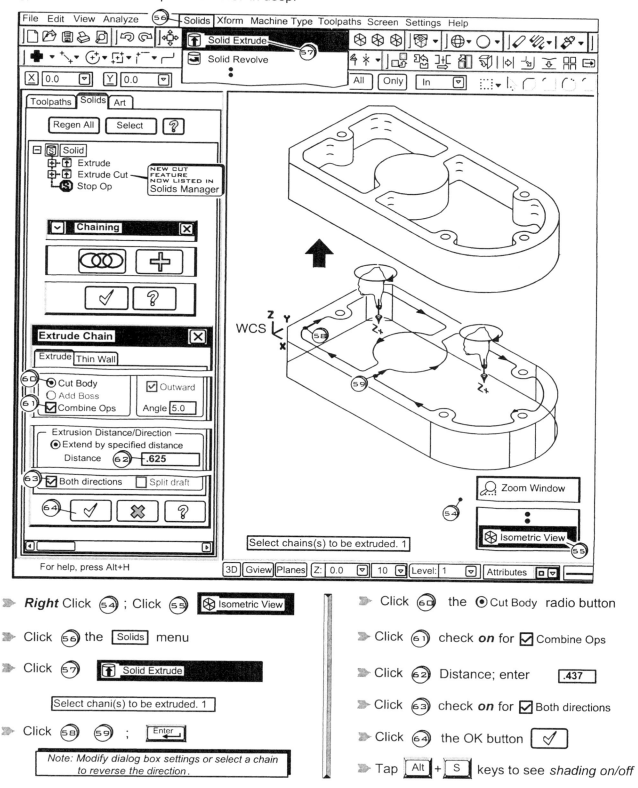

➤ **Right** Click ⑤④ ; Click ⑤⑤ 🔷 Isometric View

➤ Click ⑤⑥ the [Solids] menu

➤ Click ⑤⑦ 🔼 Solid Extrude

Select chani(s) to be extruded. 1

➤ Click ⑤⑧ ⑤⑨ ; [Enter ⏎]

Note: Modify dialog box settings or select a chain to reverse the direction.

➤ Click ⑥⓪ the ⦿ Cut Body radio button

➤ Click ⑥① check **on** for ☑ Combine Ops

➤ Click ⑥② Distance; enter [.437]

➤ Click ⑥③ check **on** for ☑ Both directions

➤ Click ⑥④ the OK button ✓

➤ Tap [Alt] + [S] keys to see *shading on/off*

h) Create the four solid .125 DIA thru holes.

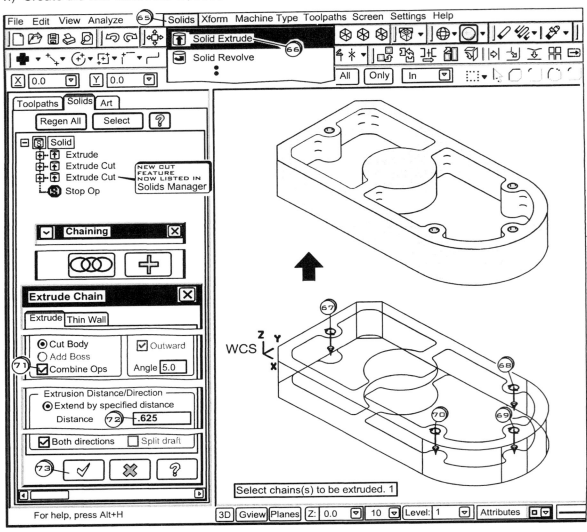

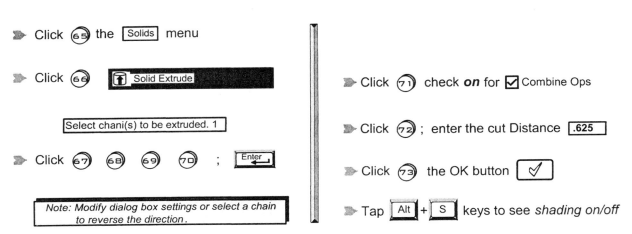

➤ Click ⑥⑤ the [Solids] menu

➤ Click ⑥⑥ [⬆ Solid Extrude]

Select chani(s) to be extruded. 1

➤ Click ⑥⑦ ⑥⑧ ⑥⑨ ⑦⓪ ; [Enter↵]

> Note: Modify dialog box settings or select a chain to reverse the direction.

➤ Click ⑦① check **on** for ☑ Combine Ops

➤ Click ⑦② ; enter the cut Distance [.625]

➤ Click ⑦③ the OK button [✓]

➤ Tap [Alt] + [S] keys to see *shading on/off*

i) Construct the .094 DIA drill thru circle.

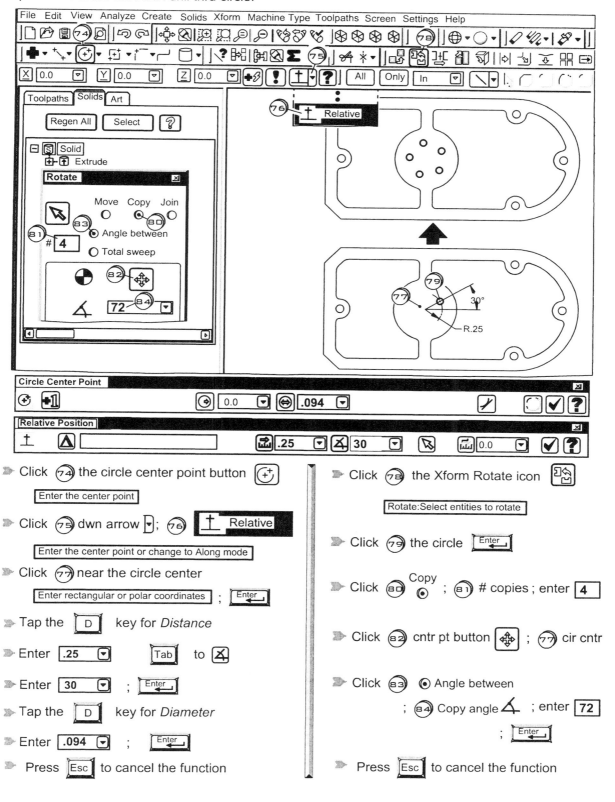

➤ Click ⑦④ the circle center point button ⊕
> Enter the center point

➤ Click ⑦⑤ dwn arrow ▼; ⑦⑥ †̲ Relative
> Enter the center point or change to Along mode

➤ Click ⑦⑦ near the circle center
> Enter rectangular or polar coordinates ; Enter◄

➤ Tap the [D] key for *Distance*

➤ Enter .25 ▼ [Tab] to ⊿

➤ Enter 30 ▼ ; Enter◄

➤ Tap the [D] key for *Diameter*

➤ Enter .094 ▼ ; Enter◄

➤ Press [Esc] to cancel the function

➤ Click ⑦⑧ the Xform Rotate icon ⊞
> Rotate:Select entities to rotate

➤ Click ⑦⑨ the circle Enter◄

➤ Click ⑧⓪ Copy ⊙ ; ⑧① # copies ; enter 4

➤ Click ⑧② cntr pt button ✛ ; ⑦⑦ cir cntr

➤ Click ⑧③ ⊙ Angle between
; ⑧④ Copy angle ⊿ ; enter 72
; Enter◄

➤ Press [Esc] to cancel the function

j) Complete the holes on the top face by extruding the five solid .094 DIA thru holes.

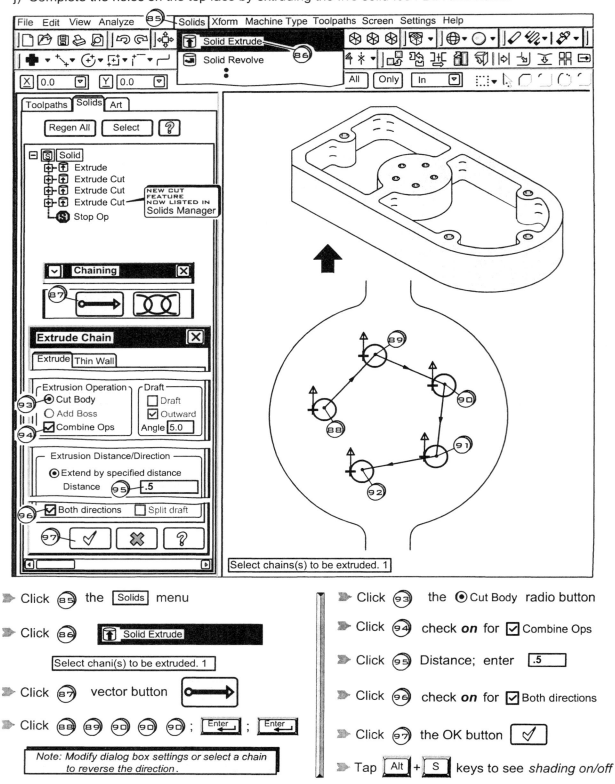

➤ Click ⑧⑤ the [Solids] menu

➤ Click ⑧⑥ [🔼 Solid Extrude]

Select chani(s) to be extruded. 1

➤ Click ⑧⑦ vector button [●—➡]

➤ Click ⑧⑧ ⑧⑨ ⑨⓪ ⑨⓪ ⑨⓪ ; [Enter◄┘] ; [Enter◄┘]

Note: Modify dialog box settings or select a chain to reverse the direction.

➤ Click ⑨③ the ⦿ Cut Body radio button

➤ Click ⑨④ check **on** for ☑ Combine Ops

➤ Click ⑨⑤ Distance; enter [.5]

➤ Click ⑨⑥ check **on** for ☑ Both directions

➤ Click ⑨⑦ the OK button [✓]

➤ Tap [Alt] + [S] keys to see *shading on/off*

k) Add the .450 DIA boss circle.

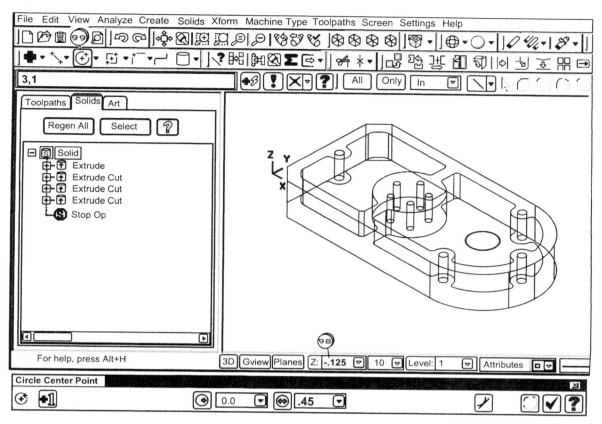

▸ Click ⊖⊜ in Z const depth; enter [**-.125**]

▸ Click ⊖⊜ the circle center point button [⊕⁺]

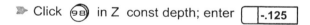

▸ Enter [**3,1**] ; [Enter ↵]

▸ Tap the [D] key for *Diameter*

▸ Enter [.45 ▾] ; [Enter ↵]

l) Extrude the solid boss x .312 in deep at a 5° taper.

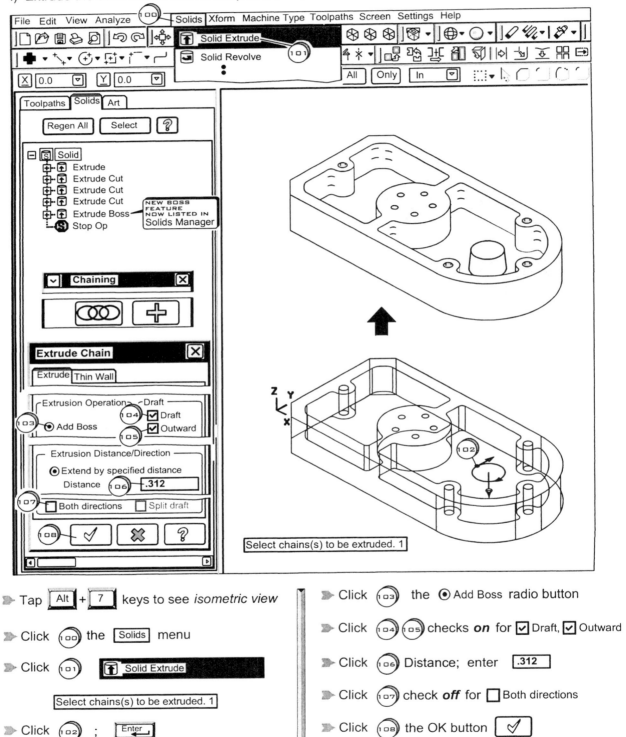

Select chains(s) to be extruded. 1

▶ Tap ⬚Alt⬚ + ⬚7⬚ keys to see *isometric view*

▶ Click ⓞ⓪ the ⬚Solids⬚ menu

▶ Click ⓞ① 🔼 Solid Extrude

　　Select chains(s) to be extruded. 1

▶ Click ⓞ② ; ⬚Enter⬚

　　Note: Modify dialog box settings or select a chain to reverse the direction.

▶ Click ⓞ③ the ⦿ Add Boss radio button

▶ Click ⓞ④ ⓞ⑤ checks *on* for ☑ Draft, ☑ Outward

▶ Click ⓞ⑥ Distance; enter ⬚.312⬚

▶ Click ⓞ⑦ check *off* for ☐ Both directions

▶ Click ⓞ⑧ the OK button ☑

▶ Tap ⬚Alt⬚ + ⬚S⬚ keys to see *shading on/off*

m) Add the R.032 in solid fillets and the .02 x 45° solid chamfers.

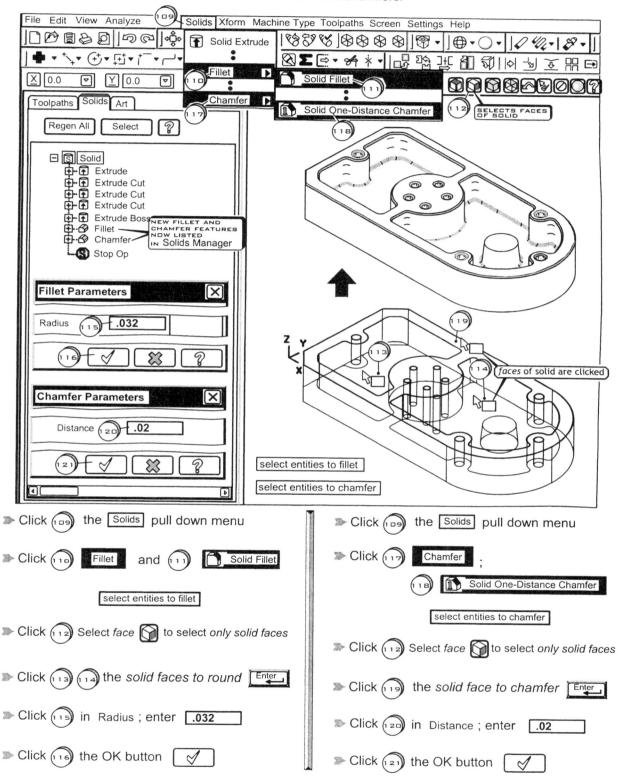

➤ Click (109) the Solids pull down menu

➤ Click (110) Fillet and (111) Solid Fillet

select entities to fillet

➤ Click (112) Select face to select *only solid faces*

➤ Click (113) (114) the *solid faces to round* Enter

➤ Click (115) in Radius ; enter .032

➤ Click (116) the OK button ✓

➤ Click (109) the Solids pull down menu

➤ Click (117) Chamfer ;
(118) Solid One-Distance Chamfer

select entities to chamfer

➤ Click (112) Select face to select *only solid faces*

➤ Click (119) the *solid face to chamfer* Enter

➤ Click (120) in Distance ; enter .02

➤ Click (121) the OK button ✓

n) Construct the .375in x .5in obround rectangle on the left face.

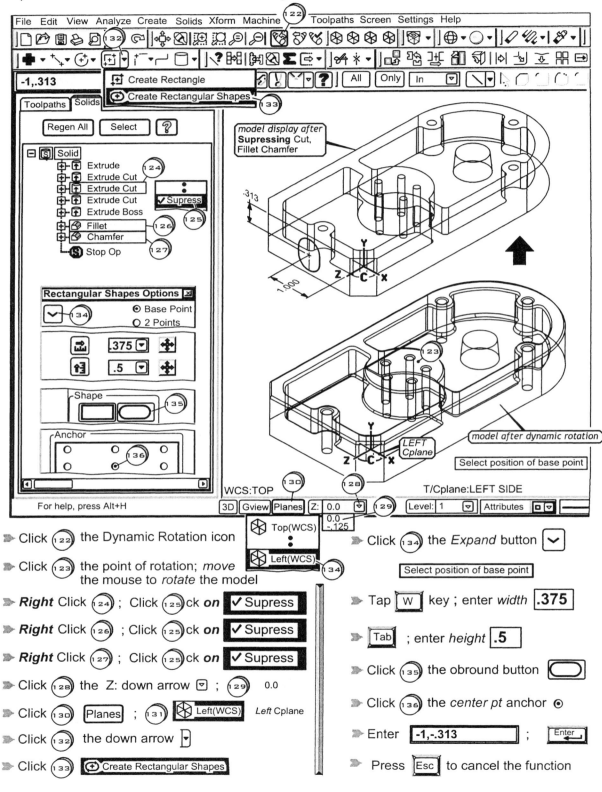

➤ Click ⒈⒉⒉ the Dynamic Rotation icon

➤ Click ⒈⒉⒊ the point of rotation; *move the mouse to rotate* the model

➤ *Right* Click ⒈⒉⒋ ; Click ⒈⒉⒌ck *on* ✓ **Supress**

➤ *Right* Click ⒈⒉⒍ ; Click ⒈⒉⒌ck *on* ✓ **Supress**

➤ *Right* Click ⒈⒉⒎ ; Click ⒈⒉⒌ck *on* ✓ **Supress**

➤ Click ⒈⒉⒏ the Z: down arrow ▽ ; ⒈⒉⒐ 0.0

➤ Click ⒈⒊⒪ **Planes** ; ⒈⒊⒈ Left(WCS) *Left* Cplane

➤ Click ⒈⒊⒉ the down arrow ▿

➤ Click ⒈⒊⒊ ⊕ Create Rectangular Shapes

➤ Click ⒈⒊⒋ the *Expand* button ▽

Select position of base point

➤ Tap W key ; enter *width* .375

➤ Tab ; enter *height* .5

➤ Click ⒈⒊⒌ the obround button ⬭

➤ Click ⒈⒊⒍ the *center pt* anchor ⊙

➤ Enter -1,.313 ; Enter↵

➤ Press Esc to cancel the function

o) Create the obround pocket in the solid model x .125 deep.

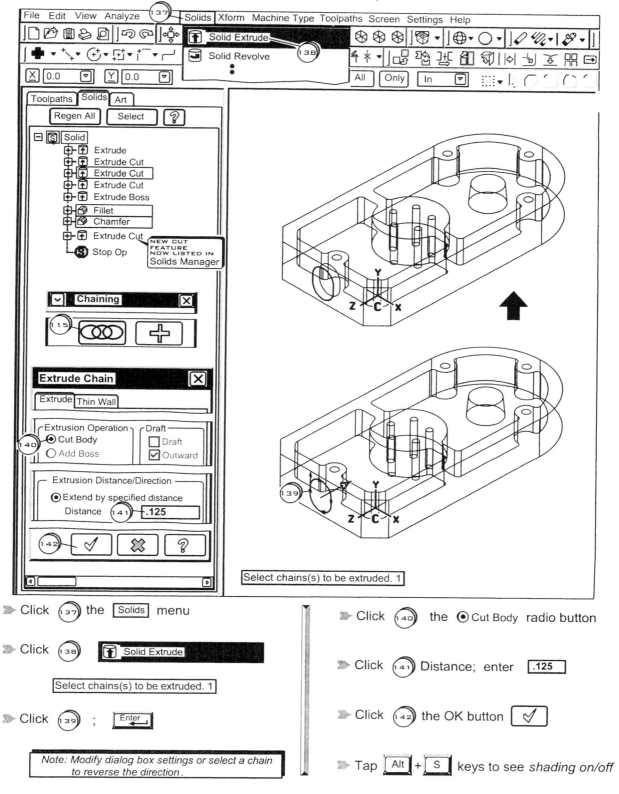

Select chains(s) to be extruded. 1

⋙ Click (137) the Solids menu

⋙ Click (138) Solid Extrude

Select chains(s) to be extruded. 1

⋙ Click (139) ; Enter

Note: Modify dialog box settings or select a chain to reverse the direction.

⋙ Click (140) the ⦿ Cut Body radio button

⋙ Click (141) Distance; enter .125

⋙ Click (142) the OK button ✓

⋙ Tap Alt + S keys to see shading on/off

p) Finish the model by creating the R.032 in solid fillets and the .02 x 45° solid chamfers.

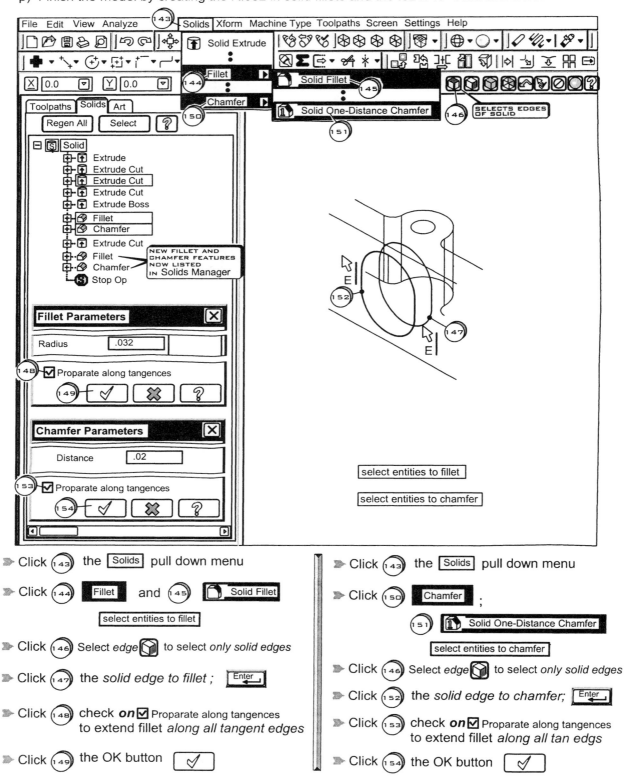

➤ Click ⒕₄₃ the [Solids] pull down menu

➤ Click ⒕₄₄ [Fillet] and ⒕₄₅ [Solid Fillet]

     [select entities to fillet]

➤ Click ⒕₄₆ Select *edge* 🔲 to select *only solid edges*

➤ Click ⒕₄₇ the *solid edge to fillet ;* [Enter]

➤ Click ⒕₄₈ check *on* ☑ Proparate along tangences
     to extend fillet *along all tangent edges*

➤ Click ⒕₄₉ the OK button [✓]

➤ Click ⒕₄₃ the [Solids] pull down menu

➤ Click ⒕₅₀ [Chamfer] ;

     ⒕₅₁ [🔳 Solid One-Distance Chamfer]

     [select entities to chamfer]

➤ Click ⒕₄₆ Select *edge* 🔲 to select *only solid edges*

➤ Click ⒕₅₂ the *solid edge to chamfer;* [Enter]

➤ Click ⒕₅₃ check *on* ☑ Proparate along tangences
     to extend fillet *along all tan edgs*

➤ Click ⒕₅₄ the OK button [✓]

q) Reset the default Tplane and Cplane back to Top and display the solid model with suppression *off all features*.

Note: All features **must be unsuppressed***(check **off**) before the part file can be saved.*

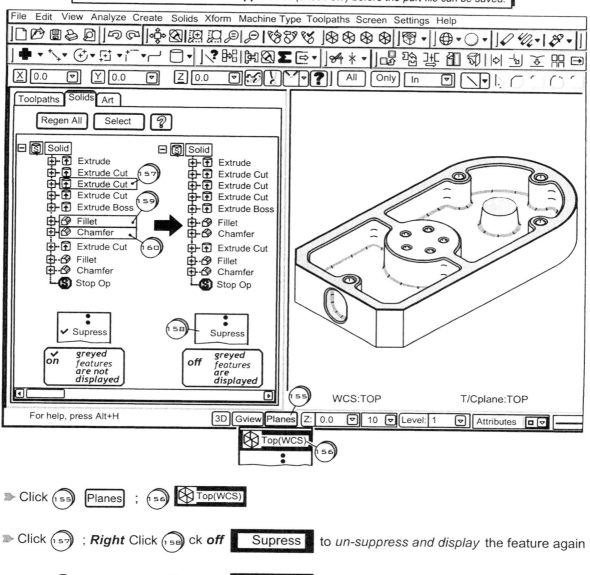

➤ Click ⓘ₅₅ [Planes] ; ⓘ₅₆ ⊗ Top(WCS)

➤ Click ⓘ₅₇ ; ***Right*** Click ⓘ₅₈ ck *off* ▮ Supress ▮ to *un-suppress and display* the feature again

➤ Click ⓘ₅₉ ; ***Right*** Click ⓘ₅₈ ck *off* ▮ Supress ▮ to *un-suppress and display* the feature again

➤ Click ⓘ₆₀ ; ***Right*** Click ⓘ₅₈ ck *off* ▮ Supress ▮ to *un-suppress and display* the feature again

➤ Tap [Alt] + [S] keys to see *shading on/off*

➤ Save the part

11-2)   File Name: **EX11-2JV** ← your initials

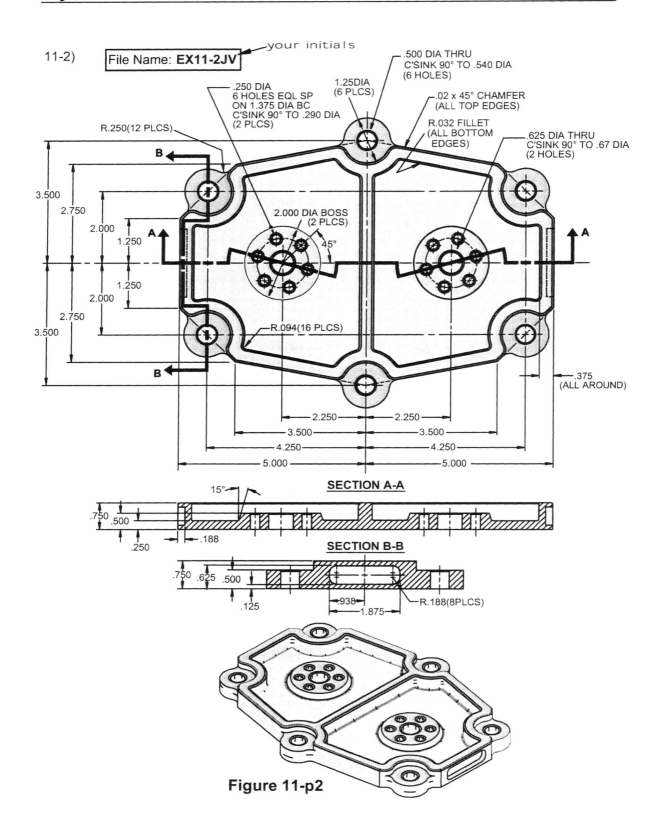

SECTION A-A

SECTION B-B

Figure 11-p2

11-2)

File Name: **EX11-2JV** ← YOUR INITIALS

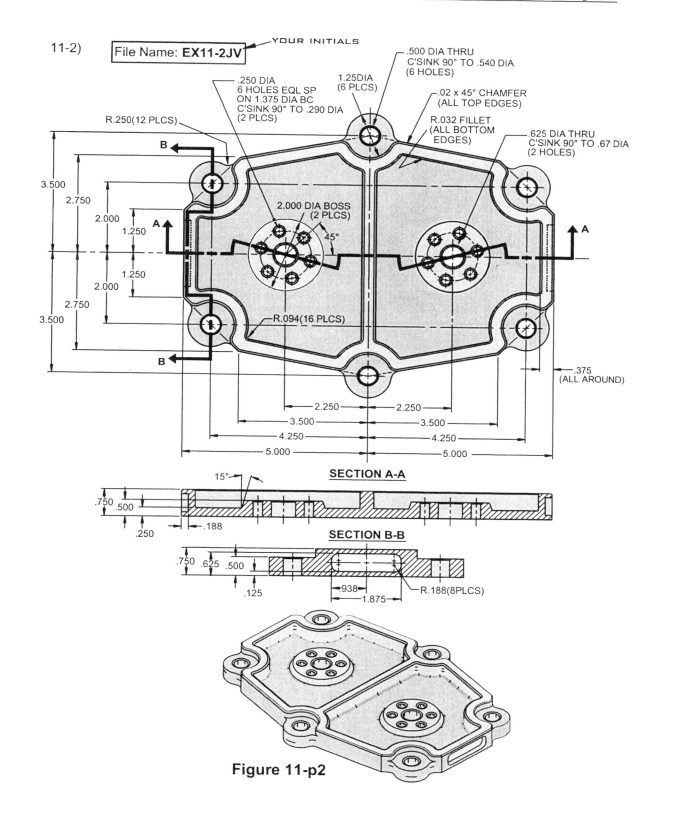

SECTION A-A

SECTION B-B

Figure 11-p2

a) Use the *Line, Parallel, Line Polar and Circle* functions to begin creating a *quarter* template of the part base

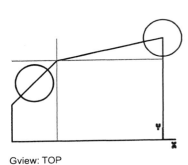

Gview: TOP

b) Use the *Xform Mirror* function to complete the preliminary shape of the base outline.

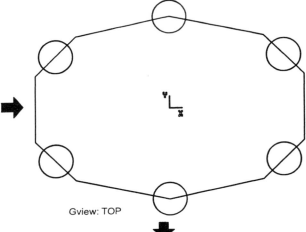

Gview: TOP

c) Trim the lines and circles.
Create the R.25 fillets.
Use the *Join entities* function to join lines A to B and C to D.

— R.250(12 PLCS)

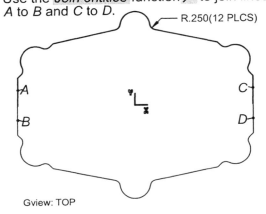

A   C

B   D

Gview: TOP

d) *Extrude Body* a Distance of .5 in

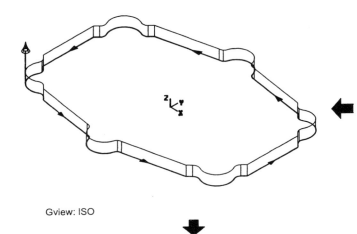

Gview: ISO

e) Create and set level 2-BASE BOSS.
Use the *Line and Circle* functions to *retrace* over the existing base outline.

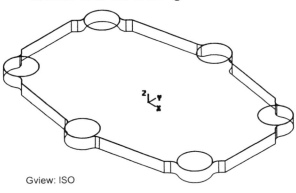

Gview: ISO

f) Turn *off* the visibility of Level 1 .
Complete the shape of the base boss outline by *trimming* and *filleting*.

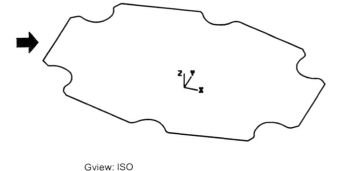

Gview: ISO

g) Turn *on* the visibility and set the Level to 1 .
Extrude Boss a Distance of .25 in.

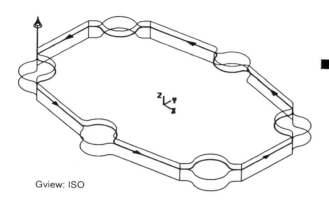

Gview: ISO

h) Use *Cut Body* to create the .5DIA
thru holes.

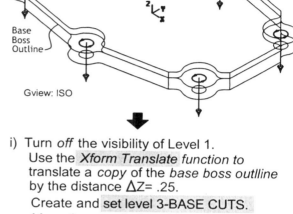

Base
Boss
Outline

Gview: ISO

j) Turn *on* the visibility of Level 1.
Use *Cut Body* to create the .5 in deep
pockets in the base.

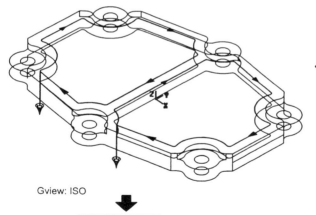

Gview: ISO

i) Turn *off* the visibility of Level 1.
Use the *Xform Translate* function to
translate a *copy* of the *base boss outlline*
by the distance ΔZ= .25.
Create and **set level 3-BASE CUTS.**
Move the copy to Level 3 and turn
off the visibility of Level 2.
Use the *Offset Contour*, *Line Parallel*
and *Fillet* functions to create the
base cut outlines

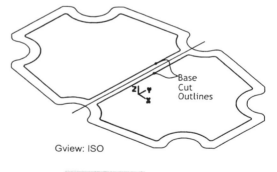

Base
Cut
Outlines

Gview: ISO

k) Set the *Z level* to .25.
Create the 2 DIA boss circles. *Extrude Boss*
a Distance of .25 in. at Draft angle 15°

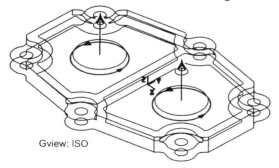

Gview: ISO

l) Set the *Z level* to .5.
Create the .250 DIA and .625 DIA circles.
Use *Cut Body* to create the drill thru holes.

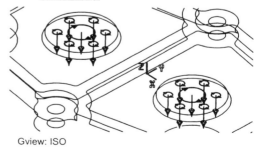

Gview: ISO

m) Click ⌈Planes⌉ ; Click ⊗ Right(WCS)
Create a polar line .375 long at 90°.
Create the 1.875 x .5 *rectangular shape
centered at the line*.

Click ⌈Planes⌉ ; Click ⊗ Top(WCS)
Use Xform Mirror to generate a *copy*
of the shape on the *left* side.

n) Use *Cut Body* to create the *left* and
right pockets.

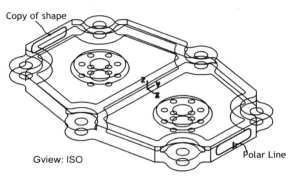

Copy of shape

Gview: ISO　　　　　　　　　Polar Line

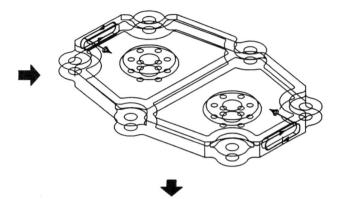

o) Complete the solid by creating the
.02 x 45° chamfers and the R.032 in
fillets on the edges of the solid
as specified in Figure 11p-2.

11-3)

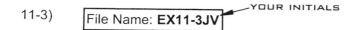

File Name: **EX11-3JV**
YOUR INITIALS

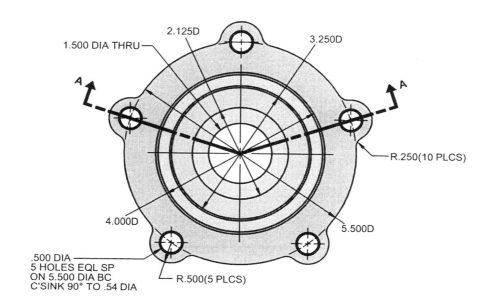

1.500 DIA THRU
2.125D
3.250D
A
A
R.250(10 PLCS)
4.000D
5.500D
.500 DIA
5 HOLES EQL SP
ON 5.500 DIA BC
C'SINK 90° TO .54 DIA
R.500(5 PLCS)

SECTION A-A

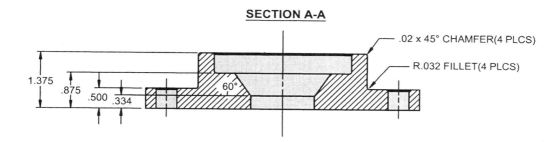

.02 x 45° CHAMFER(4 PLCS)
R.032 FILLET(4 PLCS)
1.375
.875
.500
.334
60°

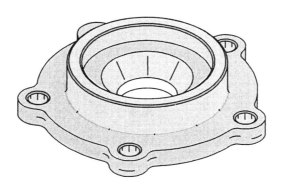

Figure 11-p3

a) Use the *Arc, Polar* function to create the R2.75 with an initial angle of 54° and final angle 126°.

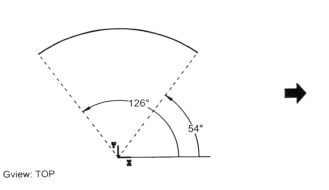

Gview: TOP

b) Add the .5DIA circle

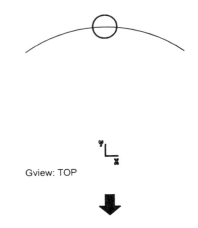

Gview: TOP

d) Use the *Xform ,Rotate* function to make *four copies* and complete the base outline.

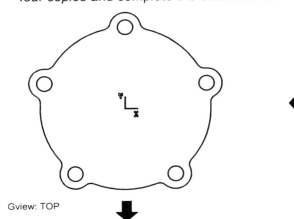

Gview: TOP

c) Create the R.5 arc and the R.25 fillets.

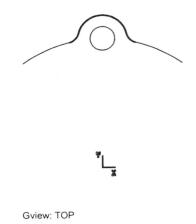

Gview: TOP

e) Use the *Extrude Body and Cut Body* functions to create the *base* of the solid model.

f) Use the *Extrude Boss* function to create the 4DIA solid cylinder boss feature.

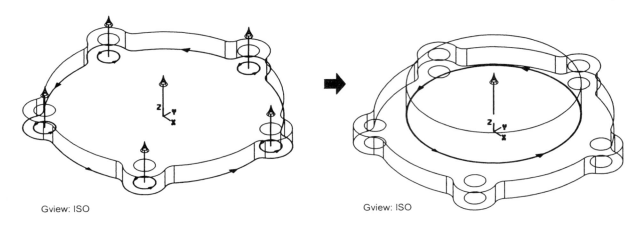

Gview: ISO   Gview: ISO

g) Click Planes ; Click ⬡ Front(WCS)

 Right Click; Click Autocursor

 Click: ⬡ Front Gview

 Click: ◐ No Hidden Wireframe

 Create the cutout profile as shown in bold.

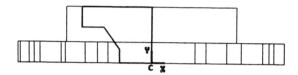

 Gview: FRONT
 Gplane: FRONT

h) Use the *Revolve Cut* function to create solid cutout by revolving and remove its geometry from the solid model.

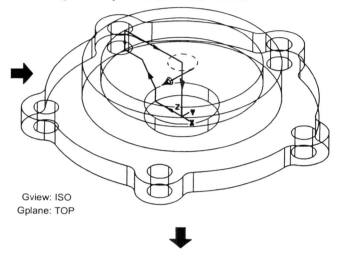

 Gview: ISO
 Gplane: TOP

i) Complete the solid by creating the .02 x 45° chamfers and the R.032 in fillets on the edges of the solid as specified in Figure 11p-3.

11-4)

File Name: **EX11-4JV**

YOUR INITIALS

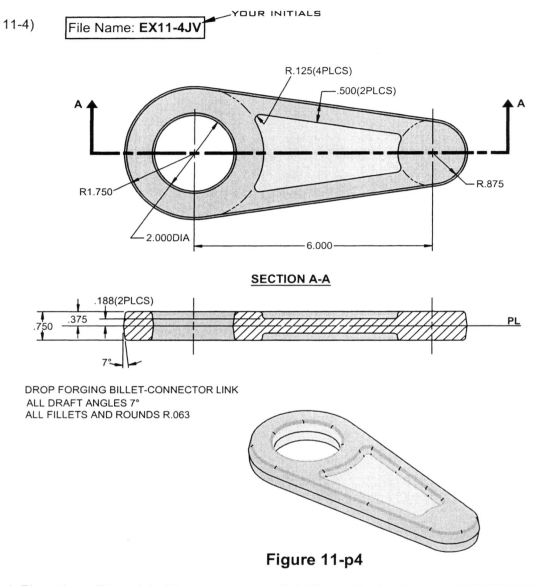

R.125(4PLCS)

.500(2PLCS)

A

A

R1.750

R.875

2.000DIA

6.000

SECTION A-A

.188(2PLCS)

.375

.750

7°

PL

DROP FORGING BILLET-CONNECTOR LINK
ALL DRAFT ANGLES 7°
ALL FILLETS AND ROUNDS R.063

Figure 11-p4

a) Place the solid model of the the connector link billet on the level named: FORGE BILLET

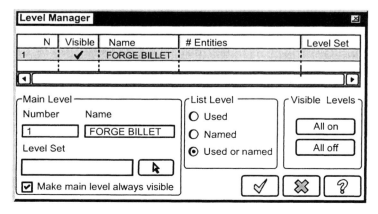

b) Create the boundary of the connecting link at the partling line **PL**

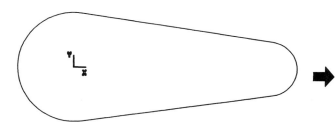

Gview: TOP

c) Use the *Extrude Body* option to extrude the *boundary* a distance of .375 in *both* directions *inward*. Enter a split draft angle of 7°.

Gview: TOP

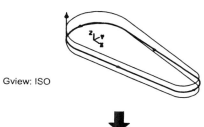

Gview: ISO

e) Change to the front view. Use the *Xform Mirror* function to create a *copy of* the pocket boundary about the parting line(PL).

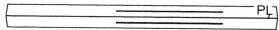

Gview: FRONT

d) Set the Z level to .188. Create the pocket *boundary*

Gview: TOP

f) Use *Cut Body* option to extrude the pocket boundaries a distance of .188 *outward*. Enter a *draft angle of 7°*.

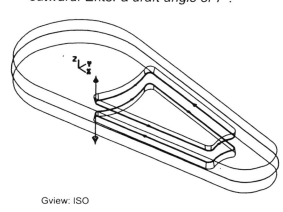

Gview: ISO

g) Set the Z level to 0. Create the 2DIA hole *boundary*

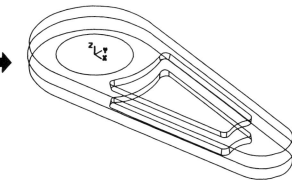

Gview: ISO

h) Use *Cut Body* option to extrude the pocket boundaries a distance of .375 *outward.* Enter a *draft angle of 7°.*

i) Create *R.063 fillets* at the *edges* of the *top and bottom faces and pockets.*

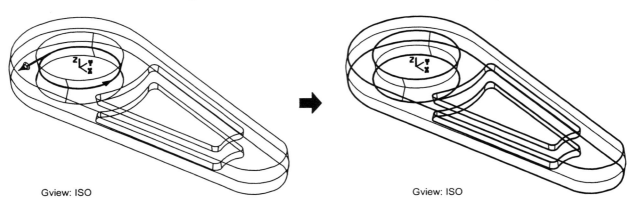

Gview: ISO   Gview: ISO

11-5)

File Name: **EX11-4JV**
YOUR INITIALS

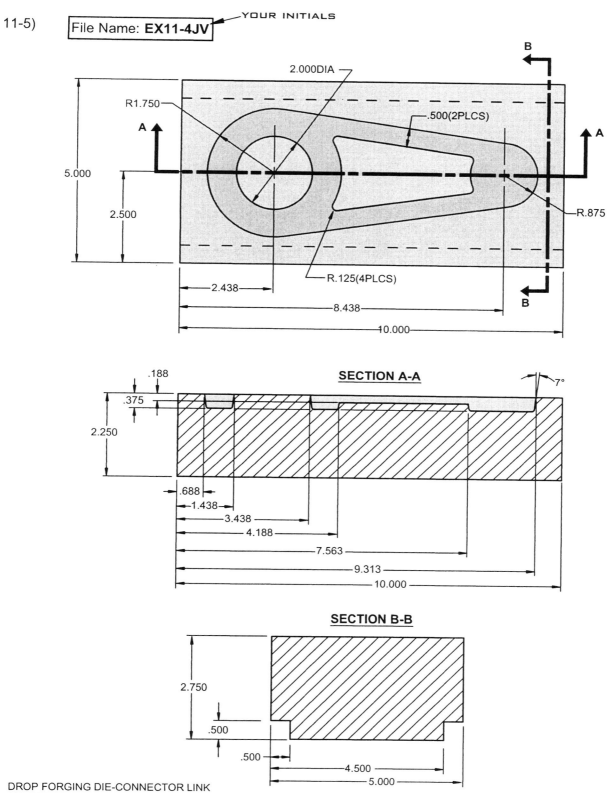

2.000DIA

R1.750

A

5.000

2.500

A

.500(2PLCS)

R.875

2.438

R.125(4PLCS)

8.438

10.000

B

B

SECTION A-A

.188

.375

2.250

7°

.688

1.438

3.438

4.188

7.563

9.313

10.000

SECTION B-B

2.750

.500

.500

4.500

5.000

DROP FORGING DIE-CONNECTOR LINK
ALL DRAFT ANGLES 7°
ALL FILLETS AND ROUNDS R.063

Figure 11-p5

a) Open the file **EX11-4-JV**. Place the solid model of the the connector link die on the level named: FORGE DIE

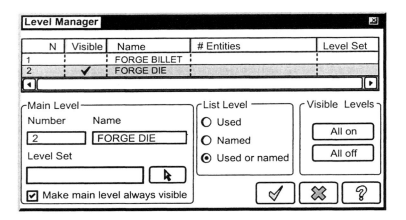

b) Select the RIGHT WCS. Create a 5 x 2.75 in, *rectangle* and two .5 x .5 in *rectangles.*

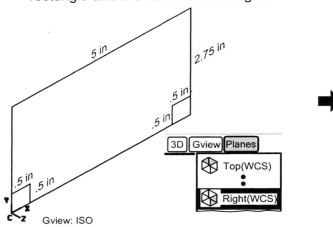

c) Use *Trim* and *Delete* functions to complete the shape of the die profile in the right view.

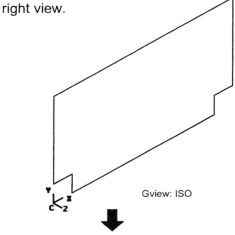

e) Select the TOP WCS.
Create a *polar line* 2.438 in long at an angle of 0°.

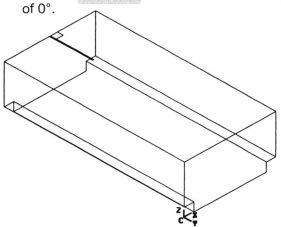

d) Use *Extrude Body* option to extrude the *die profile* a distance of 10in.

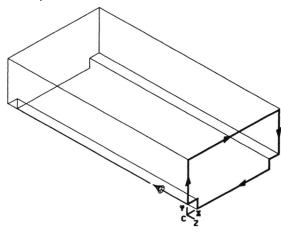

f) Turn *on* the *visibility of Level 1*. Use the **Xform, Translate** function to *move the billet into the die.*

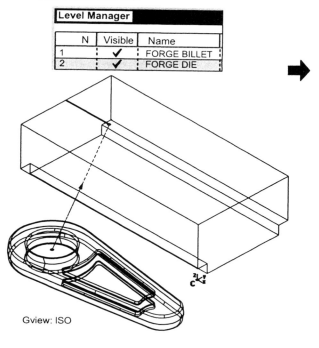

Gview: ISO

g) Click the [Solids] tab to enter the Solids Manager. Click the check **on** (✔) to *supress the fillet solid features.*

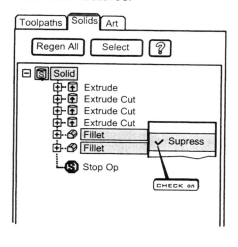

h) Select the *Boolean Remove* **non-associative function.**
Remove the billet from the die solid body.

off ☐ Keep original target solid

on ☑ Keep original tool solids

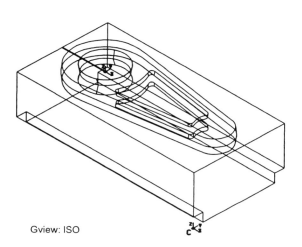

Gview: ISO

i) Turn *off the visibility of Level 1* to see the solid model of the billet die.

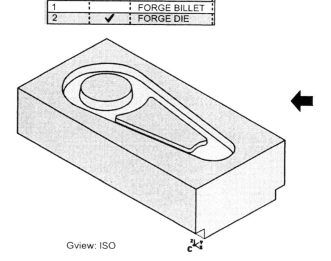

Gview: ISO

11-6)

File Name: **EX11-6**

Open the *Autocad* file **EX11-6.DWG** from the exercise CD provided with this text. Execute the *Find Features* function to rebuild the history tree of the imported brick solid.
Create any additional fillet and chamfer features in the solid as indicated in Figure 11p-6.

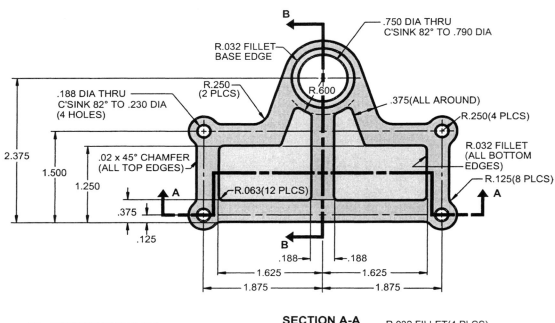

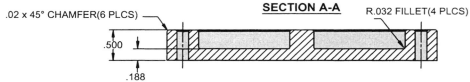

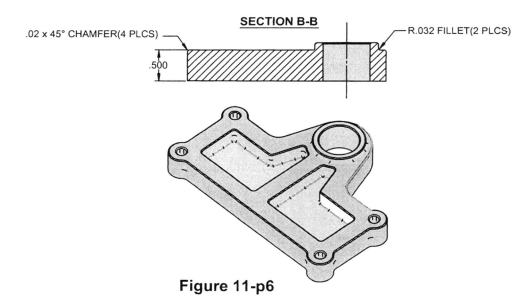

Figure 11-p6

CHAPTER - 12

EXECUTING 2D MILLING OPERATIONS ON SOLID MODELS

12-1 Chapter Objectives

After completing this chapter you will be able to:

1. Understand how to use the Solid Drill function to create drill toolpaths on a solid model.

2. Know how to create a 2D chained contour on a solid model using the Chaining dialog box: Solids mode.

3. Explain how to supress and unsupress solid features for machining and filing purposes.

The Mastercam X2 Mill package must be *active before any of the milling functions can be used.*

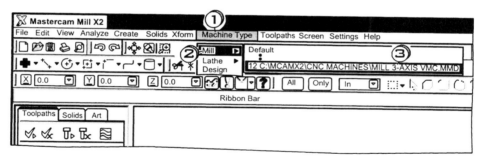

➣ Click ① the  drop down menu

➣ Click ② the **Mill** package

➣ Click ③ the  general vertical milling machine with a Fanuc type controller.

12-2 Creating Solid Drill Toolpaths

Mastercam X2 features the solid drill function to *automatically* execute the following operations:

◆ search and list all *hole sizes and Z-depths* in a solid model .

◆ create *tool paths* for drilling the holes.

◆ select *tools and drill depths* and *create the pilot drilling operation* for the holes.

◆ select *tools and drill depths* and *create the peck drilling operation* for the holes.

◆ select *tools and tap* depts and *create the tapping operation for* the holes.

The *cycle stored with each tool definition determines the drill cycle that is used in each operation.*

12-3 Activating the Solid Drilling Dialog Box

The **Solid Drilling** dialog box can be activated from either the *Toolpaths* pull down menu bar or the **2D Toolpaths** toolbar.

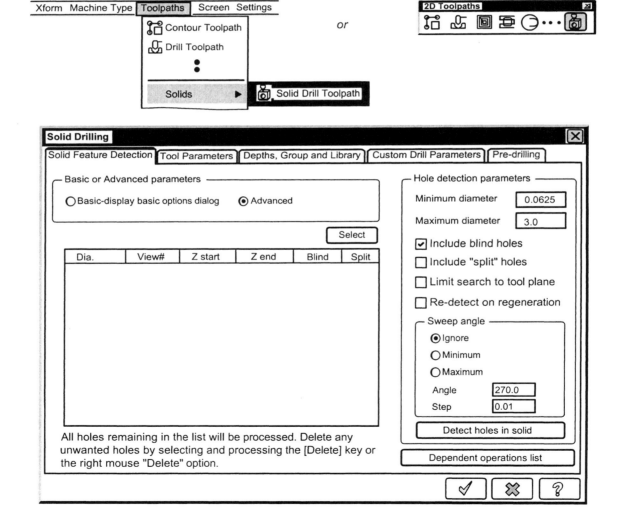

12-4 Specifying Solid Drilling Parameters

The **Solid Drilling** dialog contains parameters for filtering *automatic* hole searches in solids. For example, the operator can specify holes of a specific diameter and type(blind or thru). The operator can also specify if pilot holes are to be drilled and fine tune drill depths. The various parameters in the **Solid Drilling** dialog box will be considered in this section.

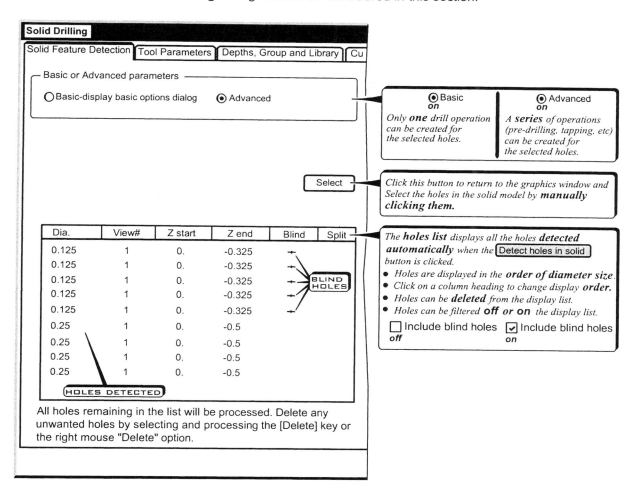

All holes remaining in the list will be processed. Delete any unwanted holes by selecting and processing the [Delete] key or the right mouse "Delete" option.

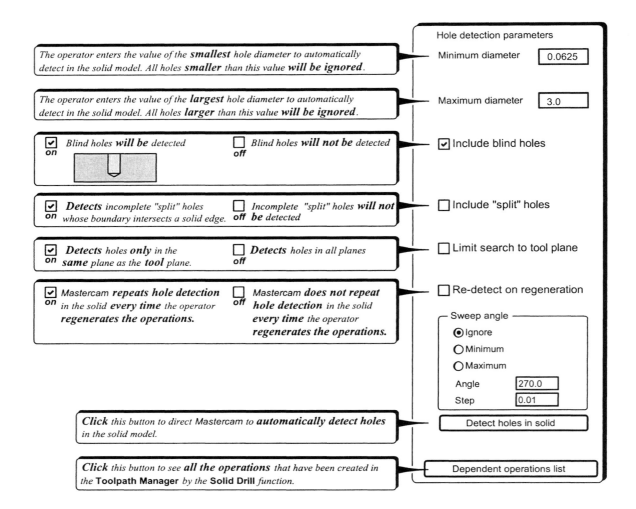

The operator enters the value of the **smallest** *hole diameter to automatically detect in the solid model. All holes* **smaller** *than this value* **will be ignored**.

The operator enters the value of the **largest** *hole diameter to automatically detect in the solid model. All holes* **larger** *than this value* **will be ignored**.

☑ *Blind holes* **will be** *detected*
on

☐ *Blind holes* **will not be** *detected*
off

☑ **Detects** *incomplete "split" holes*
on *whose boundary intersects a solid edge.*

☐ *Incomplete "split" holes* **will not**
off *be detected*

☑ **Detects** *holes* **only** *in the*
on *same plane as the* **tool** *plane.*

☐ **Detects** *holes in all planes*
off

☑ *Mastercam* **repeats hole detection**
on *in the solid* **every time** *the operator*
regenerates the operations.

☐ *Mastercam* **does not repeat**
off **hole detection** *in the solid*
every time *the operator*
regenerates the operations.

Click *this button to direct Mastercam to* **automatically detect holes** *in the solid model.*

Click *this button to see* **all the operations** *that have been created in the* **Toolpath Manager** *by the* **Solid Drill** *function.*

Hole detection parameters

Minimum diameter   0.0625

Maximum diameter   3.0

☑ Include blind holes

☐ Include "split" holes

☐ Limit search to tool plane

☐ Re-detect on regeneration

Sweep angle
◉ Ignore
○ Minimum
○ Maximum
Angle   270.0
Step   0.01

Detect holes in solid

Dependent operations list

EXAMPLE 12-1

Assume a solid model of the part shown in Figure 12-2 has been created in *Mastercam X2 Design.*

- Input the stock and material for the job

- Use the Solid Drill function to execute the operations listed in PROCESS PLAN 12-1

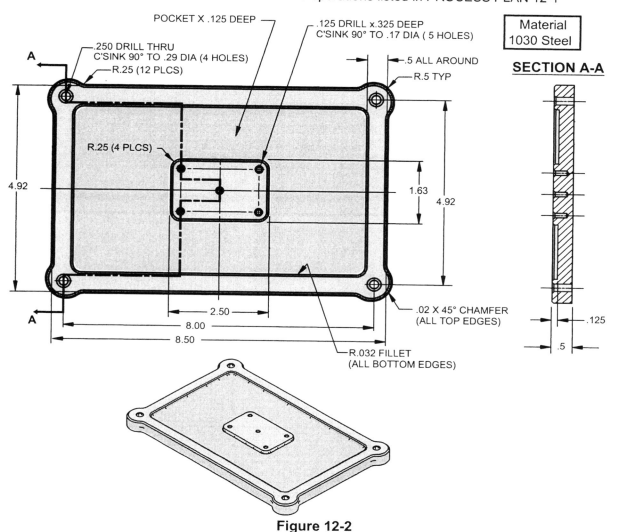

Figure 12-2

PROCESS PLAN 12-1

No.	Operation	Tooling
1	SPOT DRILL X .166 DEEP (4 HOLES)	1/8 SPOT DRILLL
2	PECK DRILL THRU(4 HOLES)	1/4 DRILL
3	C'SINK 90° TO .29 DIA (4 HOLES)	.5 CHAMFER MILL
4	SPOT DRILL X .125 DEEP (5 HOLES)	1/32 SPOT DRILL
5	PECK DRILL X .325 DEEP	1/8 DRILL
6	C'SINK 90° TO .17 DIA (5 HOLES)	.5 CHAMFER MILL

A) ENTER *Mastercam's* MACHINE GROUP PROPERTIES DIALOG BOX

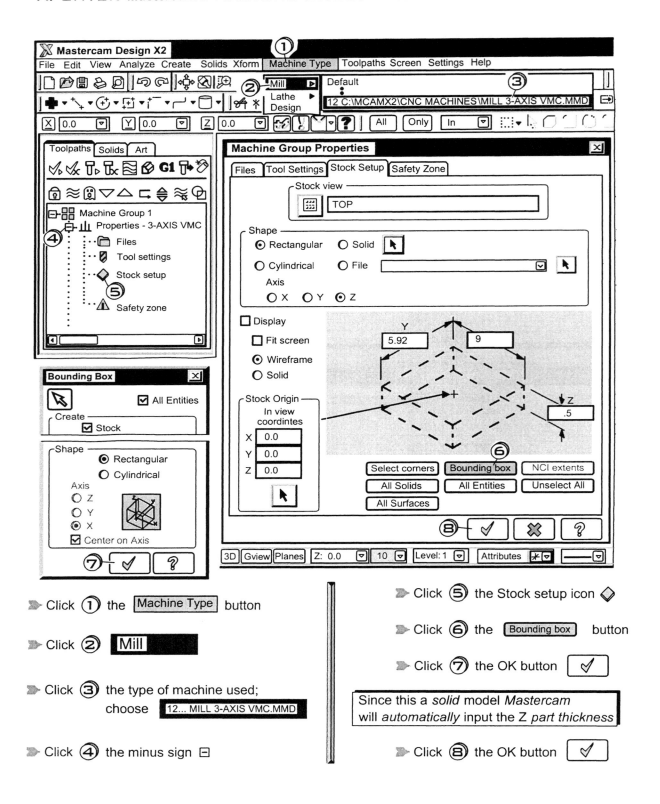

➤ Click ① the Machine Type button

➤ Click ② Mill

➤ Click ③ the type of machine used; choose 12... MILL 3-AXIS VMC.MMD

➤ Click ④ the minus sign ⊟

➤ Click ⑤ the Stock setup icon ◇

➤ Click ⑥ the Bounding box button

➤ Click ⑦ the OK button ✓

Since this a *solid* model *Mastercam* will *automatically* input the Z *part thickness*

➤ Click ⑧ the OK button ✓

B) ENTER *Mastercam's* MATERIAL MANAGER DIALOG BOX

◆ SPECIFY THE STOCK MATERIAL

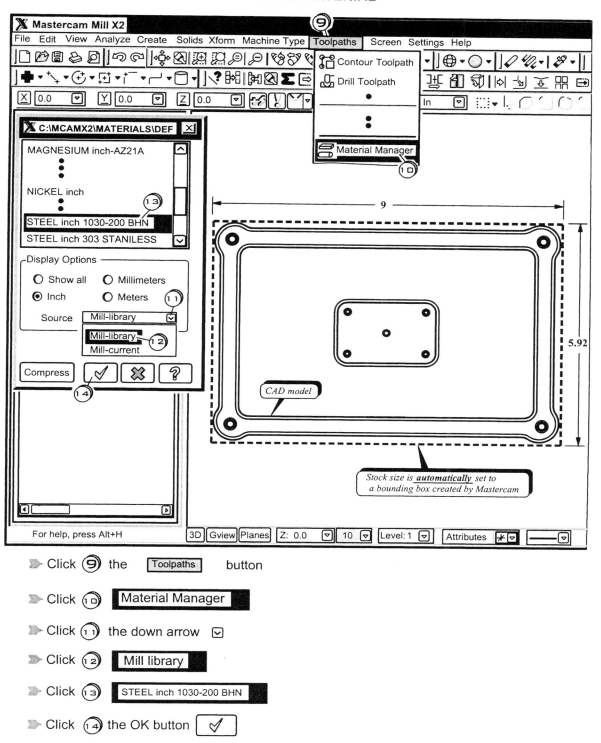

Stock size is **automatically** set to a bounding box created by Mastercam

➤ Click ⑨ the [Toolpaths] button

➤ Click ⑩ [Material Manager]

➤ Click ⑪ the down arrow ▽

➤ Click ⑫ [Mill library]

➤ Click ⑬ [STEEL inch 1030-200 BHN]

➤ Click ⑭ the OK button [✓]

C) DRILL THE .125DIA PILOT HOLES X .166 DEEP AND THE .25DIA HOLES THRU.

♦ DIRECT *Mastercam* TO AUTOMATICALLY DETECT THE HOLES TO BE DRILLED.

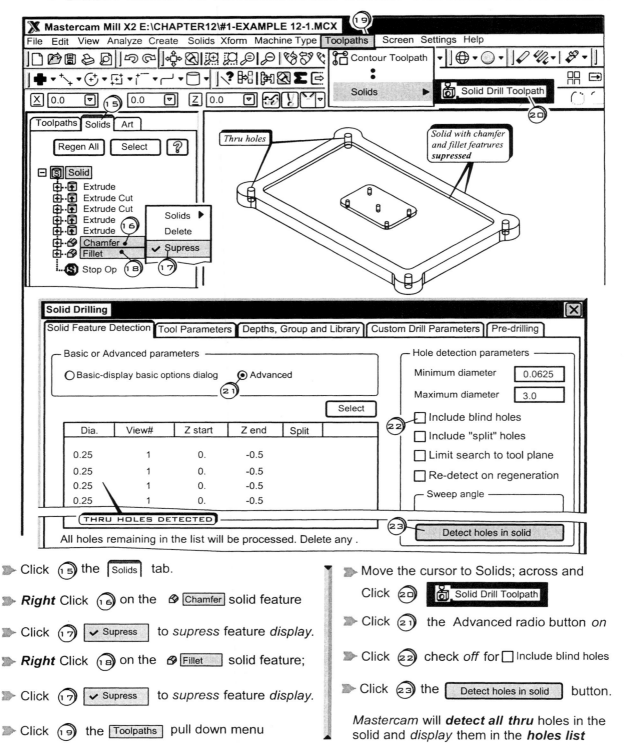

Click (15) the `Solids` tab.

Right Click (16) on the ✏ `Chamfer` solid feature

Click (17) `✓ Supress` to *supress* feature *display*.

Right Click (18) on the ✏ `Fillet` solid feature;

Click (17) `✓ Supress` to *supress* feature *display*.

Click (19) the `Toolpaths` pull down menu

Move the cursor to Solids; across and

Click (20) `Solid Drill Toolpath`

Click (21) the Advanced radio button *on*

Click (22) check *off* for ☐ Include blind holes

Click (23) the `Detect holes in solid` button.

Mastercam will **detect all thru** holes in the solid and *display* them in the **holes list**

♦ SPECIFY THE TOOL PARAMETERS FOR THE DRILLING OPERATION.

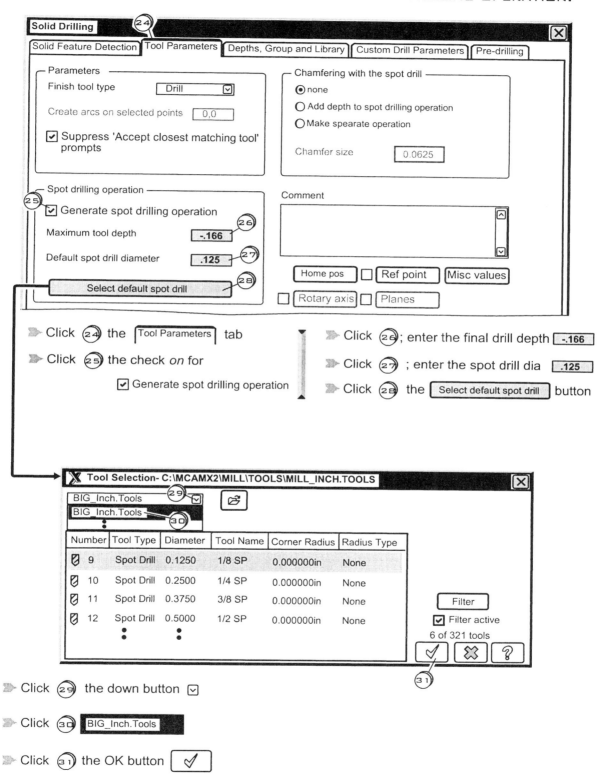

▶ Click ㉔ the [Tool Parameters] tab

▶ Click ㉕ the check *on* for

    ☑ Generate spot drilling operation

▶ Click ㉖; enter the final drill depth [-.166]

▶ Click ㉗ ; enter the spot drill dia [.125]

▶ Click ㉘ the [Select default spot drill] button

▶ Click ㉙ the down button ☑

▶ Click ㉚ BIG_Inch.Tools

▶ Click ㉛ the OK button [✓]

◆ SPECIFY PARAMETERS FOR PECK DRILLING THE .25DIA THRU HOLE.

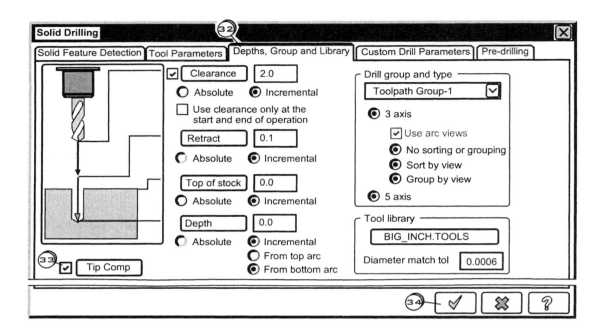

▶ Click ③② the │ Depths, Group and Library │ tab

▶ Click ③③ check *on* ☑ │ Tip Comp │ for a *thru* hole

▶ Click ③④ the OK button │ ✓ │

Mastercam will drill *all* the .125DIA
pilot holes and the .25DIA thru holes.

D) CREATE THE C'SINKS 90° TO .29DIA
◆ CREATE THE C'SINK TOOLPATH

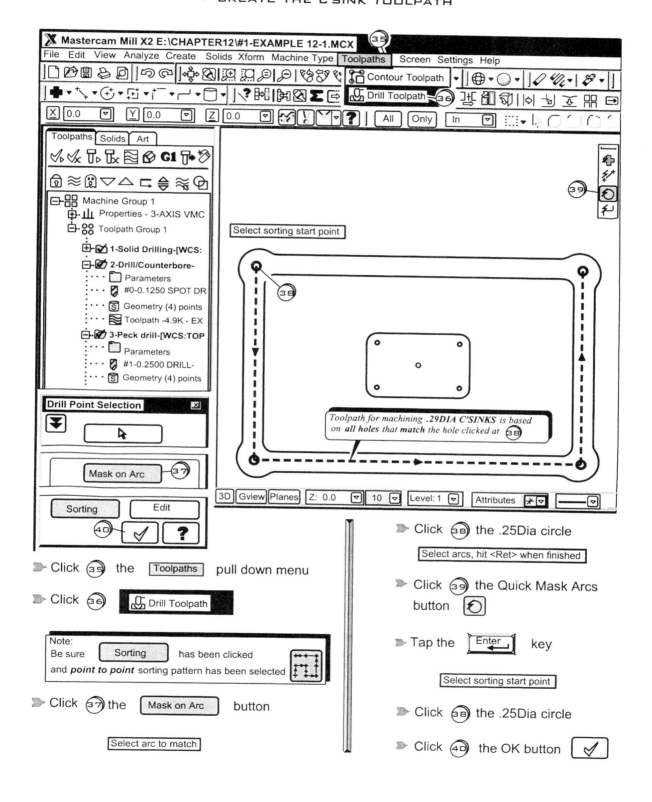

► Click ③⑤ the [Toolpaths] pull down menu

► Click ③⑥ [⚙ Drill Toolpath]

Note:
Be sure [Sorting] has been clicked
and *point to point* sorting pattern has been selected

► Click ③⑦ the [Mask on Arc] button

[Select arc to match]

► Click ③⑧ the .25Dia circle

[Select arcs, hit <Ret> when finished]

► Click ③⑨ the Quick Mask Arcs button [⟳]

► Tap the [Enter ⏎] key

[Select sorting start point]

► Click ③⑧ the .25Dia circle

► Click ④⓪ the OK button [✓]

OBTAIN THE NEEDED .5 DIA CHAMFER MILL TOOL

Mastercam will activate and display the Tool Parameters Tab. The operator can use this tab to **select a tool, set speeds and feeds** and other general toolpath parameters.

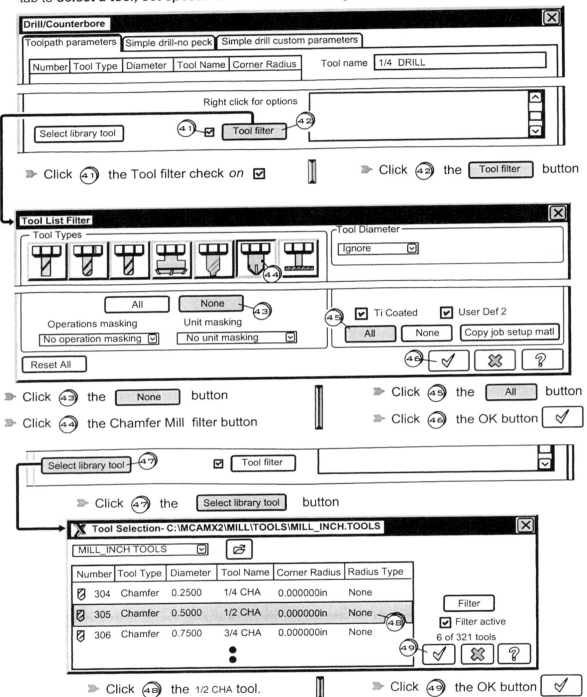

The tool will be brought from the library into your part file.

AUTO ASSIGN SPEEDS FEEDS FOR THE TOOL

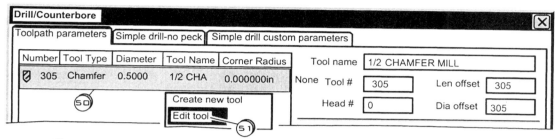

➤ Click ⑤⓪ on the .5 Dia Chamfer Mill tool

➤ **Right** Click move the cursor down and Click ⑤① [Edit tool]

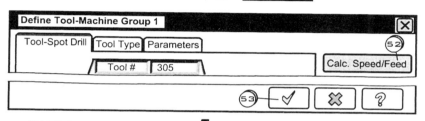

➤ Click ⑤② the [Calc. Speed/Feed] button.   ➤ Click ⑤③ the OK button ✓

◆ **ENTER THE REQUIRED C'SINK 90° TO .29 DIA MACHINING PARAMETERS**

➤ Click ⑤④ [Simple drill-no peck] tab

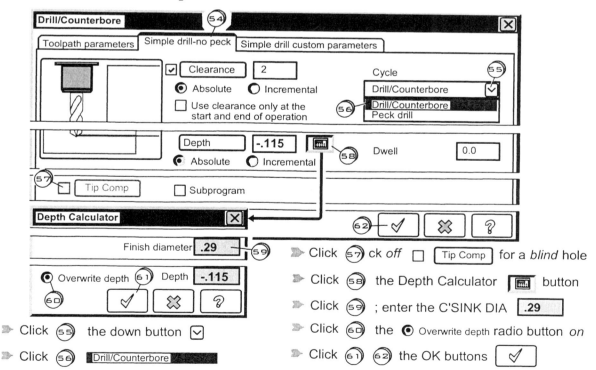

➤ Click ⑤⑦ ck *off* ☐ [Tip Comp] for a *blind* hole

➤ Click ⑤⑧ the Depth Calculator [⌨] button

➤ Click ⑤⑨ ; enter the C'SINK DIA [.29]

➤ Click ⑥⓪ the ◉ Overwrite depth radio button *on*

➤ Click ⑥① ⑥② the OK buttons ✓

➤ Click ⑤⑤ the down button ☑

➤ Click ⑤⑥ [Drill/Counterbore]

E) DRILL THE .032DIA PILOT HOLES X .125 DEEP AND THE
.125DIA HOLES X .325 DEEP.

    ◆ DIRECT *Mastercam* TO AUTOMATICALLY DETECT THE HOLES TO BE DRILLED.

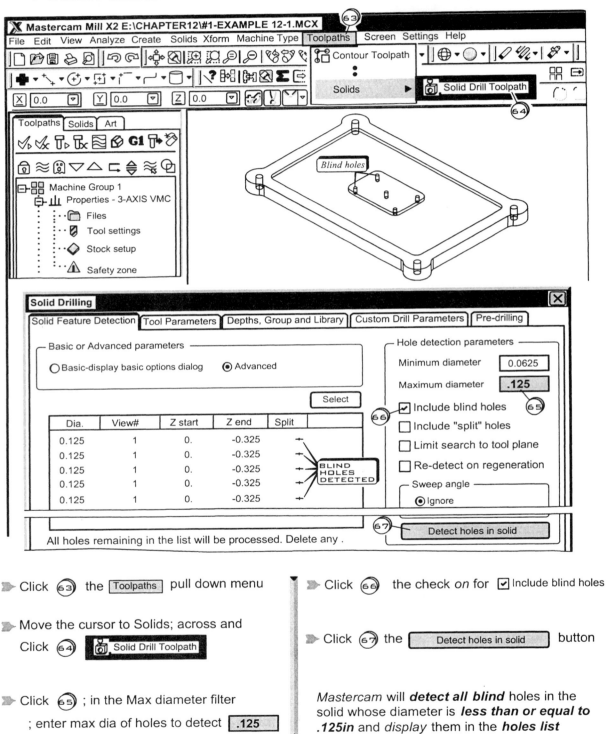

▶ Click ⑥③ the Toolpaths pull down menu

▶ Move the cursor to Solids; across and
    Click ⑥④ 📷 Solid Drill Toolpath

▶ Click ⑥⑤ ; in the Max diameter filter
    ; enter max dia of holes to detect .125

▶ Click ⑥⑥ the check *on* for ☑ Include blind holes

▶ Click ⑥⑦ the Detect holes in solid button

Mastercam will **detect all blind** holes in the
solid whose diameter is **less than or equal to**
.125in and *display* them in the **holes list**

◆ SPECIFY THE TOOL PARAMETERS FOR THE DRILLING OPERATION.

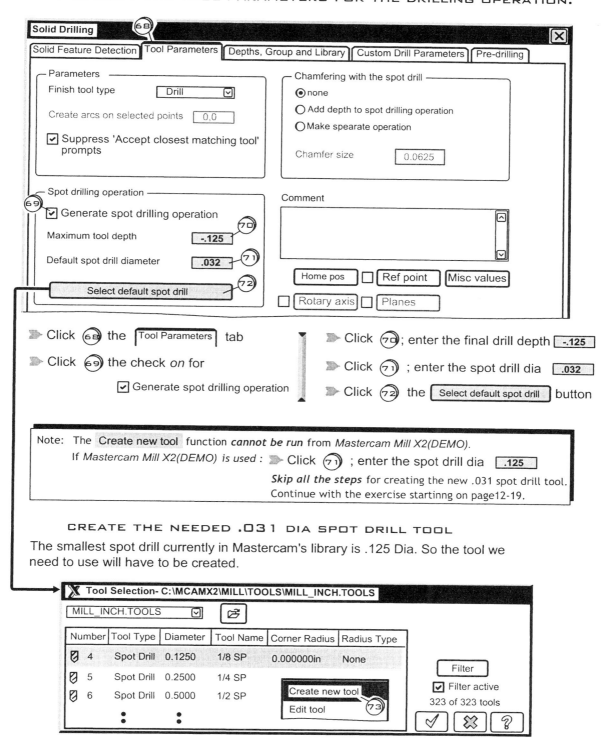

▷ Click ⑥⑧ the [Tool Parameters] tab

▷ Click ⑥⑨ the check *on* for

   ☑ Generate spot drilling operation

▷ Click ⑦⓪; enter the final drill depth [-.125]

▷ Click ⑦①; enter the spot drill dia [.032]

▷ Click ⑦② the [Select default spot drill] button

Note: The [Create new tool] function *cannot be run* from *Mastercam Mill X2(DEMO)*.

If *Mastercam Mill X2(DEMO)* is used : ▷ Click ⑦①; enter the spot drill dia [.125]

Skip all the steps for creating the new .031 spot drill tool.
Continue with the exercise startinng on page12-19.

CREATE THE NEEDED .031 DIA SPOT DRILL TOOL

The smallest spot drill currently in Mastercam's library is .125 Dia. So the tool we need to use will have to be created.

▷ ***Right*** Click move the cursor down and Click ⑦③ [Create new tool]

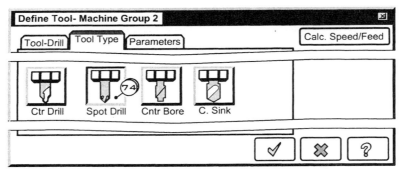

➤ Click ⑦④ the Spot Drill filter button

The tool dimensions are listed in the PROCESS PLAN 5-2 on page 5-16 as :

1/32 Spot Drill(.31 - Flute length; 1.5 - OAL)

Enter these values into the Define Tool dialog box :

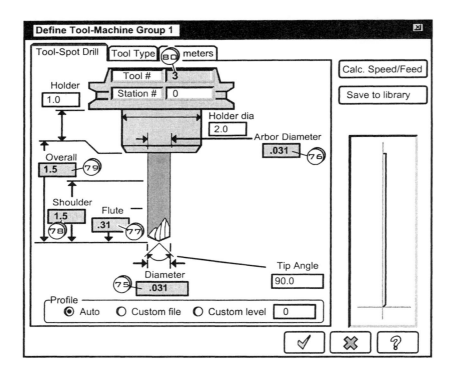

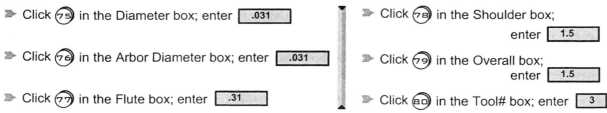

➤ Click ⑦⑤ in the Diameter box; enter **.031**

➤ Click ⑦⑥ in the Arbor Diameter box; enter **.031**

➤ Click ⑦⑦ in the Flute box; enter **.31**

➤ Click ⑦⑧ in the Shoulder box; enter **1.5**

➤ Click ⑦⑨ in the Overall box; enter **1.5**

➤ Click ⑧⓪ in the Tool# box; enter **3**

Open the Parameters Tab and open the SPOTDRIL sub-file.
Create the name 1/32 SPOT DRILLx1.5L for the new tool.

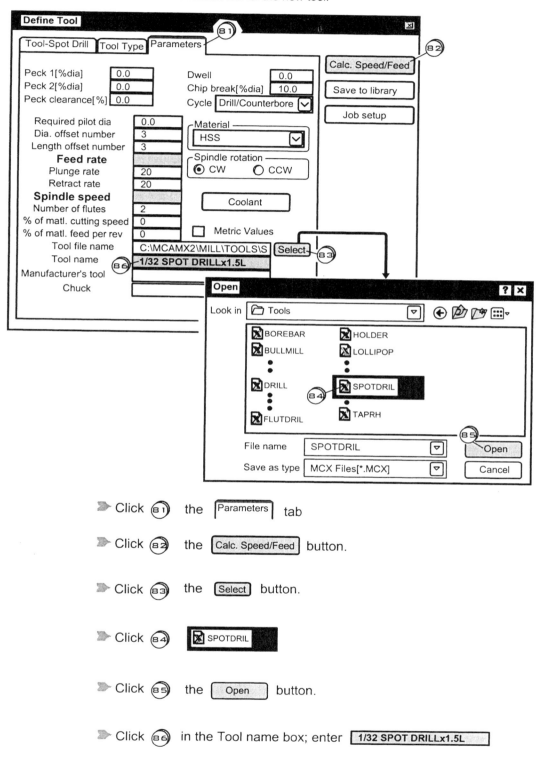

➤ Click ⓑ① the ⟨Parameters⟩ tab

➤ Click ⓑ② the ⟨Calc. Speed/Feed⟩ button.

➤ Click ⓑ③ the ⟨Select⟩ button.

➤ Click ⓑ④ ⟨ⓧ SPOTDRIL⟩

➤ Click ⓑ⑤ the ⟨Open⟩ button.

➤ Click ⓑ⑥ in the Tool name box; enter ⟨1/32 SPOT DRILLx1.5L⟩

Save the tool in the SPOTDRIL sub-file which is contained within the main tool library file MILL_INCH.Tools

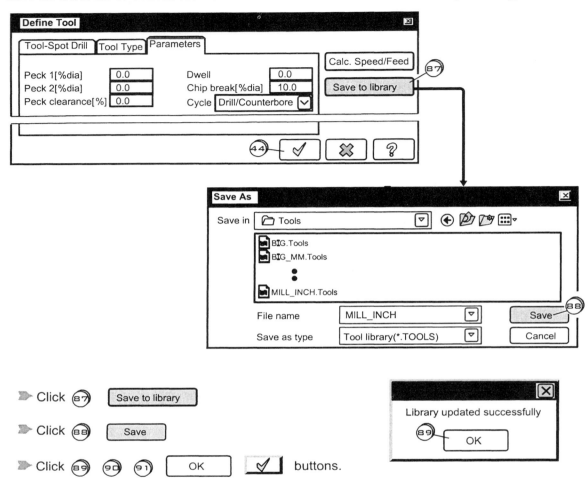

➤ Click ⑧⑦ [Save to library]

➤ Click ⑧⑧ [Save]

➤ Click ⑧⑨ ⑨⑩ ⑨① [OK] [✓] buttons.

After creating and storing the new tool in the tools library, *Mastercam* will then load it into the current job file.

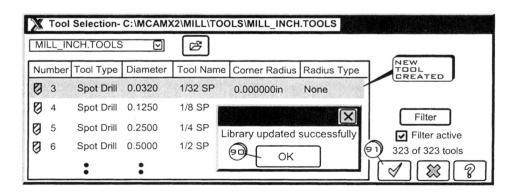

♦ SPECIFY PARAMETERS FOR PECK DRILLING THE .125DIA
HOLE X .325 DEEP.

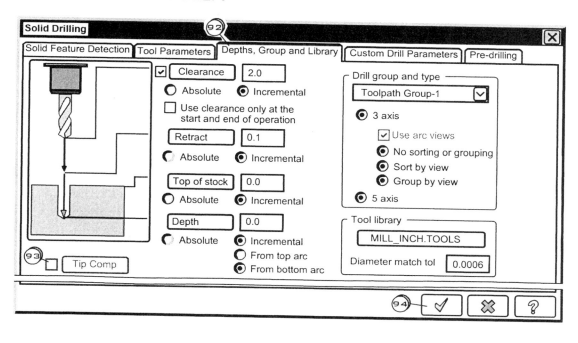

▷ Click ⓐ the ⌐Depths, Group and Library⌐ tab

▷ Click ⓐ ckeck *off* ☐ ⌐Tip Comp⌐ for a *blind* hole

▷ Click ⓐ the OK button ✓

Mastercam will drill *all* the .032DIA
pilot holes and the .125DIA x .325 deep blind holes.

F) CREATE THE C'SINKS 90° TO .17DIA
♦ CREATE THE C'SINK TOOLPATH

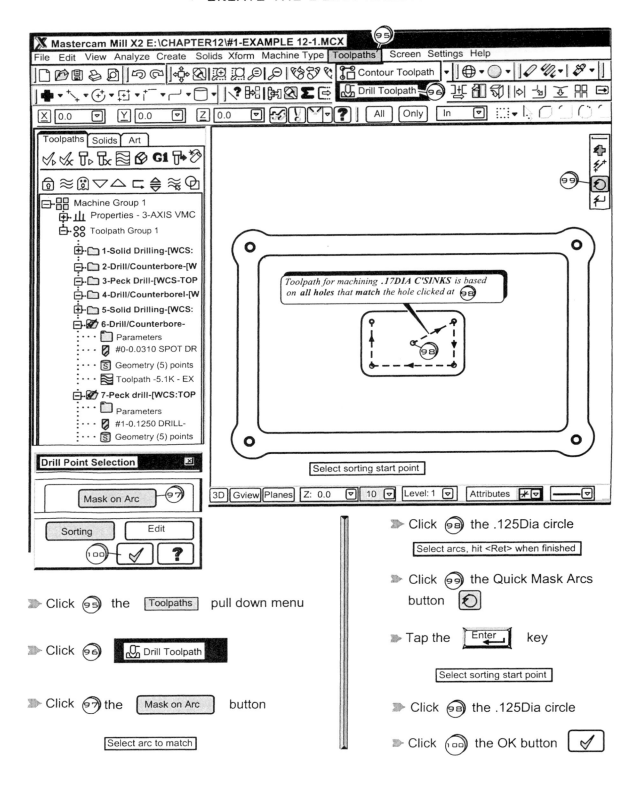

➤ Click (95) the [Toolpaths] pull down menu

➤ Click (96) [🔧 Drill Toolpath]

➤ Click (97) the [Mask on Arc] button

[Select arc to match]

➤ Click (98) the .125Dia circle

[Select arcs, hit <Ret> when finished]

➤ Click (99) the Quick Mask Arcs button 🔄

➤ Tap the [Enter] key

[Select sorting start point]

➤ Click (98) the .125Dia circle

➤ Click (100) the OK button ✓

◆ OBTAIN THE NEEDED .5 DIA CHAMFER MILL TOOL

Mastercam will activate and display the Tool Parameters Tab. The operator can use this tab to **select a tool, set speeds and feeds** and other general toolpath parameters.

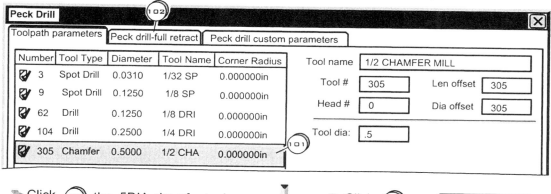

➤ Click ⟨101⟩ the .5DIA chamfer tool.   ➤ Click ⟨102⟩ the ⌐Peck drill-full retract⌐ tab.

◆ ENTER THE REQUIRED C'SINK 90° TO .17 DIA MACHINING PARAMETERS

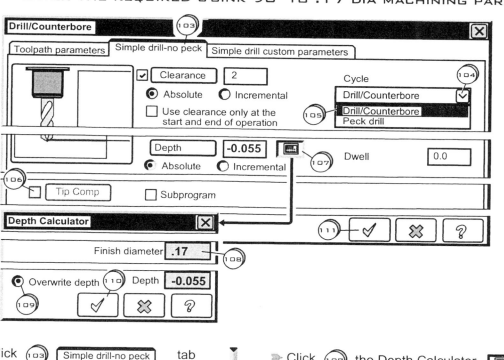

➤ Click ⟨103⟩ ⌐Simple drill-no peck⌐ tab

➤ Click ⟨104⟩ the down button ☑

➤ Click ⟨105⟩ ▇Drill/Counterbore▇

➤ Click ⟨106⟩ check *off* ☐ ⌐Tip Comp⌐

for a *blind* hole

➤ Click ⟨107⟩ the Depth Calculator ▦ button

➤ Click ⟨108⟩ ; enter the C'SINK DIA ⌐.17⌐

➤ Click ⟨109⟩ the ◉ Overwrite depth radio button *on*

➤ Click ⟨110⟩ ⟨111⟩ the OK buttons ☑

12-5 Creating a 2D Chained Contour on a Solid Model

Mastercam X2 machines a contour by moving the cutting tool along the chained curve. If a *solid model is present it will be detected automatically and the solids chaining mode button will be active*. For a *solid model a solid chain* will be created. *Mastercam* indicates a *solid chain* the **Operatios Manager** as follows: ⑤ Geometry (1) chains

12-6 Activating the Chaining Dialog Box: Solids Mode

The **Chaining** dialog box can be activated from either the *Toolpaths* pull down menu bar or the **2D Toolpaths** toolbar.

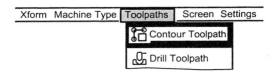

or

Chaining Dialog Box:Wireframe Mode

Chaining Dialog Box: Solids Mode

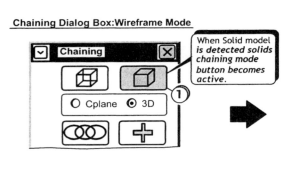

When Solid model *is detected solids chaining mode button becomes active.*

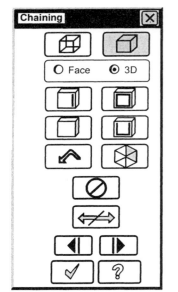

➤ Click ① the Solids Mode [▢] button

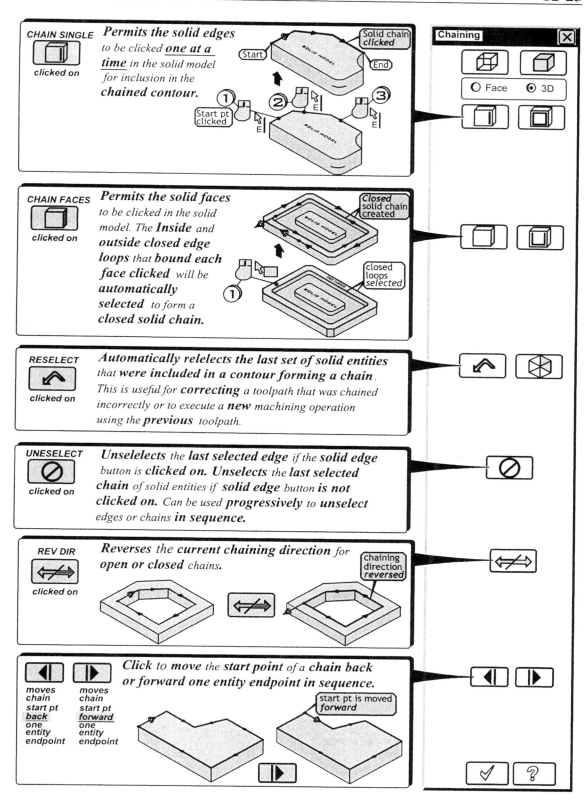

CHAIN SINGLE
clicked on

Permits the solid edges *to be clicked* <u>**one at a time**</u> *in the solid model for inclusion in the* **chained contour.**

CHAIN FACES
clicked on

Permits the solid faces *to be clicked in the solid model. The* **Inside** *and* **outside closed edge loops** *that* **bound each face clicked** *will be* **automatically selected** *to form a* **closed solid chain.**

RESELECT
clicked on

Automatically relelects the last set of solid entities *that* **were included in a contour forming a chain***. This is useful for* **correcting** *a toolpath that was chained incorrectly or to execute a* **new** *machining operation using the* **previous** *toolpath.*

UNESELECT
clicked on

Unselelects *the* **last selected edge** *if the* **solid edge** *button is* **clicked on. Unselects** *the* **last selected chain** *of solid entities if* **solid edge** *button* **is not clicked on.** *Can be used* **progressively** *to* **unselect** *edges or chains* **in sequence.**

REV DIR
clicked on

Reverses the current chaining direction *for* **open or closed** *chains.*

◀| *moves chain start pt* **back** *one entity endpoint*

|▶ *moves chain start pt* **forward** *one entity endpoint*

Click *to* **move** *the* **start point** *of a* **chain back** *or* **forward one entity endpoint in sequence.**

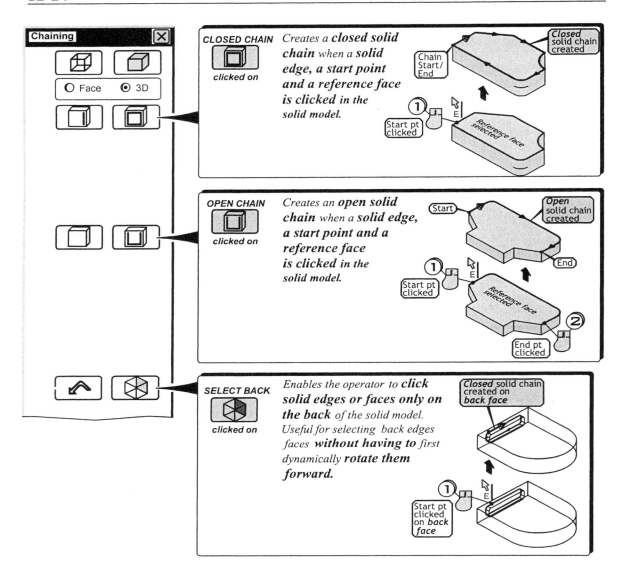

CLOSED CHAIN

clicked on

*Creates a **closed solid chain** when a **solid edge, a start point and a reference face** is clicked in the solid model.*

OPEN CHAIN

clicked on

*Creates an **open solid chain** when a **solid edge, a start point and a reference face** is clicked in the solid model.*

SELECT BACK

clicked on

*Enables the operator to **click solid edges or faces only on the back** of the solid model. Useful for selecting back edges faces **without having to** first dynamically **rotate them forward**.*

EXAMPLE 12-2

Contour machine, pocket and chamfer the part given in Example 12-1.

- Create the machining contours by chaining
- Enter the machining parameters as listed in PROCESS PLAN 12-2.

PROCESS PLAN 12-2

No.	Operation	Tooling
1	SPOT DRILL x .166 DEEP (4 HOLES)	1/8 SPOT DRILLL
2	PECK DRILL THRU(4 HOLES)	1/4 DRILL
3	C'SINK 90° TO .29 DIA (4 HOLES)	.5 CHAMFER MILL
4	SPOT DRILL x .125 DEEP (5 HOLES)	1/32 SPOT DRILL
5	PECK DRILL x .325 DEEP	1/8 DRILL
6	C'SINK 90° TO .17 DIA (5 HOLES)	.5 CHAMFER MILL
7	PROFILE x .5 DEEP LEAVE .01 FOR FINISH CUTS IN XY AND Z	1/2 FLAT END MILL
8	POCKET x .125 DEEP LEAVE .02 FOR FINISH CUTS IN XY AND Z	1/2 BULL END MILL .032 CORNER RADIUS
9	CHAMFER .02 x 45°	.5 CHAMFER MILL

A) CREATE A MACHINING CONTOUR FOR THE PROFILE OPERATION BY SOLID CHAINING

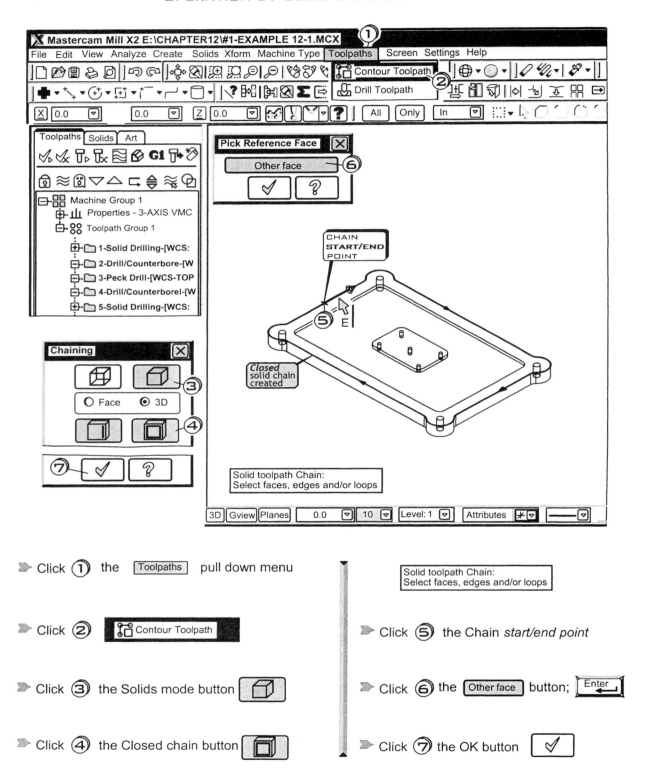

> Click ① the [Toolpaths] pull down menu

> Click ② [Contour Toolpath]

> Click ③ the Solids mode button

> Click ④ the Closed chain button

Solid toolpath Chain:
Select faces, edges and/or loops

> Click ⑤ the Chain *start/end point*

> Click ⑥ the [Other face] button; [Enter]

> Click ⑦ the OK button

B) Obtain the 1/2 End Mill tool from the tools library

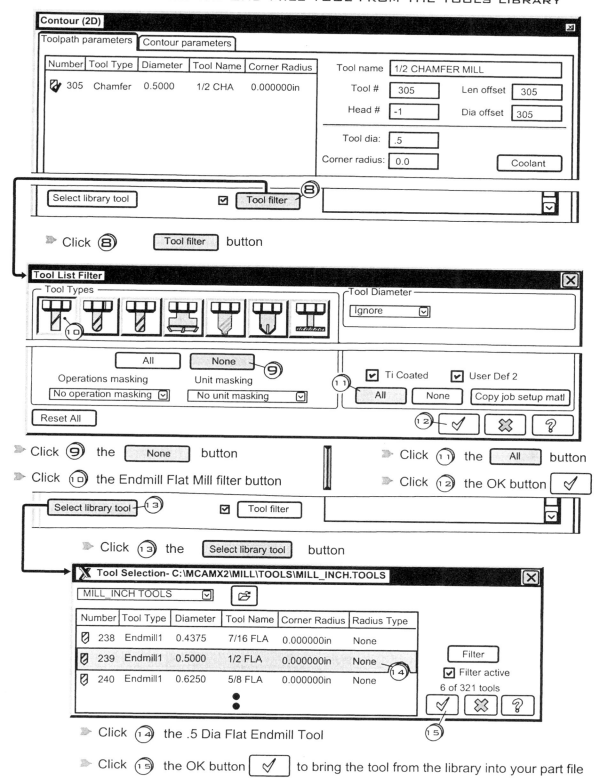

➤ Click ⑧ [Tool filter] button

➤ Click ⑨ the [None] button   ➤ Click ⑪ the [All] button

➤ Click ⑩ the Endmill Flat Mill filter button   ➤ Click ⑫ the OK button [✓]

➤ Click ⑬ the [Select library tool] button

➤ Click ⑭ the .5 Dia Flat Endmill Tool

➤ Click ⑮ the OK button [✓] to bring the tool from the library into your part file

C) AUTO ASSIGN SPEEDS FEEDS FOR THE TOOL

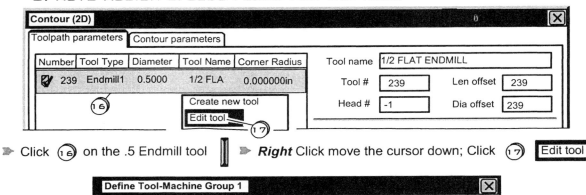

➤ Click ⑯ on the .5 Endmill tool   ➤ **Right** Click move the cursor down; Click ⑰ [Edit tool]

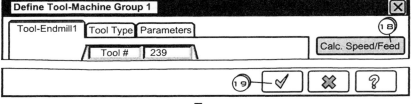

➤ Click ⑱ the [Calc. Speed/Feed] button.   ➤ Click ⑲ the OK button ✓

D) ENTER THE CONTOUR(2D) PARAMETERS LISTED IN PROCESS PLAN 12-2

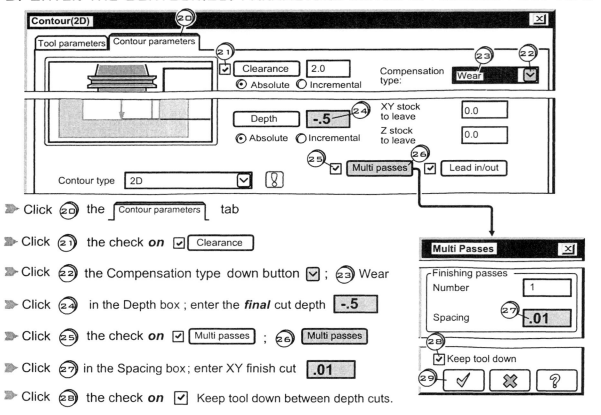

➤ Click ⑳ the [Contour parameters] tab

➤ Click ㉑ the check **on** ☑ [Clearance]

➤ Click ㉒ the Compensation type down button ☑ ; ㉓ Wear

➤ Click ㉔ in the Depth box ; enter the **final** cut depth [-.5]

➤ Click ㉕ the check **on** ☑ [Multi passes] ; ㉖ [Multi passes]

➤ Click ㉗ in the Spacing box ; enter XY finish cut [.01]

➤ Click ㉘ the check **on** ☑ Keep tool down between depth cuts.

➤ Click ㉙ the OK button [✓]

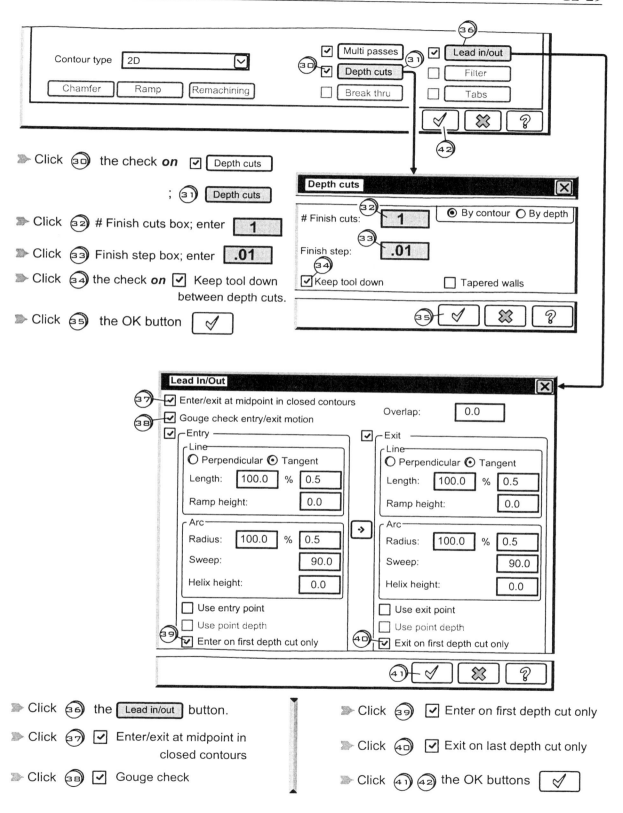

Click ③⓪ the check *on* ☑ Depth cuts

; ③① Depth cuts

Click ③② # Finish cuts box; enter 1

Click ③③ Finish step box; enter .01

Click ③④ the check *on* ☑ Keep tool down between depth cuts.

Click ③⑤ the OK button ☑

Click ③⑥ the Lead in/out button.

Click ③⑦ ☑ Enter/exit at midpoint in closed contours

Click ③⑧ ☑ Gouge check

Click ③⑨ ☑ Enter on first depth cut only

Click ④⓪ ☑ Exit on last depth cut only

Click ④① ④② the OK buttons ☑

E) CREATE A MACHINING CONTOUR FOR THE POCKETING OPERATION BY SOLID CHAINING

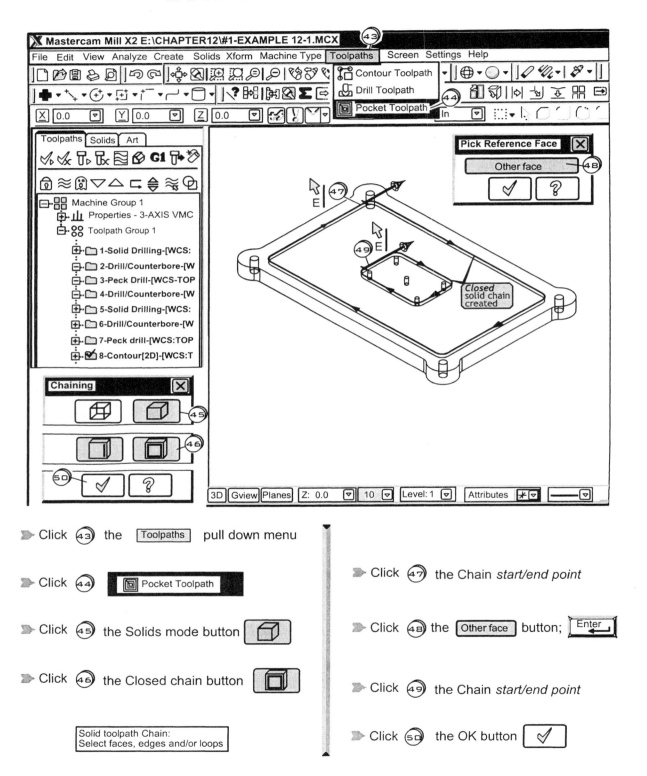

➤ Click ④③ the [Toolpaths] pull down menu

➤ Click ④④ [🔲 Pocket Toolpath]

➤ Click ④⑤ the Solids mode button [▣]

➤ Click ④⑥ the Closed chain button [▣]

Solid toolpath Chain:
Select faces, edges and/or loops

➤ Click ④⑦ the Chain *start/end point*

➤ Click ④⑧ the [Other face] button; [Enter ←]

➤ Click ④⑨ the Chain *start/end point*

➤ Click ⑤⓪ the OK button [✓]

F) OBTAIN THE 1/2 BULL END MILL TOOL FROM THE TOOLS LIBRARY

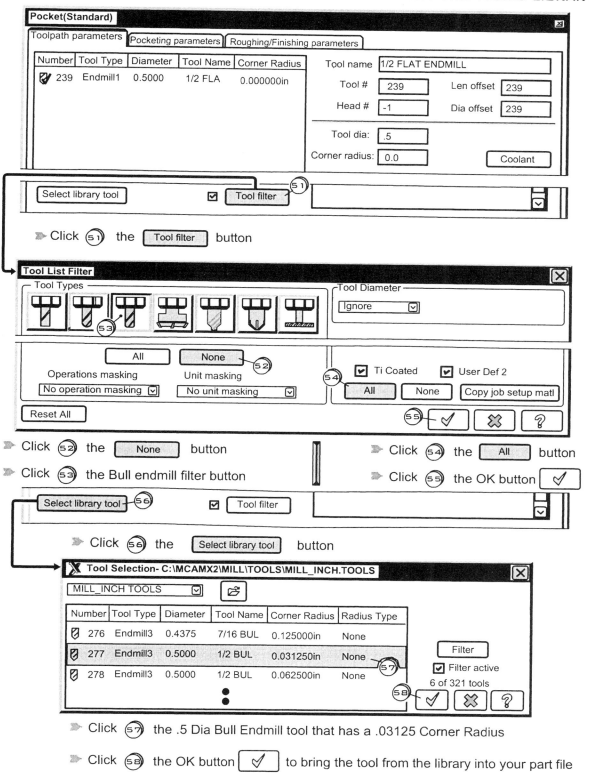

➤ Click ⑤① the [Tool filter] button

➤ Click ⑤② the [None] button

➤ Click ⑤③ the Bull endmill filter button

➤ Click ⑤④ the [All] button

➤ Click ⑤⑤ the OK button ✓

➤ Click ⑤⑥ the [Select library tool] button

➤ Click ⑤⑦ the .5 Dia Bull Endmill tool that has a .03125 Corner Radius

➤ Click ⑤⑧ the OK button ✓ to bring the tool from the library into your part file

G) AUTO ASSIGN SPEEDS FEEDS FOR THE TOOL

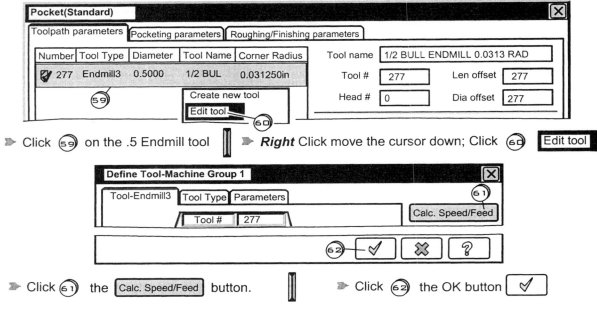

➤ Click ⑤⑨ on the .5 Endmill tool ‖ ➤ **Right** Click move the cursor down; Click ⑥⓪ [Edit tool]

➤ Click ⑥① the [Calc. Speed/Feed] button.   ‖   ➤ Click ⑥② the OK button ✓

H) ENTER THE POCKETING PARAMETERS LISTED IN PROCESS PLAN 12-2

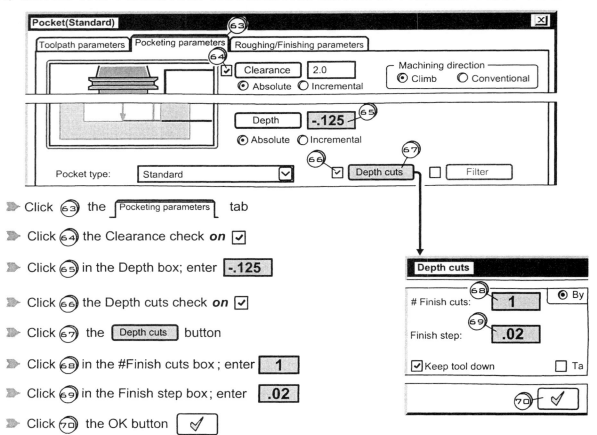

➤ Click ⑥③ the [Pocketing parameters] tab

➤ Click ⑥④ the Clearance check **on** ☑

➤ Click ⑥⑤ in the Depth box; enter [-.125]

➤ Click ⑥⑥ the Depth cuts check **on** ☑

➤ Click ⑥⑦ the [Depth cuts] button

➤ Click ⑥⑧ in the #Finish cuts box ; enter [1]

➤ Click ⑥⑨ in the Finish step box; enter [.02]

➤ Click ⑦⓪ the OK button [✓]

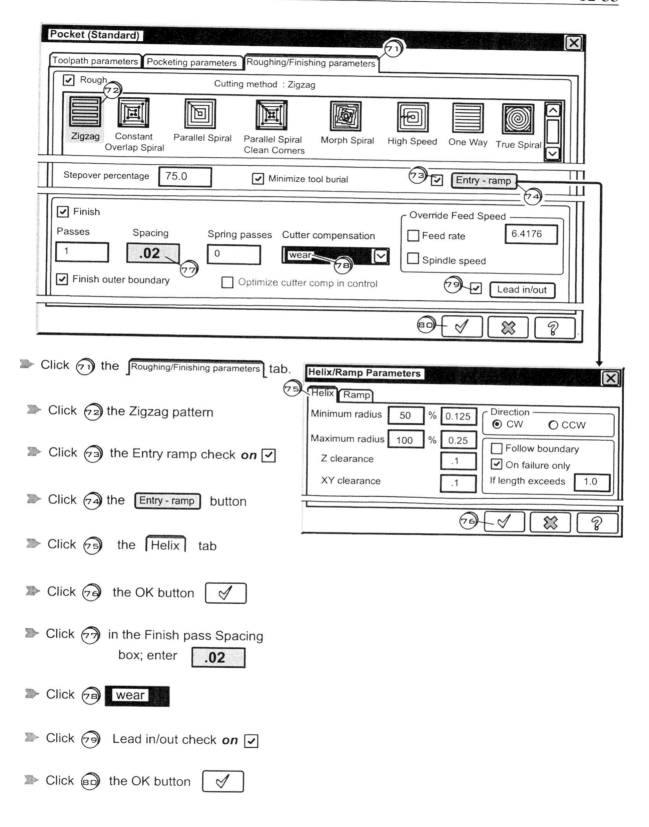

Click ⑦¹ the ⌐Roughing/Finishing parameters⌐ tab.

Click ⑦² the Zigzag pattern

Click ⑦³ the Entry ramp check *on* ☑

Click ⑦⁴ the ⌐Entry - ramp⌐ button

Click ⑦⁵ the ⌐Helix⌐ tab

Click ⑦⁶ the OK button ☑

Click ⑦⁷ in the Finish pass Spacing box; enter .02

Click ⑦⁸ **wear**

Click ⑦⁹ Lead in/out check *on* ☑

Click ⑧⁰ the OK button ☑

1) CREATE A MACHINING CONTOUR FOR THE CHAMFERING OPERATION BY SOLID CHAINING

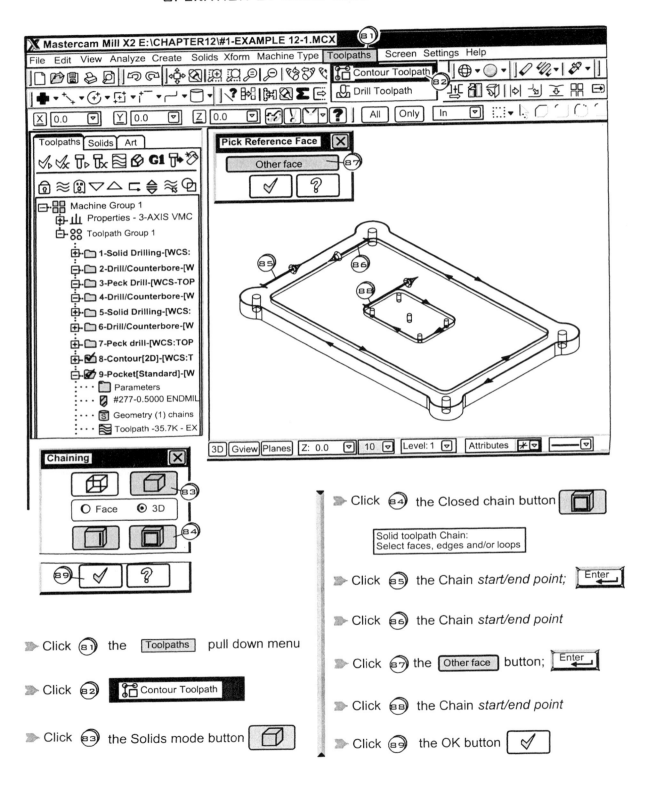

➤ Click ⑧① the [Toolpaths] pull down menu

➤ Click ⑧② [Contour Toolpath]

➤ Click ⑧③ the Solids mode button

➤ Click ⑧④ the Closed chain button

Solid toolpath Chain:
Select faces, edges and/or loops

➤ Click ⑧⑤ the Chain *start/end point;* [Enter]

➤ Click ⑧⑥ the Chain *start/end point*

➤ Click ⑧⑦ the [Other face] button; [Enter]

➤ Click ⑧⑧ the Chain *start/end point*

➤ Click ⑧⑨ the OK button

J) OBTAIN THE PREVIOUSLY LOADED 1/2 CHAMFER MILL.

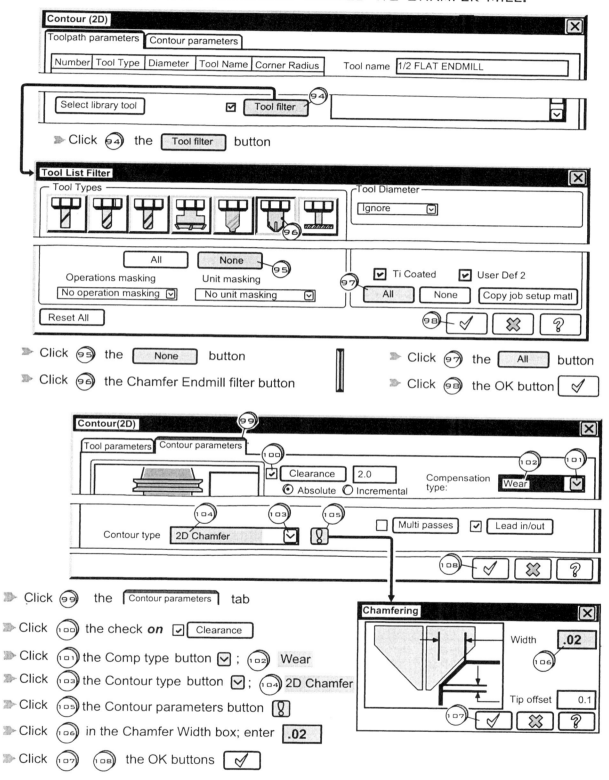

Click ⑨④ the [Tool filter] button

Click ⑨⑤ the [None] button

Click ⑨⑥ the Chamfer Endmill filter button

Click ⑨⑦ the [All] button

Click ⑨⑧ the OK button ✓

Click ⑨⑨ the [Contour parameters] tab

Click ⑩⓪ the check **on** ☑ [Clearance]

Click ⑩① the Comp type button ▽ ; ⑩② Wear

Click ⑩③ the Contour type button ▽ ; ⑩④ 2D Chamfer

Click ⑩⑤ the Contour parameters button 🔲

Click ⑩⑥ in the Chamfer Width box; enter .02

Click ⑩⑦ ⑩⑧ the OK buttons ✓

K) ALL THE SOLID MODEL FEATURES MUST BE
UNSUPRESSED BEFORE THE .MCX FILE CAN BE SAVED.

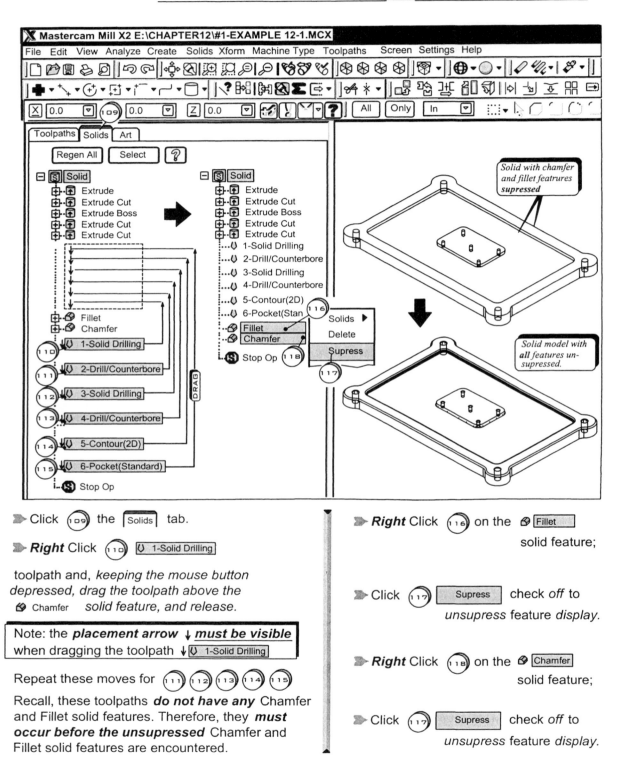

Click ⓘⓞ⑨ the ⌊Solids⌋ tab.

Right Click ⓘⓘⓞ ⌊◊ 1-Solid Drilling⌋

toolpath and, *keeping the mouse button depressed, drag the toolpath above the* ◎ Chamfer *solid feature, and release.*

> Note: the **placement arrow** ↓ **must be visible** when dragging the toolpath ↓⌊◊ 1-Solid Drilling⌋

Repeat these moves for ⓘⓘⓘ ⓘⓘⓘ②② ⓘⓘⓘ③③ ⓘⓘⓘ④④ ⓘⓘⓘ⑤⑤

Recall, these toolpaths **do not have any** Chamfer and Fillet solid features. Therefore, they **must occur before the unsupressed** Chamfer and Fillet solid features are encountered.

Right Click ⓘⓘⓘ⑥ on the ◎ ⌊Fillet⌋ solid feature;

Click ⓘⓘⓘ⑦ ⌊ Supress ⌋ check *off* to *unsupress* feature *display.*

Right Click ⓘⓘⓘ⑧ on the ◎ ⌊Chamfer⌋ solid feature;

Click ⓘⓘⓘ⑦ ⌊ Supress ⌋ check *off* to *unsupress* feature *display.*

K) REGENERATE ALL THE MACHINING OPERATIONS.

L) BACKPLOT ALL THE MACHINING OPERATIONS.

Refer to Chapter 5 Section 5-4.

L) VERIFY ALL THE MACHINING OPERATIONS.

Refer to Chapter 5 Section 5-5.

M) GENERATE THE REQUIRED WORD ADDRESS PART PROGRAM.

Refer to Chapter 5 Section 5-10.

M) SAVE THE PART FILE.

EXERCISES

Note: Use the [Alt] + [T] keys to toggle *on/off* toolpath displays for operations

12-1) a) Get the *Solid* model file **EX11-1JV** :
  ◆ from the file generated in exercise 11-1
        or

  ◆ from the CD provided at the back of this text(file is located in the folder ▭ CHAPTER12).

  b) Specify the stock size and material

  c) Create the required toolpaths for executing the operations listed in PROCESS PLAN 12P-1

  d) Backplot and verify the toolpaths

  e) Generate the word address part program   MATERIAL: 1030 STEEL

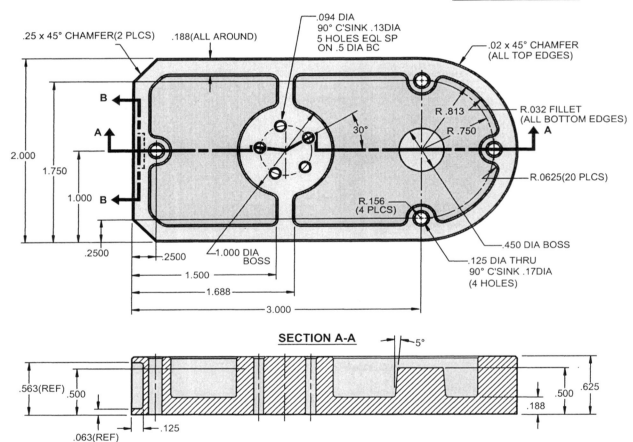

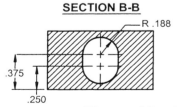

Figure 12-p1

♦ SPECIFY THE SIZE OF THE STOCK

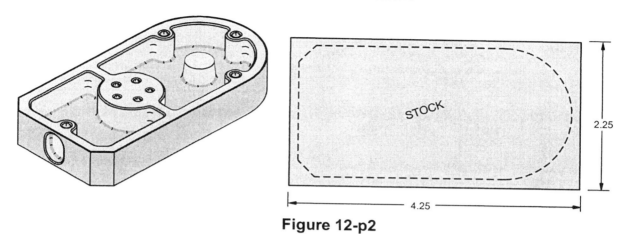

Figure 12-p2

PROCESS PLAN 12P-1

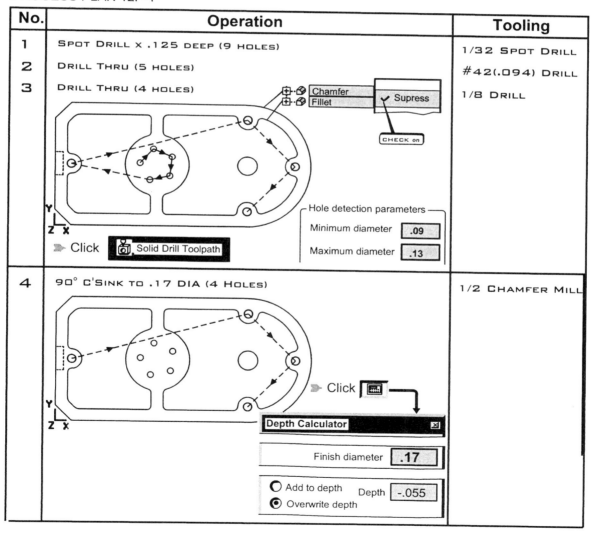

No.	Operation	Tooling
1	SPOT DRILL x .125 DEEP (9 HOLES)	1/32 SPOT DRILL
2	DRILL THRU (5 HOLES)	#42(.094) DRILL
3	DRILL THRU (4 HOLES)	1/8 DRILL
4	90° C'SINK TO .17 DIA (4 HOLES)	1/2 CHAMFER MILL

PROCESS PLAN 12P-1 (*continued*)

No.	Operation	Tooling
5	90° C'SINK TO .13 DIA (5 HOLES) 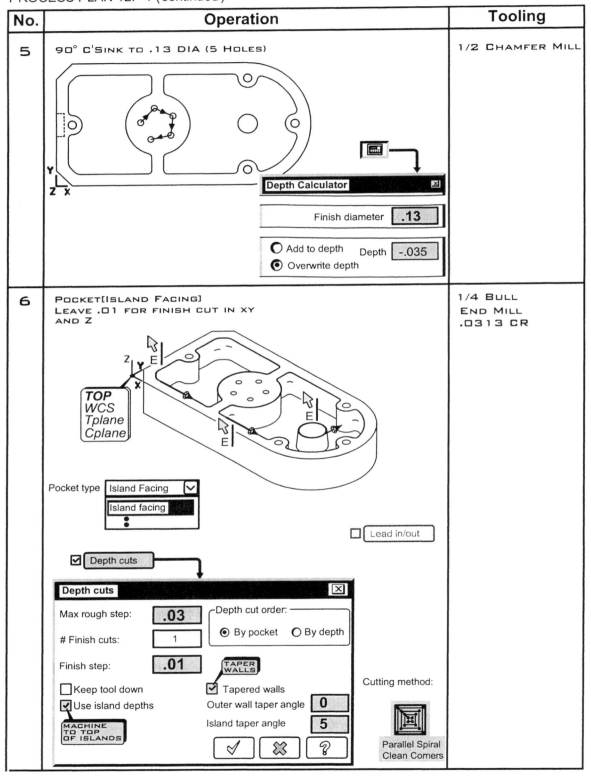	1/2 CHAMFER MILL
6	POCKET[ISLAND FACING] LEAVE .01 FOR FINISH CUT IN XY AND Z	1/4 BULL END MILL .0313 CR

PROCESS PLAN 12P-1 (*continued*)

No.	Operation	Tooling
7	POCKET[REMACHINING] LEAVE .01 FOR FINISH CUT IN XY AND Z Pocket type [Remachining ⌄] [Remachining]	1/8 BULL END MILL .0313 CR
8	ROUGH OUTSIDE X .625 DEEP LEAVE .01 FOR FINISH CUT IN XY XY stock to leave [.01] ☑ Multi passes	1/2 FLAT END MILL
9	CHAMFER INSIDE AND OUSIDE CONTOURS Contour type [2D chamfer ⌄] [Chamfer] **Chamfering** [x] Width [.02] Tip offset [0.1]	1/2 CHAMFER MILL

PROCESS PLAN 12P-1 (*continued*)

No.	Operation	Tooling
10	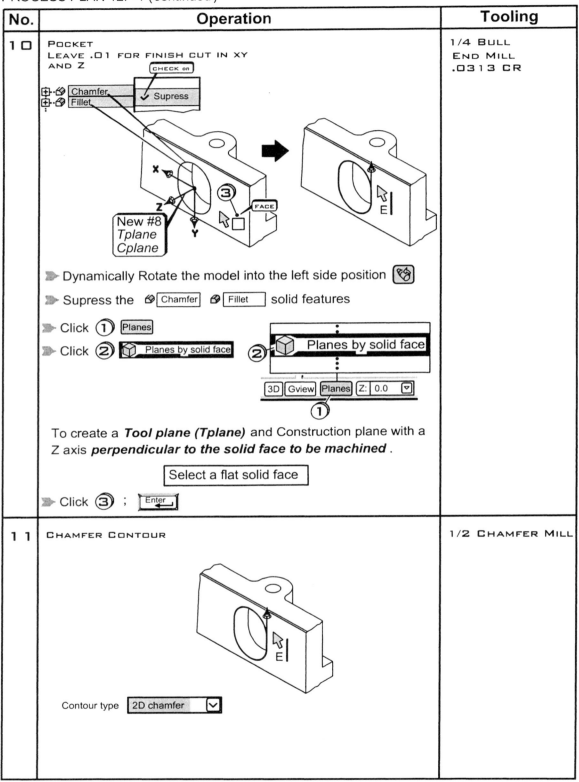 POCKET LEAVE .01 FOR FINISH CUT IN XY AND Z CHECK on Chamfer ✓ Supress Fillet New #8 *Tplane* *Cplane* ▶ Dynamically Rotate the model into the left side position ▶ Supress the ⦿ Chamfer ⦿ Fillet solid features ▶ Click ① Planes ▶ Click ② 🞐 Planes by solid face ② 🞐 Planes by solid face 3D Gview Planes Z: 0.0 ▽ ① To create a **Tool plane (Tplane)** and Construction plane with a Z axis *perpendicular to the solid face to be machined* . Select a flat solid face ▶ Click ③ ; Enter ↵	1/4 BULL END MILL .0313 CR
11	CHAMFER CONTOUR Contour type 2D chamfer ▽	1/2 CHAMFER MILL

PROCESS PLAN 12P-1 (*continued*)

RESET THE DEFAULT Tplane AND Cplane BACK TO Top

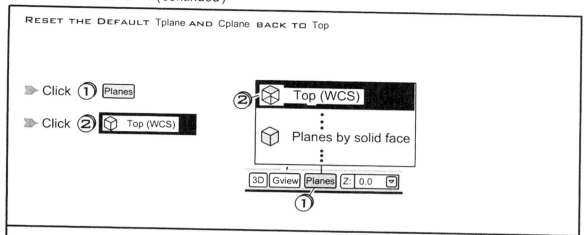

➤ Click ① Planes

➤ Click ② ◫ Top (WCS)

② ◫ Top (WCS)

⬠ Planes by solid face

[3D] [Gview] [Planes] [Z: | 0.0 | ▽]

①

UNSUPPRESS(CHECK *off*) THE SOLID FEATURES PRIOR TO SAVING THE FILE

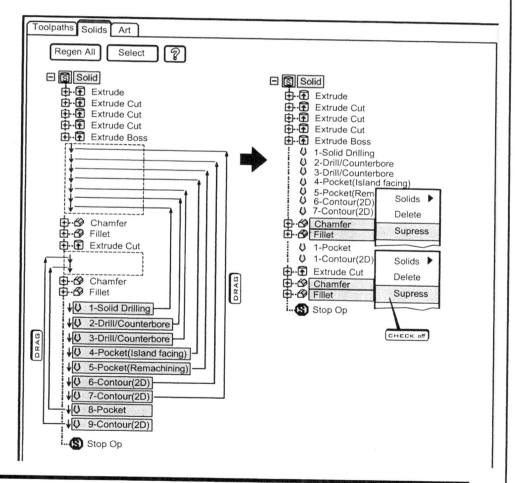

Note: All features ***must be unsuppressed*** (check ***off***)before part file can be saved.

12-2) a) Get the *Solid* model file **EX11-2JV** :
   ◆ from the file generated in exercise 11-2

b) Specify the stock size and material

c) Create the required toolpaths for executing the operations listed in PROCESS PLAN 12P-2

d) Backplot and verify the toolpaths

e) Generate the word address part program

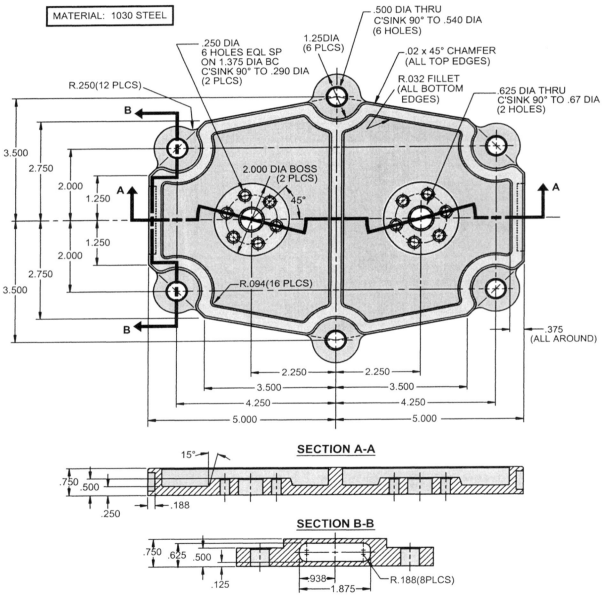

Figure 12-p3

♦ SPECIFY THE SIZE OF THE STOCK

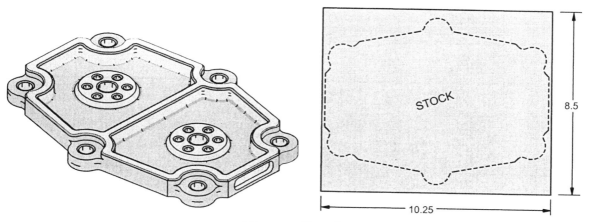

Figure 12-p4

PROCESS PLAN 12P-2

No.	Operation	Tooling
1	ROUGH OUTSIDE X .25 DEEP LEAVE .01 FOR FINISH CUT IN XY 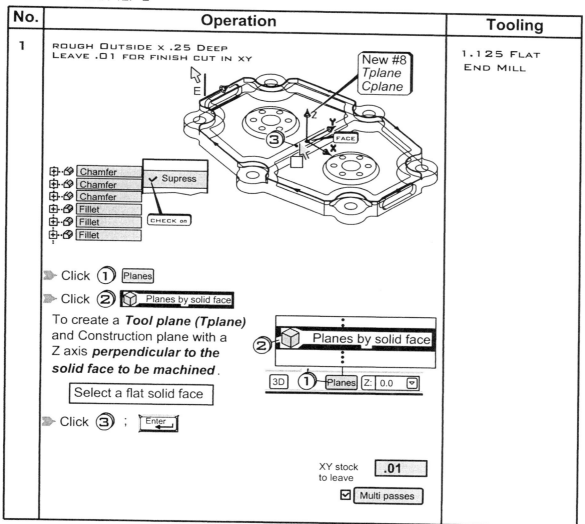	1.125 FLAT END MILL

PROCESS PLAN 12P-2 (*continued*)

No.	Operation	Tooling
2	POCKET[ISLAND FACING] LEAVE .01 FOR FINISH CUT IN XY AND Z Pocket type [Island Facing ▽] ☑ [Depth cuts] **Depth cuts** [✕] Max rough step: [.03] ┌Depth cut order: ⦿ By pocket ○ By depth # Finish cuts: [1] Finish step: [.01] [TAPER WALLS] ☐ Keep tool down ☑ Tapered walls ☑ Use island depths Outer wall taper angle [0] [MACHINE TO TOP OF ISLANDS] Island taper angle [15] [✓] [✗] [?] Cutting method: Parallel Spiral Clean Corners	1/2 BULL END MILL .0313 CR
3	POCKET[REMACHINING] LEAVE .01 FOR FINISH CUT IN XY AND Z Pocket type [Remachining ▽]	3/16 BULL END MILL .0313 CR

4
 	
5
	SPOT DRILL X .125 DEEP (12 HOLES)
	
	DRILL THRU (12 HOLES)
	
	➤ Click [🔧 Solid Drill Toolpath]
	  ┌Hole detection parameters─
	  Minimum diameter [.24]
	  Maximum diameter [.26]
	
	➤ Click ① [Planes]
	➤ Click ② [📦 Named Views]
	➤ Click ③ [NEW VIEW [8] X0. Y0. Z0.75]
		1/8 SPOT DRILL
		
		1/4 DRILL

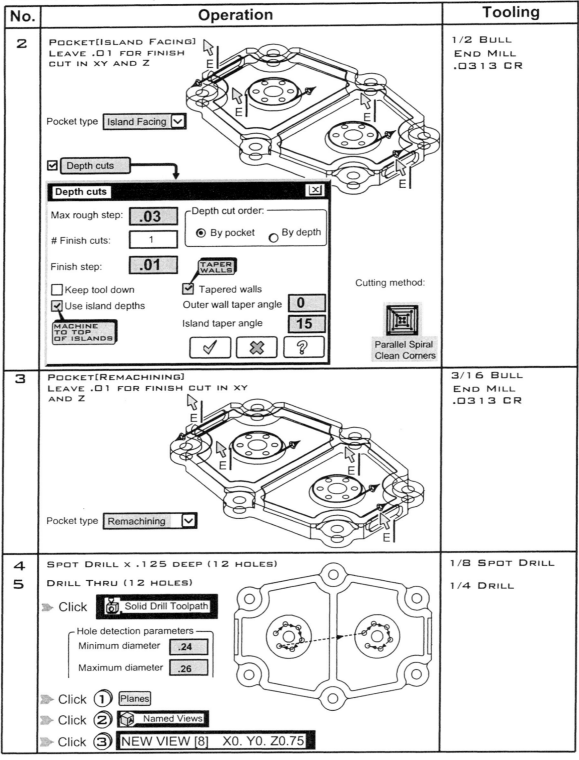

PROCESS PLAN 12P-2 (*continued*)

No.	Operation	Tooling
6	SPOT DRILL x .166 DEEP (8 HOLES)	1/4 SPOT DRILL
7	DRILL THRU (6 HOLES)	1/2 DRILL
8	DRILL THRU (2 HOLES)	5/8 DRILL

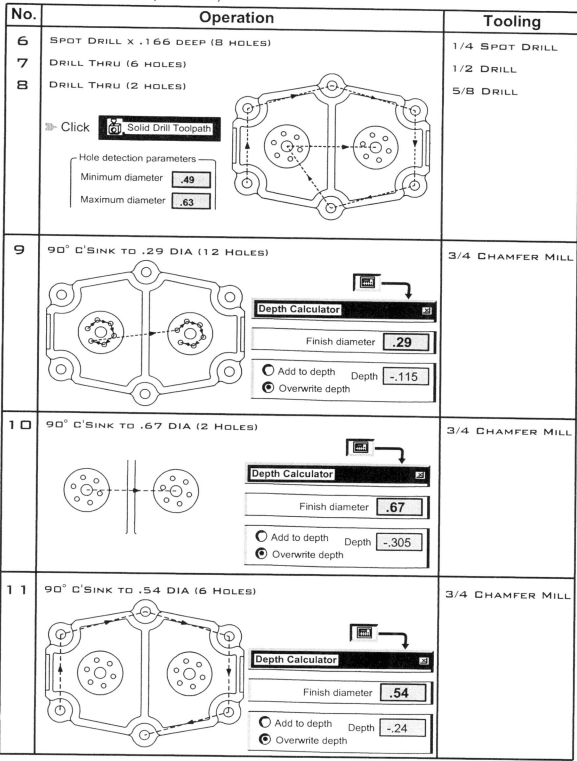

(Operation 6-8 area)

▧ Click [Solid Drill Toolpath]

Hole detection parameters
Minimum diameter **.49**
Maximum diameter **.63**

(Operation 9)

| 9 | 90° C'SINK TO .29 DIA (12 HOLES) | 3/4 CHAMFER MILL |

Depth Calculator

Finish diameter **.29**

○ Add to depth   Depth **-.115**
◉ Overwrite depth

(Operation 10)

| 10 | 90° C'SINK TO .67 DIA (2 HOLES) | 3/4 CHAMFER MILL |

Depth Calculator

Finish diameter **.67**

○ Add to depth   Depth **-.305**
◉ Overwrite depth

(Operation 11)

| 11 | 90° C'SINK TO .54 DIA (6 HOLES) | 3/4 CHAMFER MILL |

Depth Calculator

Finish diameter **.54**

○ Add to depth   Depth **-.24**
◉ Overwrite depth

PROCESS PLAN 12P-2 (*continued*)

No.	Operation	Tooling
12	CHAMFER INSIDE AND OUSIDE CONTOURS 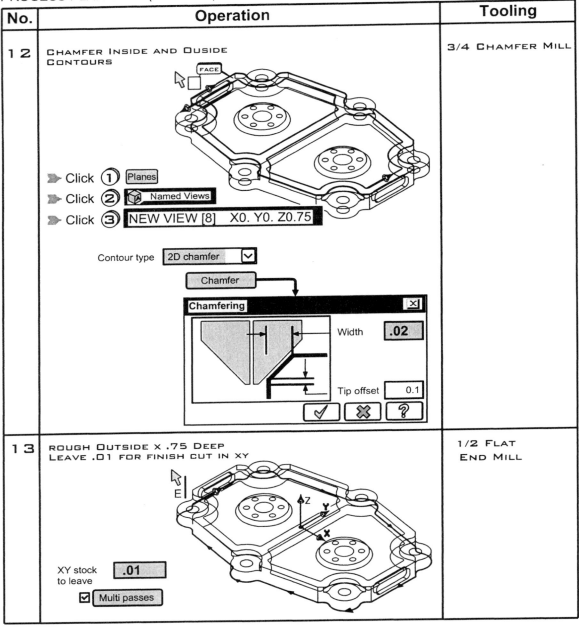 ➡ Click ① Planes ➡ Click ② Named Views ➡ Click ③ NEW VIEW [8] X0. Y0. Z0.75 Contour type 2D chamfer ▽ Chamfer **Chamfering** ▣ Width .02 Tip offset 0.1 ✓ ✗ ?	3/4 CHAMFER MILL
13	ROUGH OUTSIDE X .75 DEEP LEAVE .01 FOR FINISH CUT IN XY XY stock to leave .01 ☑ Multi passes	1/2 FLAT END MILL

PROCESS PLAN 12P-2 (*continued*)

No.	Operation	Tooling
14	POCKET LEAVE .01 FOR FINISH CUT IN XY AND Z New #9 Tplane Cplane ➤ Click ① Planes ➤ Click ② 📦 Planes by solid face To create a **Tool plane (Tplane)** and Construction plane with a Z axis *perpendicular to the solid face to be machined*. Select a flat solid face ➤ Click ③ ; Enter	3/8 FLAT END MILL
15	POCKET LEAVE .01 FOR FINISH CUT IN XY AND Z New #10 Tplane Cplane ➤ Dynamically Rotate the model into the left side position 🔄 ➤ Click ① Planes ➤ Click ② 📦 Planes by solid face To create a **Tool plane (Tplane)** and Construction plane with a Z axis *perpendicular to the solid face to be machined*. Select a flat solid face ➤ Click ③ ; Enter	3/8 FLAT END MILL

PROCESS PLAN 12P-2 (*continued*)

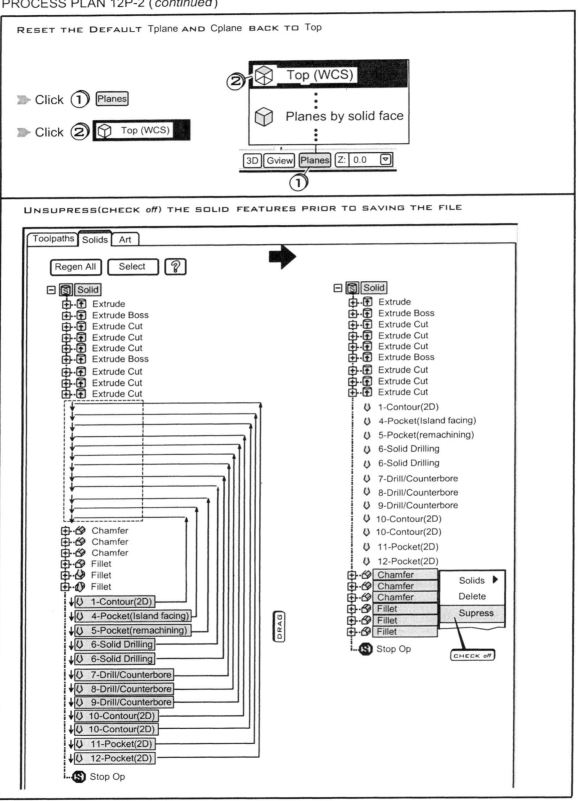

RESET THE DEFAULT Tplane AND Cplane BACK TO Top

Click ① Planes

Click ② Top (WCS)

② Top (WCS)

Planes by solid face

3D | Gview | Planes | Z: 0.0

①

UNSUPRESS(CHECK *off*) THE SOLID FEATURES PRIOR TO SAVING THE FILE

Toolpaths | Solids | Art

Regen All | Select | ?

Solid
- Extrude
- Extrude Boss
- Extrude Cut
- Extrude Cut
- Extrude Cut
- Extrude Boss
- Extrude Cut
- Extrude Cut
- Extrude Cut

- Chamfer
- Chamfer
- Chamfer
- Fillet
- Fillet
- Fillet
- 1-Contour(2D)
- 4-Pocket(Island facing)
- 5-Pocket(remachining)
- 6-Solid Drilling
- 6-Solid Drilling
- 7-Drill/Counterbore
- 8-Drill/Counterbore
- 9-Drill/Counterbore
- 10-Contour(2D)
- 10-Contour(2D)
- 11-Pocket(2D)
- 12-Pocket(2D)
- Stop Op

DRAG

Solid
- Extrude
- Extrude Boss
- Extrude Cut
- Extrude Cut
- Extrude Cut
- Extrude Boss
- Extrude Cut
- Extrude Cut
- Extrude Cut
- 1-Contour(2D)
- 4-Pocket(Island facing)
- 5-Pocket(remachining)
- 6-Solid Drilling
- 6-Solid Drilling
- 7-Drill/Counterbore
- 8-Drill/Counterbore
- 9-Drill/Counterbore
- 10-Contour(2D)
- 10-Contour(2D)
- 11-Pocket(2D)
- 12-Pocket(2D)
- Chamfer
- Chamfer
- Chamfer
- Fillet
- Fillet
- Fillet
- Stop Op

Solids ▶
Delete
Supress

CHECK *off*

12-3) a) Get the *Solid* model file **EX11-3JV** :
  ◆ from the file generated in exercise 11-3

b) Specify the stock size and material

c) Create the required toolpaths for executing the operations listed in PROCESS PLAN 12P-3

d) Backplot and verify the toolpaths

e) Generate the word address part program

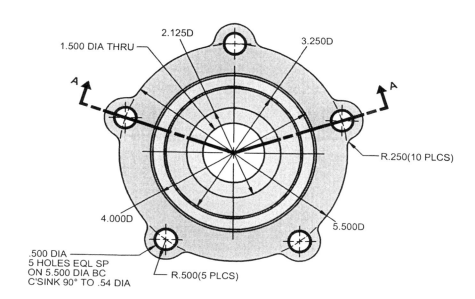

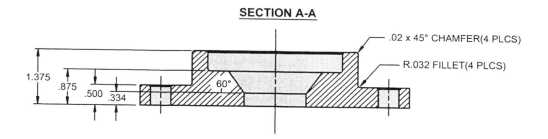

Figure 12-p5

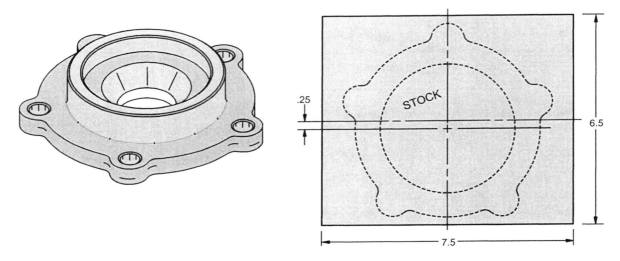

Figure 12-p6

PROCESS PLAN 12P-3

No.	Operation	Tooling
1	Pocket[Standard] x .5 deep Leave .01 for finish cut in xy and z 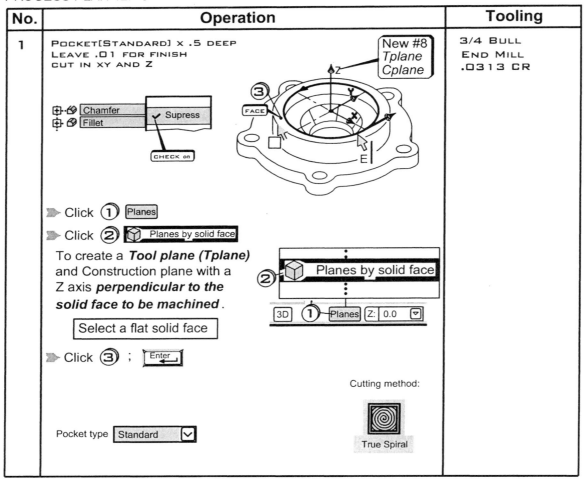	3/4 Bull End Mill .0313 CR

PROCESS PLAN 12P-3 *(continued)*

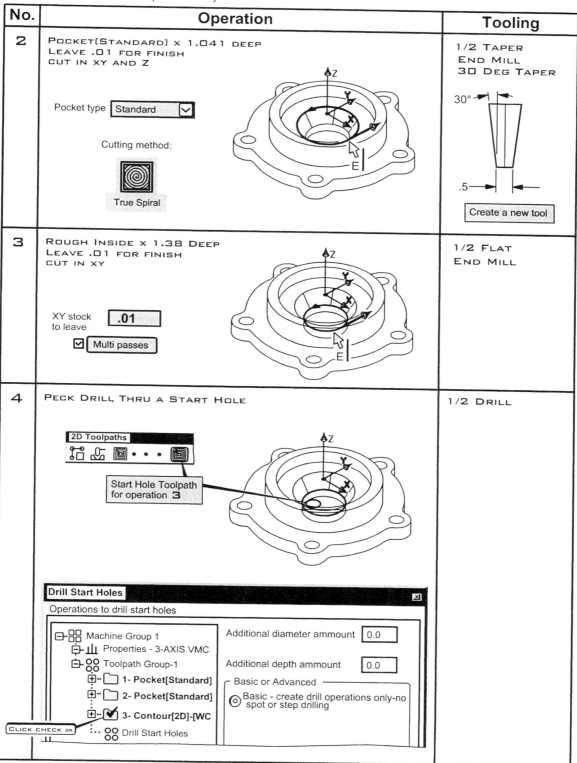

No.	Operation	Tooling
2	POCKET[STANDARD] X 1.041 DEEP LEAVE .01 FOR FINISH CUT IN XY AND Z Pocket type [Standard ▽] Cutting method: True Spiral	1/2 TAPER END MILL 30 DEG TAPER 30° .5 [Create a new tool]
3	ROUGH INSIDE X 1.38 DEEP LEAVE .01 FOR FINISH CUT IN XY XY stock to leave [.01] ☑ [Multi passes]	1/2 FLAT END MILL

4	PECK DRILL THRU A START HOLE
	
	2D Toolpaths
	
	Start Hole Toolpath for operation 3
	
	Drill Start Holes
	Operations to drill start holes
	
	⊟ Machine Group 1
	 ⊟ Properties - 3-AXIS VMC
	 ⊟ Toolpath Group-1
	  ⊞ 1- Pocket[Standard]
	  ⊞ 2- Pocket[Standard]
	  ⊞ ☑ 3- Contour[2D]-[WC
	CLICK CHECK on
	  Drill Start Holes
	
	Additional diameter ammount [0.0]
	
	Additional depth ammount [0.0]
	
	Basic or Advanced
	◉ Basic - create drill operations only-no spot or step drilling
		1/2 DRILL

PROCESS PLAN 12P-3 *(continued)*

No.	Operation	Tooling
5	POCKET[FACING] X .875 DEEP LEAVE .01 FOR FINISH CUT IN XY AND Z 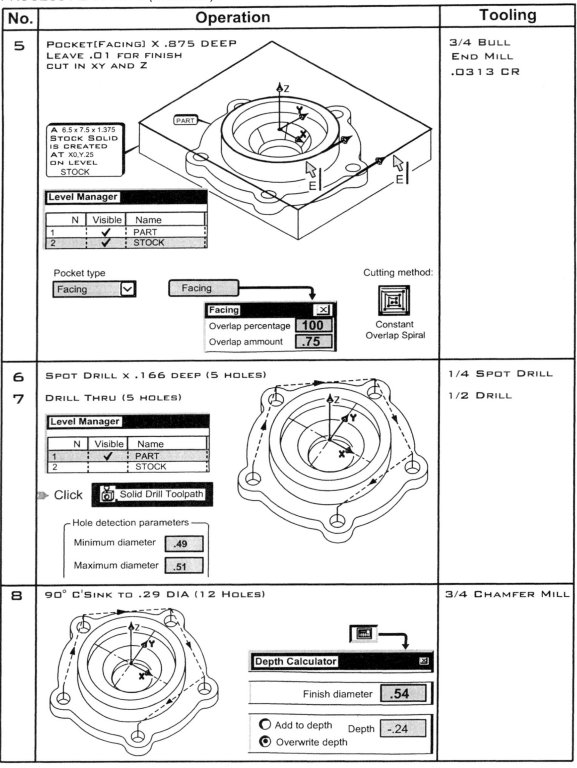	3/4 BULL END MILL .0313 CR
6 7	SPOT DRILL X .166 DEEP (5 HOLES) DRILL THRU (5 HOLES)	1/4 SPOT DRILL 1/2 DRILL
8	90° C'SINK TO .29 DIA (12 HOLES)	3/4 CHAMFER MILL

PROCESS PLAN 12P-3 *(continued)*

No.	Operation	Tooling
9	CHAMFER INSIDE AND OUSIDE CONTOURS Contour type [2D chamfer ▼] Chamfer → **Chamfering** ✕ Width [.02] Tip offset [0.1] ✓ ✗ ? 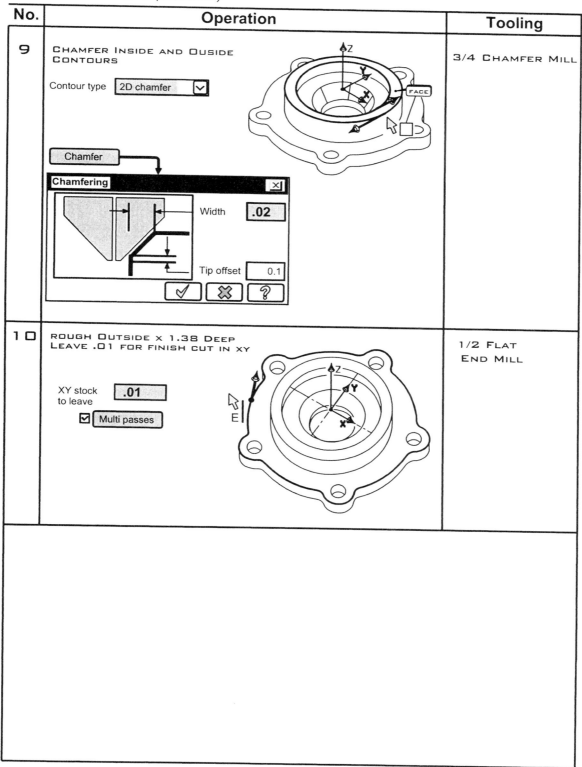	3/4 CHAMFER MILL
10	ROUGH OUTSIDE x 1.38 DEEP LEAVE .01 FOR FINISH CUT IN XY XY stock to leave [.01] ☑ Multi passes	1/2 FLAT END MILL

PROCESS PLAN 12P-3 *(continued)*

No.	Operation	Tooling

RESET THE DEFAULT Tplane AND Cplane BACK TO Top

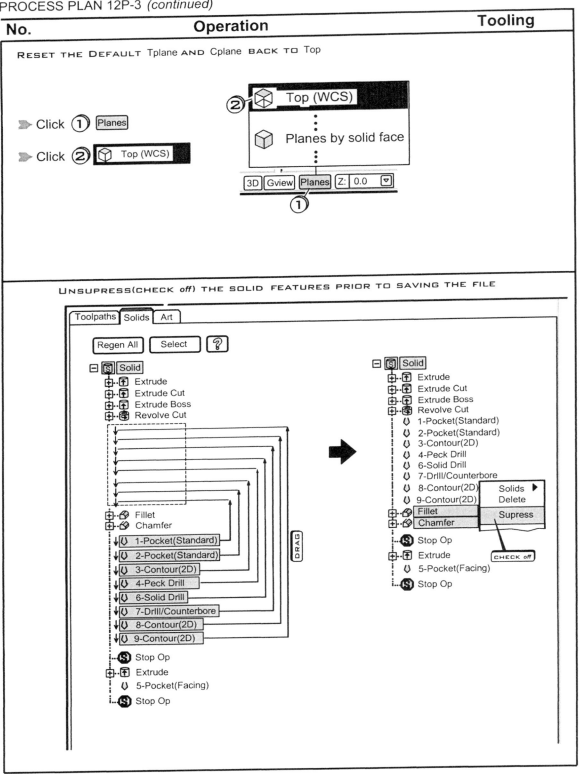

UNSUPRESS(CHECK *off*) THE SOLID FEATURES PRIOR TO SAVING THE FILE

12-4) a) Get the *Solid* model file **EX11-4JV** :
   ◆ from the file generated in exercise 11-4

b) Specify the stock size and material

c) Create the required toolpaths for executing the operations listed in PROCESS PLAN 12P-3

d) Backplot and verify the toolpaths

e) Generate the word address part program

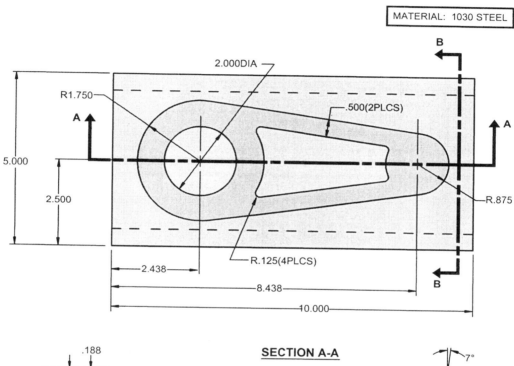

MATERIAL: 1030 STEEL

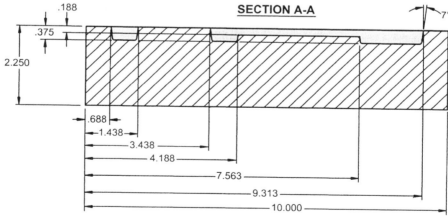

DROP FORGING DIE-CONNECTOR LINK
ALL DRAFT ANGLES 7°
ALL FILLETS AND ROUNDS R.063

Figure 12-p6

◆ SPECIFY THE SIZE OF THE STOCK

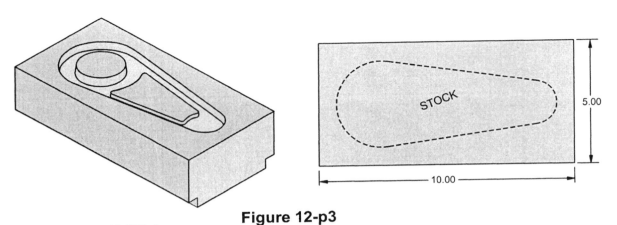

Figure 12-p3

PROCESS PLAN 12P-4

No.	Operation	Tooling
1	POCKET[ISLAND FACING] LEAVE .01 FOR FINISH CUT IN XY AND Z 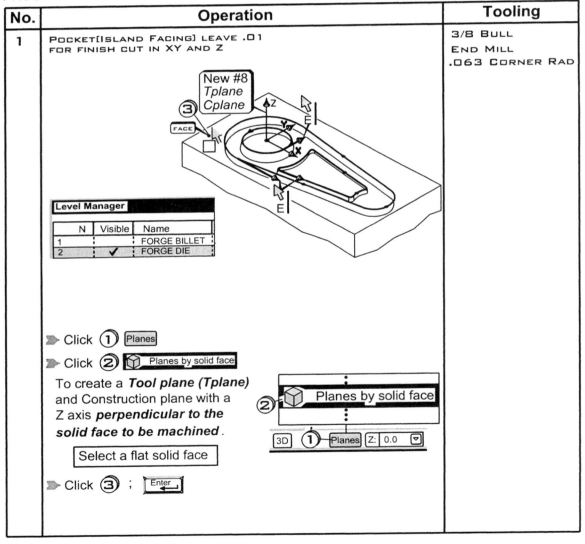	3/8 BULL END MILL .063 CORNER RAD

PROCESS PLAN 12P-4(*continued*)

No.	Operation	Tooling
	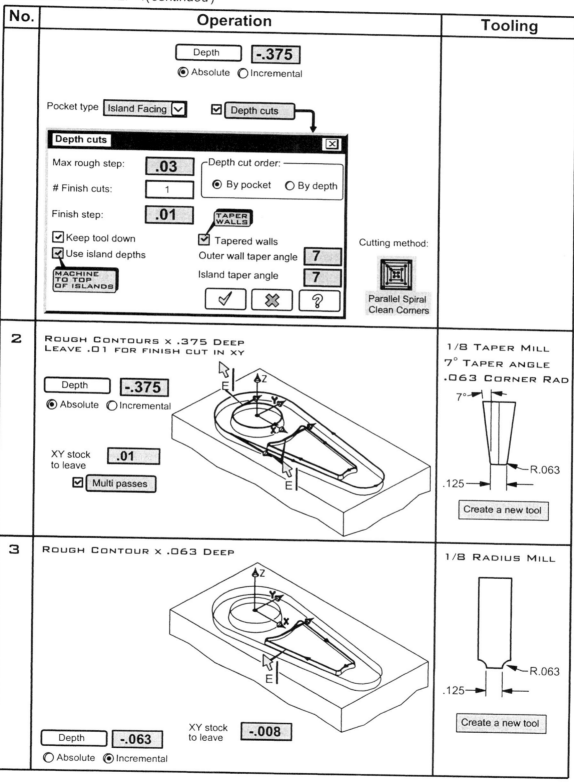	

Depth `-.375`
⦿ Absolute ○ Incremental

Pocket type `Island Facing ▾`   ☑ **Depth cuts**

Depth cuts ☒

Max rough step: `.03`

Depth cut order:
⦿ By pocket   ○ By depth

\# Finish cuts: `1`

Finish step: `.01`

TAPER WALLS

☑ Keep tool down   ☑ Tapered walls

☑ Use island depths   Outer wall taper angle `7`

MACHINE TO TOP OF ISLANDS   Island taper angle `7`

✓ ✗ ?

Cutting method:

Parallel Spiral
Clean Corners

---

2 ROUGH CONTOURS × .375 DEEP
LEAVE .01 FOR FINISH CUT IN XY

Depth `-.375`
⦿ Absolute ○ Incremental

XY stock to leave `.01`
☑ Multi passes

Tooling:
1/8 TAPER MILL
7° TAPER ANGLE
.063 CORNER RAD

7°
R.063
.125

Create a new tool

---

3 ROUGH CONTOUR × .063 DEEP

Depth `-.063`
○ Absolute ⦿ Incremental

XY stock to leave `-.008`

Tooling:
1/8 RADIUS MILL

R.063
.125

Create a new tool

PROCESS PLAN 12P-4 *(continued)*

No.	Operation	Tooling

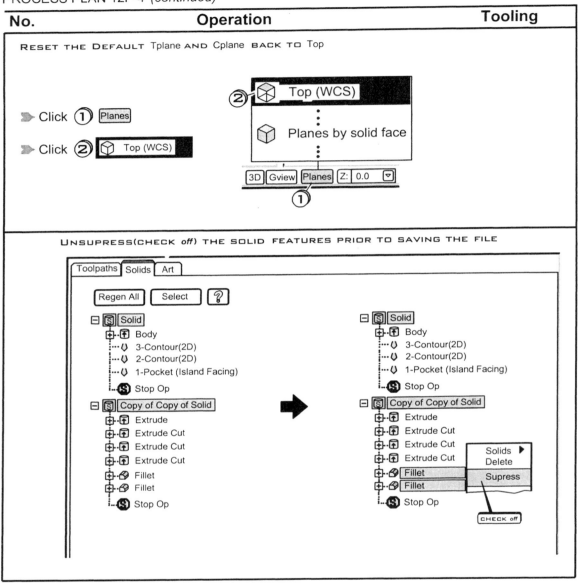

RESET THE DEFAULT Tplane AND Cplane BACK TO Top

➤ Click ① Planes

➤ Click ② ⬡ Top (WCS)

② ⬡ Top (WCS)

⬡ Planes by solid face

[3D] [Gview] [Planes] [Z: 0.0] [▽]

①

UNSUPRESS(CHECK *off*) THE SOLID FEATURES PRIOR TO SAVING THE FILE

CHAPTER - 13

FILE TRACKING AND CHANGE RECOGNITION

13-1 Chapter Objectives

After completing this chapter you will be able to:

1. Know how to use File Tracking to check for the latest version of a part file..

2. Know how to use tracking filters to build a custom list of files to track.

3. Explain the terms Original file and Incoming file.

4. Understand how to use the Change Recognition function to determine changes between the Original and Incoming files.

5. Know the process of updating the toolpaths for any operations that are affected by geometry changes in the Incoming file.

13-2 Tracking Changes in Files

Many parts in either the design or production cycle will undergo revisions and changes. *Mastercam X2* features a *File tracking* function that can check if a *newer version of the current part file exists* and display the result. In this way, the operator is can be sure that the *part file being opened contains all the latest revisions and changes*.

The File tracking fly out menu is accessible from the File menu bar.

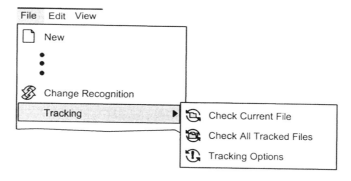

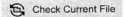 Check Current File

Opens the **File Tracking Options** dialog box and Directs Mastercam X2 to search for *newer versions of the currently loaded* part file.

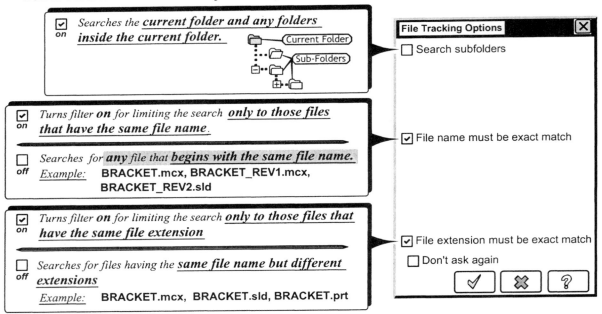

☑ *on* Searches the **current folder and any folders inside the current folder.**

Current Folder

Sub-Folders

☑ *on* Turns filter **on** for limiting the search **only to those files that have the same file name**.

☐ *off* Searches for **any file that begins with the same file name.** *Example:* BRACKET.mcx, BRACKET_REV1.mcx, BRACKET_REV2.sld

☑ *on* Turns filter **on** for limiting the search **only to those files that have the same file extension**

☐ *off* Searches for files having the **same file name but different extensions** *Example:* BRACKET.mcx, BRACKET.sld, BRACKET.prt

File Tracking Options ☒

☐ Search subfolders

☑ File name must be exact match

☑ File extension must be exact match

☐ Don't ask again

✓ ✗ ?

IF NO NEWER FILE IS FOUND

Mastercam X2 displays the message

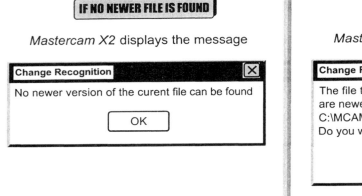

Change Recognition ☒

No newer version of the curent file can be found

OK

IF A NEWER FILE IS FOUND

Mastercam X2 displays the message

Change Recognition ☒

The file tracking tool has found multiple files that are newer than C:\MCAMX2\JVAL-MILL\EXAMPLE13-1.MCX Do you want to run change recognition now?

Yes   No

Click the Yes button to launch

Mastercam's Change Recognition function

🔁 Tracking Options

Opens the **File Tracking** dialog box which allows the operator to *create a list of files to track* as well as *set various tracking filters.*

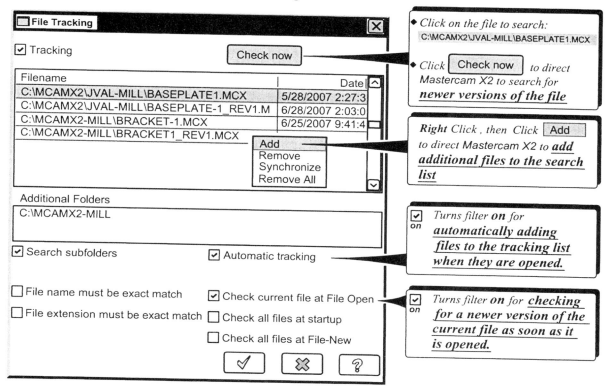

🔁 Check All Tracked Files

Opens the **File Tracking Options** dialog box and Directs *Mastercam X2* to search for *newer versions of **all the*** part files in the Options list.

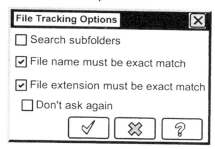

13-3 Change Recognition

Along with *File* tracking, *Mastercam X2* also has the *Change Recognition* function. Change Recognition looks at two *files and reports on their differences*. The file currently loaded, called the **Original** is *compared with* the another file called the **Incoming.**

The **Original** file **must be currently loaded before Change Recognition is launched** . The **Change Recognition** dialog box is displayed when the function is launched from either the File tracking function when *Mastercam X2* detects a *newer version* of a part file or from the File benu bar.

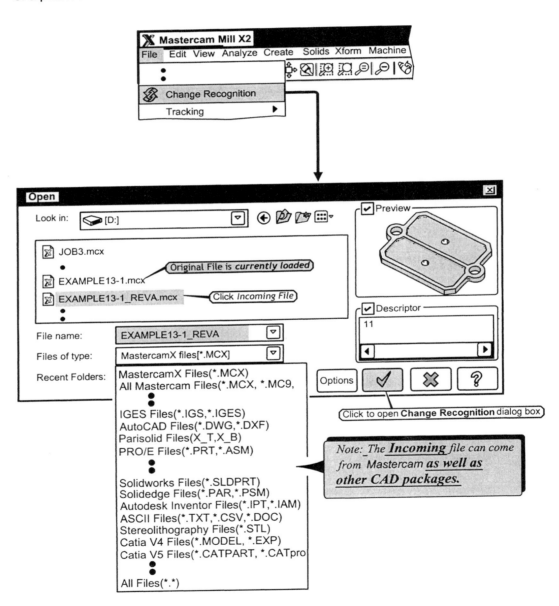

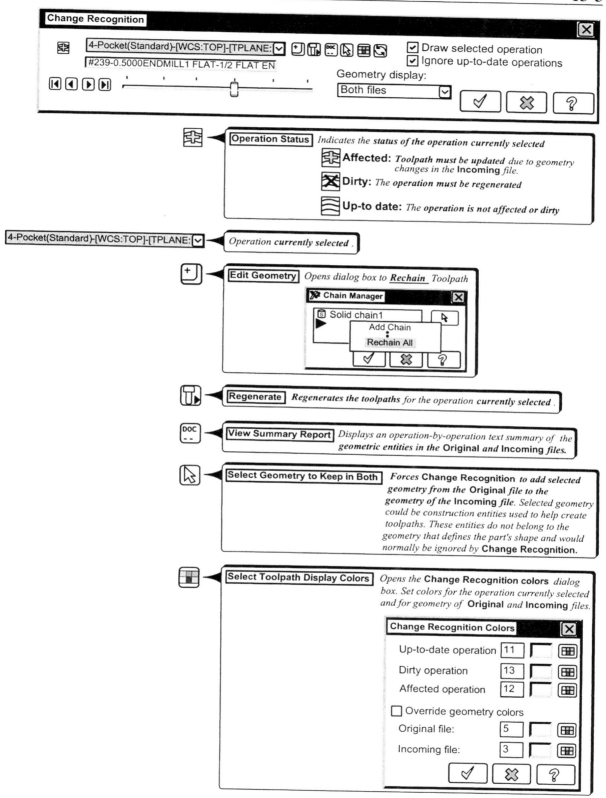

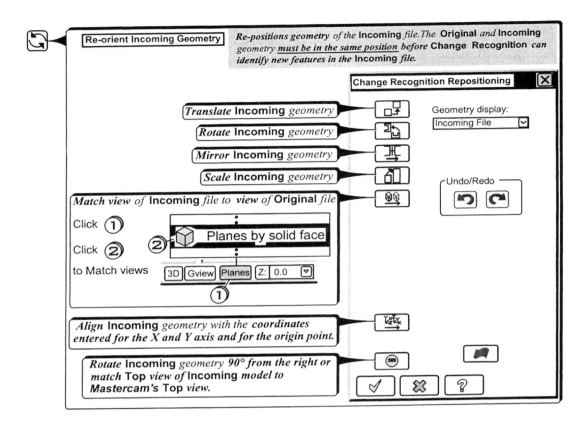

Re-orient Incoming Geometry — *Re-positions geometry of the Incoming file. The Original and Incoming geometry must be in the same position before Change Recognition can identify new features in the Incoming file.*

Change Recognition Repositioning

Translate Incoming geometry

Rotate Incoming geometry

Mirror Incoming geometry

Scale Incoming geometry

Match view of Incoming file to view of Original file

Geometry display:
Incoming File

Undo/Redo

Click ① Click ② to Match views — Planes by solid face — 3D Gview Planes Z: 0.0

Align Incoming geometry with the coordinates entered for the X and Y axis and for the origin point.

Rotate Incoming geometry 90° from the right or match Top view of Incoming model to Mastercam's Top view.

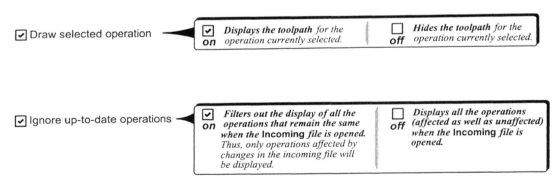

☑ Draw selected operation

☑ **on** *Displays the toolpath for the operation currently selected.*

☐ **off** *Hides the toolpath for the operation currently selected.*

☑ Ignore up-to-date operations

☑ **on** *Filters out the display of all the operations that remain the same when the Incoming file is opened. Thus, only operations affected by changes in the incoming file will be displayed.*

☐ **off** *Displays all the operations (affected as well as unaffected) when the Incoming file is opened.*

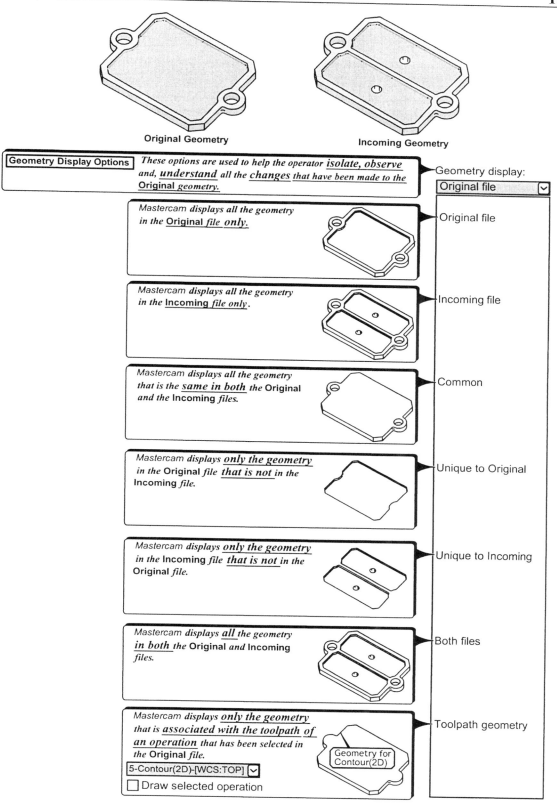

Original Geometry

Incoming Geometry

Geometry Display Options — *These options are used to help the operator isolate, observe and, understand all the changes that have been made to the Original geometry.*

Geometry display:

Original file

Mastercam displays all the geometry in the Original file only.

Original file

Mastercam displays all the geometry in the Incoming file only.

Incoming file

Mastercam displays all the geometry that is the same in both the Original and the Incoming files.

Common

Mastercam displays only the geometry in the Original file that is not in the Incoming file.

Unique to Original

Mastercam displays only the geometry in the Incoming file that is not in the Original file.

Unique to Incoming

Mastercam displays all the geometry in both the Original and Incoming files.

Both files

Mastercam displays only the geometry that is associated with the toolpath of an operation that has been selected in the Original file.

Toolpath geometry

Geometry for Contour(2D)

5-Contour(2D)-[WCS:TOP]

☐ Draw selected operation

EXAMPLE 13-1

The operations for producing the part shown Figure 13-1(left) *currently exist* in the Operations Manager of the Original file: #1-EXAMPLE13-1. Use the Change Recoginition function to *edit the existing operations and produce the Incoming file part geometry* as shown in Fig13-1(right). Note: the Incoming file is #2-EXAMPLE13-1_REVA.

Note: the DEMO version of *Mastercam X2* **does not** support the **change recognition** function

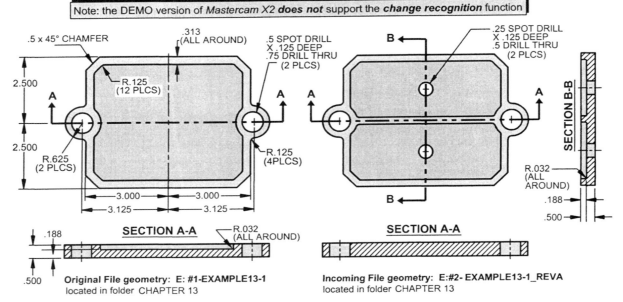

Figure 13-1

A) ENTER *Mastercam* **AND OPEN THE** ORIGINAL FILE: #1-EXAMPLE 13-1.MCX

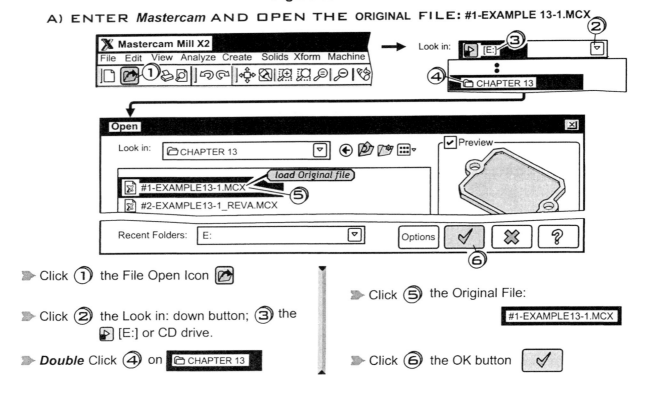

➤ Click ① the File Open Icon

➤ Click ② the Look in: down button; ③ the [E:] or CD drive.

➤ *Double* Click ④ on CHAPTER 13

➤ Click ⑤ the Original File:

#1-EXAMPLE13-1.MCX

➤ Click ⑥ the OK button

B) EXECUTE THE Change Recognition **FUNCTION AND OPEN THE**
INCOMING FILE: #2-EXAMPLE 13-1_REVA.MCX

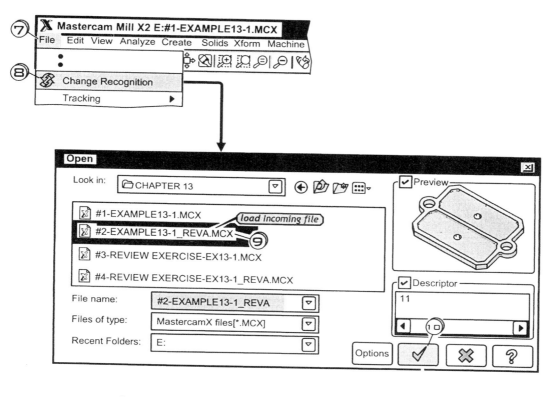

≫ Click ⑦ **File** from the menu bar

≫ Click ⑧ Change Recognition

≫ Click ⑨ the Incoming File: #2-EXAMPLE13-1_REVA

≫ Click ⑩ the OK button ✓

C) SET THE GEOMETRY DISPLAY TO Inciming File AND SUPRESS THE FILLET FEATURES.

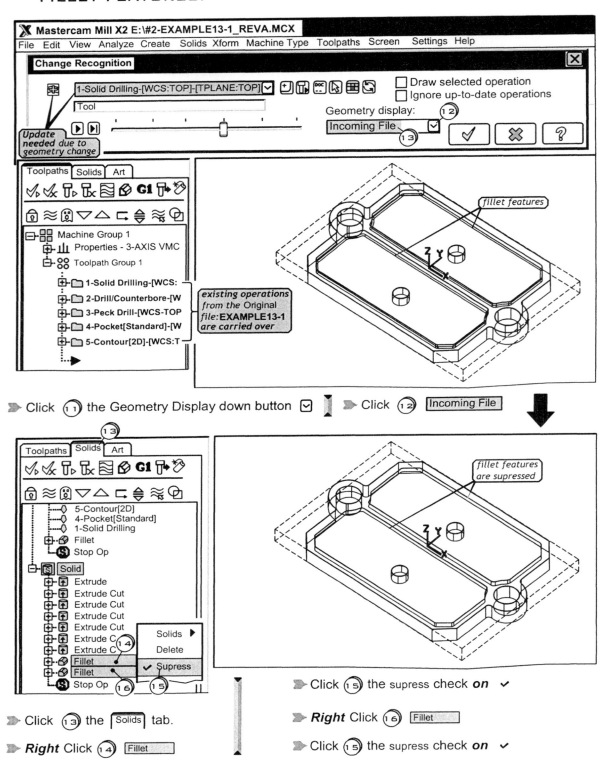

➤ Click ⑪ the Geometry Display down button ☑ ❙ ➤ Click ⑫ Incoming File

➤ Click ⑬ the Solids tab.

➤ *Right* Click ⑭ Fillet

➤ Click ⑮ the supress check *on* ✔

➤ *Right* Click ⑯ Fillet

➤ Click ⑮ the supress check *on* ✔

D) CHECK THE DIFFERENCES BETWEEN Original AND Incoming FILES.

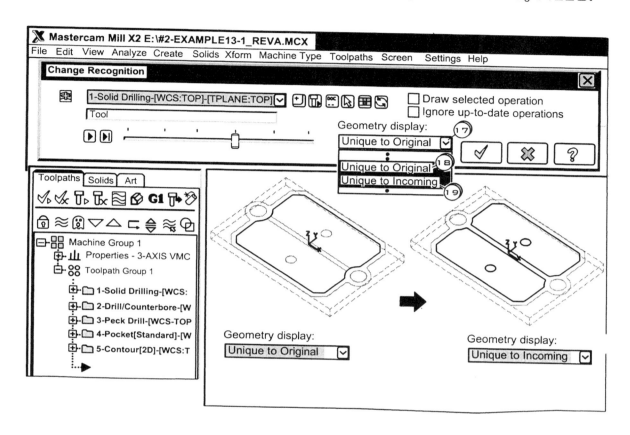

➤ Click ⑰ the Geometry Display down button ☑

➤ Click ⑱ Unique to Original

➤ Click ⑰ the Geometry Display down button ☑

➤ Click ⑲ Unique to Incoming

E) GEOMETRY CHANGES IN THE Incoming FILE REQUIRE THE
TOOLPATHS FOR OPERATION-4 TO BE UPDATED.

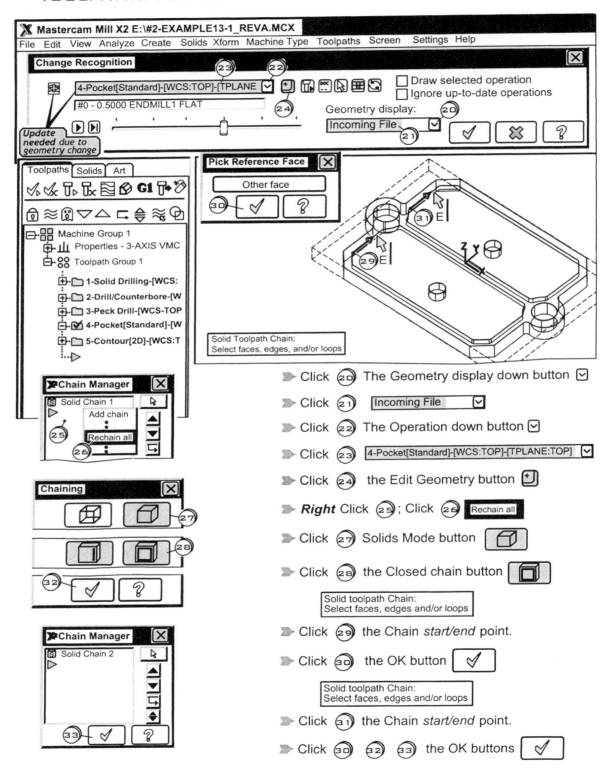

➤ Click ㉕ The Geometry display down button ☑

➤ Click ㉑   Incoming File   ☑

➤ Click ㉒ The Operation down button ☑

➤ Click ㉓   4-Pocket[Standard]-[WCS:TOP]-[TPLANE:TOP]   ☑

➤ Click ㉔   the Edit Geometry button ⊞

➤ *Right* Click ㉕ ; Click ㉖ Rechain all

➤ Click ㉗ Solids Mode button   ▱

➤ Click ㉘ the Closed chain button   ▣

Solid toolpath Chain:
Select faces, edges and/or loops

➤ Click ㉙ the Chain *start/end* point.

➤ Click ㉚   the OK button   ✓

Solid toolpath Chain:
Select faces, edges and/or loops

➤ Click ㉛ the Chain *start/end* point.

➤ Click ㉚ ㉜ ㉝ the OK buttons   ✓

F) REGENERATE THE TOOLPATH FOR OPERATION-4.

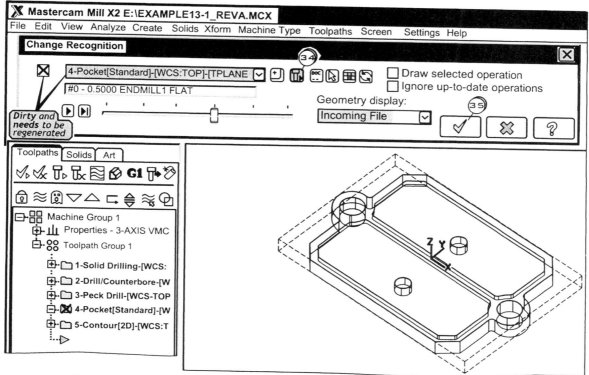

➤ Click ③④ the Regenerate button 🔲 to regenerate the toolpath for Operation-4.

G) ADD THE .25 PILOT HOLES X .188 DEEP AND THE .5 DRILL THRU HOLES.

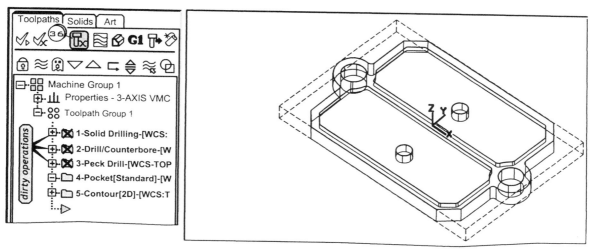

➤ Click ㉟ the **File Recognition** OK button ☑

➤ Click ㊱ the Regenerate all dirty operations button *twice*

♦ DIRECT *Mastercam* TO AUTOMATICALLY DETECT THE HOLES TO BE DRILLED.

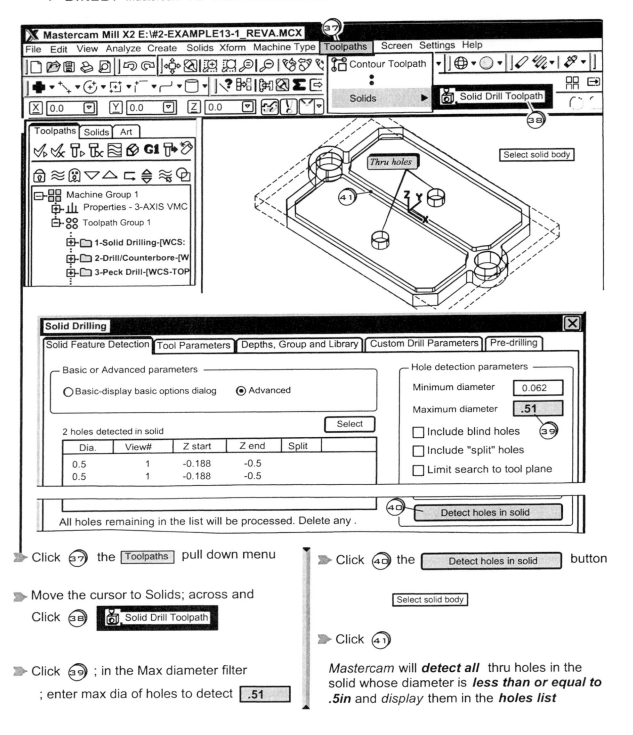

► Click ㊲ the [Toolpaths] pull down menu

► Move the cursor to Solids; across and

　Click ㊳ [🔩 Solid Drill Toolpath]

► Click ㊴ ; in the Max diameter filter

　; enter max dia of holes to detect [**.51**]

► Click ㊵ the [Detect holes in solid] button

[Select solid body]

► Click ㊶

Mastercam will **detect all** thru holes in the solid whose diameter is **less than or equal to .5in** and *display* them in the **holes list**

◆ SPECIFY THE TOOL PARAMETERS FOR THE DRILLING OPERATION.

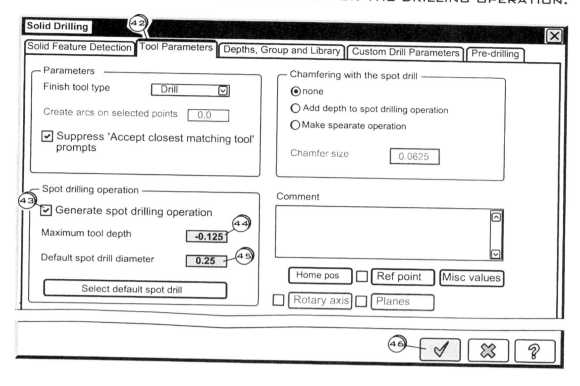

⯈ Click ㊷ the [Tool Parameters] tab

⯈ Click ㊸ the check *on* for

☑ Generate spot drilling operation

⯈ Click ㊹; enter the final drill depth [**-0.125**]

⯈ Click ㊺ ; enter the spot drill dia [**.25**]

⯈ Click ㊻ the OK button [✓]

F) ALL THE SOLID MODEL FEATURES MUST BE UNSUPRESSED BEFORE THE .MCX FILE CAN BE SAVED.

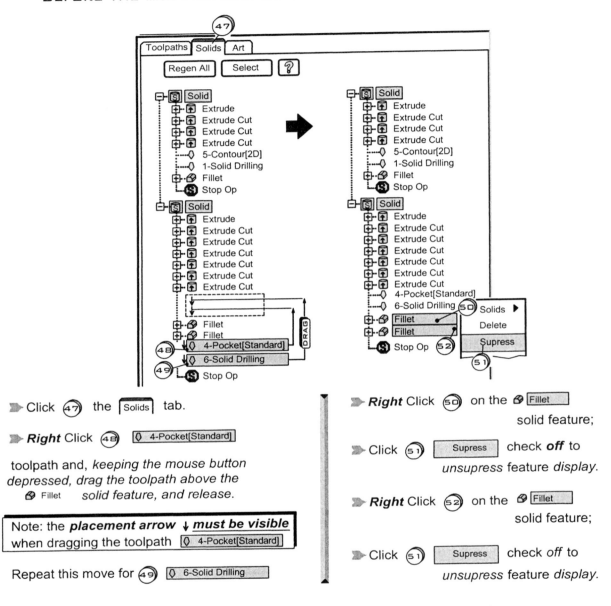

➤ Click ㊼ the ⌈Solids⌉ tab.

➤ ***Right*** Click ㊽ ⌈◊ 4-Pocket[Standard]⌉

toolpath and, *keeping the mouse button depressed, drag the toolpath above the* ⌈🖉 Fillet⌉ *solid feature, and release.*

Note: the ***placement arrow*** ↓ ***must be visible*** when dragging the toolpath ⌈◊ 4-Pocket[Standard]⌉

Repeat this move for ㊾ ⌈◊ 6-Solid Drilling⌉

➤ ***Right*** Click ㊿ on the 🖉 ⌈Fillet⌉ solid feature;

➤ Click ⑤① ⌈ Supress ⌉ check ***off*** to *unsupress* feature *display.*

➤ ***Right*** Click ⑤② on the 🖉 ⌈Fillet⌉ solid feature;

➤ Click ⑤① ⌈ Supress ⌉ check *off* to *unsupress* feature *display.*

G) SAVE THE FILE UNDER THE Incoming **FILE NAME: EXAMPLE13-1_REVA.MCX**

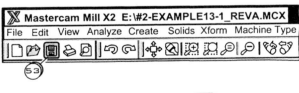

➤ Click ⑤③ the File save icon 🖫

EXERCISES

13-1) a) The operations for producing the part shown in Figure 13p-1 currently exist in the Operations Manager of the Original file **E:#3-REVIEW EXERCISE-EX13-1.MCX**, stored on the CD and located in the folder CHAPTER 13.

b) Use the Change Recognition function to edit the existing operations and produce the incoming file part geometry as shown in Figure 13p-2.
Note: the Incoming file **E:#4-REVIEW EXERCISE-EX13-1_REVA.MCX** is stored on the CD and located in the folder CHAPTER 13.

c) Backplot and Verify the regenerated toolpaths.

d) Generate the word address part program

e) Save the Incoming file with the regenerated toolpaths. *Be sure to unsupress all the solid features* before saving the file.

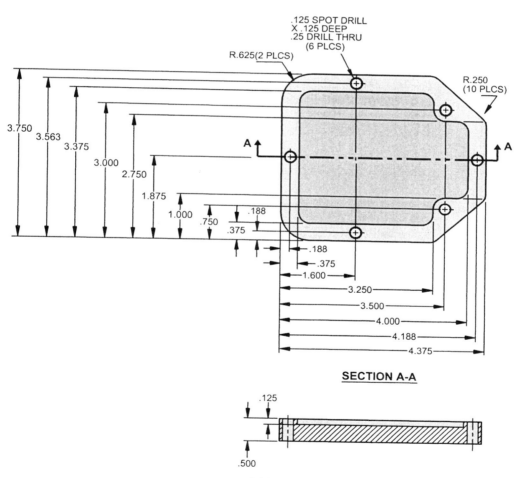

SECTION A-A

Original File geometry: #3-REVIEW EXERCISE-EX13-1.MCX

Figure 13-p1

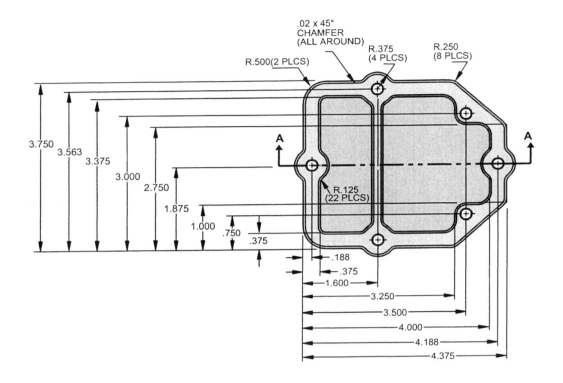

SECTION A-A

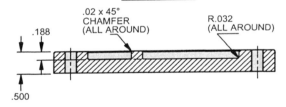

Incoming File geometry: #4-REVIEW EXERCISE-EX13-1_REVA.MCX

Figure 13-p2

APPENDIX

CUSTOMIZING *MastercamX2*

The Customize Dialog Box

Mastercam X2 features the **Customize** dialog box for creating customized tool bars, drop down menus and the graphics right click menu. Customization enables the operator to design the *Mastercam X2* interface to work optimally for specific types of applications or user preferences.

The settings for toolbars and right-click menus are stored in *Mastercam X2* files with, the extension **.mtb**

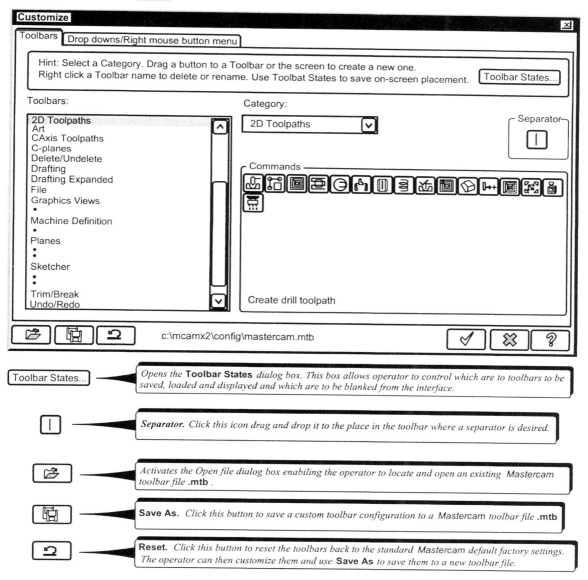

Customizing Toolbar Menus

the **Customize** dialog box enables the operator to execute the following customizing functions with respect to toolbars:

◆ Add or remove toolbar buttons

◆ Create new toolbars

◆ Rename toolbars

◆ Delete toolbars

EXAMPLE A-1

Create a custom toolbar menu with buttons for generating rectangular plates with holes arrayed in a rectangular pattern. Include the drill cycle button in the toolbar. Name the new toolbar **REC DRILL-1**

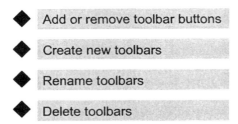

A) Enter the Customize dialog box

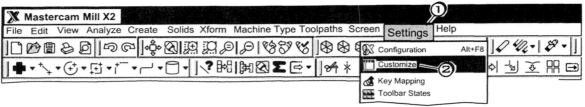

➤ **Left** Click ① the [Settings] function

➤ **Left** Click ②

B) USE THE Customize DIALOG BOX TO CREATE THE NEW TOOLBAR MENU

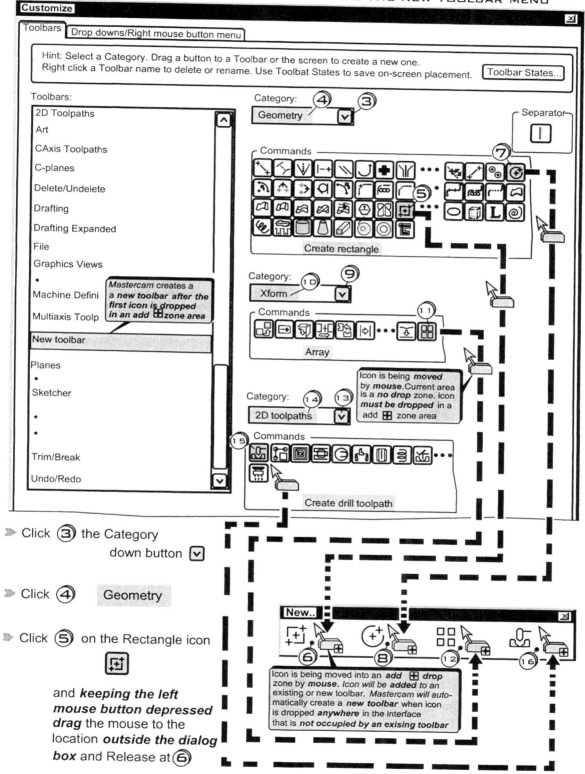

> ⇒ Click ③ the Category
>   down button ⌄

> ⇒ Click ④ Geometry

> ⇒ Click ⑤ on the Rectangle icon
>
>   and **keeping the left
>   mouse button depressed
>   drag** the mouse to the
>   location **outside the dialog
>   box** and Release at ⑥

➤ Click ⑦ on the Circle icon

and *keeping the left mouse button depressed*
drag the mouse to the location *inside the new toolbar menu*
and Release at ⑧

➤ Click ⑨ the Category
                down button ▾

➤ Click ⑩   Xform

➤ Click ⑪ on the Array icon

and *keeping the left mouse button depressed*
drag the mouse to the location *inside the new toolbar menu*
and Release at ⑫

➤ Click ⑬ the Category
                down button ▾

➤ Click ⑭   2D Toolpaths

➤ Click ⑮ on the Create drill toolpath icon

and *keeping the left mouse button depressed*
drag the mouse to the location *inside the new toolbar menu*
and Release at ⑯

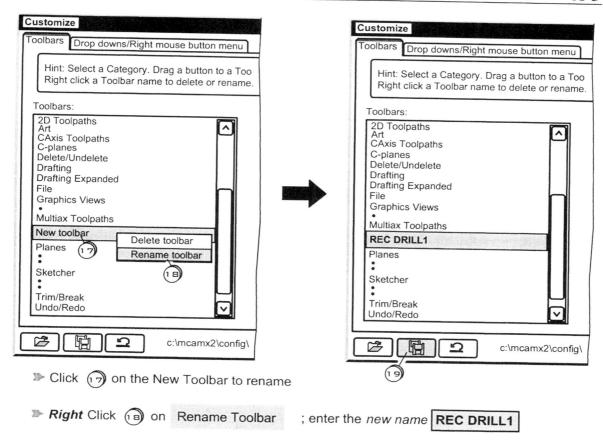

⇒ Click ⟨17⟩ on the New Toolbar to rename

⇒ **Right** Click ⟨18⟩ on Rename Toolbar ; enter the *new name* **REC DRILL1**

C) SAVE THE NEW TOOLBAR FILE

⇒ Click ⟨19⟩ the Save As [▣] button

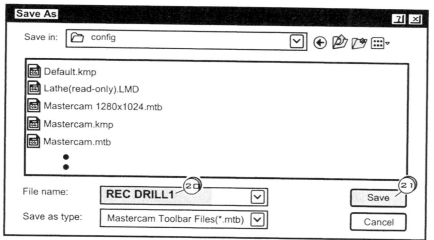

⇒ Click ⟨20⟩ in the File name box ; enter **REC DRILL1**

⇒ Click ⟨21⟩ the Save button.

Note:
The operator can *remove* an icon from a new toolbar that is *in the process of being created* by using the **Remove function** feature.

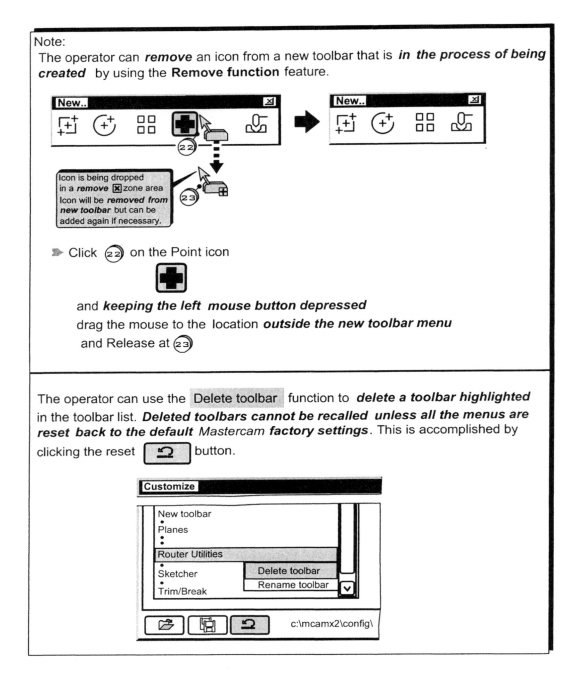

Icon is being dropped in a *remove* ⊠ zone area Icon will be *removed from new toolbar* but can be added again if necessary.

➡ Click ㉒ on the Point icon

and *keeping the left mouse button depressed*
drag the mouse to the location *outside the new toolbar menu*
and Release at ㉓

The operator can use the Delete toolbar function to *delete a toolbar highlighted* in the toolbar list. *Deleted toolbars cannot be recalled unless all the menus are reset back to the default* Mastercam *factory settings*. This is accomplished by clicking the reset ⟲ button.

The Toolbar States Dialog Box

The **Toolbar States** Dialog box dialog box is used to specify *which sets of toolbars* are to be *displayed* and which are *hidden* in the *Mastercam interface*. Additionally, the operator can create *different patterns of toolbar states*. Each pattern can be given a name and saved or deleted. Depending upon the type of job to be completed, the operator can also can also *load a previously stored toolbar configuration pattern that best fits the application*. Additionally a toolbar state can be assigned to a machine definition via the Machine Definition Manager. Upon assigning the machine definition to a machine group to create toolpaths the toolbar state would automatically load and display in the *Mastercam* window.

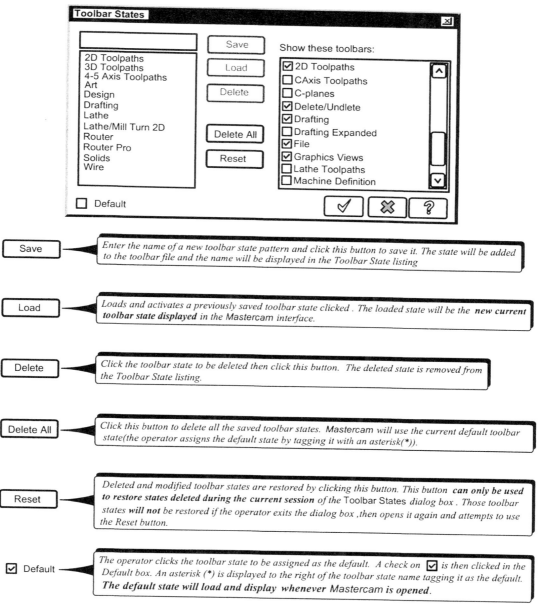

Save — Enter the name of a new toolbar state pattern and click this button to save it. The state will be added to the toolbar file and the name will be displayed in the Toolbar State listing

Load — Loads and activates a previously saved toolbar state clicked . The loaded state will be the **new current toolbar state displayed** in the Mastercam interface.

Delete — Click the toolbar state to be deleted then click this button. The deleted state is removed from the Toolbar State listing.

Delete All — Click this button to delete all the saved toolbar states. Mastercam will use the current default toolbar state(the operator assigns the default state by tagging it with an asterisk(*)).

Reset — Deleted and modified toolbar states are restored by clicking this button. This button **can only be used to restore states deleted during the current session** of the Toolbar States dialog box . Those toolbar states **will not** be restored if the operator exits the dialog box ,then opens it again and attempts to use the Reset button.

☑ **Default** — The operator clicks the toolbar state to be assigned as the default. A check on ☑ is then clicked in the Default box. An asterisk (*) is displayed to the right of the toolbar state name tagging it as the default. **The default state will load and display whenever** Mastercam **is opened.**

EXAMPLE A-2

Create a custom toolbar state for applications involving the simple drilling of holes arranged in a rectangular grid pattern. Use the new toolbar just created **REC DRILL-1**.

A) ᴇɴᴛᴇʀ ᴛʜᴇ Toolbar States ᴅɪᴀʟᴏɢ ʙᴏx

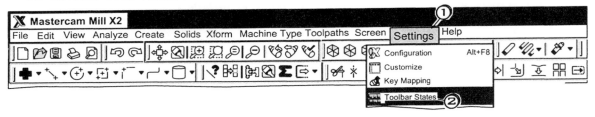

➤ **Left** Click ① the [Settings] function

➤ **Left** Click ② [Toolbar States]

B) ꜰᴏʀ ᴛʜᴇ ᴄᴜꜱᴛᴏᴍ ꜱᴛᴀᴛᴇ ꜱᴩᴇᴄɪꜰʏ ᴛʜᴇ ᴛᴏᴏʟʙᴀʀ ꜱʜᴏᴡ/ʜɪᴅᴇ ᴩᴀᴛᴛᴇʀɴ

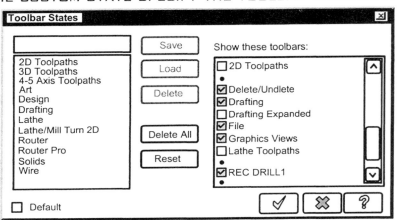

➤ Click ③ - ⑪ the check **on** ☑ for the toolbars to **show**:

③ — ☑ Delete/Undlete
④ — ☑ Drafting
⑤ — ☑ File
⑥ — ☑ Graphics Views
⑦ — ☑ REC DRILL1
⑧ — ☑ Sketcher
⑨ — ☑ Trim/Break
⑩ — ☑ Undo/Redo
⑪ — ☑ Utility

➤ Click the check **off** ☐ for all the other toolbars listed to **hide** them

C) ENTER THE NAME OF THE NEW CUSTOM TOOLBAR STATE AND SAVE IT

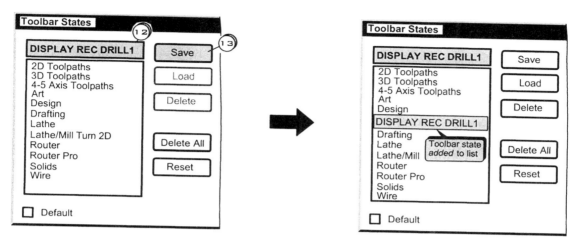

➤ Click ⑫ in the Toolbar States name box and enter: **DISPLAY REC DRILL1**

➤ Click ⑬ the [Save] button to save the new toolbar state to the toolbar file.

Mastercam will then *add* DISPLAY REC DRILL1 *to its list of toolbar states*.

D) DISPLAYING OTHER TOOLBAR STATES AND SAVING ALL WORK WITH THE CURRENT SESSION OF THE Toolbar States DIALOG BOX

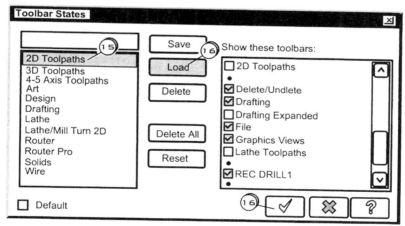

To display the 2D toolpaths Toolbar State:

➤ Click ⑮ on the toolbar state [2D Toolpaths]

➤ Click ⑯ the [Load] button to *display* the toolbar state in the *Mastercam* interface.

Important!

➤ Click ⑰ the OK [✓] button to **save all the changes made during the current session of the Toolbar States dialog box. If omitted current changes will be lost.**

Creating Customized Drop Down Menus

The **Customize** dialog box can also be used to create user defined drop down menus and the graphics right-click menu.,

EXAMPLE A-3
Repeat Example A-1. This time arrange the functions in a custom drop down menu. Include the Right-click drop down menu in the toolbar. Name the new toolbar **RT CLK-REC DRILL-1**

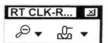

A) SPECIFY A NEW DROP DOWN MENU IS TO BE CREATED

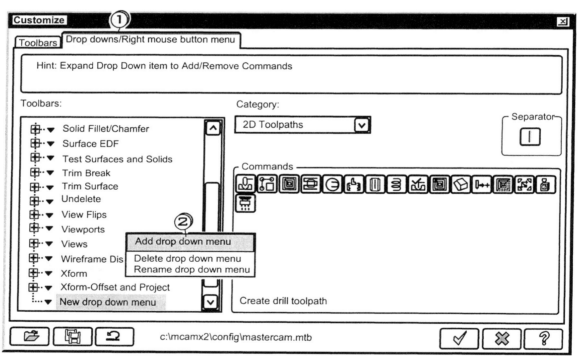

➤ Click ① the ⌐Drop downs/Right mouse button menu⌐ tab.

➤ *Right* Click; then Click ② Add drop down menu

B) USE THE Customize DIALOG BOX TO CREATE THE NEW DROP DOWN MENU

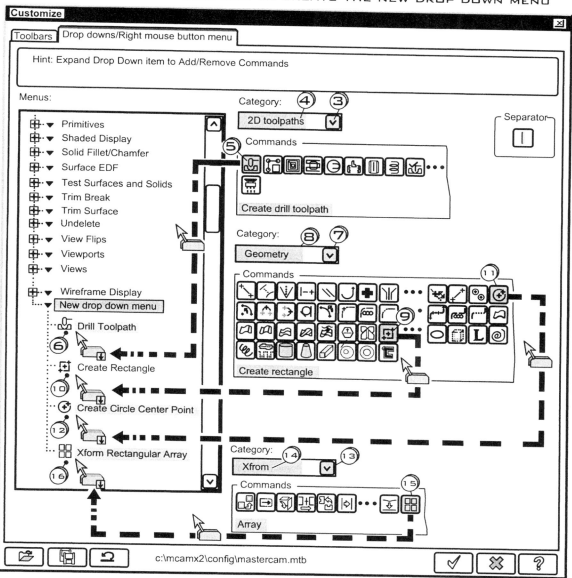

Click ③ the Category down button ☑

Click ④   2D Toolpaths

Click ⑤ on the Create drill toolpath icon

and *keeping the left mouse button depressed*
drag the mouse to the location *below the New drop down menu* and Release at ⑥

➤ Click ⑦ the Category down button ☑

➤ Click ⑧   Geometry

➤ Click ⑨ on the Rectangle icon

🔲

and *keeping the left mouse button depressed*
drag the mouse to the location *below the Drill toolpath icon* and Release at ⑩

➤ Click ⑪ on the Circle icon

⊕

and *keeping the left mouse button depressed*
drag the mouse to the location *below the Create Rectangle icon* and Release at ⑫

➤ Click ⑬ the Category down button   ☑

➤ Click ⑭   Xform

➤ Click ⑮ on the Array icon

⊞

and *keeping the left mouse button depressed*
drag the mouse to the location *below the Create circle center point icon* and Release at ⑯

C) DRAG THE NEW DROP DOWN MENUS TO THE TOOLBAR AREA IN THE *Mastercam* INTERFACE

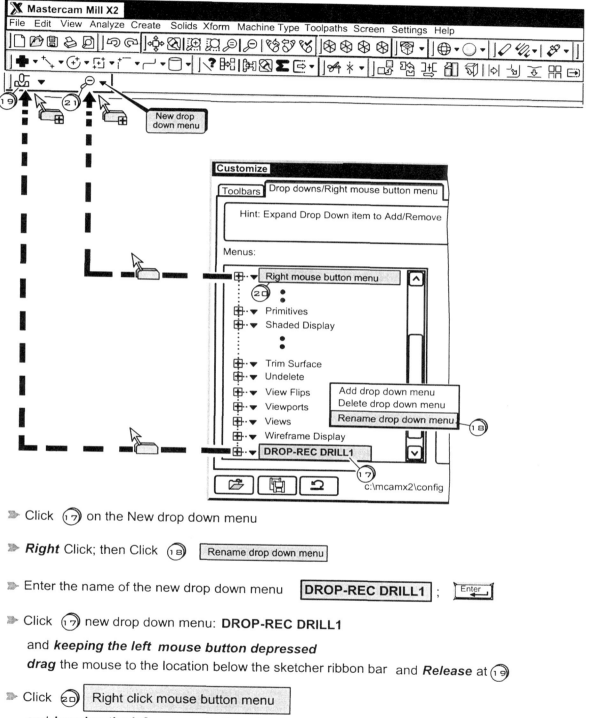

New drop down menu

Customize

Toolbars | Drop downs/Right mouse button menu

Hint: Expand Drop Down item to Add/Remove

Menus:

⊞·▼ Right mouse button menu
⊞·▼ Primitives
⊞·▼ Shaded Display
⊞·▼ Trim Surface
⊞·▼ Undelete
⊞·▼ View Flips
⊞·▼ Viewports
⊞·▼ Views
⊞·▼ Wireframe Display
⊞·▼ DROP-REC DRILL1

Add drop down menu
Delete drop down menu
Rename drop down menu

c:\mcamx2\config

➤ Click ⑰ on the New drop down menu

➤ *Right* Click; then Click ⑱   Rename drop down menu

➤ Enter the name of the new drop down menu   **DROP-REC DRILL1** ;   Enter

➤ Click ⑰ new drop down menu: **DROP-REC DRILL1**

   and *keeping the left mouse button depressed*

   drag the mouse to the location below the sketcher ribbon bar and *Release* at ⑲

➤ Click ⑳   Right click mouse button menu

   and *keeping the left mouse button depressed*

   drag the mouse to a location *inside* the new REC DRILL1 drop down menu and *Release* at ㉑

D) Name the new toolbar containing the drop down menus

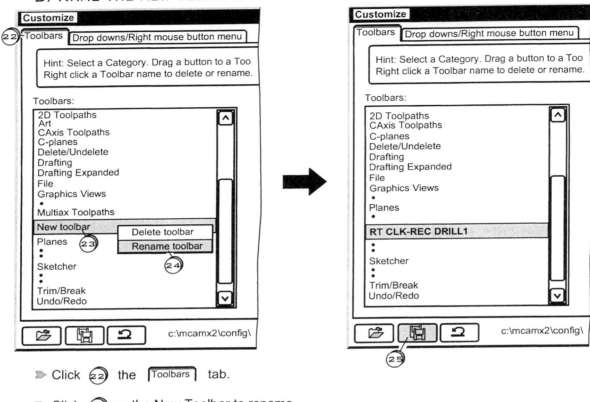

➤ Click ㉒ the [Toolbars] tab.

➤ Click ㉓ on the New Toolbar to rename

➤ *Right* Click ㉔ on | Rename Toolbar | ; enter the *new name* | RT CLK-REC DRILL1 |

E) Save the new toolbar file

➤ Click ㉕ the Save As [🖫] button

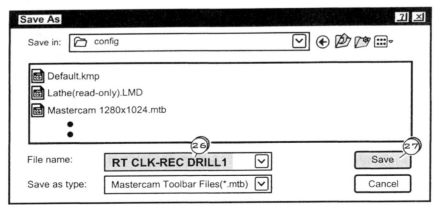

➤ Click ㉖ in the File name box ; enter | RT CLK-REC DRILL1 |

➤ Click ㉗ the [Save] button.

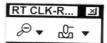

F) SAVE THE CURRENT SESSION OF THE Customize DIALOG BOX

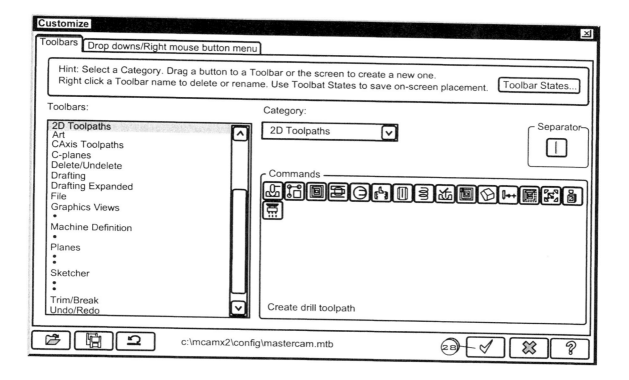

➤ Click ②Ⓑ the OK ☑ button to *save all the changes made during the current session of the Toolbar States dialog box.* *If omitted current changes will be lost.*

The Key Mapping Dialog Box

The operator can use the **Key Mapping** dialog box to assign specific *sets of **keyboard keys to Mastercam functions***. The entire mapped pattern of key sets can be saved as a custom key file with the extension **.kmp**. Many users prefer to commit the keyboard key strokes to memory and tap them rather than find and click icons in toolbars via a mouse.

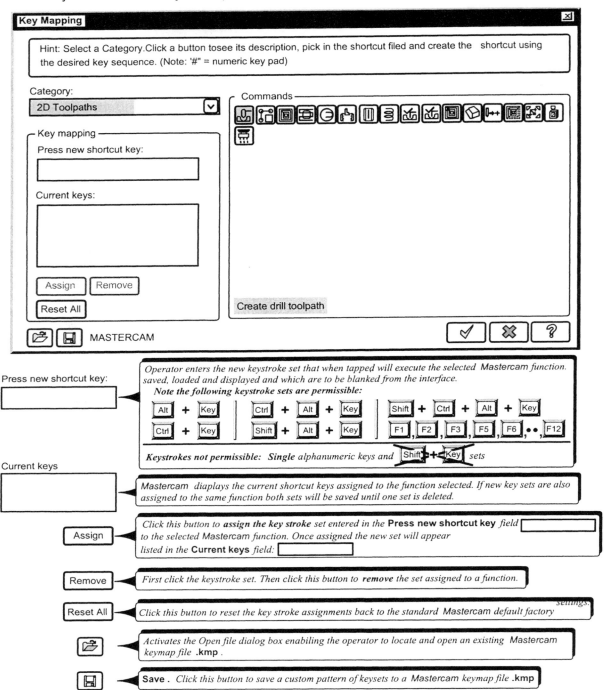

Press new shortcut key:

Operator enters the new keystroke set that when tapped will execute the selected Mastercam function. saved, loaded and displayed and which are to be blanked from the interface.
Note the following keystroke sets are permissible:

Alt + Key | Ctrl + Alt + Key | Shift + Ctrl + Alt + Key
Ctrl + Key | Shift + Alt + Key | F1, F2, F3, F5, F6, •• ,F12

Keystrokes not permissible: **Single** alphanumeric keys and Shift + Key sets

Current keys

Mastercam diaplays the current shortcut keys assigned to the function selected. If new key sets are also assigned to the same function both sets will be saved until one set is deleted.

Assign

Click this button to **assign the key stroke** set entered in the **Press new shortcut key** field to the selected Mastercam function. Once assigned the new set will appear listed in the **Current keys** field:

Remove

First click the keystroke set. Then click this button to **remove** the set assigned to a function.

Reset All

Click this button to reset the key stroke assignments back to the standard Mastercam default factory settings.

Activates the Open file dialog box enabling the operator to locate and open an existing Mastercam keymap file .kmp .

Save . Click this button to save a custom pattern of keysets to a Mastercam keymap file .kmp

EXAMPLE A-4

Create a custom pattern of key sets for generating rectangular plates with holes arrayed in a rectangular grid. Include a key set for accessing the drill cycle function . Name the new keystroke file **REC DRILL-1.**

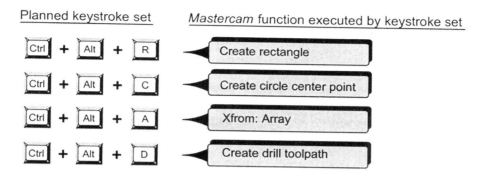

Planned keystroke set | *Mastercam* function executed by keystroke set

Ctrl + Alt + R → Create rectangle

Ctrl + Alt + C → Create circle center point

Ctrl + Alt + A → Xfrom: Array

Ctrl + Alt + D → Create drill toolpath

A) ENTER THE Key Mapping DIALOG BOX

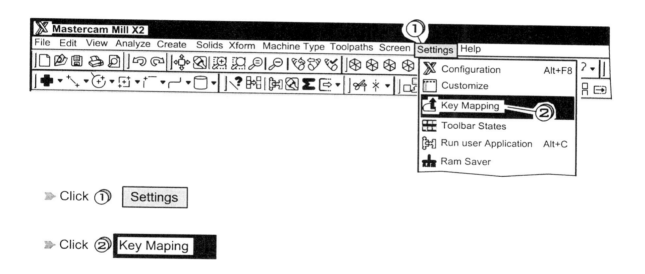

➤ Click ① Settings

➤ Click ② Key Maping

B) USE THE Key Mapping DIALOG BOX TO CREATE THE NEW PATTERN OF KEYSROKE
   SETS AND *Mastercam* FUNCTION ASSIGNMENTS.

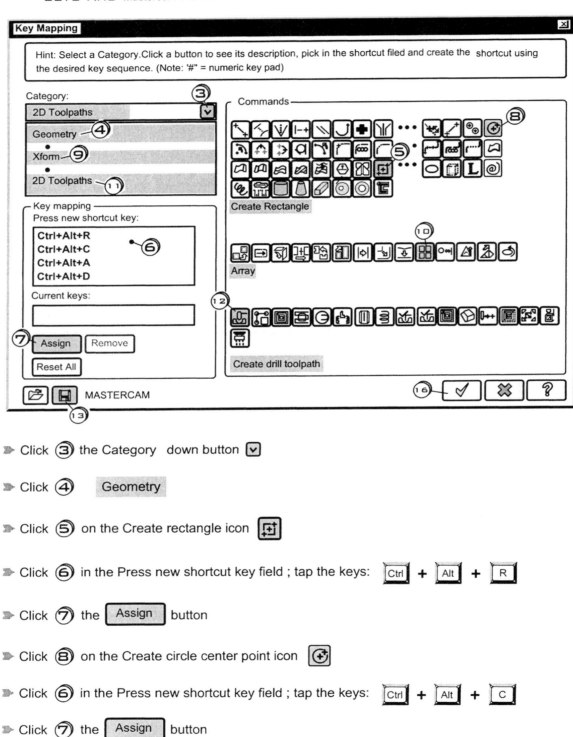

▶ Click ③ the Category down button ▼

▶ Click ④   Geometry

▶ Click ⑤ on the Create rectangle icon ▣

▶ Click ⑥ in the Press new shortcut key field ; tap the keys: Ctrl + Alt + R

▶ Click ⑦ the Assign button

▶ Click ⑧ on the Create circle center point icon ⊕

▶ Click ⑥ in the Press new shortcut key field ; tap the keys: Ctrl + Alt + C

▶ Click ⑦ the Assign button

≫ Click ③ the Category down button ☑

≫ Click ⑨ Xform

≫ Click ⑩ on the Array icon ⊞

≫ Click ⑥ in the Press new shortcut key field ; tap the keys: Ctrl + Alt + A

≫ Click ⑦ the Assign button

≫ Click ③ the Category down button ☑

≫ Click ⑪ 2D Toolpaths

≫ Click ⑫ on the Create drill toolpath icon

≫ Click ⑥ in the Press new shortcut key field ; tap the keys: Ctrl + Alt + D

≫ Click ⑦ the Assign button

C) SAVE THE NEW CUSTOM KEYSET PATTERN TO A Mastercam KEYMAP FILE

≫ Click ⑬ the Save 💾 button

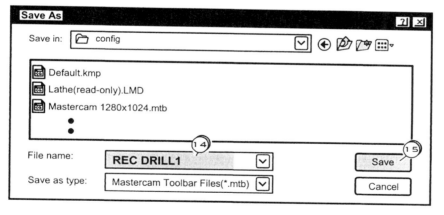

≫ Click ⑭ in the File name box ; enter **REC DRILL**1

≫ Click ⑮ the Save button.

D) SAVE THE CURRENT SESSION OF THE Key Mapping DIALOG BOX

≫ Click ⑯ the OK ✓ button to **save all the changes made during the current session of the Key Mapping dialog box. If omitted current changes will be lost.**

INDEX

Mastercam X2 Demo CD ... Fully Functional for Student Learning

Students using this text can install this fully functional version on their own PCs to:

A - Practice using the software commands contained in the examples in the text and stored on the CD

B - Copy and complete the milling exercises presented in the text and stored on the CD

RESTRICTIONS

This special demo version is not intended for commercial use and contains the following restrictions:

A - Files cannot be saved
B - Accuracy is limited to two decimal places.

To purchase a full commercial version of *Mastercam X2,* please contact

CNC Software, Inc
671 Old Post Road
Tolland, CT, 06084
Tel. 860-875-5006

Or go to *www.mastercam.com*